Average Case Analysis
of Algorithms on Sequences

Average Case Analysis of Algorithms on Sequences

WOJCIECH SZPANKOWSKI
Department of Computer Science
Purdue University

A Wiley-Interscience Publication
JOHN WILEY & SONS, INC.
New York • Chichester • Weinheim • Brisbane • Singapore • Toronto

Library of Congress Cataloging-in-Publication Data is available.
Szpankowski, Wojciech, 1952–
    Average case analysis of algorithms on sequences / Wojciech Szpankowski.
      p.   cm. – (Wiley-Interscience series in discrete mathematics and optimization)
    Includes bibliographical references and index.
    ISBN 0-471-24063-X (cloth : alk. paper)
    1. Computer algorithms. I. Title. II. Series.
  QA76.9.A43 S87 2000
  005.1—DC21   00-042253

10 9 8 7 6 5 4 3 2 1

Książke te poświęcam Moim Rodzicom,
Aleksandrze i Wincentemu,
na ich 50-cio lecie małżeństwa,
za ich nieustającą wiarę we mnie.

Contents

PART II   PROBABILISTIC AND COMBINATORIAL TECHNIQUES

PART III   ANALYTIC TECHNIQUES

Foreword

Sequences—also known as strings or words—surface in many areas of science. Initially studied by combinatorialists in relation to problems of formal linguistics, they have proved to be of fundamental importance in a large number of computer science applications, most notably in textual data processing and data compression. Indeed, designers of large internet search engines acknowledge them to be huge algorithmic factories operating on strings. In a different sphere, properties of words are essential to the processing and statistical interpretation of biological or genetic sequences. There it is crucial to discern signal from noise and to do so in a computationally efficient manner.

As its title indicates, Szpankowski's book is dedicated to the analysis of algorithms operating on sequences. First, perhaps, a few words are in order regarding analysis of algorithms. The subject was founded by Knuth around 1963 and its aim is a precise characterization of the behaviour of algorithms that operate on large ensembles of data. A complexity measure (like execution time) is fixed and there are usually a few natural probabilistic models meant to reflect the data under consideration. The analyst's task then consists in predicting the complexity to be observed. Average-case analysis focuses on expected performance; whenever possible, finer characteristics like variances, limit distributions, or large deviations should also be quantified.

For a decade or so, it has been widely recognized that average-case analyses tend to be far more informative than worst-case ones as the latter focus on somewhat special pathological configurations. Provided the randomness model is realistic, average-case complexity estimates better reflect what is encountered in practice—hence their rôle in the design, optimization, and fine tuning of algorithms. In this context, for algorithms operating on words, properties of random sequences are crucial. Their study is the central theme of the book.

"Give a man a fish and you feed him for the day; teach him to fish and you feed him for his lifetime." Following good precepts, Szpankowski's book has a largely methodological orientation. The book teaches us in a lucid and balanced fashion the two main competing tracks for analysing properties of discrete randomness. Probabilistic methods based on inequalities and approximations include moment inequalities, limit theorems, large deviations, as well as martingales and ergodic theory, all of which nicely suited to the analysis of random discrete structures. Analytic methods fundamentally rely on exact modelling by generating functions which, once viewed as transformations of the complex plane, begin to reveal their secrets. Singularity analysis, saddle point strategies, and Mellin transforms become then instrumental. As Hadamard was fond of saying, the shortest path between two truths on the real line passes through the complex domain.

Throughout the book, the methodology is put to effective use in analysing some of the major problems concerning sequences that have an algorithmic or information-theoretic nature. In fact the book starts right away with a few easily stated questions that form recurrent themes; these are relative to digital trees, data compression, string editing, and pattern-matching. A great many more problems are thoroughly analysed in later chapters, including the celebrated leader election problem of distributed computing, fast pattern matching, basic information theory, Lempel-Ziv compression, lossy compression.

Analysis of algorithms is now a mature field. The time is ripe for books like this one which treat wide fields of applications. Szpankowski's book offers the first systematic synthesis on an especially important area—algorithms on sequences. Enjoy its mathematics! Enjoy its information theory! Enjoy its multi-faceted computational aspects!

PHILIPPE FLAJOLET

Preface

An algorithm is a systematic procedure that produces in a finite number of steps the solution to a problem. The name derives from the Latin translation *Algoritmi de numero Indorum* of the 9th-century Arab mathematician al-Khwarizmi's arithmetic treatise *Al-Khwarizmi Concerning the Hindu Art of Reckoning.* The most obvious reason for *analyzing* algorithms, and data structures associated with them, is to discover their characteristics in order to evaluate their suitability for various applications or to compare them with other algorithms for the same application. Often such analyses shed light on properties of computer programs and provide useful insights of the combinatorial behaviors of such programs. Needless to say, we are interested in *good* algorithms in order to efficiently use scarce resources such as computer space and time.

Most often algorithm designs are finalized toward the optimization of the asymptotic *worst-case* performance. Insightful, elegant, and useful constructions have been set up in this endeavor. Along these lines, however, the design of an algorithm is sometimes targeted at coping efficiently with unrealistic, even pathological inputs and the possibility is neglected that a simpler algorithm that works fast on average might perform just as well, or even better in practice. This alternative solution, also called a probabilistic approach, was an important issue three decades ago when it became clear that the prospects for showing the existence of polynomial time algorithms for NP-hard problems were very dim. This fact, and the apparently high success rate of heuristic approaches to solving certain difficult problems, led Richard Karp in 1976 to undertake a more serious investigation of probabilistic algorithms. In the last decade we have witnessed an increasing interest in the *probabilistic analysis* (also called *average-case analysis*) of algorithms, possibly due to a high success rate of randomized algorithms for computational geometry, scientific visualization, molecular biology, and information theory. Nowadays worst-case and average-case analyses coexist in a friendly symbiosis, enriching each other.

The focus of this book is on *tools* and *techniques* used in the average-case analysis of algorithms, where *average case* is understood very broadly (e.g., it includes exact and limiting distributions, large deviations, variance, and higher moments). Such methods can be roughly divided into two categories: *analytic* and *probabilistic*. The former were popularized by D. E. Knuth in his monumental three volumes, *The Art of Computer Programming*, whose prime goal was to accurately predict the performance characteristics of a wide class of algorithms. Probabilistic methods were introduced by Erdős and Rényi and popularized in the book by Erdős and Spencer, *Probabilistic Methods in Combinatorics*. In general, nicely structured problems are amenable to an analytic approach that usually gives much more precise information about the algorithm under consideration. As argued by Andrew Odlyzko: "Analytic methods are extremely powerful and when they apply, they often yield estimates of unparalleled precision." On the other hand, structurally complex algorithms are more likely to be first solved by probabilistic tools that later could be further enhanced by a more precise analytic approach.

The area of analysis of algorithms (at least, the way we understand it) was born on July 27, 1963, when D. E. Knuth wrote his "Notes on Open Addressing" about hashing tables with linear probing. Since 1963 the field has been undergoing substantial changes. We see now the emergence of combinatorial and asymptotic methods that allow the classification of data structures into broad categories that are amenable to a unified treatment. Probabilistic methods that have been so successful in the study of random graphs and hard combinatorial optimization problems play an equally important role in this field. These developments have two important consequences for the analysis of algorithms: it becomes possible to predict average behavior under more *general* probabilistic models; at the same time it becomes possible to analyze much more structurally *complex* algorithms. To achieve these goals the analysis of algorithms draws on a number of branches in mathematics: combinatorics, probability theory, graph theory, real and complex analysis, number theory and occasionally algebra, geometry, operations research, and so forth.

In this book, we choose one facet of the theory of algorithms, namely, algorithms and data structures on sequences (also called strings or words) and present a detailed exposition of the analytic and probabilistic methods that have become popular in such analyses. As stated above, the focus of the book is on techniques of analysis, but every method is illustrated by a variety of specific problems that arose from algorithms and data structures on strings. Our choice stems from the fact that there has been a resurgence of interest in algorithms on sequences and their applications in computational biology and information theory.

Our choice of methods covered here is aimed at closing the gap between analytic and probabilistic methods. There are excellent books on analytic methods such as the three volumes of D. E. Knuth and the recent book by Sedgewick and Flajolet. Probabilistic methods are discussed extensively in the books by Alon and Spencer, Coffman and Lueker, and Motwani and Raghavan. However, remarkably few books have been dedicated to both analytic and probabilistic analysis of algorithms, with possible exceptions of the books by Hofri and Mahmoud.

ABOUT THIS BOOK

This is a graduate textbook intended for graduate students in computer science, discrete mathematics, information theory, applied mathematics, applied probability, and statistics. It should also be a useful reference for researchers in these areas. In particular, I hope that those who are experts in probabilistic methods will find the analytic part of this book interesting, and vice versa.

The book consists of three parts: Part I describes a class of algorithms (with associated data structures) and formulates probabilistic and analytic models for studying them. Part II is devoted to probabilistic methods, whereas Part III presents analytic techniques.

Every chapter except the first two has a similar structure. After a general overview, we discuss the method(s) and illustrate every new concept with a simple example. In most cases we try to provide proofs. However, if the method is well explained elsewhere or the material to cover is too vast (e.g., the asymptotic techniques in Chapter 8), we then concentrate on explaining the main ideas behind the methods and provide references to rigorous proofs. At the end of each chapter there is an application section that illustrates the methods discussed in the chapter in two or three challenging research problems. Needless to say, the techniques discussed in this book were selected for inclusion on exactly one account: how useful they are to solve these application problems. Naturally, the selection of these problems is very biased, and often the problem shows up in the application section because I was involved in it. Finally, every chapter has a set of exercises. Some are routine calculations, some ask for details of derivations presented in the chapter, and others are research problems denoted as ⚠ and unsolved problems marked as ▽.

Now we discuss in some detail the contents of the book. Part I has two chapters. The first chapter is on the algorithms and data structures on words that are studied in this book: We discuss digital search trees (i.e., tries, PATRICIA tries, digital search trees, and suffix trees), data compression algorithms such as Lempel-Ziv schemes (e.g., Lempel-Ziv'77, Lempel-Ziv'78, lossy extensions of Lempel-Ziv schemes), pattern matching algorithms (e.g., Knuth-Morris-Pratt and Boyer-Moore), the shortest common superstring problem, string editing problem (e.g., the longest common subsequence), and certain combinatorial optimization problems. Chapter 2 builds probabilistic models on sequences that are used throughout the book to analyze the algorithms on strings. In particular, we discuss memoryless, Markov, and mixing sources of generating random sequences.

We also review some facts from probability theory and complex analysis. We finish this chapter with an overview on special functions (e.g., the Euler gamma function and the Riemann zeta function) that are indispensable for the analytic methods of Part III.

Part II consists of four chapters. Chapter 3 is on the probabilistic and combinatorial inclusion–exclusion principle, the basic tool of combinatorial analysis. After proving the principle, we discuss three applications, namely, the depth in a trie, order statistics, and the longest aligned matching word. Chapter 4 is devoted to the most popular probabilistic tool, that of the the first and the second moment methods. We illustrate them with a variety of examples and research problems (e.g., Markov approximation of a stationary distribution and the height analysis of digital trees). In Chapter 5 we discuss both the subadditive ergodic theorem and the large deviations. We use martingale differences to derive the very powerful Azuma's inequality (also known as the method of bounded differences). Finally, in Chapter 6 we introduce elements of information theory. In particular, we use random coding technique to prove three fundamental theorems of Shannon (i.e., the source coding theorem, the channel coding theorem and the rate distortion theorem). In the applications section we turn our attention to some recent developments in data compression based on pattern matching and the shortest common superstring problem. In particular, we show that with high probability a greedy algorithm that finds the shortest common superstring is asymptotically optimal. This is of practical importance because the problem itself is NP-hard.

Part III is on analytic methods and is composed of four chapters. Chapter 7 introduces generating functions, a fundamental and the most popular analytic tool. We discuss ordinary generating functions, exponential generating functions, and Dirichlet series. Applications range from pattern matching algorithms to the Delange formula on a digital sum. Chapter 8 is the longest in this book and arguably the most important. It presents an extensive course on complex asymptotic methods. It starts with the Euler-Maclaurin summation formula, matched asymptotics and the WKB method, continues with the singularity analysis and the saddle point method, and finishes with asymptotics of certain alternating sums. In the applications section we discuss the minimax redundancy rate for memoryless sources and the limiting distribution of the depth in digital search trees. The next two chapters continue our discussion of asymptotic methods. Chapter 9 presents the Mellin transform and its asymptotic properties. Since there are good accounts on this method (cf. [132, 149]), we made this chapter quite short. Finally, the last chapter is devoted to a relatively new asymptotic method known as depoissonization. The main thrust lies in an observation that certain problems are easier to solve when a deterministic input is replaced by a Poisson process. However, nothing is for free in this world, and after solving the problem in the Poisson domain one must translate the results back to the original problem, that is, depoissonize them. We cover here almost all known depoissonization results and illustrate them with three problems: analysis of the leader election algorithm, and the depth in generalized digital search trees for memoryless and Markovian sources. The latter analysis is among the most sophisticated in this book.

PERSONAL PERSPECTIVE

I can almost pin down the exact date when I got interested in the analysis of algorithms. It was in January 1983 in Paris during a conference on performance evaluation (at that time I was doing research in performance evaluation of multiaccess protocols and queueing networks). I came from a gloomy Poland, still under martial law, to a glamorous, bright, and joyful Paris. The conference was excellent, one of the best I have ever participated in. Among many good talks one stood out for me. It was on approximate counting, by Philippe Flajolet. The precision of the analysis and the brightness (and speed) of the speaker made a lasting impression on me. I wished I could be a disciple of this new approach to the analysis of algorithms. I learned from Philippe that he was influenced by the work of D. E. Knuth. For the first time I got a pointer to the three volumes of Knuth's book, but I could not find them anywhere in Poland.

In 1984 I left Gdańsk and moved to McGill University, Montreal. I had received a paper to review on the analysis of conflict resolution algorithms. The paper was interesting, but even more exciting was a certain recurrence that amazingly had a "simple" asymptotic solution. I verified numerically the asymptotics and the accuracy was excellent. I wanted to know why. Luckily, Luc Devroye had just returned from his sabbatical and he pointed me again to Knuth's books. I found what I needed in volumes I and III. It was an illumination! I was flabbergasted that problems of this complexity could be analyzed with such accuracy. I spent the whole summer of 1984 (re)learning complex analysis and reading Knuth's books. I started solving these recurrences using the new methods that I had been learning. I was becoming a disciple of the precise analysis of algorithms.

When I moved to Purdue University in 1985, I somehow figured out that the recurrences I was studying were also useful in data structures called *tries*. It was a natural topic for me to explore since I moved from an electrical engineering department to a computer science department. I got hooked and decided to write to Philippe Flajolet, to brag about my new discoveries. In response he sent me a ton of papers of his own, solving even more exciting problems. I was impressed. In May 1987 he also sent to Purdue his best weapon, a younger colleague whose name was Philippe Jacquet. When Philippe Jacquet visited me I was working on the analysis of the so-called interval searching algorithm, which I knew how to solve but only with numerical help. Philippe got a crucial idea on how to push it using only analytic tools. We wrote our first paper [214]. Since then we have met regularly every year producing more and longer papers (cf. Chapter 10; in particular, Section 10.5.3). We have become friends.

Finally, in 1989 I *again* rediscovered the beauty of information theory after reading the paper [452] by Wyner and Ziv. There is a story associated with it. Wyner and Ziv proved that the typical length of repeated substrings found within the first n positions of a random sequence is with high probability $\frac{1}{h} \log n$, where h is the entropy of the source (cf. Section 6.5.1). They asked if this result can be extended to almost sure convergence. My work on suffix trees (cf. Section 4.2.6) paid off since I figured out that the answer is in the negative [411] (cf. also Section 6.5.1). The crucial intuition came from the work of Boris Pittel [337], who encouraged me to study it and stood behind me in the critical time when my analysis was under attack. It turned out that Ornstein and Weiss [333] proved that the Wyner and Ziv conjecture is true. To make the story short, let me say that fortunately both results were correct since we analyzed slightly different settings (i.e., I analyzed the problem in the so-called right domain of asymptotics, and Ornstein and Weiss in the left domain). The reader may read more about it in Chapter 6. Since then I have found information theory more and more interesting. Philippe Jacquet and I have even coined the term *analytic information theory* for dealing with problems of information theory by using analytic methods, that is, those in which complex analysis plays a pivotal role.

Acknowledgments

There is a long list of colleagues and friends from whom I benefited through encouragement and critical comments. They helped me in various ways during my tenure in the analysis of algorithms. All I know about analytic techniques comes from two Philippes: Jacquet and Flajolet. And vice versa, what I do not know is, of course, their fault. On a more serious note, my direct contact with Philippe Jacquet and Philippe Flajolet has influenced my thinking and they have taught me many tricks of the trade. (I sometimes wonder if I didn't become French during numerous visits to INRIA, Rocquencourt; all except for mastering the language!) Many applications in this book are results of my work with them. Needless to say, their friendship and inspirations were invaluable to me. *Merci beaucoup mes amis.*

I have been learning probabilistic techniques from masters: David Aldous, Alan Frieze, Tomasz Łuczak, and Boris Pittel. *Thanks. Thanks. Dziękuje.* Спасибо.

I thank my numerous co-authors with whom I worked over the last 15 years: David Aldous, Marc Alzina, Izydor Apostol, Alberto Apostolico, Mikhail Atallah, Luc Devroye, Jim Fill, Philippe Flajolet, Ioannis Fudos, Yann Genin, Leonidas Georgiadis, Ananth Grama, Micha Hofri, Svante Janson, Philippe Jacquet, Peter Kirschenhofer, Chuck Knessl, Guy Louchard, Tomasz Łuczak, Hosam Mahmoud, Dan Marinescu, Evaggelia Pitoura, Helmut Prodinger, Bonita Rais, Vernon Rego, Mireille Régnier, John Sadowsky, Erkki Sutinen, Jing Tang, and Leandros Tassiulas.

I also had the privilege of talking with D. E. Knuth, whose encouragement was very important to me. His influence pervades the book.

Many colleagues have read various versions of this book. I profoundly thank Luc Devroye, Michael Drmota, Leonidas Georgiadis, Philippe Jacquet, Svante Janson, Chuck Knessl, Yiannis Kontoyiannis, Guy Louchard, John Kieffer, Gabor Lugosi, Kihong Park, Helmut Prodinger, Yuri Reznik, and Erkki Sutinen. I am particularly in debt to Leonidas Georgiadis, who read the entire book and gave me many valuable comments.

As with every large project, I am sure I did not avoid mistakes and typos. I will try to keep errata on my home page www.cs.purdue.edu/people/spa. Readers' help in eliminating remaining inaccuracies will be greatly appreciated. Please send comments to spa@cs.purdue.edu.

This book was written over the last four years. I have received invaluable help from staff and faculty of the Department of Computer Sciences at Purdue University. I am grateful to the National Science Foundation, which supported me over the last 15 years. The last three chapters were written during my sabbatical at Stanford University during fall 1999. I thank Tom Cover for being a wonderful host and his students for welcoming me into their weekly seminar.

Philippe Flajolet kindly agreed to write a Foreword to this book. I thank him from the bottom of my heart. Philippe Jacquet has been entertaining me for more than ten years with his good humor and comic sketches. He generously agreed to contribute ten comic illustrations to this book; one for each chapter. If only for these sketches, the book should be on your shelf!

Finally, I am in debt to my family for putting up with me through these years and helping me, in their own way, to continue this project. My wife, Mariola, designed the cover. I have much to thank her for, including some twenty years with me.

Wojtek Szpankowski

West Lafayette, Indiana, 1997–1998
Stanford, California, 1999

Part I

PROBLEMS ON WORDS

1

Data Structures and Algorithms on Words

In this book we choose one facet of the theory of algorithms, namely data structures and algorithms on sequences (strings, words) to illustrate probabilistic, combinatorial, and analytic techniques of analysis. In this chapter, we briefly describe some data structures and algorithms on words (e.g., tries, PATRICIA tries, digital search trees, pattern matching algorithms, Lempel-Ziv schemes, string editing, and shortest common superstring) that are used extensively throughout the book to illustrate the methods of analysis.

Data structures and algorithms on sequences have experienced a new wave of interest due to a number of novel applications in computer science, communications, and biology. Among others, these include dynamic hashing, partial match retrieval of multidimensional data, searching and sorting, pattern matching, conflict resolution algorithms for broadcast communications, data compression, coding, security, genes searching, DNA sequencing, genome maps, double digest problem, and so forth. To satisfy these diversified demands various data structures were proposed for these algorithms. Undoubtedly, the most popular data structures in algorithms on words are digital trees [3, 269, 305] (e.g., tries, PATRICIA, digital search trees), and in particular suffix trees [3, 17, 77, 375, 383, 411, 412]. We discuss various digital trees and introduce several parameters characterizing them that we shall study throughout the book.

The importance of digital trees stems from their abundance of applications in other problems such as data compression (Section 1.2), pattern matching (Section 1.3), and the shortest common superstring problem (Section 1.4). These problems recently became very important due to the need for an efficient storage and transmission of multimedia, and possible applications to DNA sequencing.

Graphs and directed acyclic graphs (DAG) also find several applications in problems on strings. In particular, we consider the *edit distance* problem and its variants (Section 1.5). Finally, we close this chapter with a brief discussion of certain class of optimization problems on graphs that find applications for algorithms on sequences (Section 1.6).

1.1 DIGITAL TREES

We start our discussion with a brief review of the **digital trees**. The most basic digital tree, known as a **trie** (from re*trie*val), is defined first, and then other digital trees (such as PATRICIA, digital search trees and suffix trees) are described in terms of the trie.

The primary purpose of a trie is to store a set $\mathcal{X}$ of strings (words, sequences), say $\mathcal{X} = \{X^1, \ldots, X^n\}$. Throughout the book, the terms strings, words and sequences are used interchangeably. Each string is a finite or infinite sequence of symbols taken from a finite alphabet $\mathcal{A} = \{\omega_1, \ldots, \omega_V\}$ of size $V = |\mathcal{A}|$. We use a generic notation $\mathcal{D}_n$ for all digital trees built over a set $\mathcal{X}$ of n strings. A string will be stored in a leaf of the trie. The trie over $\mathcal{X}$ is built recursively as follows: For $|\mathcal{X}| = 0$, the trie is, of course, empty. For $|\mathcal{X}| = 1$, $trie(\mathcal{X})$ is a single node. If $|\mathcal{X}| > 1$, $\mathcal{X}$ is split into V subsets $\mathcal{X}_1, \mathcal{X}_2, \ldots, \mathcal{X}_V$ so that a string is in $\mathcal{X}_j$ if its first symbol is ω_j. The tries $trie(\mathcal{X}_1), trie(\mathcal{X}_2), \ldots, trie(\mathcal{X}_V)$ are constructed in the same way except that at the kth step, the splitting of sets is based on the kth symbol. They are then connected from their respective roots to a single node to create $trie(\mathcal{X})$. Figure 1.1 illustrates such a construction. Observe

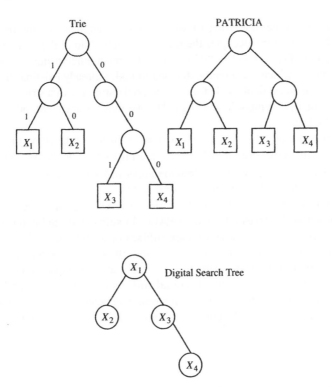

Figure 1.1. A trie, Patricia trie and a digital search tree (DST) built from the following four strings $X^1 = 11100\ldots$, $X^2 = 10111\ldots$, $X^3 = 00110\ldots$, and $X^4 = 00001\ldots$.

that all strings are stored in external nodes (shown as boxes in Figure 1.1) while internal nodes are branching nodes used to direct strings to their destinations (i.e., external nodes). When a new string is inserted, the search starts at the root and proceeds down the tree as directed by the input symbols (e.g., symbol "0" in the input string means move to the right and "1" means proceed to the left as shown in Figure 1.1).

There are many possible variations of the trie. One such variation is the **b-trie**, in which a leaf is allowed to hold as many as b strings (cf. [145, 305, 411]). The b-trie is particularly useful in algorithms for extendible hashing in which the capacity of a page or other storage unit is b. A second variation of the trie, the **PATRICIA trie** (*P*ractical *A*lgorithm *T*o *R*etrieve *I*nformation *C*oded *I*n *A*lphanumeric) eliminates the waste of space caused by nodes having only one branch. This is done by collapsing one-way branches into a single node (Figure 1.1). In a **digital search tree** (DST), shown also in Figure 1.1, strings are directly stored in nodes so that external nodes are eliminated. More precisely, the root contains the first string (however, in some applications the root is left empty). The next string occupies the right or the left child of the root depending on whether its first symbol is "0" or "1". The remaining strings are stored in available nodes which are directly attached to nodes already existing in the tree. The search for an available node follows the prefix structure of a string as in tries. That is, if the next symbol in a string is "0" we move to the right, otherwise we move to the left.

As in the case of a trie, we can consider an extension of the above digital trees by allowing them to store up to b strings in an (external) node. We denote such digital trees as $\mathcal{D}_n^{(b)}$, but we often drop the upper index when $b = 1$ is discussed. Figure 1.1 illustrates these definitions for $b = 1$.

One of the most important example of tries and PATRICIA tries are **suffix trees** and **compact suffix trees** (also called PAT). In suffix trees and compact suffix trees, the words stored in these trees are suffixes of a given string $X = x_1 x_2 \ldots$; that is, the word $X^j = x_j x_{j+1} x_{j+2} \ldots$ is the jth suffix of X which begins at the jth position of X. Thus a suffix tree is a trie and a compact suffix tree is a PATRICIA trie in which the words are all suffixes of a *given* string. Clearly, in this case the strings X^j for $j = 1, \ldots, n$ strongly depend on each other while in a trie the strings of $\mathcal{X}$ might be completely independent. A suffix tree is illustrated in Figure 1.2.

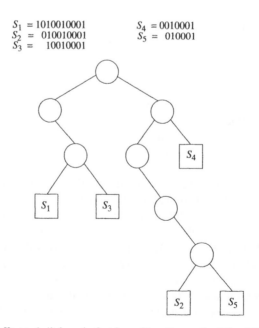

$$S_1 = 1010010001 \qquad S_4 = 0010001$$
$$S_2 = 010010001 \qquad S_5 = 010001$$
$$S_3 = 10010001$$

Figure 1.2. Suffix tree built from the first five suffixes $S_1, \ldots, S_5$ of $X = 0101101110 \ldots$.

Certain characteristics of tries and suffix trees are of primary importance. We define them below.

Definition 1.1 (Digital Trees Parameters) *Let us consider a b-digital tree $\mathcal{D}_n^{(b)}$ built over n strings and capable of storing up to b strings in an (external) node.*

(i) *The mth depth $D_n^{(b)}(m)$ is the length of a path from the root of the digital tree to the (external) node containing the mth string.*

(ii) *The typical depth $D_n^{(b)}$ is the depth of a randomly selected string, that is,*

$$\Pr\{D_n^{(b)} \le k\} = \frac{1}{n} \sum_{m=1}^{n} \Pr\{D_n^{(b)}(m) \le k\}. \tag{1.1}$$

(iii) *The (external) path length $L_n^{(b)}$ is the sum of all depths, that is,*

$$L_n^{(b)} = \sum_{m=1}^{n} D_n^{(b)}(m). \tag{1.2}$$

(iv) *The height $H_n^{(b)}$ is the length of the longest depth, that is,*

$$H_n^{(b)} = \max_{1 \le m \le n} \{D_n^{(b)}(m)\}. \tag{1.3}$$

(v) *The fill-up level $F_n^{(b)}$ is the maximal full level in the tree $\mathcal{D}_n^{(b)}$, that is, $\mathcal{D}_n^{(b)}$ is a full tree up to level $F_n^{(b)}$ but not on the level $F_n^{(b)} + 1$. In other words, on levels $0 \le i \le F_n^{(b)}$ there are exactly V^i nodes (either external or internal) in the tree, but there are less than $V^{F_n^{(b)}+1}$ nodes on level $F_n^{(b)} + 1$.*

(vi) *The shortest depth $s_n^{(b)}$ is the length of the shortest path from the root to an external node, that is,*

$$s_n^{(b)} = \min_{1 \le m \le n} \{D_n^{(b)}(m)\}. \tag{1.4}$$

(Observe that $F_n^{(b)} \le s_n^{(b)}$.)

(vii) *The size $S_n^{(b)}$ is the number of (internal) nodes in $\mathcal{D}_n^{(b)}$.*

(viii) *The kth average profile $\bar{B}_n^{(b)}(k)$ is the average number of strings stored on level k of $\mathcal{D}_n^{(b)}$.*

These parameters can be conveniently represented in another way that reveals combinatorial relationships between strings stored in such digital trees. We start with the following definition.

Definition 1.2 **(Alignments)** *For the set of strings $\mathcal{X} = \{X^1, X^2, \ldots, X^n\}$, the alignment $C_{i_1 \ldots i_{b+1}}$ between $b + 1$ strings $X^{i_1}, \ldots, X^{i_{b+1}}$ is the length of the longest common prefix of all these $b + 1$ strings.*

To illustrate this definition and show its usefulness to the analysis of digital trees, we consider the following example.

Example 1.1: *Illustration of the Self-Alignments* Let us consider the suffix tree shown in Figure 1.2. Define $C = \{C_{ij}\}_{i,j=1}^5$ for $b = 1$ as the self-alignment matrix which is shown explicitly below:

$$C = \begin{bmatrix} \star & 0 & 2 & 0 & 0 \\ 0 & \star & 0 & 1 & 4 \\ 2 & 0 & \star & 0 & 0 \\ 0 & 1 & 0 & \star & 1 \\ 0 & 4 & 0 & 1 & \star \end{bmatrix}.$$

Observe that we can express the parameters of Definition 1.1 in terms of the self-alignments C_{ij} as follows:

$$D_n(1) = \max_{2 \leq j \leq 5} \{C_{1j}\} + 1 = 3,$$

$$H_n = \max_{1 \leq i < j \leq n} \{C_{ij}\} + 1 = 5,$$

$$s_n = \min_{1 \leq i \leq n} \max_{1 \leq j \leq n} \{C_{ij}\} + 1 = 2$$

(since $b = 1$ we drop b from the above notations).

Certainly, similar relationships hold for tries, but not for PATRICIA tries and digital search trees. In the latter case, however, one can still express parameters of the trees in terms of the alignments matrix C. For example, the depth of the fourth string $D_4(4)$ can be expressed as follows:

$$D_4(4) = \max\{\min\{C_{41}, D_4(1)\}, \min\{C_{42}, D_4(2)\}, \min\{C_{43}, D_4(3)\}\}.$$

This is a bit too complicated to be of any help.   ∎

The above example suggests that there are relatively simple relationships between parameters of a trie and the alignments. Indeed, this is the case as the theorem below shows. The reader is asked to provide a formal proof in Exercise 1.

Theorem 1.3 *In a trie the following holds:*

$$D_n^{(b)}(i_{b+1}) = \max_{1 \leq i_1, \ldots, i_b \leq n} \{C_{i_1 \ldots i_{b+1}}\} + 1, \tag{1.5}$$

$$H_n^{(b)} = \max_{1 \le i_1,\ldots,i_{b+1} \le n} \{C_{i_1 \ldots i_{b+1}}\} + 1, \tag{1.6}$$

$$D_n^{(b)}(n+1) = \max_{1 \le i_1,\ldots,i_b \le n} \{C_{i_1 \ldots i_b, n+1}\} + 1, \tag{1.7}$$

$$s_n^{(b)} = \min_{1 \le i_{b+1} \le n} \{D_n^{(b)}(i_{b+1})\} = \min_{1 \le i_{b+1} \le n} \max_{1 \le i_1,\ldots,i_b \le n} \{C_{i_1 \ldots i_{b+1}}\} + 1 \tag{1.8}$$

for any $1 \le i_1, \ldots, i_{b+1} \le n$, *where* $D_n^b(n+1) = I_n$ *is the depth of insertion.*

In passing, we should mention that the above combinatorial relationships find applications in problems not directly related to digital trees. We shall meet them again in Section 6.5.2 when analyzing the shortest common superstring problem described in Section 1.4.

The digital trees are used very extensively in this book as illustrative examples. We analyze the height of tries, PATRICIA tries, and suffix trees in Chapter 4. Recurrences arising in the analysis of digital trees are discussed in Chapter 7, while the typical depth of digital search trees is studied in Chapters 8–10.

1.2 DATA COMPRESSION: LEMPEL-ZIV ALGORITHMS

Source coding is an area of information theory (see Chapter 6) that deals with problems of optimal data compression. The most successful and best-known data compression schemes are due to Lempel and Ziv [460, 461]. Efficient implementation of these algorithms involves digital trees. We describe here some aspects of the Lempel-Ziv schemes and return to them in Chapter 6.

We start with some definitions. Consider a sequence $\{X_k\}_{k=1}^{\infty}$ taking values in a finite alphabet $\mathcal{A}$ (e.g., for English text the cardinality $|\mathcal{A}| = 26$ symbols, while for an image $|\mathcal{A}| = 256$). We write X_m^n to denote $X_m, X_{m+1} \ldots X_n$. We encode X_1^n into a *binary* (compression) code $\mathscr{C}_n$, and the decoder produces the reproduction sequence $\hat{X}_1^n$ of X_1^n. More precisely, a code $\mathscr{C}_n$ is a function $\phi : \mathcal{A}^n \to \{0, 1\}^*$. On the decoding side, the decoder function $\psi : \{0, 1\}^* \to \mathcal{A}^n$ is applied to find $\hat{X}_1^n = \psi(\phi(X_1^n))$. Let $\ell(\mathscr{C}_n(X_1^n))$ be the length of the code $\mathscr{C}_n$ (in bits) representing X_1^n. Then the *bit rate* is defined as

$$r(X_1^n) = \frac{\ell(\mathscr{C}_n(X_1^n))}{n}.$$

For example, for text $r(X_1^n)$ is expressed in bits per symbol, while for image compression in bits per pixel or in short bpp.

We shall discuss below two basic Lempel-Ziv schemes, namely the so-called Lempel-Ziv'77 (LZ77) [460] and Lempel-Ziv'78 (LZ78) [461]. Both schemes are examples of **lossless** compression; that is, the decoder can recover exactly the encoded sequence. A number of interesting problems arise in lossy extensions of the Lempel-Ziv schemes. In the **lossy** data compression, discussed below, some information can be lost during the encoding.

1.2.1 Lempel-Ziv'77 Algorithm

The Lempel-Ziv algorithm partitions or parses a sequence into phrases that are similar in some sense. Depending on how such a parsing is encoded we have different versions of the algorithm. However, the basic idea is to find the longest prefix of yet uncompressed sequence that occurs in the already compressed sequence.

More specifically, let us assume that the first n symbols X_1^n are given to the encoder and the decoder. This initial string is sometimes called the "database string" or the "training sequence." Then we search for the longest prefix $X_{n+1}^{n+\ell}$ of X_{n+1}^{∞} that is repeated in X_1^n, that is,

Let I_n be the largest ℓ such that $X_{n+1}^{n+\ell} = X_m^{m+\ell-1}$ for some prescribed range of m and ℓ.

Depending on the range of m and ℓ, we can have different versions of the LZ77 scheme. In the original LZ77 algorithm [460], m and ℓ were restricted to a window size W and "lookahead" buffer B, that is, $n - W + 1 \le m \le n$ and $\ell \le B$. This implementation is sometimes called the *sliding window* LZ77. In the *fixed database* (FDLZ) version [452, 453], one sets $1 \le m \le W$ and $m - 1 + \ell \le W$; that is, the database sequence is fixed and the parser always looks for matches inside such a fixed substring X_1^W. Such a scheme is sometimes called the Wyner-Ziv scheme [452]. Finally, in the *growing database* version the only restriction is that $1 \le m \le n$ (i.e., the database consists of the last n symbols).

In general, the code built for LZ77 consists of the triple (m, ℓ, char) where char is the symbol $X_{m+\ell}$. Since the pointer to m needs $\log_2 n$ bits, the length ℓ could be coded in $O(\log I_n)$ bits and char requires $\log |\mathcal{A}|$ bits, the code length of a *phrase* is $\log_2 n + O(\log I_n) + \log_2 |\mathcal{A}|$ bits. In Exercise 5 we propose a formula for the code length of the FDLZ, while in Exercise 6 the reader is asked to find the code length for all other versions of the Lempel-Ziv schemes.

The heart of all versions of the Lempel-Ziv schemes is the algorithm that finds the longest prefix of length I_n that occurs in the database string of length n. It turns out that the suffix tree discussed in Section 1.1 can be used to efficiently find such a prefix. Indeed, let us consider a sequence $X = 1010010001\dots$, and assume X_1^4 is the database string. The suffix tree built over X_1^4 is shown in Figure 1.3. Let us now look for I_4, that is, the longest prefix of X_5^{∞} that occurs (starts) in the database X_1^4. In the growing database implementation it is X_5^8 since it is equal to

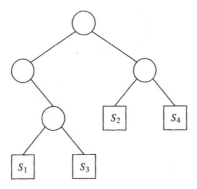

Figure 1.3. Suffix tree built from the first four suffixes of $X = 1010010001\ldots$.

X_2^5. This can be seen by inserting the fifth suffix of X into the suffix tree from Figure 1.3—which actually leads to the suffix tree shown in Figure 1.2.

1.2.2 Lempel-Ziv'78 Algorithm

The Lempel-Ziv'78 (LZ78) is a *dictionary-based* scheme that partitions a sequence into phrases (blocks) of variable sizes such that a new block is the shortest substring not seen in the past as a phrase. Every such phrase is encoded by the index of its prefix appended by a symbol, thus LZ78 code consists of pairs (`pointer, symbol`). A phrase containing only one symbol is coded with the index equal to zero.

Example 1.2: *The Lempel-Ziv'78 and Its Code* Consider the string $X_1^{14} = ababbbabbaaaba$ over the alphabet $\mathcal{A} = \{a, b\}$, which is parsed and coded as follows:

Phrase No:	1	2	3	4	5	6	7
Sequence:	(a)	(b)	(ab)	(bb)	(abb)	(aa)	(aba)
Code:	0a	0b	1b	2b	3b	1a	3a

Observe that we need $\lceil \log_2 7 \rceil$ bits to code a phrase, and two bits to code a symbol, so in total for 7 phrases we need 28 bits.   ■

The most time consuming part of the algorithm is finding the next phrase, that is, searching the dictionary. However, this can be speeded up by using a digital search tree to build the dictionary. For example, the string 11001010001000100 is parsed into (1)(10)(0)(101)(00)(01)(000)(100), and this process is represented in Figure 1.4 using the digital search tree structure. In this case, however, we

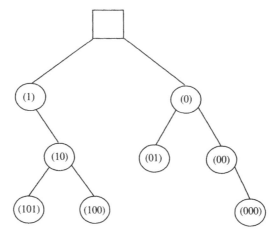

Figure 1.4. A digital tree representation of the Lempel-Ziv parsing for the string 11001010001000100.

leave the root empty (or we put an empty phrase into it). To show that the root is different from other nodes we draw it in Figure 1.4 as a square. All other phrases of the Lempel-Ziv parsing algorithm are stored in internal nodes (represented in the figure as circles). When a new phrase is created, the search starts at the root and proceeds down the tree as directed by the input symbols exactly in the same manner as in the digital tree construction (cf. Section 1.1). The search is completed when a branch is taken from an existing tree node to a new node that has not been visited before. Then the edge and the new node are added to the tree. The phrase is just a concatenation of symbols leading from the root to this node, which also stores the phrase.

We should observe differences between digital search trees discussed in Section 1.1 and the one described above. For the Lempel-Ziv scheme we consider a word of *fixed length*, say n, while before we dealt with *fixed number of strings*, say m, resulting in a digital tree consisting of exactly m nodes. Looking at Figure 1.4, we conclude that the number of nodes in the associated digital tree is equal to the number of phrases generated by the Lempel-Ziv algorithm.

1.2.3 Extensions of Lempel-Ziv Schemes

Finally, we shall discuss two extensions of Lempel-Ziv schemes, namely *generalized* Lempel-Ziv'78 and *lossy* Lempel-Ziv'77. Not only are these extensions useful from a practical point of view (cf. [11, 29, 299, 399, 361]), but they are also a source of interesting analytical problems. We return to them in Chapters 6, 9, and 10.

Generalized Lempel-Ziv'78   Let us first consider the **generalized** Lempel-Ziv'78 scheme. It is known that the original Lempel–Ziv scheme does not cope very well with sequences containing a long string of repeated symbols (i.e., the associated digital search tree is a skewed one with a long path). To somewhat remedy this situation, Louchard, Szpankowski and Tang [299] introduced a generalization of the Lempel–Ziv parsing scheme that works as follows: Fix an integer $b \geq 1$. The algorithm parses a sequence into phrases such that the next phrase is the *shortest* phrase seen in the past by *at most* $b - 1$ phrases ($b = 1$ corresponds to the original Lempel–Ziv algorithm). It turns out that such an extension of the Lempel-Ziv algorithm protects against the propagation of errors in the dictionary (cf. [399, 361]).

Example 1.3:   *Generalized Lempel-Ziv'78*   Consider the sequence

$$\alpha\beta\alpha\beta\beta\alpha\beta\alpha\beta\alpha\alpha\alpha\alpha\alpha\alpha\alpha\gamma$$

over the alphabet $\mathcal{A} = \{\alpha, \beta, \gamma\}$. For $b = 2$ it is parsed as follows:

$$(\alpha)(\beta)(\alpha)(\beta)(\beta\alpha)(\beta\alpha)(\beta\alpha\alpha)(\alpha\alpha)(\alpha\alpha)(\alpha\alpha\alpha)(\gamma)$$

that has seven *distinct* phrases and eleven phrases. The code for this new algorithm consists: (i) either of (`pointer`, `symbol`) when `pointer` refers to the *first* previous occurrence of the prefix of the phrase and `symbol` is the value of the last symbol of this phrase; (ii) or just (`pointer`) if the phrase has occurred previously (i.e., it is the second or the third or . . . the bth occurrence of this phrase). For example, the code for the previously parsed sequence is for $b = 2$: $0\alpha0\beta122\alpha33\alpha1\alpha55\alpha0\gamma$ (e.g., the phrase (2α) occurs for the first time as a new phrase, hence (2) refers to the second distinct phrase appended by α, while code (5) represents a phrase that has its second occurrence as the fifth distinct phrase). Observe that this code is of length 47 bits since there are eleven phrases each requiring up to $\lceil \log_2 7 \rceil = 3$ bits and seven symbols need 14 additional bits (i.e., $47 = 11 \cdot 3 + 7 \cdot 2 = 47$). The original LZ78 code needs 54 bits. We saved 7 bits! But, the reader may verify that the same sequence requires only 46 bits for $b = 3$ (so only one additional bit is saved), while for $b = 4$ the bit count increases again to 52.   ∎

The above example suggests that $b = 3$ is (at least local) optimum for the above sequence. Can one draw similar conclusions "on average" for a typical sequence (i.e., generated randomly)? This book is intended to provide tools to analyze such problems.

As for the original Lempel-Ziv algorithm, the most time-consuming part of the construction is to generate a new phrase. An efficient way of accomplishing this

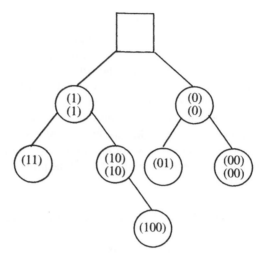

Figure 1.5. A 2-digital search tree representation of the generalized Lempel–Ziv parsing for the string 1100101000100010011.

is by means of generalized digital search trees, introduced in Section 1.1, namely b-digital search tree (b-DST). We recall that in such a digital tree one is allowed to store up to b strings in a node. In Figure 1.5 we show the 2-DST constructed from the sequence 1100101000100010011.

Lossy Extension of Lempel-Ziv'77   We now discuss another extension, namely a **lossy** Lempel-Ziv'77 scheme. In such a scheme in the process of encoding some information is lost. To control this loss, one needs a measure of fidelity $d(\cdot, \cdot)$ between two sequences. For example, the Hamming distance is defined as

$$d_n(X_1^n, \hat{X}_1^n) = \frac{1}{n} \sum_{i=1}^{n} d_1(X_i, \hat{X}_i)$$

where $d_1(X_i, \hat{X}_i) = 0$ for $X_i = \hat{X}_i$ and 1 otherwise. In the square error distortion we set $d(X_i, \hat{X}_i) = (X_i - \hat{X}_i)^2$.

Let us now fix $D > 0$. In the lossy LZ77, we consider the longest prefix of the uncompressed file that approximately (within distance D) occurs in the database sequence. More precisely, the quantity I_n defined in Section 1.2.1 becomes in this case:

Let I_n be the largest K such that a prefix of X_{n+1}^{∞} of length K is within distance D from X_i^{i-1+K} for some $1 \le i \le n - K + 1$, that is, $d(X_i^{i-1+K}, X_{n+1}^{n+K}) \le D$.

Not surprisingly, the bit rate of such a compression scheme depends on the probabilistic behavior of I_n. We shall analyze it in Chapter 6. The reader is also referred to [91, 278, 303, 403, 439].

1.3   PATTERN MATCHING

There are various kinds of patterns occurring in strings that are important to locate. These include squares, palindromes, and specific patterns. For example, in computer security one wants to know if a certain pattern (i.e., a substring, or even better a subsequence) appears (too) frequently in an audit file (text) since this may indicate an intrusion. In general, pattern matching involves a pattern H and a text string T. One is asked to determine the existence of H within T, the first occurrence of H, the number of occurrences or the location of all occurrences of H.

Two well-known pattern matching algorithms are the Knuth-Morris-Pratt (KMP) algorithm and the Boyer-Moore (BM) algorithm [3, 77]. In this section we focus on the former. The efficiency of these algorithms depends on how quickly one determines the location of the next matching attempt provided the previous attempt was unsuccessful. The key observation here is that following a mismatch at, say the kth position of the pattern, the preceding $k - 1$ symbols of the pattern and their structure give insight as to where the next matching attempt should begin. This idea is used in the KMP pattern matching algorithm and is illustrated in the following example and Figure 1.6.

Example 1.4:   *The Morris–Pratt Algorithm*   We now consider a simplified version of the KMP algorithm, namely that of the Morris–Pratt pattern matching algorithm. Let $H_1^6 = 011010$ and the text string $T_1^{10} = 1011011011$, as shown in

$$
\begin{array}{ll}
H & 0\ 1\ 1\ 0\ 1\ 0 \\
T & 1\ 0\ 1\ 1\ 0\ 1\ 1\ 0\ 1\ 1 \\
& \text{(a) first attempt}
\end{array}
$$

$$
\begin{array}{ll}
H & 0\ 1\ 1\ 0\ 1\ 0 \\
T & 1\ 0\ 1\ 1\ 0\ 1\ 1\ 0\ 1\ 1 \\
& \text{(b) second attempt}
\end{array}
$$

$$
\begin{array}{ll}
H & 0\ 1\ 10\ 1\ 0 \\
T & 1\ 0\ 1\ 1\ 0\ 1\ 1\ 0\ 1\ 1 \\
& \text{(c) third attempt}
\end{array}
$$

Figure 1.6. Comparisons made by the Morris–Pratt pattern matching algorithm

Figure 1.6. When attempting to match P with T, we proceed from left to right, comparing each symbol. No match is made with the first symbol of each, so the pattern H is moved one position to the right. On the second attempt, the sixth symbol of H does not match the text, so this attempt is halted and the pattern H is shifted to the right. Notice that it is not fruitful to begin matching at either the third or fourth position of T since the suffix 01 of the so far matched pattern $H_1^5 = 01101$ is equal to the prefix 01 of H_1^5. Thus the next matching attempt begins at the fifth symbol of T.   ∎

Knowing how far to shift the pattern H is the key to both the KMP and the BM algorithms. Therefore, the pattern H is preprocessed to determine the shift. Let us assume that a mismatch occurs when comparing T_l with H_k. Then some alignment positions can be disregarded without further text-pattern comparisons. Indeed, let $1 \le i \le k$ be the largest integer such that $H_{k-i}^{k-1} = H_1^i$, that is, i is the longest prefix of H that is equal to the suffix of H^{k-1} of length i. Then positions $l - k + 1, l - k + 2, \ldots, l - i + 1$ of the text do not need to be inspected and the pattern can be shifted by $k - i$ positions (as already observed in Figure 1.6). The set of such i can be known by a preprocessing of H.

There are different variants of the classic Knuth-Morris-Pratt algorithm [272] that differ by the way one uses the information obtained from the mismatching position. We formally define two variants, and provide an example. They can be described formally by assigning to them the so-called shift function S that determines by how much the pattern H can be shifted before the next comparison at $l + 1$ is made. We have:

Morris-Pratt variant:

$$S = \min\{k : \ \min\{s > 0 : H_{1+s}^{k-1} = H_1^{k-1-s}\}\} \ ;$$

Knuth-Morris-Pratt variant:

$$S = \min\{k : \ \min\{s : \ H_{1+s}^{k-1} = H_1^{k-1-s} \text{ and } H_k \ne H_{k-s}\}\}$$

There are several parameters of pattern matching algorithms that either determine their performance or shed some light on their behaviors. For example, the efficiency of an algorithm is characterized by its complexity, defined below.

Definition 1.4

(i) *For any pattern matching algorithm that runs on a given text T and a given pattern H, let $M(l, k) = 1$ if the lth symbol T_l of the text is compared by the algorithm to the kth symbol H_k of the pattern, and $M(l, k) = 0$ otherwise.*

(ii) *For a given pattern matching algorithm the partial complexity function $C_{r,n}$ is defined as*

$$C_{r,s}(H_1^m, T_1^n) = \sum_{l \in [r,s], k \in [1,m]} M[l, k]$$

where $1 \leq r < s \leq n$. For $r = 1$ and $s = n$ the function $C_{1,n} := C_n$ is called the **complexity** *of the algorithm.*

We illustrate some of the notions just introduced in the following example.

Example 1.5: *Illustration to Definition 1.4* Let $H = abacabacabab$ and $T = abacabacabaaa$. The first mismatch occurs for $M(12, 12)$. The comparisons performed from that point are:

1. **Morris-Pratt variant:**

   $(12, 12); (12, 8); (12, 4); (12, 2); (12, 1); (13, 2); (13, 1)$,

   where the text character is compared in turn with pattern characters (b, c, c, b, a, b, a) with the alignment positions $(1, 5, 9, 11, 12, 12, 13)$.

2. **Knuth-Morris-Pratt variant:**

   $(12, 12); (12, 8); (12, 2); (12, 1); (13, 2); (13, 1)$,

   where the text character is compared in turn with pattern characters (b, c, b, a, b, a) with the alignment positions $(1, 5, 11, 12, 12, 13)$.   ∎

It is interesting to observe that the subset $\{1, 5, 12\}$ appears in all variants. We will see that these positions share a common property of "unavoidability" explored in Section 5.5.2. We shall also discuss the pattern matching problem in Section 7.6.2.

1.4 SHORTEST COMMON SUPERSTRING

Various versions of the **shortest common superstring** (SCS) problem play important roles in data compression and DNA sequencing. In fact, in laboratories DNA sequencing (cf. [285, 445]) is routinely done by sequencing large numbers of relatively short fragments and then heuristically finding a short common superstring. The problem can be formulated as follows: given a collection of strings, say $X^1, X^2, \ldots, X^n$ over an alphabet $\mathcal{A}$, find the shortest string Z such that each of X^i appears as a substring (a consecutive block) of Z. In DNA sequencing, another formulation of the problem may be of even greater interest. We call it an *approximate* SCS and one asks for a superstring that contains *approximately* (e.g., in the Hamming distance sense) the original strings X^1, X^2, ..., X^n as substrings.

More precisely, suppose $X = x_1 x_2 \ldots x_r$ and $Y = y_1 y_2 \ldots y_s$ are strings over the same finite alphabet $\mathcal{A}$. We also write $|X|$ for the length of X. We define their *overlap* $o(X, Y)$ by

$$o(X, Y) = \max\{j : y_i = x_{r-j+i}, 1 \le i \le j\}.$$

If $X \ne Y$ and $k = o(X, Y)$, then

$$X \cdot Y = x_1 x_2 \ldots x_r y_{k+1} y_{k+2} \ldots y_s.$$

is the superstring of X and Y, where $\cdot$ is the concatenation operation. Let $\mathcal{S}$ be a set of all superstrings built over the strings X^1, ..., X^n. Then,

$$O_n^{\text{opt}} = \sum_{i=1}^{n} |X_i| - \min_{Z \in \mathcal{S}} |Z|$$

is the *optimal overlap* in the shortest common superstring.

Example 1.6:   *Common Superstring and Its Graph Representation*   Let us consider the following five strings: $X^1 = abaaab$, $X^2 = aabaaaa$, $X^3 = aababb$, $X^4 = bbaaba$, and $X^5 = bbbb$. We first find C_{ij} that represents the length of the longest suffix of X^i that is also equal to a prefix of X^j. In our case

$$C = \begin{bmatrix} \star & 3 & 3 & 1 & 1 \\ 1 & \star & 2 & 0 & 0 \\ 0 & 0 & \star & 2 & 2 \\ 3 & 4 & 4 & \star & 0 \\ 0 & 0 & 0 & 2 & \star \end{bmatrix}$$

Let now $\mathcal{G}(C)$ be a weighted digraph built on the set of strings $\{X^1, \ldots, X^5\}$ with weights C_{ij}. This digraph is shown in Figure 1.7. Observe that the optimal (maximum) Hamiltonian path in $\mathcal{G}(C)$ determines the maximum overlap between strings X^1, ..., X^5. One can construct the shortest common superstring, which in our case is $Z = abaaababbbbaabaaaa$ and $O_n^{\text{opt}} = 3 + 2 + 2 + 4 = 11$.   ■

From the above example, we should conclude that computing the shortest common superstring is as hard as finding the longest Hamiltonian path; hence the problem is NP-hard. Thus, constructing a good approximation to SCS is of

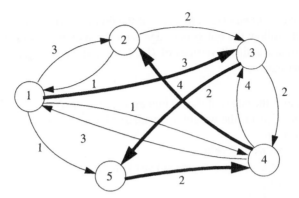

Figure 1.7. The digraph $\mathcal{G}(C)$ from Example 1.6. Optimal Hamiltonian path (starting at node 4 is shown in bold.

prime interest. It has been shown recently that a greedy algorithm can compute in $O(n \log n)$ time a superstring that in the worst case is only β times (where $2 \leq \beta \leq 4$) longer than the shortest common superstring [53, 427]. Often, one is interested in maximizing total overlap of SCS using a greedy heuristic and showing that such a heuristic produces an overlap O_n^{gr} that approximates well the optimal overlap O_n^{opt}.

A generic greedy algorithm for the SCS problem can be described as follows (cf. [53, 157, 427]: Its input is the n strings $X^1, X^2, \ldots, X^n$. It outputs a string Z which is a superstring of the input.

Generic greedy algorithm

1. $I \leftarrow \{X^1, X^2, X^3, \ldots, X^n\}$; $O_n^{gr} \leftarrow 0$;
2. **repeat**
3.   choose $X, Y \in I$; $Z = X \oplus Y$;
4. $I \leftarrow (I \setminus \{X, Y\}) \cup \{Z\}$;
5. $O_n^{gr} \leftarrow O_n^{gr} + o(X, Y)$;
6. **until** $|I| = 1$

Different variants of the above generic algorithm can be envisioned by interpreting appropriately the "choose" statement in Step 3 above:

RGREEDY: In Step 3, X is the string Z produced in the previous iteration, while Y is chosen in order to maximize $o(X, Y) = o(Z, Y)$. Our initial choice for X is X^1. Thus, in RGREEDY we have one "long" string Z which grows by addition of strings at the *right hand end*. In terms of the digraph $\mathcal{G}(C)$, one starts from the vertex X_1 and follow the out-going edge with the maximum weight $C_{1,j}$. This process continues until all vertices of $\mathcal{G}(C)$ are visited, disregarding cycles of length smaller than n.

GREEDY : Sort the edges of $\mathcal{G}_n(C)$ into $e_1, e_2, \ldots, e_N$, $N = n^2$ so that $C(e_i) \geq C(e_{i+1})$ where $C(e_i)$ is the weight assigned to edge e_i; $S_G \leftarrow \emptyset$;
For $i = 1$ to N do:
if $S_G \cup \{e_i\}$ contains in $\mathcal{G}$ neither
(i) a vertex of outdegree or indegree at least 2 in S_G,
(ii) a directed cycle,
then $S_G \leftarrow S_G \cup \{e_i\}$.
On termination S_G contains the $n - 1$ edges of a Hamilton path of $\mathcal{G}$ and corresponds to a superstring of $X^1, X^2, \ldots, X^n$. The selection of an edge weight (X^i, X^j) corresponds to overlapping X^i to the left of X^j.

MGREEDY : In Step 3 choose X, Y in order to maximize $o(X, Y)$. If $X \neq Y$ proceed as in GREEDY. If $X = Y$, then $I \leftarrow I \setminus \{X\}$, O_n^{gr} is not incremented, and $C \leftarrow C \cup \{X\}$ where the set C is initially empty. On termination we add the final string left in I to C.

MGREEDY : sort the edges $\mathcal{G}$ into $e_1, e_2, \ldots, e_N$, $N = n^2$ so that $w(e_i) \geq w(e_{i+1})$; $S_{MG}, C \leftarrow \emptyset$;
For $i = 1$ to N do:
if $S_{MG} \cup \{e_i\}$ contains no vertex of outdegree or indegree at least 2 in S_{MG},
then $S_{MG} \leftarrow S_{MG} \cup \{e_i\}$.
If e_i closes a cycle, **then** $C \leftarrow C \cup \{e_i\}$.
    On termination the edges of S_{MG} form a collection of vertex disjoint cycles $C_1, C_2, \ldots, C_t$. Each C_j contains one edge f_j which is a member of C and f_j is a lowest weight edge of C_j. Let $\mathcal{P}_j = C_j - f_j$. The catenation of paths $\mathcal{P}_1, \mathcal{P}_2, \ldots, \mathcal{P}_t$ define a superstring of the input.
    We shall analyze the shortest common superstring in Chapters 4 and 5.

1.5 STRING EDITING PROBLEM

The string editing problem arises in many applications, notably in text editing, speech recognition, machine vision and molecular sequence comparison (cf. [445]). Algorithmic aspects of this problem have been studied rather extensively in the past (cf. [19, 323, 375, 445]). In fact, many important problems on words are special cases of string editing, including the *longest common subsequence* problem (cf. [77, 67, 375]). In the following, we review the string editing problem and its relationship to the longest path problem in a grid graph.

Let Y be a string consisting of ℓ symbols over the alphabet $\mathcal{A}$. There are three operations that can be performed on a string, namely *deletion* of a symbol, *insertion* of a symbol, and *substitution* of one symbol for another symbol in $\mathcal{A}$. With each operation is associated a *weight* function. We denote by $W_I(y_i)$, $W_D(y_i)$ and $W_Q(x_i, y_j)$ the weight of insertion and deletion of the symbol $y_i \in \mathcal{A}$, and substitution of x_i by $y_j \in \mathcal{A}$, respectively. An *edit script* on Y is any sequence of edit operations, and the total weight is the sum of weights of the edit operations.

The **string editing** problem deals with two strings, say Y of length ℓ (for ℓong) and X of length s (for $s$hort), and consists of finding an edit script of minimum (maximum) total weight that transforms Y into X. The maximum (minimum) weight is called the *edit distance from X to Y*, and it is also known as the Levenshtein distance. In molecular biology, the Levenshtein distance is used to measure similarity (homogeneity) of two molecular sequences, say DNA sequences (cf. [375]).

The string edit problem can be solved by the standard dynamic programming method. Let $C_{\max}(i, j)$ denote the maximum weight of transforming the prefix of X of size i into the prefix of Y of size j. Then (cf. [19, 323, 445])

$$C_{\max}(i, j) = \max\{C_{\max}(i - 1, j - 1) + W_Q(x_i, y_j), C_{\max}(i - 1, j) + W_D(x_i), C_{\max}(i, j - 1) + W_I(y_j)\}$$

for all $1 \leq i \leq \ell$ and $1 \leq j \leq s$. We compute $C_{\max}(i, j)$ row by row to finally obtain the total cost $C_{\max} = C_{\max}(\ell, s)$ of the maximum edit script. A similar procedure works for the minimum edit distance.

The edit distance can be conveniently represented on a grid graph. The key observation is to note that interdependency among the partial optimal weights $C_{\max}(i, j)$ induces an $\ell \times s$ grid-like directed acyclic graph, called a *grid graph*. In such a graph vertices are points in the grid and edges go only from (i, j) to its neighboring points, namely, $(i, j + 1)$, $(i + 1, j)$ and $(i + 1, j + 1)$. A horizontal edge from $(i - 1, j)$ to (i, j) carries the weight $W_I(y_j)$; a vertical edge from $(i, j - 1)$ to (i, j) has weight $W_D(x_i)$; and finally a diagonal edge from $(i - 1, j - 1)$ to (i, j) is weighted according to $W_Q(x_i, y_j)$. Figure 1.8 shows an example of such an edit graph. The edit distance is the longest (shortest) path from the point $O = (0, 0)$ to $E = (\ell, s)$.

Finally, we should mention that by properly selecting the weights of W_I, W_D and W_Q we can model several variations of the string editing problem. For example, in the standard setting the deletion and insertion weights are identical, and usually constant, while the substitution weight takes two values, one (high) when matching between a letter of X and a letter of Y occurs, and another value (low)

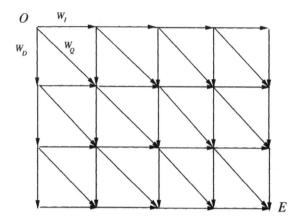

Figure 1.8. Example of a grid graph of size $\ell = 4$ and $s = 3$.

in the case of a mismatch (e.g., in the *Longest Common Subsequence* problem [67, 375], one sets $W_I = W_D = 0$, and $W_Q = 1$ when a matching occurs, and $W_Q = -\infty$ for a mismatch).

We shall analyze the edit distance problem in Chapter 5.

1.6   OPTIMIZATION PROBLEMS

In the previous two sections we have seen that combinatorial optimization problems on graphs often arise in the design and analysis of algorithms and data structures on sequences. Undoubtedly, these problems are also a source of exciting and interesting probabilistic problems. Therefore, in this final section we present a fairly general description of optimization problems.

We consider a class of optimization problems that can be formulated as follows: Let n be an integer (e.g., number of vertices in a graph, size of a matrix, number of strings in a digital tree, etc.), and S_n a set of objects (e.g., set of vertices, elements of a matrix, strings, etc). We shall investigate the behavior of the optimal values $Z_{\max}(S_n)$ and $Z_{\min}(S_n)$ defined as follows

$$Z_{\max}(S_n) = \max_{\alpha \in \mathcal{B}_n} \left\{ \sum_{i \in S_n(\alpha)} w_i(\alpha) \right\} \quad , \quad Z_{\min}(S_n) = \min_{\alpha \in \mathcal{B}_n} \left\{ \sum_{i \in S_n(\alpha)} w_i(\alpha) \right\},$$

$$(1.9)$$

and

$$Z_{\max}(\mathcal{S}_n) = \max_{\alpha \in \mathcal{B}_n} \left\{ \min_{i \in \mathcal{S}_n(\alpha)} w_i(\alpha) \right\} \quad , \quad Z_{\min}(\mathcal{S}_n) = \min_{\alpha \in \mathcal{B}_n} \left\{ \max_{i \in \mathcal{S}_n(\alpha)} w_i(\alpha) \right\},$$

(1.10)

where $\mathcal{B}_n$ is a set of all feasible solutions, $\mathcal{S}_n(\alpha)$ is a set of objects from $\mathcal{S}_n$ belonging to the αth feasible solution, and $w_i(\alpha)$ is the weight assigned to the ith object in the αth feasible solution. We often write $Z_{\max}$ and $Z_{\min}$ instead of $Z_{\max}(\mathcal{S}_n)$ and $Z_{\min}(\mathcal{S}_n)$, respectively.

For example, in the traveling salesman problem, $\mathcal{B}_n$ represents the set of all Hamiltonian paths in a graph built over n vertices, $\mathcal{S}_n(\alpha)$ is the set of edges belonging to the αth Hamiltonian path, and $w_i(\alpha)$ is the length (weight) of the ith edge. Traditionally, formulation (1.9) is called the *optimization problem with sum-objective function*, while (1.10) is known as either the *capacity optimization problem* ($Z_{\max}$) or the *bottleneck optimization problem* ($Z_{\min}$).

Certainly, combinatorial optimization problems arise in many areas of science and engineering. Among others are the capacity and bottleneck assignment problem (see Exercise 3.11) the (bottleneck and capacity) quadratic assignment problem, the minimum spanning tree, the minimum weighted k-clique problem, geometric location problems, and some others not directly related to optimization (but of interest to us), such as the height and depth of digital trees (see Sections 4.2.4–4.2.6), the maximum queue length, and hashing with lazy deletion (cf. [6]), pattern matching (cf. [28]), the shortest common superstring, and the longest common subsequence, discussed in the previous subsections.

We return to the general optimization optimization problem in Chapter 5 (see Exercise 5.17).

1.7 EXERCISES

1.1 Prove (1.5)–(1.8) of Theorem 1.3.

1.2 Establish combinatorial relationships between PATRICIA and digital trees parameters and the alignments $\{C_{ij}\}_{ij=1}^n$, as we did in Theorem 1.3 for tries.

1.3 Parse and build the Lempel-Ziv'77 and Lempel-Ziv'78 codes for the following sequence: $X = ababbbaabbbabbbbababbb$. How will the parsing and the code change when generalized LZ78 is used with $b = 4$.

1.4 Fix an integer $b \geq 1$. Generalize the Lempel-Ziv'77 scheme in a manner similar to the way we generalized the Lempel-Ziv'78 scheme in Section 1.2.3.

1.5 Consider the FDLZ, that is, the fixed database version of the Lempel-Ziv scheme. Let the database sequence be denoted as $\hat{X}_1^n$ and the source sequence be X_1^M. The source sequence is partitioned into phrases of length $I_1, I_2, \ldots, I_K$ where K is the total number of phrases (i.e., $I_1 + \cdots + I_K = M$). Show that the compression ratio $r_n(X_1^M)$ can be expressed as

$$r_n(X_1^M) = \frac{1}{M} \sum_{i=1}^{K} \log_2 n + O(\log I_i).$$

1.6 Given a sequence of length n, derive a closed-form formula for the code lengths of the LZ77, LZ78 and the generalized-LZ78 schemes in terms of the number of phrases M_n, the size of the alphabet V, and the parameter b.

1.7 Construct the compression code for $X = ababbbaabbbabbbbbabbbb$ using the lossy LZ77 with $D = 0.3$ assuming the Hamming distance.

1.8 Compute the complexity C_n introduced in Definition 1.4 for $H = abab$ and $T = abbbbabababbbbab$ when the KMP algorithm is used to find all occurrences of H in T.

1.9 Construct the optimal and the three greedy heuristics RGREEDY, GREEDY, and MGREEDY for the shortest common superstring of $X^1 = abababab$, $X^2 = bbbbbbb$, $X^3 = aaaaaaaaa$, $X^4 = babbbaaa$ and $X^5 = bbaabbaabbaa$.

1.10 Find the longest common subsequence of $X = abcccabbbbbabab$ and $Y = cabbbabababb$ over $\mathcal{A} = \{a, b, c\}$.

1.11 Determine the feasible set $\mathcal{B}_n$ and set of objects $\mathcal{S}_n$ for the shortest common superstring problem and the longest common subsequence problem.

2

Probabilistic and Analytical Models

We first discuss several probabilistic models that are used throughout this book. Then we briefly review some basic facts from probability theory (e.g., types of stochastic convergence) and complex analysis (e.g., Cauchy's residue theorem). We conclude with a brief discussion of certain special functions (e.g., Euler's gamma function, and Riemann's zeta function) that we shall use throughout the book.

Data structures and algorithms on strings are used in a variety of applications, ranging from information theory, telecommunications, wireless communications, approximate pattern matching, molecular biology, game theory, coding theory, and source coding to stock market analysis. It is often reasonable, and sometimes *sine qua non*, to assume that strings are generated by a random source of known or unknown statistics. Applications often dictate what is a reasonable set of probabilistic assumptions. For example, it is inappropriate to postulate that suffixes of a (single) sequence are independent, but experiments support the claim that bits in a hashing address form an independent sequence, while bases in a DNA sequence should be modeled by a Markov chain. We propose a few basic (generic) probabilistic models that one often encounters in the analysis of problems on words. We use them throughout the book. Finally, we review some facts from probability theory and complex analysis, and finish this chapter with a brief discussion of special functions that often arise in the analysis of algorithms.

2.1 PROBABILISTIC MODELS OF STRINGS

Throughout this book, we shall deal with sequences of discrete random variables. We write $\{X_k\}_{k=1}^{\infty}$ for a one-sided infinite sequence of random variables; however, we often abbreviate it as X provided it is clear from the context that we are talking about a sequence, not a single variable. We assume that the sequence $\{X_k\}_{k=1}^{\infty}$ is defined over a finite alphabet $\mathcal{A} = \{\omega_1, \ldots, \omega_V\}$ of size V. A partial sequence is denoted as $X_m^n = (X_m, \ldots, X_n)$ for $m < n$. When more than one sequence is analyzed (as in the digital trees) we either use an upper index to denote sequences: $X^1 = \{X_k^1\}_{k=1}^{\infty}, \ldots, X^m = \{X_k^m\}_{k=1}^{\infty}$ for m sequences; or we write $X(1) = \{X_k(1)\}_{k=1}^{\infty}, \ldots, X(m) = \{X_k(m)\}_{k=1}^{\infty}$. Finally, we shall always assume that a probability measure exists, and we write $P(X_1^n) = \Pr\{X_k = x_k, 1 \le k \le n, x_k \in \mathcal{A}\}$ for the probability mass, where we use lowercase letters for a realization of a stochastic process.

Sequences are generated by information sources, usually satisfying some constraints. We also call them probabilistic models. We start with the most elementary source, namely the **memoryless source**.

(B) Memoryless Source

Symbols of the alphabet $\mathcal{A} = \{\omega_1, \ldots, \omega_V\}$ occur independently of one another; thus $X = X_1 X_2 X_3 \ldots$ can be described as the outcome of an infinite sequence of Bernoulli trials in which $\Pr\{X_j = \omega_i\} = p_i$ and $\sum_{i=1}^{V} p_i = 1$. If $p_1 = p_2 = \ldots = p_V = 1/V$, then the model is called *symmetric* or the source is *unbiased*; otherwise, the model is *asymmetric* or the source is *biased*. In this book, we often consider *binary alphabet* $\mathcal{A} = \{0, 1\}$ with p being the probability of "0" and $q = 1 - p$ the probability of "1".

When one deals with many strings (e.g., when building a digital tree), it is necessary to say whether the strings are dependent.

In many cases, assumption (B) is not very realistic. For instance, if the strings are words from the English language, then there is certainly a dependence between consecutive letters. For example, h is much more likely to follow an s than a b. When this is the case, assumption (B) can be replaced by:

(M) Markov Source

There is a Markovian dependency between consecutive symbols in a string; that is, the probability $p_{ij} = \Pr\{X_{k+1} = \omega_j | X_k = \omega_i\}$ describes the conditional probability of sampling symbol ω_j immediately after symbol ω_i. We denote by $\mathsf{P} = \{p_{ij}\}_{i,j=1}^{V}$ the transition matrix, and by $\pi = (\pi_1, \ldots, \pi_V)$ the stationary vector satisfying $\pi\mathsf{P} = \pi$. In general, X_{k+1} may depend on the last r symbols, and then we have rth order Markov chains (however, in this book we mostly deal with $r = 1$).

In information theory **Markov sources** are usually denoted as S_k (i.e., set of states). A generalization of Markov sources is the so-called **finite state source** (cf. [163]) or **hidden Markov source** (cf. [75]) in which the output string X_k is a function of a Markov source S_k, that is, $X_k = f(S_k)$ for some function $f(\cdot)$. Finally, if in a finite state source the current state, say S_k, is computable from the current source sample X_k and previous state S_{k-1} (i.e., $S_k = g(S_{k-1}, X_k)$ for some function g), then the source is called the **unifilar source** (cf. [163]).

There are two further generalizations of Markovian sources, namely the **mixing source** and the **stationary ergodic source**, that are very useful in practice, especially for dealing with problems of data compression or molecular biology when one expects long dependency among symbols of a string.

(MX) (Strongly) ψ-Mixing Source

Let $\mathbb{F}_m^n$ be a σ-field generated by $\{X_k\}_{k=m}^{n}$ for $m \leq n$. The source is called *mixing*, if there exists a bounded function $\psi(g)$ such that for all $m, g \geq 1$ and any two events $A \in \mathbb{F}_1^m$ and $B \in \mathbb{F}_{m+g}^{\infty}$ the following holds

$$(1 - \psi(g))\Pr\{A\}\Pr\{B\} \leq \Pr\{AB\} \leq (1 + \psi(g))\Pr\{A\}\Pr\{B\}. \qquad (2.1)$$

If, in addition, $\lim_{g \to \infty} \psi(g) = 0$, then the source is called *strongly* mixing.

In words, model (MX) postulates that the dependency between $\{X_k\}_{k=1}^{m}$ and $\{X_k\}_{k=m+g}^{\infty}$ is getting weaker and weaker as g becomes larger (note that when the sequence $\{X_k\}$ is i.i.d., then $\Pr\{AB\} = \Pr\{A\}\Pr\{B\}$). The "quantity" of dependency is characterized by $\psi(g)$ (cf. [49, 58, 117]). Occasionally, we write the ψ-mixing condition in an equivalent form as follows: There exist constants $c_1 \leq c_2$ such that

$$c_1 \Pr\{A\}\Pr\{B\} \leq \Pr\{AB\} \leq c_2 \Pr\{A\}\Pr\{B\} \qquad (2.2)$$

for all $m, g \geq 1$. In some derivations, we shall require that $c_1 > 0$ which will impose further restrictions on the process (see Theorem 2.1 below).

A weaker mixing condition, namely ϕ-mixing is defined as follows:

$$-\phi(g) \leq \Pr\{B|A\} - \Pr\{B\} \leq \phi(g), \qquad \Pr\{A\} > 0, \qquad (2.3)$$

provided $\phi(g) \to 0$ as $g \to \infty$. In general, strongly ψ-mixing implies ϕ-mixing condition but not *vice versa* (cf. [58]). In this book, we shall mostly work with (strongly) ψ-mixing condition.

The following result shows that Markov sources are special cases of strongly mixing sources.

Theorem 2.1 *If X_k is a finite-state irreducible stationary aperiodic Markov chain, then X_k is a strongly ψ-mixing process, hence also ϕ-mixing.*

If, in addition, the transition probabilities $p_{ij} > 0$ for all $i, j \in \mathcal{A}$, then the Markov chain is ψ-mixing with the constant $c_1 > 0$ in (2.2).

Proof. We shall follow here Karlin and Ost [229]. Define two events

$$A = \{X_{t_1} = i_1, X_{t_2} = i_2, \ldots, X_{t_m} = i_m\} \qquad t_1 < t_2 < \cdots < t_m,$$

$$B_g = \{X_{s_1+g} = j_1, X_{s_2+g} = j_2, \ldots, X_{s_r+g} = j_r\} \quad t_m \leq s_1 < s_2 < \cdots < s_r$$

for any $g > 0$. Without loss of generality, we assume that $\Pr\{A\} > 0$ and $\Pr\{B_g\} > 0$. By Markov property

$$\Pr\{A \cap B_g\} = \Pr\{B_g|X_{s_1+g} = j_1\}\Pr\{X_{s_1+g} = j_1|X_{t_m} = i_m\}\Pr\{A\}$$

$$= \Pr\{A\}\Pr\{B_g\}\frac{\Pr\{X_{s_1+g} = j_1|X_{t_m} = i_m\}}{\Pr\{X_{s_1+g} = j_1\}}. \qquad (2.4)$$

Since X_k is an irreducible aperiodic Markov chain over a finite alphabet, it is ergodic, and since it is also a stationary chain its stationary distribution is $\Pr\{X_t = i\}$ for every $t \geq 0$. We also know from the Perron-Frobenius theorem (see Table 4.1 in Chapter 4) that the transition probability $\Pr\{X_{s_1+g} = j_1|X_{t_m} = i_m\}$ converges exponentially fast to $\Pr\{X_{s_1+g} = j_1\}$. That is, there exists $\rho < 1$ (which is related to the second largest eigenvalue of the transition matrix of the

Markov chain) such that

$$\left| \frac{\Pr\{X_{s_1+g} = j_1 | X_{t_m} = i_m\}}{\Pr\{X_{s_1+g} = j_1\}} - 1 \right| = O(\rho^g) \tag{2.5}$$

as $g \to \infty$. This and (2.4) imply that

$$(1 - O(\rho^g))\Pr\{A\}\Pr\{B_g\} \leq \Pr\{A \cap B_g\} \leq (1 + O(\rho^g))\Pr\{A\}\Pr\{B_g\},$$

which proves the first part of the theorem.

To show that $c_1 > 0$ when $p_{ij} > 0$ for all $i, j \in \mathcal{A}$ observe that by (2.4)

$$\Pr\{A \cap B_g\} = C_{ij}(g)\Pr\{A\}\Pr\{B_g\}$$

where

$$C_{ij}(g) = \frac{\Pr\{X_{s_1+g} = j | X_{t_m} = i\}}{\Pr\{X_{s_1+g} = j\}}.$$

Clearly, $\Pr\{X_{s_1+g} = j\} > 0$ for all $j \in \mathcal{A}$. Furthermore, if $p_{ij} > 0$ for all $i, j \in A$, then $C_{ij}(1) > 0$ and hence $C_{ij}(g) > 0$ for all $g > 1$. Setting $c_1 = \min_{i,j \in \mathcal{A}} \inf_{g \geq 1}\{C_{ij}(g)\}$ and using (2.5), we establish the announced result. ∎

The most general probabilistic model is the stationary ergodic source in which we only assume that the sequence is stationary and ergodic.

(S) Stationary and Ergodic Source

The sequence $\{X_k\}_{k=1}^{\infty}$ of letters from a finite alphabet is a *stationary and ergodic* sequence of random variables.

In the stationary model, the probability mass $\Pr\{X_1^n = x_1^n\}$ does not depend on time-shift, that is, if m is an integer, then for every n and m the following holds $\Pr\{X_{1+m}^{n+m} = x_1^n\} = \Pr\{X_1^n = x_1^n\}$. It turns out that that a stationary distribution may be approximated by a k order Markov distribution (see Section 4.2.1). Such an approximation becomes asymptotically accurate as $k \to \infty$ (see Theorem 6.33 in Chapter 6). In passing we should point out that memoryless and Markov sources may be either stationary or not. In this book, we shall always work with stationary models, unless stated otherwise.

Recently, B. Vallée introduced in [431] (cf. also [68]) new *probabilistic dynamical sources* that are based on theory of dynamical systems. The basic idea is to assign an infinite sequence to a real number x in the interval $[0, 1]$ (e.g., the binary expansion of x, the continued fraction expansion of x). A probabilistic behavior of such sources is induced by selecting the initial x according to a given density function.

2.2 REVIEW OF PROBABILITY

Probability theory is a basic tool in the average-case analysis of algorithms. We shall learn in this book a considerable number of probabilistic tricks that provide insights into algorithmic behavior. In this section, we briefly review some elementary definitions and results from probability theory. The reader is referred to Billingsley [49], Durrett [117], Feller [122, 123], and Shiryayev [389] for more detailed discussions.

In this book, we mostly deal with discrete random variables. Let X_n denote the value of a parameter of interest depending on n (e.g., depth in a suffix tree or a trie built over n strings). We write $\Pr\{X_n = k\}$ for the probability mass of X_n. The expected value or the mean $\mathbf{E}[X_n]$ and the variance $\mathbf{Var}[X_n]$ are computed in a standard way as:

$$\mathbf{E}[X_n] = \sum_{k=0}^{\infty} k \Pr\{X_n = k\},$$

$$\mathbf{Var}[X_n] = \sum_{k=0}^{\infty} (k - \mathbf{E}[X_n])^2 \Pr\{X_n = k\}.$$

We also write $\mathbf{E}[X^r] = \sum_{k=0}^{\infty} k^r \Pr\{X_n = k\}$ for the rth moment of X_n. Finally, we write $I(A)$ for the *indicator function*, that is, $I(A) = 1$ when the event A occurs, and zero otherwise.

Below, we first discuss some standard inequalities for expectations, and then review types of stochastic convergence.

2.2.1 Some Useful Inequalities

We review here some inequalities that play a considerable role in probability theory; they are often used in the probabilistic analysis of algorithms.

Markov's Inequality   *For a nonnegative function $g(\cdot)$ and a random variable X the following holds*

$$\Pr\{g(X) \geq t\} \leq \frac{\mathbf{E}[g(X)]}{t} \tag{2.6}$$

for any t > 0. Indeed, we have the following chain of obvious inequalities

$$\mathbf{E}[g(X)] \geq \mathbf{E}[g(X)I(g(X) \geq t)] \geq t\mathbf{E}[I(g(X) \geq t)] = t\mathrm{Pr}\{g(X) \geq t\},$$

where we recall that $I(A)$ is the indicator function of the event A.

Chebyshev's Inequality If one replaces $g(X)$ by $|X - \mathbf{E}[X]|^2$ and t by t^2 in the Markov inequality, then

$$\mathrm{Pr}\{|X - \mathbf{E}[X]| > t\} \leq \frac{\mathbf{Var}[X]}{t^2}, \tag{2.7}$$

which is known as Chebyshev's inequality.

Schwarz's Inequality (also called **Cauchy-Schwarz**) *Let X and Y be such that* $\mathbf{E}[X^2] < \infty$ *and* $\mathbf{E}[Y^2] < \infty$. *Then*

$$\mathbf{E}[|XY|]^2 \leq \mathbf{E}[X^2]\mathbf{E}[Y^2], \tag{2.8}$$

where throughout the book we shall write $\mathbf{E}[X]^2 := (\mathbf{E}[X])^2$.

Jensen's Inequality *Let* $f(\cdot)$ *be a downward convex function, that is, for* $\lambda \in (0, 1)$

$$\lambda f(x) + (1 - \lambda)f(y) \geq f(\lambda x + (1 - \lambda)y).$$

Then

$$f(\mathbf{E}[X]) \leq \mathbf{E}[f(X)], \tag{2.9}$$

with the equality holding when $f(\cdot)$ *is a linear function.*

Minkowski's Inequality *If* $\mathbf{E}[|X|^p] < \infty$, *and* $\mathbf{E}[|Y|^p] < \infty$, *then* $\mathbf{E}[|X + Y|]^p < \infty$ *and*

$$\mathbf{E}[|X + Y|^p]^{1/p} \leq \mathbf{E}[|X|^p]^{1/p} + \mathbf{E}[|Y|^p]^{1/p} \tag{2.10}$$

for $1 \leq p < \infty$.

Hölder's Inequality *Let* $0 \leq \theta \leq 1$, *and* $\mathbf{E}[X] < \infty$, $\mathbf{E}[Y] < \infty$. *Then*

$$\mathbf{E}[X^\theta Y^{1-\theta}] \leq \mathbf{E}[X]^\theta E[Y]^{1-\theta}. \tag{2.11}$$

Inequality on Means Let $(p_1, p_2, \ldots, p_n)$ be a probability vector (i.e., $\sum_{i=1}^{n} p_i = 1$) and $(a_1, \ldots, a_n)$ any vector of positive numbers. The *mean of order* $b \neq 0$ $(-\infty \leq b \leq \infty)$ is defined as

$$M_n(b) := \left(\sum_{i=1}^{n} p_i a_i^b \right)^{\frac{1}{b}}.$$

The *inequality on means* asserts that $M_n(b)$ is a nondecreasing function of b, that is,

$$r < s \qquad \Rightarrow \qquad M_n(r) \leq M_n(s), \tag{2.12}$$

where the equality holds if and only if $a_1 = a_2 = \cdots = a_n$. Furthermore, **Hilbert's inequality on means** states

$$\lim_{b \to -\infty} M_n(b) = \min\{a_1, \ldots, a_n\}, \tag{2.13}$$

$$\lim_{b \to \infty} M_n(b) = \max\{a_1, \ldots, a_n\}. \tag{2.14}$$

We shall see in Chapter 4 that Markov's inequality and Chebyshev's inequality constitute a foundation for the so-called first and second moment methods. These two simple methods are probably the most often used probabilistic tools in the analysis of algorithms.

2.2.2 Types of Stochastic Convergence

It is important to know various ways that random variables may converge. Let $X_n = \{X_n, n \geq 1\}$ be a sequence of random variables, and let their distribution functions be $F_n(x)$, respectively. A good, easy-to-read account on various types of convergence can be found in Shiryayev [389].

The first notion of convergence of a sequence of random variables is known as **convergence in probability.** The sequence X_n converges to a random variable X *in probability*, denoted $X_n \to X$ (pr.) or $X_n \overset{pr}{\to} X$, if for any $\epsilon > 0$

$$\lim_{n \to \infty} \Pr\{|X_n - X| < \epsilon\} = 1.$$

It is known that *if* $X_n \overset{pr}{\to} X$, *then* $f(X_n) \overset{pr}{\to} f(X)$ *provided* f *is a continuous function* (cf. [49, 117]). The reader is asked to prove this fact in Exercise 6.

Note that convergence in probability does not say that the difference between X_n and X becomes very small. What converges here is the *probability* that the difference between X_n and X becomes very small. It is, therefore, possible, although unlikely, for X_n and X to differ by a significant amount and for such differences to occur infinitely often. A stronger kind of convergence that does not allow such behavior is called **almost sure convergence** or **strong convergence**. This convergence assures that *the set of sample points for which X_n does not converge to X has probability zero*. In other words, a sequence of random variables X_n converges to a random variable X *almost surely*, denoted $X_n \to X$ (a.s.) or $X_n \overset{(a.s.)}{\to} X$, if for any $\epsilon > 0$,

$$\lim_{N \to \infty} \Pr\{\sup_{n \geq N} |X_n - X| < \epsilon\} = 1.$$

From this formulation of almost sure convergence, it is clear that if $X_n \to X$ (a.s.), the probability of infinitely many large differences between X_n and X is zero. As the term strong implies, almost sure convergence implies convergence in probability.

A simple sufficient condition for almost sure convergence can be inferred from the *Borel-Cantelli lemma* presented below.

Lemma 2.2 (Borel-Cantelli) *If for every $\epsilon > 0$ $\sum_{n=0}^{\infty} \Pr\{|X_n - X| > \epsilon\} < \infty$, then $X_n \overset{a.s.}{\to} X$.*

Proof. It follows directly from the following chain of inequalities

$$\Pr\{\sup_{n \geq N} |X_n - X| \geq \epsilon\} = \Pr\{\bigcup_{n \geq N} (|X_n - X| \geq \epsilon)\} \leq \sum_{n \geq N} \Pr\{|X_n - X| \geq \epsilon\} \to 0.$$

The inequality above is a consequence of the fact that the probability of a sum of events is smaller than the sum of the probability of the events. The last convergence is a consequence of our assumption that $\sum_{n=0}^{\infty} \Pr\{|X_n - X| > \epsilon\} < \infty$. ∎

A third type of convergence is defined on the distribution functions $F_n(x)$. The sequence of random variables X_n **converges in distribution** or **converges in law** to the random variable X, denoted $X_n \overset{d}{\to} X$ if

$$\lim_{n \to \infty} F_n(x) = F(x) \tag{2.15}$$

for each point of continuity of $F(x)$. One can prove that the above definition is equivalent to the following: $X_n \overset{d}{\to} X$ if

$$\lim_{n \to \infty} E[f(X_n)] = E[f(X)] \tag{2.16}$$

for all bounded continuous functions f. We shall return to this type of convergence in Chapter 8.

The next type of convergence is the **convergence in mean of order** p or **convergence in** L^p which postulates that $\mathbf{E}[|X_n - X|^p] \to 0$ as $n \to \infty$. We write it as $X_n \overset{L^p}{\to} X$. Finally, we introduce **convergence in moments** for which $\lim_{n\to\infty} \mathbf{E}[X_n^p] = \mathbf{E}[X^p]$ for any $p \geq 1$.

We now describe relationships (implications) between various type of convergence. The reader is referred to [49, 117, 389] for a proof.

Theorem 2.3   *We have the following implications:*

$$X_n \overset{a.s.}{\to} X \Rightarrow X_n \overset{pr}{\to} X, \tag{2.17}$$

$$X_n \overset{L^p}{\to} X \Rightarrow X_n \overset{pr}{\to} X, \tag{2.18}$$

$$X_n \overset{pr}{\to} X \Rightarrow X_n \overset{d}{\to} X \tag{2.19}$$

$$X_n \overset{L^p}{\to} X \Rightarrow \mathbf{E}[X_n^p] \to \mathbf{E}[X^p]. \tag{2.20}$$

No other implications hold in general.

It is easy to devise an example showing that convergence in probability does not imply convergence in mean (e.g., take $X_n = n$ with probability $1/n$ and zero otherwise). To obtain convergence in mean from the convergence in probability one needs somewhat stronger conditions. For example, if $|X_n| \leq Y$ and $\mathbf{E}[Y] < \infty$, then by the *dominated convergence theorem* (cf. [49, 117]) we know that convergence in probability implies convergence in mean. To generalize it, one introduces the so called **uniform integrability**. *It is said that a sequence* $\{X_n, \ n \geq 1\}$ *is uniformly integrable if*

$$\sup_{n\geq 1} \mathbf{E}[|X_n| I(|X_n| > a)] \to 0 \tag{2.21}$$

when $a \to \infty$. The above is equivalent to

$$\lim_{a\to\infty} \sup_{n\geq 1} \int_{|x|>a} x\, dF_n(x) = 0.$$

Then the following is true (cf. [49, 117]): *if X_n is uniformly integrable, then $X_n \overset{pr}{\to} X$ implies $X_n \overset{L^p}{\to} X$.*

2.3 REVIEW OF COMPLEX ANALYSIS

Much of the necessary complex analysis involves the use of Cauchy's integral formula and Cauchy's residue theorem. Here, we informally recall a few facts from analytic functions, and then discuss the above two theorems. We shall return to them in Part III. For precise definitions and formulations the reader is referred to many excellent books such as Henrici [195], Hille [196], Remmert [363] or Titchmarsh [424]. We shall follow here Flajolet and Sedgewick [149].

The main topic of this section is **analytic function**, its definitions, properties, and a few applications. Analytic functions can be characterized by one of three equivalent ways: by convergent series, by differentiability property, and by integrals vanishing on cycles. In discussing these three definitions, we introduce the necessary concepts from complex analysis.

Convergent Series. A function $f(z)$ of a complex variable z is analytic at point $z = a$ if it has a convergent series representation in a neighborhood of $z = a$, that is,

$$f(z) = \sum_{n \geq 0} f_n (z - a)^n, \qquad z \in B(a, r),$$

where $B(a, r)$ is a ball with center a and radius $r > 0$. From this definition, one immediately concludes that if $f(z)$ and $g(z)$ are analytic, then $f(z) + g(z)$, $f(z)g(z)$, and $f(z)/g(z)$ with $g(z) \neq 0$ are analytic, too. Furthermore, the above implies that if $f(z)$ is analytic at $z = a$, then there is a disk called the *disk of convergence* such that the series representing $f(z)$ is convergent inside this disk and divergent outside the disk. The radius of this disk is called *radius of convergence*. In Chapter 7 we discuss this in depth; in particular, we prove Hadamard's theorem (see Theorem 7.1), which gives the radius of convergence of a series in terms of its coefficients.

Holomorphic Function. Equivalent name for analytic functions is holomorphic. A holomorphic function $f(z)$ has a derivative at a point $z = a$ defined as

$$\frac{df(z)}{dz} = \lim_{\Delta z \to 0} \frac{f(z + \Delta z) - f(z)}{\Delta z}$$

that does not depend on the way Δz goes to zero. Riemann proved that a holomorphic function is analytic, that is, it has a local convergent series representation.

Integrals. Finally, we discuss a very important concept of complex integrals. We should first define a *simply connected* domain as an open connected set having the property that any simple closed path (cf. [363]) can be continuously deformed to a point inside this set. It is a basic fact of complex analysis that *if $f(z)$ is an analytic function on an open simply connected set, then*

$$\oint f(z) := \int_C f(z) = 0$$

along any closed path C inside this set. This fact actually is equivalent to analyticity of $f(z)$. It is worth knowing that one can bound the integrals as follows:

$$\left| \int_C f(z) \right| \leq |C| \max_C \{|f(z)|\} \tag{2.22}$$

where $|C|$ is the path length of the curve C. Finally, we cite one more result that plays a significant role in the analysis (cf. [363]).

Theorem 2.4   (i) *Let C be a piecewise continuous differentiable path and Ω a region. If $g(w, z)$ is continuous on $C \times \Omega$ and for every $w \in C$ the function $g(w, z)$ is analytic in Ω, then the function*

$$h(z) = \int_C g(\xi, z)d\xi$$

is analytic in Ω.

(ii) *The function*

$$h(z) = \int_a^\infty f(t, z)dt$$

is an analytic function of z at all points z if: (i) the integral converges; (ii) $f(z, t)$ is an analytic function of z and t; (iii) $\frac{\partial f(z,t)}{\partial t}$ is a continuous function of both variables; (iv) the integral $\int_a^\infty \frac{\partial f(z,t)}{\partial z}dt$ converges uniformly.

Example 2.1:   *A Simple Integral*   In complex analysis a circular integral over z^{-n} for $n = 0, 1, \ldots$ plays a very special role. Observe that this function is not analytic at $z = 0$. It is easy to see that

$$\oint z^{-n}dz = 2\pi i \int_0^1 e^{-2\pi i(n-1)}dt = \begin{cases} 2\pi i & \text{for } n = 1 \\ 0 & \text{otherwise} \end{cases} \tag{2.23}$$

where we substituted $z = e^{2\pi i t}$. We shall often use this simple fact throughout the book.   ∎

Meromorphic Functions and Residues.   A quotient of two analytic functions gives a *meromorphic function* that is analytic everywhere but a set of points called poles, where the denominator vanishes. More formally, a meromorphic function $f(z)$ can be represented in a neighborhood of $z = a$ with $z \neq a$ by the Laurent series as:

$$f(z) = \sum_{n \geq -M} f_n(z-a)^n$$

for some integer M. If the above holds with $f_{-M} \neq 0$, then it is said that $f(z)$ has a *pole* of order M at $z = a$.

An important tool frequently used in the complex analysis is the **residue.** The residue of $f(z)$ at a point a is the coefficient at $(z-a)^{-1}$ in the Laurent expansion of $f(z)$ around a, and it is denoted as

$$\text{Res}[f(z); \; z = a] := f_{-1} = \lim_{z \to a}(z - a)f(z).$$

There are many simple rules to evaluate residues and the reader can find them in any standard book on complex analysis (e.g., [195, 363]). For example, if $f(z)$ and $g(z)$ are analytic around $z = a$, then

$$\text{Res}[\frac{f(z)}{g(z)}; \; z = a] = \frac{f(a)}{g'(a)}, \qquad g(a) = 0, \quad g'(a) \neq 0; \qquad (2.24)$$

if $g(z)$ is not analytic at $z = a$, then

$$\text{Res}[f(z)g(z); \; z = a] = f(a)\text{Res}[g(z); \; z = a]. \qquad (2.25)$$

Evaluating residues of multiple poles is much more computationally involved. Actually, the easiest way is to use the `series` command in MAPLE that produces a series development of a function. The residue is simply the coefficient at $(z - a)^{-1}$. For example, the following session of MAPLE computes the series expansion of $f(z) = 1/(1 - 2^z)^2$ at $z = 0$:

```
> series(1/(1-2^z)^2, z=0, 4);
```

$$\frac{1}{\ln(2)^2} z^{-2} - \frac{1}{\ln(2)} z^{-1} + \frac{5}{12} + O(z)$$

From the above we see that $\text{Res}[f(z); \; z = 0] = \frac{1}{\ln 2}$.

Residues are very important in evaluating contour integrals as demonstrated by the following theorem.

Theorem 2.5 (Cauchy Residue Theorem) *If $f(z)$ is analytic within and on the boundary of a simple closed curve C except at a finite number of poles $a_1, a_2, \ldots, a_N$ inside of C having residues $\text{Res}[f(z); \; z = a_1], \ldots, \text{Res}[f(z); \; z =$*

a_N], *respectively, then*

$$\frac{1}{2\pi i} \int_C f(z)dz = \sum_{j=1}^{N} \text{Res}[f(z); \; z = a_j],$$

where the curve C is traversed counterclockwise.

Sketch of Proof. Let us assume there is only one pole at $z = a$. Since $f(z)$ is meromorphic, it has the Laurent expansion, which after integration over C leads to

$$\int_C f(z)dz = \sum_{n \geq 0} f_n \int_C (z-a)^n + f_{-1} \int_C \frac{dz}{z-a} = 2\pi i \text{Res}[f(z), z = a],$$

since the first integral is zero by the calculation of Example 2.1.   ∎

The Cauchy residue theorem can be used to prove the next *very* important result, namely **Cauchy's Integral Theorem**. In Part III of this book, it will be the most often used paradigm.

Theorem 2.6   (Cauchy Coefficient Formula)   *Let $f(z)$ be analytic inside a simply connected region with C being a closed curve oriented counterclockwise that encircles the origin $z = 0$. Then*

$$f_n := [z^n]f(z) = \frac{1}{2\pi i} \oint f(z) \frac{dz}{z^{n+1}}. \tag{2.26}$$

Also, the following holds

$$f^{(k)}(z) = \frac{k!}{2\pi i} \oint \frac{f(w)dw}{(w-z)^{k+1}} \tag{2.27}$$

where $f^{(k)}(z)$ is the kth derivative of $f(z)$.

Proof. These formulas are direct consequences of Cauchy's residue theorem. Indeed,

$$f_n := [z^n]f(z) = \text{Res}[f(z)z^{-n-1} : z = 0] = \frac{1}{2\pi i} \oint f(z) \frac{dz}{z^{n+1}}.$$

Throughout this book, we shall write $[z^n]f(z)$ for the coefficient of $f(z)$ at z^n.   ∎

We finish this section with two examples that illustrate what we have learned so far. The second example is particularly worth studying since we shall do this type of calculation quite often in Part III.

Example 2.2: *Liouville's Theorem* We prove the following theorem of Liouville: *If a function $f(z)$ is analytic in the whole complex plane $\mathbb{C}$ and $f(z) \leq B$ for some constant B, then $f(z)$ must be constant.* Let us first assume that $f(z) = B$. Then by Example 2.1 we show that $f_n = 0$ for $n \geq 1$. Conversely, let $|f(z)| \leq B$. Then $f_n = O(R^{-n})$ for *any* R (it doesn't matter how large) by (2.26) and (2.22), so f_n must be zero except for $n = 0$. ∎

Example 2.3: *Expansion of Meromorphic Functions* Let $f(z)$ be a meromorphic for $|z| < R$ with isolated poles at $a_1, \ldots, a_m$. Then

$$f_n := [z^n] f(z) = - \sum_{j=1}^{m} \text{Res}[f(z) z^{-n-1}, z = a_j] + O(R^{-n}).$$

Indeed, let us compute the following integral

$$I_n = \frac{1}{2\pi i} \oint_R f(z) \frac{dz}{z^{n+1}},$$

where the integration is along a circle of radius R and center at the origin. Observe that the circle of radius R contains all the poles of the function in addition to the pole at $z = 0$. On one hand, by our estimate (2.22) this integral is $|I_n| = O(R^{-n})$ (since $f(z)$ is analytic on the circle of radius R). On the other hand, inside the circle of radius R there are poles at $z = 0$ and $z = a_j$. The pole at $z = 0$ contributes f_n by *Cauchy's coefficient formula* while the poles at $z = a_j$ contribute $\text{Res}[f(z) z^{-n-1}, z = a_j]$, which leads to our result. We shall return to these estimates in Section 8.3.1. ∎

2.4 SPECIAL FUNCTIONS

In this subsection, we review properties of two special functions: **Euler's gamma function** and **Riemann's zeta function**. Both are used very frequently in the average-case analysis of algorithms. We shall meet them quite often in Part III of this book. There are many excellent books treating these functions in depth. The reader might inspect a new book of Temme [422], an excellent account on special functions and asymptotics by Olver [331], and an old, traditional (but still in use) book of Whittaker and Watson [448]. The new book [14] by Andrews, Askey and Roy is also excellent. Here we will follow Temme [422]. A full list of properties for these special functions can be found in the handbook of Abramowitz and Stegun [2] and the three-volume opus on transcendental functions [35] by H. Bateman.

2.4.1 Euler's Gamma Function

A desire to generalize $n!$ to the complex plane led Euler to introduce one of the most useful special functions, namely, the *gamma function*. It is defined as

$$\Gamma(z) = \int_0^\infty t^{z-1} e^{-t} dt, \quad \Re(z) > 0. \tag{2.28}$$

To see that the above integral generalized $n!$, let us integrate it by parts. We obtain

$$\Gamma(z+1) = -\int_0^\infty t^z d\left(e^{-t}\right) = z\Gamma(z). \tag{2.29}$$

Observe now that $\Gamma(1) = 1$, and then $\Gamma(n+1) = n!$ for n natural, as desired.

Analytic Continuation. We now analytically continue the gamma function to the whole complex plane. We first extend the definition to $-1 < \Re(z) < 0$ by considering (2.29) and writing

$$\Gamma(z) = \frac{\Gamma(z+1)}{z}, \quad -1 < \Re(z) < 0.$$

Since $\Gamma(z+1)$ is well defined for $-1 < \Re(z) < 0$ (indeed, $\Re(z+1) > 0$), we can enlarge the region of definition to the strip $-1 < \Re(z) < 0$. However, at $z = 0$ there is a pole whose residue is easy to evaluate; that is,

$$\text{Res}[\Gamma(z); z = 0] = \lim_{z \to 0} z\Gamma(z) = 1.$$

Now we can further extend to $-2 < \Re(z) < -1$ by applying (2.29) twice to get

$$\Gamma(z) = \frac{\Gamma(z+2)}{z(z+1)}, \quad -2 < \Re(z) < -1.$$

Observe that

$$\text{Res}[\Gamma(z); z = -1] = \lim_{z \to -1} (z+1)\Gamma(z) = -1.$$

In general, let us assume we have already defined $\Gamma(z)$ up to the strip $-n < \Re(z) < -n+1$. Then, extension to $-n-1 < \Re(z) < -n$ is obtained by

$$\Gamma(z) = \frac{\Gamma(z+n+1)}{z(z+1)\cdots(z+n)}.$$

The residue at $z = -n$ becomes

$$\text{Res}[\Gamma(z); z = -n] = \lim_{z \to -n} (z + n)\Gamma(z) = \frac{(-1)^n}{n!} \tag{2.30}$$

for all $n = 0, -1, \ldots$. The above formula is the most useful, and certainly responsible for many applications of the gamma function to discrete mathematics and the analysis of algorithms. The reader should take a good look at it.

Beta Function. Closely related to the gamma function is the *beta function* defined as

$$B(w, z) = \int_0^1 t^{w-1}(1 - t)^{z-1}dt, \quad \Re(w) > 0 \text{ and } \Re(z) > 0. \tag{2.31}$$

The reader is asked to prove in Exercise 11 the following formula that relates the gamma and the beta functions:

$$B(w, z) = \frac{\Gamma(w)\Gamma(z)}{\Gamma(w + z)}. \tag{2.32}$$

Beta function is often used to prove properties of the gamma function (e.g., see Exercises 14 and 16).

Infinite Products. In 1856 Weierstrass defined the gamma function as

$$\frac{1}{\Gamma(z)} = ze^{\gamma z} \prod_{n=1}^{\infty} \left[\left(1 + \frac{z}{n}\right) e^{-z/n} \right],$$

where the constant γ is known as Euler's constant defined as

$$\gamma = \lim_{n \to \infty} \left(\sum_{k=1}^{n} \frac{1}{k} - \ln(n + 1) \right) = 0.5772157 \ldots .$$

Using this definition of γ, and some algebraic manipulation the reader is asked in Exercise 12 to derive the following Euler's product representation:

$$\Gamma(z) = \lim_{n \to \infty} \frac{n!n^z}{z(z + 1) \cdots (z + n)}.$$

In Exercise 13 we suggest showing that the above representation is equivalent to the integral formula (2.29) for $\Re(z) > 0$.

Gamma Function on Imaginary Line. Several applications of the gamma function follow from its desirable behavior for purely imaginary variable; that is, for

$z = iy$ where $y \in \mathbb{R}$. Let us start with the well-known *reflection formula* of the gamma function (see Exercise 14):

$$\Gamma(z)\Gamma(1 - z) = \frac{\pi}{\sin \pi z}, \quad z \notin \mathbb{Z}. \tag{2.33}$$

Set now $z = iy$ for $y \in \mathbb{R}$ and use (2.29) to get

$$\Gamma(iy)\Gamma(-iy) = \frac{\pi}{-iy \sin \pi iy}.$$

But since $-i \sin iy = \sinh y = (e^y - e^{-y})/2$, we finally derive

$$\Gamma(iy)\Gamma(-iy) = |\Gamma(iy)|^2 = \frac{\pi}{y \sinh \pi y},$$

where we also use the fact that $\Gamma(-iy)$ is conjugate to $\Gamma(iy)$, that is, $\Gamma(-iy) = \overline{\Gamma(iy)}$ (in general, $\Gamma(\bar{z}) = \overline{\Gamma(z)}$). A consequence of the above identity is the following asymptotic property of the gamma function

$$|\Gamma(iy)| \sim \sqrt{\frac{2\pi}{|y|}} e^{-\pi|y|/2} \quad \text{as} \quad |y| \to \pm\infty, \tag{2.34}$$

where $f(z) \sim g(z)$ if $\lim_{z \to \infty} \frac{f(z)}{g(z)} = 1$ (see Section 8.1.1 for a precise definition). The above shows that the gamma function decays exponentially fast on the imaginary line. In general, using (2.29) one can show that

$$|\Gamma(x + iy)| \sim \sqrt{2\pi}|y|^{x - \frac{1}{2}} e^{-\pi|y|/2} \quad \text{as} \quad |y| \to \pm\infty \tag{2.35}$$

for any $x \in \mathbb{R}$.

Asymptotic Expansions.   In many applications Stirling's asymptotic formula for $n!$ proves to be extremely useful. Not surprisingly then, the same is true for asymptotic expansion of the gamma function. It can be proved [422, 448] that (see Example 8.8.3 in Chapter 8)

$$\Gamma(z) = \sqrt{2\pi} z^{z - \frac{1}{2}} e^{-z} \left(1 + \frac{1}{12z} + \frac{1}{288z^2} + \cdots \right)$$

for $z \to \infty$ when $|\arg(z)| < \pi$.

The above asymptotic is helpful in deriving another approximation for large z. Let us consider the ratio of $\Gamma(z + a)$ and $\Gamma(z + b)$ for large z. To see how it behaves, consider

$$\Gamma^*(z) = \frac{\Gamma(z)}{\sqrt{2\pi}\, z^{z-\frac{1}{2}} e^{-z}},$$

which tends to one as $z \to \infty$. Then,

$$\frac{\Gamma(z+a)}{\Gamma(z+b)} = z^{a-b} \frac{\Gamma^*(z+b)}{\Gamma^*(z+b)} Q(z, a, b),$$

where $Q(z, a, b) = 1 + O(1/z)$ as $z \to \infty$. Hence, the above ratio is approximately z^{a-b}. More precisely

$$z^{b-a} \frac{\Gamma(z+a)}{\Gamma(z+b)} = 1 + \frac{(a-b)(a+b-1)}{2z} + O(1/z^2) \qquad (2.36)$$

as $z \to \infty$ along any curve joining $z = 0$ and $z = \infty$ providing $z \ne -a, -a - 1, \ldots,$ and $z \ne -b, -b - 1, \ldots$. A full asymptotic expansion of this ratio can be found in Exercise 16 and in Temme [422].

Psi Function. The derivative of the logarithm of the gamma function plays an important role in the theory and applications of special functions. It is known as the *psi function* and is defined as:

$$\psi(z) = \frac{d}{dz} \ln \Gamma(z) = \frac{\Gamma'(z)}{\Gamma(z)}.$$

Using Weierstrass's product form of the gamma function, one can derive the following

$$\psi(z) = -\gamma + \sum_{n=0}^{\infty} \left(\frac{1}{n+1} - \frac{1}{z+n} \right), \quad z \ne 0, -1, -2, \ldots. \qquad (2.37)$$

From the above, we conclude that the psi function possesses simple poles at all nonpositive integers, and

$$\text{Res}[\psi(z); z = -n] = \lim_{z \to -n} (z+n)\psi(z) = -1, \quad n \in \mathbb{N}. \qquad (2.38)$$

Laurent's Expansions. As we observed above, the gamma function and the psi function do have simple poles at all nonpositive integers. Thus one can expand these functions around $z = -n$ using the Laurent's series. The following is known (cf. [422]):

$$\Gamma(z) = \frac{(-1)^n}{n!} \frac{1}{(z+n)} + \psi(n+1) \qquad (2.39)$$

$$+ \frac{1}{2}(z+n)\left(\pi^2/3 + \psi^2(n+1) - \psi'(n+1)\right) + O((z+n)^2),$$

$$\psi(z) = \frac{-1}{(z+n)} + \psi(m+1) + \sum_{k=2}^{\infty} \left((-1)^n \zeta(n) + \sum_{i=1}^{k} i^{-k}\right)(z+n)^{k-1}, \quad (2.40)$$

where $\zeta(z)$ is the Riemann zeta function defined in Section 2.4.2. In particular,

$$\Gamma(z) = \frac{1}{z} - \gamma + O(z), \tag{2.41}$$

$$\Gamma(z) = \frac{-1}{z+1} + \gamma - 1 + O(z+1). \tag{2.42}$$

We shall use the above formulas quite often in Part III (see Section 7.6.3).

2.4.2   Riemann's Zeta Function

We discuss here the Riemann zeta function $\zeta(z)$. This is the most fascinating special function that still hides from us its beautiful properties (e.g., the Riemann conjecture concerning zeros of $\zeta(z)$). We uncover only the tip of the iceberg. The reader is referred to Titchmarsh and Heath-Brown [425] for more in-depth discussion. The Riemann zeta function is defined as

$$\zeta(z) = \sum_{n=1}^{\infty} \frac{1}{n^z}, \quad \Re(z) > 1. \tag{2.43}$$

The *generalized* zeta function $\zeta(z, a)$ (also known as Hurwitz zeta function) is defined as

$$\zeta(z, a) = \sum_{n=0}^{\infty} \frac{1}{(n+a)^z}, \quad \Re(z) > 1,$$

where $a \neq 0, -1, -2, \ldots$ is a constant. It is evident that $\zeta(z, 1) = \zeta(z)$.

Integral Representation and Analytical Continuation.   To analytically continue zeta function to the whole complex plane we use an integral representation. We start with a simple observation that

$$\frac{1}{n^z} = \frac{1}{\Gamma(z)} \int_0^{\infty} t^{z-1} e^{-nt} dt,$$

which follows from the definition of the gamma function and the substitution $w = nt$. Summing all $n \geq 1$, and interchanging summation and integration (which is allowed due to uniform convergence), we obtain

$$\zeta(z) = \frac{1}{\Gamma(z)} \int_0^\infty \frac{t^{z-1}}{e^t - 1} dt, \quad \Re(z) > 1. \tag{2.44}$$

It can be proved [422, 448] that the above expression can be further transformed into

$$\zeta(z) = \frac{\Gamma(1-z)}{2\pi i} \int_{-\infty}^{(0+)} \frac{t^{z-1}}{e^{-t} - 1} dt, \tag{2.45}$$

where the integral $\int_{-\infty}^{(0+)}$ is known as the Hankel integral (the reader is referred to Table 8.3 in Chapter 8 for a precise definition and properties). Thus $\zeta(z)$ may not be analytic where $\Gamma(1-z)$ is not analytic, that is, at $z = 1, 2, \ldots$. But we know already that $\zeta(z)$ is analytic for $\Re(z) > 1$; hence $z = 1$ is the only singularity of $\zeta(z)$.

Finally, the *reflection formula for zeta function* due to Riemann relates the gamma function and the zeta function (cf. [448]):

$$\zeta(s)\Gamma\left(\frac{s}{2}\right)\pi^{-s/2} = \zeta(1-s)\Gamma\left(\frac{1-s}{2}\right)\pi^{-(1-s)/2}, \tag{2.46}$$

which is true for all s where $\zeta(s)$ and $\Gamma(s)$ functions are defined.

Zeta Function on Imaginary Line. The zeta function $\zeta(z)$ behaves differently from the gamma function on imaginary lines, which has important consequences for certain asymptotics that we shall learn in Chapters 7 and 8. Let now $z = x+iy$, where $x, y \in \mathbb{R}$, and we inspect the behavior of $\zeta(z)$ for fixed x and $y \to \pm\infty$. The following result is proved in Whittaker and Watson [448]

$$\zeta(z) = \begin{cases} O(|y|^{\frac{1}{2}-x}) & \text{for } x < 0 \\ O(|y|^{\frac{1-x}{2}}) & \text{for } 0 < x < 1 \\ O(1) & \text{for } x > 1 \end{cases} \tag{2.47}$$

as $y \to \pm\infty$. Thus unlike the gamma function, $\zeta(z)$ function has a polynomial growth along the imaginary axis. It is worth remembering the above relationships.

Laurent's Expansion. By (2.45), we know that $\zeta(z)$ has only one pole at $z = 1$. It can be proved that its Laurent series around $z = 1$ is

$$\zeta(z) = \frac{1}{z-1} + \sum_{k=0}^\infty \frac{(-1)^k \gamma_k}{k!} (z-1)^k, \tag{2.48}$$

where γ_k are the so–called Stieltjes constants for $k \geq 0$ defined as

$$\gamma_k = \lim_{m \to \infty} \left(\sum_{i=1}^{m} \frac{\ln^k i}{i} - \frac{\ln^{k+1} m}{k+1} \right).$$

In particular, $\gamma_0 = \gamma = 0.577215\ldots$ is the Euler constant, and $\gamma_1 = -0.072815\ldots$.
From the above (or directly from (2.45)), we conclude that

$$\text{Res}[\zeta(z); z = 1] = 1.$$

We shall use the $\zeta(z)$ function quite extensively in Part III. We will learn more about this fascinating function. Among other things, we will relate the value of the zeta function on nonnegative integers to the so-called Bernoulli numbers discussed in Chapter 8 (Table 8.1). Stay tuned!

2.5   EXTENSIONS AND EXERCISES

2.1   Prove the following formula for the $(k + 1)$st moment of a discrete non-negative random variable X:

$$E[X^{k+1}] = \sum_{m \geq 0} \Pr\{X > m\} \sum_{i=0}^{k} (m + 1)^{k-i} m^i.$$

In particular,

$$E[X] = \sum_{m \geq 0} \Pr\{X > m\}.$$

2.2   Prove that (2.15) and (2.16) are equivalent.

2.3   Construct counterexamples to every reverse implication from Theorem 2.3.

2.4   Prove Theorem 2.3.

2.5   Prove the following result: *If X_n is a sequence of nonnegative random variables such that $X_n \overset{a.s.}{\to} X$ and $E[X_n] \to E[X]$, then $X_n \overset{L^1}{\to} X$.*

2.6   Prove that if $X_n \overset{pr}{\to} X$, then $f(X_n) \overset{pr}{\to} f(X)$ provided f is a continuous function.

2.7   Prove (2.24)–(2.25).

2.8   Compute the residue of $f(z) = \Gamma(z)(1 - 2^z)^{-1}$ at $z = 0$.

2.9 Extend Liouville's theorem to polynomial functions, that is, prove that *if $f(z)$ is of at most polynomial growth, that is, $|f(z)| \le B|z|^r$ for some $r > 0$, then it is a polynomial.*

2.10 Estimate the growth of the coefficients $f_n = [z^n] f(z)$ of

$$f(z) = \frac{1}{(1-z)(1-2^z)}$$

using the argument from Example 2.3.

2.11 Prove formula (2.32).
Hint. Compute the integral

$$I(w, z) = \int_0^\infty \int_0^\infty x^{2w-1} y^{2z-1} e^{-(x^2+y^2)} dx dy$$

in two different ways (cf. [422]).

2.12 Show that the Weierstrass product formula for the gamma function implies the Euler product form representation for the gamma function.
Hint. Use the definition of the Euler constant.

2.13 Prove that the product formulas for the gamma function are equivalent to the integral formula (2.28).
Hint. Compute

$$\int_0^\infty \left(1 - \frac{t}{n}\right)^n t^{z-1} dt$$

in two different manners (cf. [422]).

2.14 Prove the *reflection formula* (2.33) for the gamma function.
Hint. Compute

$$B(z, 1-z) = \int_0^\infty \frac{s^{z-1}}{1+s} ds$$

in two ways (cf. [422]).

2.15 Prove the growth (2.47) of $\zeta(z)$ function on imaginary lines for $0 < x < 1$. In general, prove that: *if $\sum_{n \ge 0} a_n < \infty$, then the series*

$$f(z) = \sum_{n=1}^\infty \frac{a_n}{n^z}$$

grows like $|f(x + iy)| = O(|y|^{1-x})$ *for* $0 < x < 1$ *as* $y \to \pm\infty$ (cf. [424]).

2.16 ⚠ In this exercise we extend the asymptotic formula (2.36) for $\Gamma(z + a)/\Gamma(z + b)$. Prove the following asymptotic expansion for $\Re(b - a) > 0$ (cf. [422]):

$$\frac{\Gamma(z + a)}{\Gamma(z + b)} \sim z^{a-b} \sum_{n=0}^{\infty} c_n \frac{\Gamma(b - a + n)}{\Gamma(b - a)} \frac{1}{z^n} \quad \text{as} \quad z \to \infty, \qquad (2.49)$$

where

$$c_n = (-1)^n \frac{B_n^{a-b+1}(a)}{n!},$$

and $B_n^{(w)}(x)$ are the so-called *generalized Bernoulli polynomials* defined as

$$e^{xz}\left(\frac{z}{e^z - 1}\right)^w = \sum_{n=0}^{\infty} \frac{B_n^{(w)}(x)}{n!} z^n \quad |z| < 2\pi.$$

Hint. Express the ratio $\Gamma(z + a)/\Gamma(z + b)$ in terms of the beta function, that is,

$$\frac{\Gamma(z + a)}{\Gamma(z + b)} = \frac{B(z + a, b - a)}{\Gamma(b - a)}.$$

Then use the integral representation of the beta function to show that

$$\frac{\Gamma(z + a)}{\Gamma(z + b)} = \frac{1}{\Gamma(b - a)} \int_0^\infty u^{b-a-1} e^{-zu} f(u)\,du,$$

where

$$f(u) = e^{-au}\left(\frac{1 - e^{-u}}{u}\right)^{b-a-1}$$

Part II

PROBABILISTIC AND COMBINATORIAL TECHNIQUES

3

Inclusion-Exclusion Principle

The inclusion-exclusion principle is one of the oldest methods in combinatorics, number theory, discrete mathematics, and probabilistic analysis. It allows us to compute either the probability that exactly r events occur out of n events (probabilistic inclusion-exclusion principle) or the number of objects that belong exactly to r sets out of n possibly intersecting sets (combinatorial inclusion-exclusion principle). In this chapter, we derive a general form of the inclusion-exclusion principle, and discuss three applications: depth in a trie, order statistics, and the longest aligned word.

In combinatorics we often count the number of objects that satisfy certain properties or belong to certain sets. The (combinatorial) principle of inclusion and exclusion is very helpful. Imagine the following situation: Sets A and B are subset of S, and we wish to count the number of objects that belong neither to A nor to B. Certainly it is not $|S| - |A| - |B|$, since the objects in $A \cap B$ have been subtracted twice. The correct answer is $|S| - |A| - |B| + |A \cap B|$, as one may expect.

In probability, we often need to compute the probability that at least one event out of n events $A_1, \ldots, A_n$ does occur. The events $A_1, \ldots, A_n$ might be dependent and not disjoint. For example, we know that $\Pr\{A \cup B\} = \Pr\{A\} + \Pr\{B\} - \Pr\{A \cap B\}$. In this chapter we generalize these examples and discuss some applications of the inclusion-exclusion principle.

Finally, a word about notation. Throughout the book, we shall write $\log(\cdot)$ for logarithm of an unspecified base. We shall also use $\ln(\cdot)$ for natural logarithm and $\lg := \log_2$ for binary logarithm.

3.1 PROBABILISTIC INCLUSION-EXCLUSION PRINCIPLE

Let $A_1, \ldots, A_n$ be n events which may be intersecting and dependent. We want to compute the probability P_r that exactly r events occur in a random trial. If we write $\bar{A}_i$ for the complement of A_i, then naturally

$$P_r = \sum_{1 \leq j_1 < \cdots < j_r \leq n} \Pr\{A_{j_1} \ldots A_{j_r} \bar{A}_{j_{r+1}} \ldots \bar{A}_{j_n}\}, \qquad (3.1)$$

where $j_1, \ldots, j_n$ is a permutation of $(1, 2, \ldots, n)$. To simplify further derivations, let us define $J_r = \{(j_1, \ldots, j_r) : 1 \leq j_1 < \cdots < j_r \leq n\}$ and for $k \geq 1$

$$S_k := \sum_{J_k} \Pr\{A_{j_1} \ldots A_{j_k}\} \qquad (3.2)$$

with $S_0 = 1$. Observe that

$$S_k = \sum_{J_k} \Pr\{A_{j_1} \ldots A_{j_k}\} = \sum_{J_k} \Pr\{A_{j_1} \ldots A_{j_k}(A_{j_{k+1}} \cup \bar{A}_{j_{k+1}}) \ldots (A_{j_n} \cup \bar{A}_{j_n})\}$$

$$= \sum_{r=k}^{n} \binom{r}{k} \sum_{J_r} \Pr\{A_{j_1} \ldots A_{j_r} \bar{A}_{j_{r+1}} \ldots \bar{A}_{j_n}\}$$

$$= \sum_{r=k}^{n} \binom{r}{k} P_r,$$

where P_r is defined in (3.1).

But we are interested in P_r rather than in S_k, so we need to invert the above. This can be done as follows:

$$\sum_{k=r}^{n}(-1)^{r+k}\binom{k}{r}S_k = \sum_{k=r}^{n}\sum_{s=k}^{n}(-1)^{r+k}\binom{k}{r}\binom{s}{k}P_s$$

$$= \sum_{s=r}^{n}\left(\sum_{k=r}^{s}(-1)^{k-r}\binom{k}{r}\binom{s}{k}\right)P_s.$$

However, as can be easily verified,

$$\sum_{k=r}^{s}(-1)^{k-r}\binom{k}{r}\binom{s}{k} = \sum_{l=0}^{s-r}(-1)^l\binom{s-r}{l} = \begin{cases} 1 & \text{if } r = s \\ 0 & \text{otherwise.} \end{cases}$$

We conclude that

$$P_r = \sum_{k=r}^{n}(-1)^{r+k}\binom{k}{r}S_k.$$

The above equation is an example of the *inversion formula*.

Thus we have just proved the *principle of inclusion and exclusion*. Observe that we did not make any assumption regarding the events $A_1, \ldots, A_n$.

Theorem 3.1 (Inclusion-Exclusion Principle) *Let $A_1, \ldots, A_n$ be events in an arbitrary probability space, and let P_r be the probability of exactly r of them to occur as defined precisely in (3.1). Then*

$$P_r = \sum_{k=r}^{n}(-1)^{r+k}\binom{k}{r}\sum_{1\le j_1 <\cdots< j_k \le n}\Pr\{A_{j_1}\ldots A_{j_k}\} \qquad (3.3)$$

$$= \sum_{k=r}^{n}(-1)^{r+k}\binom{k}{r}S_k,$$

where S_k is explicitly defined in (3.2).

We shall discuss some applications of the inclusion-exclusion principle in the applications section of this chapter. However, for convenience we provide some examples that illustrate the method.

Example 3.1: *Computing a Distribution Through Its Moments* This example is adopted from Bollobás [55]. Let X be a random variable defined on $\{0, 1, \ldots, n\}$, and let $\mathbf{E}_k[X] = \mathbf{E}[X(X-1)\cdots(X-k+1)]$ be the kth factorial moment of X. We shall prove that

$$\Pr\{X = r\} = \frac{1}{r!} \sum_{k=r}^{n} (-1)^{r+k} \frac{\mathbf{E}_k[X]}{(k-r)!}.$$

Indeed, let $A_i = \{X \geq i\}$ for all $i = 1, \ldots, n$. We now evaluate $S_k = \sum_{J_k} \Pr\{\bigcap_{j \in J_k} A_j\}$ where J_k is defined above. Observe that

$$S_1 = \sum_{j=1}^{n} \Pr\{X \geq j\} = \sum_{i=1}^{n} i \Pr\{X = i\} = \mathbf{E}[X],$$

$$S_2 = \sum_{i=1}^{n} \sum_{j=i+1}^{n} \Pr\{X \geq j\} = \sum_{j=2}^{n} (j-1)\Pr\{X \geq j\}$$

$$= \sum_{k=2}^{n} \Pr\{X = k\} \sum_{j=1}^{k-1} j = \frac{\mathbf{E}_2[X]}{2}.$$

In Exercise 1 we ask the reader to prove the following easy generalization:

$$S_k = \sum_{1 \leq j_1 < \cdots < j_k \leq n} \Pr\{X \geq j_1, \ldots, X \geq j_k\} = \frac{1}{k!} \mathbf{E}_k[X]. \qquad (3.4)$$

But $\{X = r\}$ is equivalent to the event that exactly r of A_i occur, since $\{X = r\} = A_1 A_2 \ldots A_r \bar{A}_{r+1} \ldots \bar{A}_n = P_r$ (other events contributing to the probability P_r are empty). By Theorem 3.1 and (3.4) we prove the announced result. ∎

As mentioned above, we did not make any assumptions regarding A_i, and A_i can live on arbitrary spaces. In fact, such a generalization of the inclusion-exclusion principle is due to Fréchet. Its usefulness is illustrated in the next example.

Example 3.2: *Laplace's Formula on the Sum of I.I.D. Uniform Distributions*
Let $X_1, \ldots, X_n$ be independent random variables uniformly distributed on the interval $(0, 1)$. We prove Laplace's formula (cf. [373]), namely,

$$\Pr\{X_1 + \cdots + X_n < x\} = \frac{1}{n!} \sum_{k=0}^{n} (-1)^k \binom{n}{k} (x-k)_+^n,$$

where $x_+ = \max\{0, x\}$. Observe that the distribution of $X_1 + \cdots + X_n$ is equal to the volume of the following set

$$\Omega = \{(x_1, \ldots, x_n) : x_1 + \cdots + x_n < x, \ 0 \leq x_i \leq 1, \ i = 1, \ldots, n\}.$$

To compute this we use the Fréchet formula (i.e., Theorem 3.1 applied to $\mathbb{R}^n$ space). Define the simplex

$$\widetilde{\Omega} = \{(x_1, \ldots, x_n) : x_1 + \cdots + x_n < x, \ x_i \geq 0, \ i = 1, \ldots, n\},$$

and $A_i = \widetilde{\Omega} \cap \{x_i > 1\}$. Clearly, the volume $V(\Omega)$ is equal to the volume of $\widetilde{\Omega}$ minus those parts that do not belong to the unit cube $[0, 1]^n$. But, this can be estimated by applying the inclusion-exclusion principle with a condition that none of A_i occurs. By setting $r = 0$ in Theorem 3.1, we obtain

$$\Pr\{X_1 + \cdots + X_n < x\} = V(\Omega) = \sum_{k=0}^{n} (-1)^k S_k,$$

where $S_0 = V(\widetilde{\Omega})$ and

$$S_k = \sum_{1 \leq j_1 < \cdots < j_k \leq n} V(A_{j_1}, \ldots, A_{j_k}).$$

In the above, $V(A_{j_1}, \ldots, A_{j_k})$ is the volume of the part of $\widetilde{\Omega}$ that possesses properties $A_{j_1}, \ldots, A_{j_k}$, that is, $x_{j_1} > 1, \ldots, x_{j_k} > 1$. To compute $V(A_{j_1}, \ldots, A_{j_k})$ we must set k variables in the equation $x_1 + \cdots + x_n = x$ to 1; hence

$$V(A_{j_1}, \ldots, A_{j_k}) = \frac{(x - k)_+^n}{n!}.$$

This completes the proof of Laplace's formula. ∎

In many situations, one needs to estimate the probability that *at least* one event occurs. This is also known as the *inclusion-exclusion principle*, and in fact it follows directly from Theorem 3.1. Indeed,

$$\Pr\{\bigcup_{i=1}^{n} A_i\} = \sum_{r=1}^{n} P_r$$

$$= \sum_{r=1}^{n} \sum_{k=r}^{n} (-1)^{r+k} \binom{k}{r} S_k = \sum_{k=1}^{n} (-1)^k S_k \sum_{r=1}^{k} \binom{k}{r} (-1)^r$$

$$= \sum_{k=1}^{n} (-1)^k S_k [(1 - 1)^k - 1] = \sum_{k=1}^{n} (-1)^{k+1} S_k.$$

We summarize this finding in the next corollary.

Corollary 3.2 *For arbitrary events the following holds:*

$$\Pr\{\bigcup_{i=1}^{n} A_i\} = \sum_{k=1}^{n} (-1)^{k+1} \sum_{J_k} \Pr\{\bigcap_{j\in J_k} A_j\} = \sum_{k=1}^{n} (-1)^{k+1} S_k, \qquad (3.5)$$

where $J_k = \{1 \leq j_1 < \cdots < j_k \leq n\}$.

It is indeed quite rare to be able to compute all the probabilities involved in the inclusion-exclusion formula. Therefore, often one must retreat to inequalities. The most simple, yet still very powerful, is the following Boole's inequality

$$\Pr\{\bigcup_{i=1}^{n} A_i\} \leq \sum_{i=1}^{n} \Pr\{A_i\}, \qquad (3.6)$$

which can be easily proved by induction. However, we can do much better. The reader is asked in Exercise 2 to use mathematical induction to prove the following Bonferroni's inequalities.

Lemma 3.3 (Bonferroni's Inequalities) *For every even integer $m \geq 0$ we have*

$$\sum_{j=1}^{m} (-1)^{j+1} \sum_{1 \leq t_1 < \cdots < t_j \leq n} \Pr\{A_{t_1} \ldots A_{t_j}\} \leq \Pr\{\bigcup_{i=1}^{n} A_i\}$$

$$\leq \sum_{j=1}^{m+1} (-1)^{j+1} \sum_{1 \leq t_1 < \cdots < t_j \leq n} \Pr\{A_{t_1} \ldots A_{t_j}\}.$$

In Section 3.3.3 we use Bonferroni's inequalities to derive the limiting distribution of the longest aligned match between two sequences.

3.2 COMBINATORIAL INCLUSION-EXCLUSION PRINCIPLE

In this section, we count the number of objects (properties) that belong exactly to r (possibly intersecting) subsets. Imagine the following scenario: There are N objects in a set X whose subsets are $A_1, \ldots, A_n$ of cardinalities $N_1, \ldots, N_n$, respectively. We are interested in counting the number, $N(r)$, of objects that belong exactly to r subsets, where $r = 0, 1, \ldots, n$. Let $N_{j_1,\ldots,j_k} = |A_{j_1} \cap \cdots \cap A_{j_k}|$, that is, $N_{j_1,\ldots,j_k}$ is the cardinality of the intersection of the subsets $A_{j_1}, \ldots, A_{j_k}$ where $1 \leq j_1 < \cdots < j_k \leq n$. The following, also known as the *combinatorial inclusion-exclusion principle*, is a direct consequence of Theorem 3.1.

Theorem 3.4   *For any n and $0 \leq r \leq n$, the number $N(r)$ of objects that belong to exactly r subsets $A_1, \ldots, A_n$ of a set X of cardinality N is*

$$N(r) = \sum_{k=r}^{n} (-1)^{r+k} \binom{k}{r} \sum_{1 \leq j_1 < \cdots < j_k \leq n} N_{j_1, \ldots, j_k} \tag{3.7}$$

$$= \sum_{k=r}^{n} (-1)^{r+k} \binom{k}{r} W_k,$$

where $W_k = \sum_{1 \leq j_1 < \cdots < j_k \leq n} N_{j_1, \ldots, j_k}$.

Proof. We derived the above directly from the probabilistic inclusion-exclusion principle. It suffices to assume that all objects in X are uniformly distributed; hence the probability of a subset A of X is $\Pr\{A\} = |A|/N$. Then the probability P_r of Theorem 3.1 becomes $N(r)/N$, and (3.7) follows immediately from (3.3). ∎

■

We illustrate the combinatorial inclusion-exclusion principle in a few examples. More can be found in the exercises at the end of this chapter. We start with a simple enumeration of surjections.

Example 3.3:   *Number of Surjections*   Let X be a set of cardinality n, and $Y = \{y_1, \ldots, y_m\}$. We want to enumerate the number of surjections from X on Y, that is, the functions from X to Y whose image is Y. Let us call this number s_{nm}. Clearly, $s_{nm} = n!$ if $n = m$ and $s_{nm} = 0$ if $m > n$. Let A_i be the subset of all functions $f : X \to Y$ such that $y_i \notin f(X)$. In terms of Theorem 3.4 we need to find $N(0) = s_{nm}$, that is, the number of functions that *do not* belong to any of A_i. But $|A_{j_1} \cap \cdots \cap A_{j_k}| = (m - k)^n$, thus by (3.7) with $r = 0$ we obtain

$$s_{nm} = \sum_{k=0}^{m} (-1)^k \binom{m}{k} (m - k)^n.$$

Interestingly enough, we also proved that

$$\sum_{k=0}^{n} (-1)^k \binom{n}{k} (n - k)^n = n!,$$

$$\sum_{k=0}^{m} (-1)^k \binom{m}{k} (m - k)^n = 0 \quad \text{for} \quad m > n.$$

Such combinatorial relationships are sometimes quite cumbersome to prove. Counting may help!   ■

Let us now apply Theorem 3.4 to count sequences satisfying certain properties. In those problems the **Möbius function** plays a pivotal role. It is defined as follows: Consider an integer n that has a unique factorization as a product of prime powers, that is,

$$n = p_1^{e_1} p_2^{e_2} \cdots p_r^{e_r}, \tag{3.8}$$

where p_i are distinct prime numbers and $e_i \geq 1$. We define the Möbius function $\mu(n)$ as $\mu(1) = 1$ and for $n > 1$

$$\mu(n) = \begin{cases} 0 & \text{if any power } e_i > 1 \\ (-1)^r & \text{if } e_1 = \cdots = e_r = 1. \end{cases} \tag{3.9}$$

The Möbius function finds applications in counting problems that involve divisors of n. Let us agree to write $d|n$ when d is a divisor of n. Observe that

$$\sum_{d|n} \mu(d) = \begin{cases} 1 & \text{if } n = 1 \\ 0 & \text{otherwise,} \end{cases} \tag{3.10}$$

where the summation is over all divisors d of n. Indeed, to prove the above, let n have the factorization (3.8). By the definition of the Möbius function we have

$$\sum_{d|n} \mu(d) = \sum_{i=0}^{r} \binom{r}{i}(-1)^i = (1-1)^r = 0,$$

since one finds $\binom{r}{i}$ divisors d consisting of i distinct primes involved in the factorization of n.

This leads us to the following interesting *inversion formula*.

Theorem 3.5 (Möbius Inversion Formula) *Let $f(n)$ and $g(n)$ be functions defined for a positive integer n that satisfy*

$$f(n) = \sum_{d|n} g(d).$$

Then

$$g(n) = \sum_{d|n} \mu(d) f\left(\frac{n}{d}\right). \tag{3.11}$$

Proof. By (3.10) we have

$$\sum_{d|n} \mu(d) f\left(\frac{n}{d}\right) = \sum_{d|n} \mu\left(\frac{n}{d}\right) f(d)$$

$$= \sum_{d|n} \mu\left(\frac{n}{d}\right) \sum_{d'|d} g(d') = \sum_{d'|n} g(d') \sum_{m|(n/d')} \mu(m)$$

$$= g(n)$$

as desired. ∎

An inversion formula like the above is very useful in counting and proving combinatorial identities. We shall return to it in Chapter 7 where we use analytic tools to explore them (see Section 7.5.2). We finish this section with an application of Theorem 3.5.

Example 3.4: *Primitive Sequences* Let us count the number of binary sequences of length n that are primitive, that is, they are *not* expressible as a concatenation of some identical smaller sequences. For example, 010101 is not a primitive sequence while 011010 is a primitive one. Let $f(n)$ be the number of such primitive sequences. There are 2^n all binary strings, and every one is uniquely expressible as a concatenation of n/d identical *primitive* strings of length d, where d is a divisor of n. Thus for every n

$$2^n = \sum_{d|n} f(d).$$

Then by the Möbius inversion formula we have

$$f(n) = \sum_{d|n} \mu\left(\frac{n}{d}\right) 2^d$$

for any $n \geq 0$. ∎

3.3 APPLICATIONS

We now tackle some problems in which the inclusion-exclusion principle plays a significant role. We start with deriving the generating function for the depth in a trie, then deal with order statistics, and finish with deriving the asymptotic distribution for the longest aligned word in two randomly generated strings. The last problem is by far the most challenging.

3.3.1   Depth in a Trie

In Section 1.1 we discussed tries built over n binary strings $X^1, \ldots, X^n$. We assume that those strings are generated by a binary memoryless source with "0" occurring with probability p and "1" with probability $q = 1 - p$. We are interested here in the depth $D_n(m)$ as defined in Definition 1.1. By Definition 1.2, C_{ij} is the length of the longest string that is a prefix of X^i and X^j. By Theorem 1.3

$$D_n(m) = \max_{1 \le i \ne m \le n} \{C_{i,m}\} + 1. \tag{3.12}$$

Certainly, the alignments C_{ij} are dependent random variables even for the memoryless source.

Our goal is to compute the generating function of the depth $D_n := D_n(1)$, that is, $D_n(u) = \sum_{k=0}^{\infty} \Pr\{D_n = k\} u^k = \mathbf{E}[u^{D_n}]$. Observe that

$$\Pr\{D_n > k\} = \Pr\{\bigcup_{i=2}^{n} A_{i,1}\},$$

where $A_{ij} = \{C_{ij} \ge k\}$. Thus one can apply the principle of inclusion-exclusion, as presented in Corollary 3.2, to obtain

$$\Pr\{D_n > k\} = \Pr\{\bigcup_{i=2}^{n} [C_{i,1} \ge k]\}$$

$$= \sum_{r=1}^{n-1} (-1)^{r+1} \binom{n-1}{r} \Pr\{C_{2,1} \ge k, \ldots, C_{r+1,1} \ge k\},$$

since the probability $\Pr\{C_{2,1} \ge k, \ldots, C_{r+1,1} \ge k\}$ does not depend on the choice of strings (i.e., it is the same for any r-tuple of selected strings). Moreover, it can be easily explicitly computed. Indeed, we find

$$\Pr\{C_{2,1} \ge k, \ldots, C_{r+1,1} \ge k\} = (p^{r+1} + q^{r+1})^k,$$

since r independent strings must agree on the first k symbols, and the probability that they agree on a symbol at a given position is $p^{r+1} + q^{r+1}$. Thus

$$\sum_{k=0}^{n} \Pr\{D_n > k\} u^k = \sum_{r=1}^{n-1} (-1)^{r+1} \binom{n-1}{r} \frac{1}{1 - u(p^{r+1} + q^{r+1})}.$$

In Exercise 4 we ask the reader to prove that

$$\sum_{k=0}^{\infty} \Pr\{D_n > k\} u^k = \frac{1 - \mathbf{E}[u^{D_n}]}{1 - u},$$

which finally leads to

$$D_n(u) = 1 + \sum_{r=1}^{n-1} (-1)^r \binom{n-1}{r} \frac{1 - u}{1 - u(p^{r+1} + q^{r+1})}$$

for $|u| < 1$.

3.3.2 Order Statistics

Let $Y_1, \ldots, Y_m$ be a sequence of random variables. We often need to compute the maximum or minimum values of such a sequence. In some problems (e.g., combinatorial optimization problems discussed in Section 1.6) we are also interested in the rth greatest value of $Y_1, \ldots, Y_m$, where $r = 1, \ldots, m$. More precisely, let $\min_{1 \le i \le m}\{Y_i\} := Y_{(1)} \le Y_{(2)} \le \cdots \le Y_{(m)} := \max_{1 \le i \le m}\{Y_i\}$. We call $Y_{(r)}$ the rth *order statistic* of $Y_1, \ldots, Y_m$.

The inclusion-exclusion principle can be used to find the probabilistic behavior of order statistics. Indeed, let $A_i = \{Y_i > x\}$, then $\{Y_{(m)} > x\} = \{\max\{Y_1, \ldots, Y_m\} > x\} = \bigcup_{i=1}^{m} A_i$. More interestingly, we can express the rth order statistic as follows

$$\{Y_{(r)} > x\} = \bigcup_{1 \le j_1 < \cdots < j_{m-r+1} \le m} \bigcap_{i=1}^{m-r+1} A_{j_i}, \tag{3.13}$$

since for $\{Y_{(r)} > x\}$ to hold it is required that at least $m - r + 1$ variables out of $Y_1, \ldots, Y_m$ are greater than x. Indeed, for $r = m$ we have $\{Y_{(m)} > x\} = \{\max_{1 \le i \le m}\{Y_i\} > x\} = \bigcup_{i=1}^{m}\{Y_i > x\}$, and for $r = m - 1$ we obtain

$$\{Y_{(m-1)} > x\} = \{Y_1 > x \ \& \ Y_2 > x \text{ or} \ldots \text{or } Y_1 > x \ \& \ Y_m > x$$

$$\text{or} \ldots \text{or } Y_{m-1} > x \ \& \ Y_m > x\}$$

and the general case is given in (3.13).

The next finding presents a simple result on the maximum of a set of random variables.

Theorem 3.6 *Let $Y_1, Y_2, \ldots, Y_m$ be a sequence of random variables with distribution functions $F_1(y), F_2(y), \ldots, F_m(y)$, respectively, and let $R_i(y) = \Pr\{Y_i \ge y\}$. Define $Y_{(m)} = \max_{1 \le i \le m}\{Y_i\}$.*

(i) *If a_m is the smallest solution of*

$$\sum_{k=1}^{m} R_k(a_m) = 1, \tag{3.14}$$

then

$$\mathbf{E}[Y_{(m)}] \le a_m + \sum_{k=1}^{m} \sum_{j=a_m}^{\infty} R_k(j).$$

(ii) *If the distribution function $F_i(y)$ of Y_i satisfies for all $1 \le i \le m$ the following two conditions*

$$F_i(y) < 1 \quad \text{for all} \ \ y < \infty, \tag{3.15}$$

and

$$\lim_{y \to \infty} \sup_{i} \frac{1 - F_i(cy)}{1 - F_i(y)} = 0 \quad \text{for} \ \ c > 1, \tag{3.16}$$

then $Y_{(m)}/a_m \le 1$ in probability (pr.), that is, for any $\varepsilon > 0$

$$\lim_{m \to \infty} \Pr\{Y_{(m)} \ge (1 + \varepsilon)a_m\} = 0,$$

where a_m solves (3.14).

(iii) *If $Y_1, \ldots, Y_m$ are independently and identically distributed with common distribution function $F(\cdot)$, then $Y_{(m)} \sim a_m$ (pr.), that is, for any $\varepsilon > 0$*

$$\lim_{m \to \infty} \Pr\{(1 - \varepsilon)a_m \le Y_{(m)} \le (1 + \varepsilon)a_m\} = 1,$$

where a_m is a solution of (3.14) that in this case becomes $mR(a_m) = 1$.

Proof. For (i) observe that for any a

$$Y_{(m)} \le a + \sum_{k=1}^{m} [Y_k - a]^+,$$

where $t^+ = \max\{0, t\}$. Since $[Y_k - a]^+$ is a nonnegative random variable, then $\mathbf{E}[Y_k - a]^+ = \int_a^\infty R_k(y)dy$ (assuming for simplicity that Y_k is a continuous random variable), thus

$$\mathbf{E}[Y_{(m)}] \le a + \sum_{k=1}^{m} \int_a^\infty R_k(x)dx.$$

Minimizing the right-hand side of the above with respect to a yields part (i). For part (ii) we use (3.13) for $r = m$, and after a simple application of (3.6), we arrive at

$$\Pr\{Y_{(m)} > x\} = \Pr\{\bigcup_{i=1}^{m} Y_i > x\} \leq \sum_{i=1}^{m} R_i(x).$$

Let now $x = (1+\varepsilon)a_m$ where $\varepsilon > 0$ and a_m be defined by (3.14). Observe that by (3.15) $a_m \to \infty$. Then, (3.16) with $c = 1+\varepsilon$ implies $R((1+\varepsilon)a_m) = o(1)R(a_m)$; hence due to (3.14)

$$\Pr\{Y_{(m)} \geq (1 + \varepsilon)a_m\} = o(1) \sum_{i=1}^{m} R(a_m) \to 0.$$

Part (iii) can be proved using the additional assumption about the independence of $Y_1, \ldots, Y_m$. The reader is asked in Exercise 9 to provide details of the proof. ∎

Finally, we present a simple asymptotic result concerning the behavior of $Y_{(r)}$ for the so-called exchangeable random variables. A sequence $\{Y_i\}_{i=1}^{m}$ is *exchangeable* if for any k-tuple $\{j_1, \ldots, j_k\} \subset \{1, \ldots, m\}$ the following holds:

$$\Pr\{Y_{j_1} < y_{j_1}, \ldots, Y_{j_k} < y_{j_k}\} = \Pr\{Y_1 < y_1, \ldots, Y_k < y_k\},$$

that is, the joint distribution depends only on the *number* of variables (cf. [161]).

Theorem 3.7 *Let $\{Y_i\}_{i=1}^{m}$ be exchangeable random variables with distribution $F(\cdot)$ having property (3.15), and let $R_r(x) = \Pr\{Y_1 > x, \ldots, Y_r > x\}$ satisfy for all $c > 1$*

$$\lim_{x \to \infty} \frac{R_{m-r+1}(cx)}{R_{m-r+1}(x)} = 0 \tag{3.17}$$

uniformly over all $r = 1, \ldots, m$. Define $a_m^{(r)}$ as the smallest solution of

$$\binom{m}{r-1} R_{m-r+1}(a_m^{(r)}) = 1.$$

Then $Y_{(r)} \leq a_m^{(r)}$ (pr.) as $m \to \infty$.

Proof. We use (3.13) with $x = (1 + \varepsilon)a_m^{(r)}$. By (3.6) and the assumption of exchangebility, we obtain

$$\Pr\{Y_{(r)} > (1+\varepsilon)a_m^{(r)}\} \le \binom{m}{m-r+1} R_{m-r+1}((1+\epsilon)a_m^{(r)})$$

$$= \binom{m}{r-1} R_{m-r+1}(a_m^{(r)})o(1) \to 0,$$

where the last equation follows from (3.17).   ∎

The reader is asked in Exercises 9–11 to extend the above results and apply them to some combinatorial optimization problems.

3.3.3   Longest Aligned Word

We report here some results of Karlin and Ost [229]. Imagine two sequences of length n, say $X = (X_1, \ldots, X_n)$ and $Y = (Y_1, \ldots, Y_n)$. We assume that both are generated by memoryless sources over a binary alphabet, say *Bernoulli*(p_1, p_2) and *Bernoulli*(r_1, r_2), respectively. We are interested in the length W_n of the *longest aligned match word*, that is, a word as long as possible that is common to both strings and starts at the same position in both strings. More precisely:

$$W_n := \max\{k : \exists_{1 \le t \le n} \, X_t = Y_t, \ldots, X_{t+k-1} = Y_{t+k-1} \text{ and } X_{t+k} \ne Y_{t+k}\}.$$

To see how this problem is related to the topic of this section, let us observe that $W_n = \max_k\{A(n, k)\}$, where

$A(n, k) = \{\text{There exists an aligned match word of length } k \text{ in both sequences}\}.$

Then $\Pr\{W_n \ge k\} = \Pr\{A(n, k)\}$ and

$$A(n, k) = \bigcup_{t=1}^{n} A_t(k),$$

where $A_t(k)$ is the event signifying a k-word match starting at position $1 \le t \le n$. Thus, the principle of inclusion-exclusion (cf. Corollary 3.2) can be used to evaluate $\Pr\{A(n, k)\}$, and hence $\Pr\{W_n \ge k\}$.

We start with the formulation of the result of Karlin and Ost [229]. Let $P = p_1 r_1 + p_2 r_2$ be the probability of a match at a given position. The probability of a match of length at least k starting at any position is equal to P^k. For convenience we also define $Q = P^{-1}$. We prove below that the maximal aligned match W_n is concentrated around

$$k = \frac{\log n}{\log(P^{-1})} + x + 1 = \log_Q n + x + 1,$$

where x is a real number. We denote by $\widetilde{x}$ such x (that may depend on n) that $k = \log_Q n + \widetilde{x} + 1$ is an integer.

Theorem 3.8   (Karlin and Ost, 1987)   *Under the above assumptions*

$$\lim_{n \to \infty} \Pr\{W_n \leq \log_Q n + \widetilde{x}\} = \exp\left(-(1 - P)P^{\widetilde{x}+1}\right). \qquad (3.18)$$

If x is a fixed real number, then the limiting distribution of W_n does not exist, but one can claim

$$\liminf_{n \to \infty} \Pr\{W_n \leq \log_Q n + x\} = \exp\left(-(1 - P)P^x\right) \qquad (3.19)$$

$$\limsup_{n \to \infty} \Pr\{W_n \leq \log_Q n + x\} = \exp\left(-(1 - P)P^{x+1}\right) \qquad (3.20)$$

where $Q^{-1} = P = p_1 r_1 + p_2 r_2$ is the probability of a match at a given position.

Before we proceed with the proof, some explanations are in order. The limiting distribution of W_n does not exist for general real x, but an asymptotic distribution may be found that resembles the *extreme distribution* $\exp(-e^{-x})$ (cf. Galambos [161]). Indeed, we shall prove below that for large n

$$\lim_{n \to \infty} \Pr\{W_n \leq \log_Q n + x\} \sim \exp\left(-(1 - P)P^{x+1-\langle \log_Q n + x \rangle}\right),$$

where $\langle \log n \rangle = \log n - \lfloor \log n \rfloor$. This fractional part $\langle \log_Q n \rangle$ behaves very erratically. We shall study its property in Section 8.2.3. The asymptotic distribution is presented in Figure 3.1, where the staircase-like function $\exp(-(1 - P)P^{x+1-\langle \log_Q n + x \rangle})$ is plotted together with the lower extreme distribution $\exp\left(-(1 - P)P^x\right)$ and the upper extreme distribution $\exp\left(-(1 - P)P^{x+1}\right)$. We should point out, however, that such a behavior is not a surprise, since W_n as a discrete order statistic is expected not to have a limiting distribution. Already in 1970 Anderson [12] observed that the limiting distribution of the maximum of *discrete* i.i.d. random variables may not exist.

Now we are ready to prove the Karlin and Ost result. Throughout this derivation we assume that

$$k = \frac{\log n}{\log(P^{-1})} + \widetilde{x} + 1 = \log_Q n + \widetilde{x} + 1, \qquad (3.21)$$

where $\widetilde{x}$ is such that k is an integer. As observed above, we need to estimate $\Pr\{A(n, k)\}$, and after applying Corollary 3.2 we can write

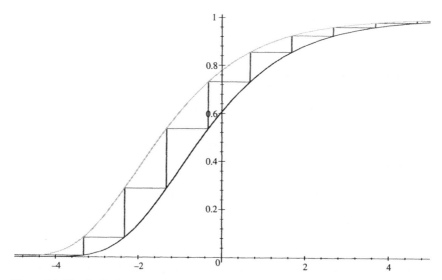

Figure 3.1. The distribution function of $W_n - \log_2 20$ with two continuous extremes for binary symmetric Bernoulli models.

$$\Pr\{A(n, k)\} = \sum_{m=1}^{n}(-1)^{m+1} \sum_{1 \leq t_1 < \ldots < t_m \leq n} \Pr\{A_{t_1} \ldots A_{t_m}\} = \sum_{s=1}^{n}(-1)^{m+1} S_m(n),$$

where $S_m(n) = \sum_{1 \leq t_1 < \ldots < t_m \leq n} \Pr\{A_{t_1} \ldots A_{t_m}\}$. We write below A_t for $A_t(k)$.

The plan for establishing the limit of $\Pr\{A(n, k)\}$ is as follows: We shall prove that for k as in (3.21) every term $S_m(n)$ converges to a function, say $a_m(\tilde{x})$ as $n \to \infty$; that is, for $m = 1, 2, \ldots$

$$\lim_{n \to \infty} S_m(n) = \lim_{n \to \infty} \sum_{1 \leq t_1 < \ldots < t_m \leq n} \Pr\{A_{t_1} \ldots A_{t_m}\} = a_m(\tilde{x}). \qquad (3.22)$$

By Bonferroni's inequalities for every even m

$$\liminf_{n \to \infty} \Pr\{A(n, k)\} \geq \sum_{i=1}^{m}(-1)^{i+1} a_i(\tilde{x}),$$

$$\limsup_{n \to \infty} \Pr\{A(n, k)\} \leq \sum_{i=1}^{m+1}(-1)^{i+1} a_i(\tilde{x}).$$

Next, we shall show that the alternating series

$$\sum_{i=1}^{\infty} (-1)^{i+1} a_i(\tilde{x}) = a(\tilde{x})$$

converges uniformly for bounded x, which would imply that

$$\lim_{n \to \infty} \Pr\{W_n \le \log_Q n + \tilde{x}\} = 1 - a(\tilde{x}).$$

We now compute $\lim_{n \to \infty} S_m(n)$ for every m. Let us start with $m = 1$. Observe that $\Pr\{A_t\} = P^k$; thus

$$\lim_{n \to \infty} S_1(n) = \lim_{n \to \infty} n P^{\log_Q n + \tilde{x} + 1} = P^{\tilde{x}+1} := \phi(\tilde{x}).$$

This was quite easy! The next step requires a little more work. Let

$$S_2(n) = \sum_{t_1 < t_2} \Pr\{A_{t_1} A_{t_2}\} = \frac{1}{2} \sum_{|t_2 - t_1| \le k} \Pr\{A_{t_1} A_{t_2}\} + \frac{1}{2} \sum_{|t_2 - t_1| > k} \Pr\{A_{t_1} A_{t_2}\}$$

$$:= T_1(n) + T_2(n).$$

By independence $T_2(n)$ can be easily computed, and we arrive at

$$\lim_{n \to \infty} T_2(n) = \frac{1}{2} \lim_{n \to \infty} (S_1(n))^2 = \frac{1}{2} \phi^2(\tilde{x}).$$

To assess $T_1(n)$, we must take into account overlapping parts. Let the two overlapping words of length k be such that the first word starts at position t and the overlapping begins at position $t + j$ for $1 \le j \le k$. Let C_2 denote the extended match of length $k + j$. Then $\Pr\{C_2 > k + j\} = P^{k+j}$, and hence

$$T_1(n) = \sum_{t=1}^{n} \sum_{j=1}^{k} P^{k+j} = n P^k \left(\frac{P}{1-P} \right) (1 - P^k).$$

In summary, for k as in (3.21)

$$\lim_{n \to \infty} S_2(n) = \frac{1}{2} (\phi(\tilde{x}))^2 + \left(\frac{P}{1-P} \right) \phi(\tilde{x})$$

since $(1 - P^k) = (1 - P^{\tilde{x}+1}/n) \to 1$ as $n \to \infty$.

Let us now consider a general m and compute $S_m(n)$. We partition $m = n_1 + 2n_2 + 3n_3 + \cdots + m n_m$ so that $S_m^{(n_1, n_2, \ldots, n_m)}(n)$ denotes the cumulation of probabilities $\Pr\{A_{t_1}, \ldots, A_{t_m}\}$ of joint events which consist of n_1 matches of nonoverlapping k-words, n_2 pairs of overlapping k-word matches which are mutually independent, n_3 groups of overlapping three k-word matching which are disjoint between the different triples, etc., and n_m (necessarily at most 1) successive overlapping k-word matches. Then, $S_m(n) = \sum_{\mathcal{P}(n_1,\ldots,n_m)} S_m^{(n_1,n_2,\ldots,n_m)}(n)$ where $\mathcal{P}(n_1,\ldots,n_m)$ is the set of all integer partitions such that $m = n_1 + 2n_2 + 3n_3 + \cdots + mn_m$. It is not too difficult to observe that

$$S_m^{(n_1,n_2,\ldots,n_m)}(n) = \frac{1}{n_1! n_2! \cdots n_m!} \prod_{i=1}^{m} \left(nP^k \left(\frac{P}{1-P} \right)^{m-1} (1 - P^k)^{m-1} \right)^{n_i}.$$

Indeed, consider n_i groups of overlapping i word matches of length k. Since any permutation of these words leads to the same probability, we see the term $1/n_i!$ in the above. Let a group of i k-word matches start at position $1 \leq t \leq n$ and the overlaps begin at positions $t + j_1, \ldots, t + j_m$, respectively. Such an extended C_m match has the following probability:

$$\Pr\{C_m > k + j_1 + \cdots + j_m\} = \sum_{j_1=1}^{k} \cdots \sum_{j_m=1}^{k} P^{k+j_1+\cdots+j_m}$$

$$= P^k \left(\frac{P}{1-P} \right)^{m-1} (1 - P^k)^{m-1}.$$

Summing over all $1 \leq t \leq n$, and noting that the n_i groups are independent, we obtain the desired formula. Thus, for k as in (3.21) we arrive at

$$\lim_{n \to \infty} S_m^{(n_1,n_2,\ldots,n_m)}(n) = \frac{1}{n_1! n_2! \cdots n_m!} \prod_{i=1}^{m} s_i^{n_i},$$

where $s_i = \phi(\tilde{x})(P/(1 - P))^{i-1} = P^{\tilde{x}+1}(P/(1 - P))^{i-1}$ and finally,

$$a_m(\tilde{x}) = \lim_{n \to \infty} S_m(n) = \sum_{n_1+2n_2+\cdots+mn_m=m} \frac{1}{n_1! n_2! \cdots n_m!} \prod_{i=1}^{m} s_i^{n_i}. \qquad (3.23)$$

To complete the proof, we need to sum up the alternating series

$$\sum_{m=1}^{\infty} (-1)^{m+1} a_m(\tilde{x}).$$

For this we need one result from generating functions, namely,

$$\sum_{m=0}^{\infty} \sum_{n_1+2n_2+\cdots mn_m=m} \frac{x^m}{n_1! n_2! \cdots n_m!} \prod_{i=1}^{m} c_i^{n_i} = \exp \left(\sum_{i=1}^{\infty} c_i x^i \right), \qquad (3.24)$$

provided the series $\sum_{i=1}^{\infty} c_i x^i$ converges. The above formula is a simple application of the convolution formula of exponential generating functions, and will be discussed in Chapter 7. In fact, we prove it in Example 7.5. (This is the point where analytic and probabilistic analyses meet and need each other to solve the problem at hand.) Knowing this, we can proceed as follows:

$$\sum_{m=1}^{\infty} (-1)^{m+1} a_m(\widetilde{x})$$

$$= \sum_{m=1}^{\infty} (-1)^{m+1} \sum_{n_1+2n_2+\cdots+mn_m=m} \frac{1}{n_1! n_2! \cdots n_m!} \left(\frac{P}{1-P} \right)^{\sum_{i=1}^{m} n_i (i-1)}$$

$$\times (\phi(\widetilde{x}))^{\sum_{i=1}^{m} n_i} = \sum_{m=1}^{\infty} (-1)^{m+1} \left(\frac{P}{1-P} \right)^m$$

$$\times \sum_{n_1+2n_2+\cdots+mn_m=m} \frac{1}{n_1! n_2! \cdots n_m!} \left(\frac{1-P}{P} \phi(\widetilde{x}) \right)^{\sum_{i=1}^{m} n_i}$$

$$= 1 - \exp\left(-(1-P) P^{\widetilde{x}+1} \right),$$

where the last equality comes from identity (3.24) with $x = -P/(1-P)$ and $c_i = (1-P)\phi(\widetilde{x})/P$. The above identities are true only for $P/(1-P) < 1$, that is, for $P < \frac{1}{2}$. But with the help of analytic continuation (see Chapter 8) we can extend their validity to the whole interval $0 < P < 1$. This proves (3.18) of Theorem 3.8. To establish the second part of Theorem 3.8, we observe that for any real x we can write $\Pr\{W_n \geq \log_Q + x + 1\} = \Pr\{W_n \geq \lfloor \log_Q + x + 1 \rfloor\}$ and $\lfloor \log_Q + x + 1 \rfloor = \log_Q + x + 1 + \langle \log_Q + x \rangle$.

3.4 EXTENSIONS AND EXERCISES

3.1 Prove (3.4), that is,

$$S_k = \sum_{1 \leq j_1 < \cdots < j_k \leq n} \Pr\{X \geq j_1, \ldots, X \geq j_k\} = \frac{1}{k!} \mathbf{E}_k[X].$$

3.2 Prove Lemma 3.3 using mathematical induction.

3.3 (Bell's Inequality) Prove the following useful inequality of Bell:

$$\Pr\{A \cap B\} \leq \Pr\{A \cap C\} + \Pr\{B \cap \bar{C}\},$$

where A, B, C are events and $\bar{C}$ is the complementary event to C. The above inequality is used in quantum physics to establish the Einstein-Podolsky-Rosen paradox.

3.4 Prove that for any random variable X over nonnegative integers

$$\sum_{k=0}^{n} \Pr\{X > k\}u^k = \frac{1 - \mathbf{E}[u^X]}{1 - u}$$

for $|u| < 1$.

3.5 Prove that if X is a nonnegative random variable with finite moments and for all m

$$\lim_{k \to \infty} \frac{\mathbf{E}_k[X]k^m}{k!} = 0,$$

then

$$\Pr\{X = r\} = \frac{1}{r!} \sum_{k=r}^{\infty} (-1)^{r+k} \frac{\mathbf{E}_k[X]}{(k-r)!},$$

where, as before, $\mathbf{E}_k[X]$ is the kth factorial moment of X.

3.6 Enumerate the number of binary circular sequences, where two sequences obtained by a rotation are considered the same.

3.7 Let $n = p_1^{e_1} p_2^{e_2} \cdots p_r^{e_r}$ be a positive integer, and denote by $\phi(n)$ the number of integers $1 \le k \le n$ such that $\gcd(n, k) = 1$. (The function ϕ is known as the Euler function.) Prove that

$$\phi(n) = n \sum_{k=1}^{n} (-1)^{k+1} \sum_{1 \le j_1 < \ldots < j_k \le n} \frac{1}{p_{j_1} \cdots p_{j_k}} = n \prod_{i=1}^{n} \left(1 - \frac{1}{p_i}\right),$$

and

$$\sum_{d|n} \phi(d) = n.$$

Conclude that

$$\phi(n) = n \sum_{d|n} \frac{\mu(d)}{d}$$

for all positive integers n.

3.8 How many positive integers less than 1000 have no factor between 1 and 10?

3.9 Prove part (iii) of Theorem 3.6. Then establish the following refinement: Let $Y_1, \ldots, Y_m$ be i.i.d. and let a_m be the smallest solution of $m(1 - \Pr\{Y_1 > a_m\}) = 1$. Then

$$\lim_{m \to \infty} \Pr\{\max\{Y_1, \ldots, Y_m\} = \lfloor a_m \rfloor \text{ or } \lfloor a_m \rfloor + 1\} = 1.$$

Hint. Verify that

$$\lim_{m \to \infty} (\Pr\{\max\{Y_1, \ldots, Y_m\} < x\} - \exp(n(1 - \Pr\{Y_1 > x\}))) = 0$$

for real x.

3.10 (Aldous, Hofri, and Szpankowski, 1992) Prove the following result: *If $\{X_k\}_{k=1}^{\infty}$ are i.i.d. Poisson distributed random variables with mean ρ, then for large enough integers a and n*

$$\Pr\{\max_{1 \le k \le n} X_k < a\} - \exp(-ne^{-\rho}\rho^a/a!) \to 0 \quad \text{as} \quad n, a \to \infty$$

and

$$\Pr\{\max_{1 \le k \le n} X_k = \lfloor a_n \rfloor + 1 \text{ or } \lfloor a_n \rfloor\} = 1 - O(1/a_n) \to 1 \quad \text{as } n \to \infty.$$

For large n the sequence $\{a_n\}$ satisfies

$$a_n \sim \frac{\log n - \rho}{\log(\log n - \rho) - \log \rho} \sim \frac{\log n}{\log \log n},$$

where a_n is defined as the smallest solution of the equation

$$n \cdot \frac{\gamma(a_n, \rho)}{\Gamma(a_n)} = 1.$$

In the above $\gamma(x, \rho) \equiv \int_0^{\rho} t^{x-1} e^{-t} dt$ is the incomplete gamma function, and $\Gamma(x) = \gamma(x, \infty)$ is the gamma function discussed in Section 2.4.

3.11 ⚠ Let $W = \{w_{i,j}\}_{i,j=1}^n$ be a matrix of *weights* that are i.i.d. distributed according to a common strictly continuous and increasing distribution $F(\cdot)$. Consider the following *bottleneck assignment problem*:

$$Z_{\min} = \min_{\sigma \in B_n} \{\max_{1 \le i \le n} w_{i,\sigma(i)}\},$$

and *capacity assignment problem*

$$V_{\max} = \max_{\sigma \in B_n} \{ \min_{1 \le i \le n} w_{i, \sigma(i)} \},$$

where B_n is a set all permutations of $\{1, 2, \ldots, n\}$. Prove that

$$\lim_{n \to \infty} \frac{Z_{(d)}}{F^{-1}(\log n / n)} = 1 \quad \text{(pr.)}$$

$$\lim_{n \to \infty} \frac{V_{(d)}}{F^{-1}(1 - \log n / n)} = 1 \quad \text{(pr.)},$$

where $F^{-1}(\cdot)$ denotes the inverse function of the distribution $F(\cdot)$.

3.12 ⚠(Karlin and Ost, 1987) Extend Theorem 3.8 to Markov sources. More generally, compute the asymptotic distribution of the longest aligned word match found in r or more sequences among s sequences generated by Markov sources.

3.13 ⚠(Karlin and Ost, 1987) Compute the limiting distribution of the *longest common word* (not necessary aligned) common to r or more of s sequences generated by memoryless and Markov sources.

4

The First and Second Moment Methods

In the analysis of algorithms, we are usually interested in time complexities, storage requirements, or the value of a particular parameter characterizing the algorithm. These quantities are *random* (nondeterministic) since either the input is assumed to vary or the algorithm itself may make random decisions (e.g., in randomized algorithms). We often are satisfied with average values; however, for many applications this might be too simplistic. In many instances these quantities of interest, though random, behave in a very deterministic manner for large inputs. The *first and second moment methods* are the most popular probabilistic tools used to derive such relationships. We discuss them in this chapter together with some interesting applications.

Imagine that X_n is a random variable representing a quantity of interest (e.g., the number of pattern occurrences in a random text of length n). Usually, it is not difficult to compute the average $\mathbf{E}[X_n]$ of X_n but it is equally easy to come up with examples that show how poor such a measure of variability of X_n could be. But often we can discover refined information about X_n (e.g., that with high probability $X_n \sim \mathbf{E}[X_n]$ as $n \to \infty$). In this case, the random variable X_n converges (in probability or almost surely) to a *deterministic* value $a_n = \mathbf{E}[X_n]$ as n becomes larger (see Example 4.2).

Consider the following example: Let $M_n = \max\{C_1, \ldots, C_n\}$ where C_i are dependent random variables (e.g., M_n may represent the longest path in a digital tree). Again, M_n varies in a random fashion but for large n we shall find out that $M_n \sim a_n$ where a_n is a (deterministic) sequence (see Example 4.1). More precisely, for every $\varepsilon > 0$ the probability of $(1-\varepsilon)a_n \leq M_n \leq (1+\varepsilon)a_n$ becomes closer and closer to 1; thus according to the definition from Section 2.2.2 we say that $M_n \overset{pr}{\to} a_n$ or $M_n \to a_n$ (pr.).

These and other problems can be investigated by two simple probabilistic techniques called the **first and second moment methods**. They resemble Boole's and Bonferroni's inequalities, but instead of bounding the probability of a union of events by marginal probabilities they use the first and the second moments. We also briefly discuss the **fourth moment method**.

Here is the plan for the chapter: We first discuss some theoretical underpinnings of the methods which we illustrate through a few examples. Then we present several applications such as a Markov approximation of a stationary distribution, evaluate the number of primes dividing a given number, and finally estimate the height in tries, PATRICIA tries, digital search trees, and suffix trees.

4.1 THE METHODS

Let us start by recalling Markov's inequality (2.6) from Chapter 2: For a nonnegative random variable X, we can bound the probability that X exceeds $t > 0$ as follows: $\Pr\{X \geq t\} \leq \mathbf{E}[X]/t$. If in addition, X is an integer-valued random variable, after setting $t = 1$ we obtain

$$\Pr\{X > 0\} \leq \mathbf{E}[X], \tag{4.1}$$

which is the *first moment method*. We can derive it in another manner (without using the Markov inequality). Observe that

$$\Pr\{X > 0\} = \sum_{k=1}^{\infty} \Pr\{X = k\} \leq \sum_{k=0}^{\infty} k\Pr\{X = k\} = \mathbf{E}[X].$$

The above implies Boole's inequality (3.6). Indeed, let A_i $(i = 1, \ldots, n)$ be events, and set $X = I(A_1) + \cdots + I(A_n)$ where, as before, $I(A) = 1$ if A occurs, and zero otherwise. Boole's inequality (3.6) follows.

In a typical application of (4.1), we expect to show that $\mathbf{E}[X] \to 0$, and hence that $X = 0$ occurs **with high probability** (**whp**). Throughout, we shall often write **whp** instead of the more formal "convergence in probability." We illustrate the first moment method in a simple example.

Example 4.1: *Maximum of Dependent Random Variables* Let $M_n = \max\{X_1, \ldots, X_n\}$ where X_i are dependent but identically distributed random variables with the distribution function $F(\cdot)$. We assume that the distribution function is such that for all $c > 1$

$$1 - F(cy) = \delta(y)(1 - F(y)),$$

where $\delta(y) \to 0$ as $y \to \infty$. We shall prove that **whp** $M_n/a_n \leq 1$ where a_n is the smallest solution of the following equation:

$$n(1 - F(a_n)) = 1,$$

provided $a_n \to \infty$ as $n \to \infty$. Indeed, observe that $\{M_n > x\} = \bigcup_{i=1}^{n} A_i$ where $A_i = \{X_i > x\}$. Thus by the first moment method,

$$\Pr\{M_n > x\} \leq n\mathbf{E}[I(A_1)] = n(1 - F(x)).$$

Let us set now $x = (1 + \varepsilon)a_n$ for any $\varepsilon > 0$, where a_n is defined as above. Then

$$\Pr\{M_n > (1+\varepsilon)a_n\} \leq n(1 - F((1+\varepsilon)a_n)) = \delta(a_n)n(1 - F(a_n)) = \delta(a_n) \to 0,$$

since $a_n \to \infty$. Thus $M_n \leq (1 + \varepsilon)a_n$ **whp**, as desired (see also Section 3.3.2). ∎

The first moment method uses only the knowledge of the average value of X to bound the probability. A better estimate of the probability can be obtained if one knows the variance of X. Indeed, Chebyshev already noticed that (cf. (2.7))

$$\Pr\{|X - \mathbf{E}[X]| \geq t\} \leq \frac{\mathbf{Var}[X]}{t^2}.$$

Setting in the Chebyshev inequality $t = |\mathbf{E}[X]|$ we arrive at

$$\Pr\{X = 0\} \leq \frac{\mathbf{Var}[X]}{\mathbf{E}[X]^2}. \tag{4.2}$$

Indeed, we have

$$\Pr\{X = 0\} \le \Pr\{X \, (X - 2\mathbf{E}[X]) \ge 0\} = \Pr\{|X - \mathbf{E}[X]| \ge |\mathbf{E}[X]|\} \le \frac{\mathbf{Var}[X]}{\mathbf{E}[X]^2}.$$

Inequality (4.2) is known as (Chebyshev's version of) the *second moment method*.

We now derive a refinement of (4.2). Shepp [385] proposed to apply Schwarz's inequality (2.8) to obtain the following:

$$\mathbf{E}[X]^2 = \mathbf{E}[I \, (X \ne 0)X]^2 \le \mathbf{E}[I \, (X \ne 0)]\mathbf{E}[X^2] = \Pr\{X \ne 0\}\mathbf{E}[X^2],$$

which leads to a refined version of the second moment inequality, namely,

$$\Pr\{X = 0\} \le \frac{\mathbf{Var}[X]}{\mathbf{E}[X^2]} \le \frac{\mathbf{Var}[X]}{\mathbf{E}[X]^2}. \tag{4.3}$$

Actually, another formulation of Shepp's inequality due to Chung and Erdős is quite useful. Consider $X_n = I(A_1) + \cdots + I(A_n)$ for a sequence of events $A_1, \ldots, A_n$, and observe that $\{X_n > 0\} = \bigcup_{i=1}^n A_i$. Rewriting (4.3) as

$$\Pr\{X_n > 0\} \ge \frac{\mathbf{E}[X_n]^2}{\mathbf{E}[X_n^2]},$$

we obtain, after some simple algebra,

$$\Pr\{\bigcup_{i=1}^n A_i\} \ge \frac{\left(\sum_{i=1}^n \Pr\{A_i\}\right)^2}{\sum_{i=1}^n \Pr\{A_i\} + \sum_{i \ne j} \Pr\{A_i \cap A_j\}}. \tag{4.4}$$

In a typical application, if we are able to prove that $\mathbf{Var}[X_n]/\mathbf{E}[X_n^2] \to 0$, then $X_n \overset{pr}{\to} \mathbf{E}[X_n]$ as $n \to \infty$, that is, for any $\varepsilon > 0$

$$\lim_{n \to \infty} \Pr\{|X_n - \mathbf{E}[X_n]| \ge \varepsilon\mathbf{E}[X_n]\} = 0.$$

We present below two simple examples delaying a discussion of a more sophisticated problem till the applications section.

Example 4.2: *Pattern Occurrences in a Random Text*   Let H be a given string (pattern) of length m, and T be a random string of length n. Both strings are over a finite alphabet, and we assume that $m = o(n)$. Our goal is to find a typical (in a probabilistic sense) behavior of the number of the pattern occurrences in T. We denote this by O_n. Is it true that $O_n \sim \mathbf{E}[O_n]$ **whp**? We compute the mean $\mathbf{E}[O_n]$ and the variance $\mathbf{Var}[O_n]$. Let I_i be a random variable equal to 1 if the pattern H occurs at position i, and 0 otherwise. Clearly $O_n = I_1 + I_2 + \cdots + I_{n-m+1}$, and hence

$$E[O_n] = \sum_{i=1}^{n-m+1} E[I_i] = (n - m + 1)P(H),$$

where $P(H) = \Pr\{T_i^{i+m-1} = H\}$ for some $1 \leq i \leq n - m + 1$, that is, the probability that the pattern H occurs at position i of the text T. The variance is also easy to compute since

$$\begin{aligned}
\text{Var}[O_n] &= \sum_{i=1}^{n-m+1} E[I_i^2] + \sum_{1 \leq i < j \leq n-m+1} \text{Cov}[I_i I_j] \\
&= \sum_{i=1}^{n-m+1} E[I_i] + \sum_{1 \leq i < j \leq n-m+1} \text{Cov}[I_i I_j],
\end{aligned}$$

where

$$\text{Cov}[I_i I_j] = E[I_i I_j] - E[I_i]E[I_j].$$

Let us compute the covariance $\text{Cov}[I_i I_j]$. Observe that $\text{Cov}[I_i I_j] = 0$ for $|j - i| > m$, and otherwise $\text{Cov}[I_i I_j] \leq E[I_i I_j] \leq E[I_i] = P(H)$. Thus

$$\text{Var}[O_n] \leq (n - m + 1)P(H) + \sum_{|j-i| \leq m} E[I_i] \leq (n - m + 1)P(H) + 2m^2 P(H).$$

In view of the above and Chebyshev's inequality, we obtain for any $\varepsilon > 0$

$$\begin{aligned}
\Pr\{|O_n/E[O_n] - 1| > \varepsilon\} &\leq \frac{(n - m + 1 + 2m^2)P(H)}{\varepsilon^2(n - m + 1)^2 P^2(H)} \\
&\leq \frac{1}{\varepsilon^2(n - m + 1)P(H)} + \frac{2m^2}{\varepsilon^2(n - m + 1)^2 P(H)} \to 0,
\end{aligned}$$

provided $m = o(n)$, which is a natural assumption to make since H is *given*. This proves that $O_n/E[O_n] \to 1$ **whp**. A fuller discussion of this problem can be found in Section 7.6.2. ∎

In the next example, we consider a *deterministic* problem and use randomized technique to solve it. It is adopted from Alon and Spencer [9].

Example 4.3: *Erdős Distinct Sum Problem*   Consider a set of positive integers $\{x_1, \ldots, x_k\} \subset \{1, \ldots, n\}$. Let $f(n)$ denote the maximal k such that there exists a set $\{x_1, \ldots, x_k\}$ with *distinct* sums. The simplest set with distinct sums is

$\{2^i : i \leq \log_2 n\}$, which also shows that $f(n) \geq 1 + \lfloor \log_2 n \rfloor$. Erdős asked to prove that

$$f(n) \leq \log_2 n + C$$

for some constant C. We shall show that

$$f(n) \leq \log_2 n + \frac{1}{2} \log_2 \log_2 n + O(1).$$

Indeed, let $\{x_1, \ldots, x_k\}$ be a distinct sum set, and let B_i be a random variable taking values 0 and 1 with equal probability $\frac{1}{2}$. Consider

$$X = B_1 x_1 + \cdots + B_k x_k.$$

Certainly,

$$\mathbf{E}[X] = \frac{x_1 + \cdots + x_k}{2},$$

$$\mathbf{Var}[X] = \frac{x_1^2 + \cdots + x_k^2}{4} \leq \frac{n^2 k}{4}.$$

Using Chebyshev's inequality with $t = \lambda \sqrt{\mathbf{Var}[X]}$ we obtain, after reversing the inequality,

$$1 - \frac{1}{\lambda^2} \leq \mathrm{Pr}\{|X - \mathbf{E}[X]| \leq \lambda n \sqrt{k}/2\}$$

for all $\lambda > 1$. On the other hand, we should observe that, due to the assumption that $\{x_1, \ldots, x_k\}$ is a distinct sum set, the probability that X has a particular value is equal either to 0 or 2^{-k}. Since there are $\lambda n \sqrt{k}$ values within $|X - \mathbf{E}[X]| \leq \lambda n \sqrt{k}/2$, we obtain

$$\mathrm{Pr}\{|X - \mathbf{E}[X]| \leq \lambda n \sqrt{k}/2\} \leq 2^{-k} \lambda n \sqrt{k}.$$

Comparing the above two inequalities leads to

$$n \geq \frac{2^k}{\sqrt{k}} \frac{1 - \lambda^{-2}}{\lambda}.$$

Since $2^{f(n)} \leq nk$, we find

$$f(n) \leq \log_2 n + \frac{1}{2} \log_2 \log_2 n + O(1),$$

which completes the proof of the claim.   ∎

Finally, we discuss the *fourth moment method*. Here is a simple derivation. Using Hölder's inequality (2.11) we have

$$\mathbf{E}[Y^2] = \mathbf{E}[|Y|^a|Y|^{2-a}] \le \mathbf{E}[|Y|^{ap}]^{1/p}\mathbf{E}[|Y|^{(2-a)q}]^{1/q}$$

where $0 < a < 2$ and $1/p + 1/q = 1$. Set now $a = 2/3$, $p = 3/2$ (so that $1/p = 2/3$ and $1/q = 1/3$) to conclude that

$$\mathbf{E}[Y^2] \le \mathbf{E}[|Y|]^{2/3}\mathbf{E}[|Y^4|]^{1/3},$$

which yields the fourth moment inequality (cf. [43])

$$\mathbf{E}[|Y|] \ge \frac{\mathbf{E}[Y^2]^{3/2}}{\mathbf{E}[Y^4]^{1/2}}. \tag{4.5}$$

We illustrate the fourth moment method in a simple example, while the reader is encouraged to read Berger [43] for considerably deeper analysis.

Example 4.4: *Discrepancy of a Sum of Random Variables* As in [69], consider $Y = X_1 + \cdots + X_n$ where $X_i \in \{-1, 0, 1\}$ are i.i.d. and $\Pr\{X_i = -1\} = \Pr\{X_i = +1\} = p$ while $\Pr\{X_i = 0\} = 1 - 2p$. Actually, we allow p to vary with n provided $np \ge 1$. We wish to find a lower bound for $\mathbf{E}[|Y|]$. Observe that $\mathbf{E}[Y^2] = np$ and

$$\mathbf{E}[Y^4] = \sum_{i=1}^{n}\sum_{j=1}^{n}\sum_{k=1}^{n}\sum_{l=1}^{n}\mathbf{E}[X_iX_jX_kX_l].$$

When all indices i, j, k, l are different, then clearly $\mathbf{E}[X_iX_jX_kX_l] = 0$. In fact, $\mathbf{E}[X_iX_jX_kX_l]$ is not vanishing *only if* all four indices are the same or two of them are equal to one value and the other two equal to another value. But there are only $O(n^2)$ such indices, thus $\mathbf{E}[Y^4] = \Theta(n^2p^2)$, and by the fourth moment inequality we prove that

$$\mathbf{E}[|Y|] = \Omega(\sqrt{np}).$$

In fact, one can prove that $\mathbf{E}[|Y|] = \Theta(\sqrt{np})$. ∎

We summarize this section by presenting all the results derived so far.

Theorem 4.1 *Let X be a random variable.*

First Moment Method If X is nonnegative integer-valued, then

$$\Pr\{X > 0\} \le \mathbf{E}[X].$$

Chebyshev's Second Moment Method *For arbitrary X we have*

$$\Pr\{X = 0\} \leq \frac{\text{Var}[X]}{\text{E}[X]^2};$$

Shepp's Second Moment Method

$$\Pr\{X = 0\} \leq \frac{\text{Var}[X]}{\text{E}[X^2]}.$$

Chung and Erdős' Second Moment Method *For any set of events $A_1, \ldots, A_n$*

$$\Pr\{\bigcup_{i=1}^{n} A_i\} \geq \frac{\left(\sum_{i=1}^{n} \Pr\{A_i\}\right)^2}{\sum_{i=1}^{n} \Pr\{A_i\} + \sum_{i \neq j} \Pr\{A_i \cap A_j\}}.$$

Fourth Moment Method *If X possess the first four moments, then*

$$\text{E}[|X|] \geq \frac{\text{E}[X^2]^{3/2}}{\text{E}[X^4]^{1/2}}.$$

4.2 APPLICATIONS

It is time to apply what we have learned so far to some interesting problems. We start this section with a kth order Markov approximation of a stationary distribution (see Section 4.2.1). Then we shall present the Hardy and Ramanujan result [190] concerning the number of primes dividing a large integer n (see Section 4.2.2), and finally we consider the height in digital trees and suffix trees (see Sections 4.2.3–4.2.6). In Chapter 6 we shall discuss other applications of information theoretical flavor in which the methods of this chapter are used.

4.2.1 Markov Approximation of a Stationary Distribution

In information theory and other applications, one often approximates a stationary process by a kth order Markov chain. Such an approximation can be easily derived through the Markov inequality; hence the first moment method. We discuss this next. The following presentation is adopted from Cover and Thomas [75].

Let $\{X_k\}_{k=-\infty}^{\infty}$ be a two-sided stationary process with the underlying distribution $P(x_1^n) = \Pr\{X_1^n = x_1^n\}$. We consider the *random* variable $P(X_0^{n-1})$ and estimate its probabilistic behavior through a kth order Markov approximation. The latter is defined as follows:

$$P^k(X_0^{n-1}) := P(X_0^{k-1}) \prod_{i=k}^{n-1} P(X_i | X_{i-k}^{i-1})$$

for fixed k.

Let us evaluate the ratio $P^k(X_0^{n-1})/P(X_0^{n-1})$ by the first moment method. First of all, we have

$$\mathbf{E}\left[\frac{P^k(X_0^{n-1})}{P(X_0^{n-1})}\right] = \sum_{x_0^{n-1}} P(x_0^{n-1}) \frac{P^k(x_0^{n-1})}{P(x_0^{n-1})}$$

$$= \sum_{x_0^{n-1}} P^k(x_0^{n-1}) \le 1.$$

By Markov's inequality, for any $a_n \to \infty$, we have

$$\Pr\left\{\frac{1}{n} \log \frac{P^k(X_0^{n-1})}{P(X_0^{n-1})} \ge \frac{1}{n} \log a_n\right\} = \Pr\left\{\frac{P^k(X_0^{n-1})}{P(X_0^{n-1})} \ge a_n\right\} \le \frac{1}{a_n}.$$

Taking $a_n = n^2$, we conclude by the Borel–Cantelli lemma that the event

$$\frac{1}{n} \log \frac{P^k(X_0^{n-1})}{P(X_0^{n-1})} \ge \frac{1}{n} \log a_n$$

occurs infinitely often with probability zero. Hence,

$$\limsup_{n\to\infty} \frac{1}{n} \log \frac{P^k(X_0^{n-1})}{P(X_0^{n-1})} \le 0 \quad (\text{a.s.}).$$

Actually, we can also find an upper bound on $P(X_0^{n-1})$. Let us consider $P(X_0^{n-1}|X_{-\infty}^{-1})$ as an approximation of $P(X_0^{n-1})$. We shall again use Markov's inequality, so we must first evaluate

$$\mathbf{E}\left[\frac{P(X_0^{n-1})}{P(X_0^{n-1}|X_{-\infty}^{-1})}\right] = \mathbf{E}_{X_{-\infty}^{-1}}\left[\mathbf{E}_{X_0^{n-1}}\left[\frac{P(X_0^{n-1})}{P(X_0^{n-1}|X_{-\infty}^{-1})}\bigg| X_{-\infty}^{-1}\right]\right]$$

$$= \mathbf{E}_{X_{-\infty}^{-1}}\left[\sum_{x_0^{n-1}} \frac{P(x_0^{n-1})}{P(x_0^{n-1}|X_{-\infty}^{-1})} P(x_0^{n-1}|X_{-\infty}^{-1})\right] \le 1.$$

By the same argument as above we conclude that

$$\limsup_{n \to \infty} \frac{1}{n} \log \frac{P(X_0^{n-1})}{P(X_0^{n-1}|X_{-\infty}^{-1})} \leq 0 \quad \text{(a.s.)}.$$

In summary, we formulate the following lemma (cf. [8, 75]).

Lemma 4.2 **(Algoet and Cover, 1988 Sandwich Lemma)** *Let P be a stationary measure and P^k its kth order Markov approximation, as defined above. Then*

$$\limsup_{n \to \infty} \frac{1}{n} \log \frac{P^k(X_0^{n-1})}{P(X_0^{n-1})} \leq 0 \quad \text{(a.s.)}, \tag{4.6}$$

$$\limsup_{n \to \infty} \frac{1}{n} \log \frac{P(X_0^{n-1})}{P(X_0^{n-1}|X_{-\infty}^{-1})} \leq 0 \quad \text{(a.s.)} \tag{4.7}$$

for large n.

In fact, the proof of the above lemma allows us to conclude the following: Given $\varepsilon > 0$, there exists N_ε such that for $n > N_\varepsilon$ with probability at least $1 - \varepsilon$ we have

$$\frac{1}{n^2} P^k(X_0^{n-1}) \leq P(X_0^{n-1}) \leq n^2 P(X_0^{n-1}|X_{-\infty}^{-1}).$$

The sandwich lemma was proved by Algoet and Cover [8], who used it to establish the existence of the entropy rate for general stationary processes. We shall return to it in Chapter 6.

4.2.2 Excursion into Number Theory

We present below Turán's proof of the Hardy and Ramanujan result concerning the number, $v(n)$, of primes dividing a number chosen randomly between 1 and n. We prove that $v(n)$ is close to $\ln \ln n$ as $n \to \infty$. We use the second moment method, thus we again solve a deterministic problem with a probabilistic tool. We shall follow the exposition of Alon and Spencer [9].

Throughout this section let p denote a prime number. We need two well-known results from number theory (cf. [190, 423]):

$$\sum_{p \leq x} \frac{1}{p} = \ln \ln x + C + o(1), \tag{4.8}$$

$$\pi(x) = \frac{x}{\ln x}(1 + o(1)), \tag{4.9}$$

where C is a constant $C = 0.26149\ldots$, and $\pi(x)$ denotes the number of primes smaller than x.

We now choose x randomly from the set $\{1, \ldots, n\}$. For prime p we set $I_p = 1$ if $p|x$, and zero otherwise. We ask for the number of prime factors $v(x)$ of x. Observe that

$$X := \sum_{p \le n} I_p = v(x).$$

Since x can be chosen on n different ways, and in $\lfloor n/p \rfloor$ cases it will be divisible by p, we easily find that

$$E[I_p] = \frac{\lfloor n/p \rfloor}{n} = \frac{1}{p} + O\left(\frac{1}{n}\right),$$

and by (4.8) we also have

$$E[X] = \sum_{p \le n}\left(\frac{1}{p} + O\left(\frac{1}{n}\right)\right) = \ln\ln n + O(1).$$

Now we bound the variance

$$\mathbf{Var}[X] = \sum_{p \le n}\mathbf{Var}[I_p] + \sum_{p \ne q \le n}\mathbf{Cov}[I_p I_q] \le E[X] + \sum_{p \ne q \le n}\mathbf{Cov}[I_p I_q],$$

since $\mathbf{Var}[I_p] \le E[I_p]$. In the above p and q are two distinct primes. Observe that $I_p I_q = 1$ if and only if $p|x$ and $q|x$, which further implies that $pq|x$. In view of this we have

$$\mathbf{Cov}[I_p I_q] = E[I_p I_q] - E[I_p]E[I_q] = \frac{\lfloor n/(pq) \rfloor}{n} - \frac{\lfloor n/p \rfloor}{n}\frac{\lfloor n/q \rfloor}{n}$$

$$\le \frac{1}{pq} - \left(\frac{1}{p} - \frac{1}{n}\right)\left(\frac{1}{q} - \frac{1}{n}\right)$$

$$\le \frac{1}{n}\left(\frac{1}{p} + \frac{1}{q}\right).$$

Then by (4.9)

$$\sum_{p \ne q}\mathbf{Cov}[I_p I_q] \le \frac{2\pi(n)}{n}\sum_{p \le n}\frac{1}{p} \le \frac{2n(\ln\ln n + O(1))}{n\ln n} = O\left(\frac{\ln\ln n}{\ln n}\right) \to 0.$$

Finally, by Chebyshev's inequality with $t = \omega(n)\sqrt{\ln\ln n}$, where $\omega(n) \to \infty$ arbitrarily slowly, we arrive at the main result

$$\Pr\{|v(x) - \ln\ln n| > \omega(n)\sqrt{\ln\ln n}\} \leq \frac{c}{\omega(n)},$$

where c is a constant. This is summarized in the theorem below.

Theorem 4.3   (Hardy and Ramanujan, 1920; Turán, 1934)   *For $\omega(n) \to \infty$ arbitrarily slowly, the number of $x \in \{1, \ldots, n\}$ such that*

$$|v(x) - \ln\ln n| > \omega(n)\sqrt{\ln\ln n}$$

is $O(n/\omega(n)) = o(n)$.

4.2.3   Height in Tries

In the following subsections, we shall analyze various digital trees, namely tries, PATRICIA tries, and digital search trees (DST), and their generalizations such as b-tries. As we discussed in Section 1.1, these digital trees are built over a set $\mathcal{X} = \{X^1, \ldots, X^n\}$ of n independent strings of (possibly) infinite length. Each string X^i is generated by either a memoryless source or a Markovian source. We concentrate on estimating the typical length of the height in such digital trees. We recall that the height is the longest depth in a tree.

In this subsection, we consider a trie built over n strings generated independently by memoryless and Markov sources. The reader may recall that in Theorem 1.3 we express the height H_n in terms of the alignments C_{ij} as

$$H_n = \max_{1 \leq i < j \leq n} \{C_{ij}\} + 1, \tag{4.10}$$

where C_{ij} is defined as the length of the longest common prefix of strings X^i and X^j. Throughout this section, we again apply the first moment method to bound the height from above, and the Chung-Erdős second moment method to establish a lower bound.

The results of the following subsections are mostly adopted from Devroye [94, 99, 101], Pittel [337], and Szpankowski [410, 411, 412]. However, the reader is advised to study also a series of papers by Arratia and Waterman [21, 22, 23, 25] on similar topics.

Memoryless Source.   We now assume that every string from $\mathcal{X}$ is generated by a memoryless source. To simplify the analysis, we only consider a binary alphabet $\mathcal{A} = \{0, 1\}$; hence the source outputs an independent sequence of 0's and 1's with probabilities p and $q = 1 - p$, respectively. Observe that $P_2 = p^2 + q^2$ is the probability that two independently generated strings agree on a given position (i.e., either both symbols are 0 or both are 1). We also write $Q_2 = 1/P_2$ to simplify our exposition. Clearly, the alignment C_{ij} has a geometric distribution, that is, for all $i, j \in \{1, \ldots, n\}$ and $k \geq 0$

$$\Pr\{C_{ij} = k\} = P_2^k(1 - P_2)$$

and hence $\Pr\{C_{ij} \geq k\} = P_2^k$.

We now derive a typical behavior of H_n. We start with an upper bound. By the first moment method (or equivalently the Boole inequality), for any nonnegative integer k

$$\Pr\{H_n > k\} = \Pr\{\max_{1 \leq i < j \leq n} \{C_{ij}\} \geq k\} \leq n^2 \Pr\{C_{ij} \geq k\},$$

since the number of pairs (i, j) is bounded by n^2. Set now $k = \lfloor 2(1 + \varepsilon) \log_{Q_2} n \rfloor$ for any $\varepsilon > 0$. Then

$$\Pr\{H_n > 2(1 + \varepsilon) \log_{Q_2} n\} \leq \frac{n^2}{n^{2(1+\varepsilon)}} = \frac{1}{n^{2\varepsilon}} \to 0.$$

Thus $H_n/(2 \log_{Q_2} n) \leq 1$ (pr.).

We now match the above upper bound and prove that $H_n/(2 \log_{Q_2} n) = 1$ (pr.) by showing that $\Pr\{H_n > 2(1 - \varepsilon) \log_{Q_2} n\} \to 1$ for any $\varepsilon > 0$. We set below $k = \lfloor 2(1 - \varepsilon) \log_{Q_2} n \rfloor$. We use the Chung-Erdős formulation of the second moment method with $A_{ij} = \{C_{ij} \geq k\}$. Observe that the sums in (4.4) are over pairs (i, j). Let us define

$$S_1 = \sum_{1 \leq i < j \leq n} \Pr\{A_{ij}\},$$

$$S_2 = \sum_{(i,j) \neq (l,m)} \Pr\{A_{ij} \cap A_{lm}\},$$

where the summation in S_2 is over all pairs $1 \leq i, j \leq n, 1 \leq l, m \leq n$ such that $(i, j) \neq (l, m)$. Then by (4.4)

$$\Pr\{H_n \geq k\} \geq \frac{S_1^2}{S_1 + S_2}.$$

The following is obvious for $n \geq 2$

$$S_1 = \sum_{1 \leq i < j \leq n} \Pr\{A_{ij}\} = \frac{1}{2} n(n - 1) P_2^k \geq \frac{n^2}{4} P_2^k.$$

The sum S_2 is a little harder to deal with. We must consider two cases:

(i) all indices i, j, l, m are different,
(ii) either $i = l$ (i.e., we have (i, j) and (i, m)) or $j = m$ (so we have (i, j) and (l, j)).

Let us split $S_2 = S_2' + S_2''$ such that S_2' is the summation over all different indices (as in (i) above), and S_2'' covers the latter case. Notice that C_{ij} and C_{lm} are independent when the indices are different, hence

$$S_2' \leq \frac{n^4}{4} P_2^{2k}.$$

To evaluate S_2'' we must compute the probability $\Pr\{C_{ij} \geq k, C_{im} \geq k\}$. But, as in Section 3.3.1, we easily see that

$$\Pr\{C_{ij} \geq k, C_{i,m} \geq k\} = (p^3 + q^3)^k = P_3^k,$$

since the probability of having the same symbol at a given position for *three* strings X_i, X_j and X_m is equal to $P_3 = p^3 + q^3$. But, there are no more than $n^3/6$ pairs with one common index, thus

$$S_2'' \leq n^3 (p^3 + q^3)^k.$$

In summary,

$$S_2 = \sum_{(i,j),(l,m)} \Pr\{C_{ij} \geq k, \ C_{lm} \geq k\} \leq \frac{n^4}{4} P_2^{2k} + n^3 (p^3 + q^3)^k.$$

To proceed further, we need to bound the sum S_2''. We first prove a useful inequality in the following lemma.

Lemma 4.4 *For all $s \geq t > 0$ the following holds*

$$(p^s + q^s)^{1/s} \leq (p^t + q^t)^{1/t}, \tag{4.11}$$

where $0 < q = 1 - p < 1$.

Proof. Let $f(x) = (p^x + q^x)^{1/x}$ for $x > 0$. Then

$$f'(x) = \frac{f(x)}{x} \left(\frac{p^x \ln p + q^x \ln q}{p^x + q^x} - \frac{\ln f(x)}{x} \right).$$

For $p \geq q$ define $a = q/p$. For $x \geq 1$ we proceed as follows:

$$\frac{p^x \ln p + q^x \ln q}{p^x + q^x} - \frac{\ln f(x)}{x} \leq \frac{1}{1+a^x} \ln p + \frac{1}{1+a^{-x}} \ln q - \frac{1}{x} \ln p$$

$$\leq \frac{1}{1+a^{-x}} (\ln q - \ln p) + \left(1 - \frac{1}{x}\right) \ln p$$

$$\leq 0.$$

Thus $f'(x) < 0$ for $x \geq 1$ and $f(x)$ is a decreasing function. This can be easily extended to $0 \leq x < 1$ so the lemma is proved for all $x \geq 0$. In Figure 4.1 we plot the function $f(s) = (p^s + q^s)^{1/s}$ for $s = 2$ and $s = 3$.   ∎

The rest is easy: An application of the Chung-Erdős formula (4.4) le⌐ ' to

$$\Pr\{H_n > k\} \geq \frac{1}{1/S_1 + S_2'/S_1^2 + S_2''/S_1^2}$$

$$\geq \frac{1}{4n^{-2} P_2^{-k} + 1 + 16(p^3 + q^3)^k/(n P_2^{2k})}$$

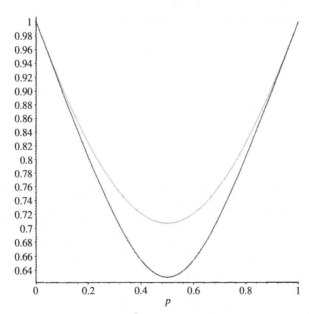

Figure 4.1. Function $(p^3 + q^3)^{\frac{1}{3}}$ (lower curve) plotted against $(p^2 + q^2)^{\frac{1}{2}}$.

$$\geq \frac{1}{1 + 4n^{-2\varepsilon} + 16n^{-1}P_2^{-k/2}}$$

$$\geq \frac{1}{1 + 4n^{-2\varepsilon} + 16n^{-\varepsilon}} = 1 - O(1/n^\varepsilon) \to 1,$$

where the third inequality follows from $k = 2(1 - \varepsilon)\log_{Q_2} n$, and the last one is a consequence of (4.11). Thus we have shown that $H_n/(2\log_{Q_2} n) \geq 1$ (pr.), which completes our proof of

$$\Pr\{2(1 - \varepsilon)\log_{Q_2} n \leq H_n \leq 2(1 + \varepsilon)\log_{Q_2} n\} = 1 - O(n^{-\varepsilon}) \qquad (4.12)$$

for any $\varepsilon > 0$, where $Q_2^{-1} = P_2 = p^2 + q^2$. In passing we should observe that the above implies the convergence in probability of $H_n/\log_{Q_2} n \to 1$, but the rate of convergence is too slow to use the Borel-Cantelli lemma and to conclude almost sure convergence of $H_n/\log_{Q_2} n$. Nevertheless, at the end of this section we provide an argument (after Kingman [240] and Pittel [337]) that allows us to justify such a convergence (see Theorem 4.7 for a complete statement of the result).

Memoryless Source for b-Tries. We now generalize the above to **b-tries** in which every (external) node is capable of storing up to b strings. As in Theorem 1.3, we express the height H_n in terms of the alignments $C_{i_1,\dots,i_{b+1}}$ as

$$H_n = \max_{1 \leq i_1 < \cdots < i_{b+1} \leq n} \{C_{i_1,\dots,i_{b+1}}\} + 1,$$

where $C_{i_1,\dots,i_{b+1}}$ is the length of the longest common prefix of strings $X^{i_1}, \dots, X^{i_{b+1}}$. We follow the same route as for $b = 1$, so we only outline the proof. First of all, observe that

$$\Pr\{C_{i_1,\dots,i_{b+1}} \geq k\} = P_{b+1}^k,$$

where $P_{b+1} = p^{b+1} + q^{b+1}$ represents the probability of a match in a *given* position of $b + 1$ strings $X^{i_1}, \dots, X^{i_{b+1}}$. As before, we write $Q_{b+1} = 1/P_{b+1}$. Our goal is to prove that $H_n \sim (b+1)\log_{Q_{b+1}} n$ (pr.). To simplify the analysis, we let $\mathcal{D}_{\mathbf{i}} = \{\mathbf{i} = (i_1, \dots, i_{b+1}) : i_k \neq i_l \text{ whenever } k \neq l\}$ and $A_{\mathbf{i}} = \{C_{i_1,\dots,i_{b+1}} \geq k\}$.

The upper bound is an easy application of the first moment method. We have for $k = \lfloor (1 + \varepsilon)(b + 1)\log_{Q_{b+1}} n \rfloor$

$$\Pr\{H_n > k\} = \Pr\{\max_{\mathcal{D}_{\mathbf{i}}} \{C_{i_1,\dots,i_{b+1}}\} \geq k\}$$

$$\leq n^{b+1} \Pr\{C_{i_1,\ldots,i_{b+1}} \geq (1+\varepsilon)(b+1)\log_{Q_{b+1}} n\}$$

$$= \frac{1}{n^{(b+1)\varepsilon}} \to 0,$$

which is the desired result.

The lower bound is slightly more intricate due to complicated indexing. But in principle it is nothing else than the second moment method, as in the case $b = 1$. Let us denote by S_1 and S_2 the two sums appearing in the Chung–Erdős inequality. The first sum can be estimated as follows:

$$S_1 = \sum_{\mathcal{D}_i} \Pr\{A_i\} = \binom{n}{b+1} P_{b+1}^k \geq \frac{(n-b)^{b+1}}{(b+1)!} P_{b+1}^k.$$

The second sum S_2 we upper bound by

$$S_2 = \sum_{\mathcal{D}_i \neq \mathcal{D}_j} \Pr\{A_i \cap A_j\}$$

$$\leq \frac{1}{[(b+1)!]^2} n^{2(b+1)} P_{b+1}^{2k} + \sum_{i=1}^{b} \binom{b+1}{i} n^{2(b+1)-i} P_{2(b+1)-i}^k.$$

Using the Chung-Erdős formula we obtain

$$\Pr\{H_n > k\}$$

$$\geq \frac{1}{1/S_1 + S_2/S_1^2}$$

$$\geq \frac{1}{\frac{(b+1)!}{(n-b)^{b+1} P_{b+1}^k} + \left(1 + \frac{b}{n-b}\right)^{2(b+1)} + \sum_{i=1}^{b} \binom{b+1}{i}[(b+1)!]^2 \frac{n^{2(b+1)-i}}{(n-b)^{2(b+1)}} \frac{P_{2(b+1)-i}^k}{P_{b+1}^{2k}}}.$$

Setting now $k = \lfloor (1-\varepsilon)(b+1)\log_{Q_{b+1}} n \rfloor$ and using inequality (4.11) we finally arrive at

$$\Pr\{H_n > (1-\varepsilon)(b+1)\log_{Q_{b+1}} n\} \geq \frac{1}{c_1 n^{(b+1)\varepsilon} + 1 + \sum_{i=1}^{b} c_2(b) n^{i\varepsilon}}$$

$$\geq \frac{1}{1 + c(b)n^\varepsilon} = 1 - O(n^{-\varepsilon}),$$

where $c_1, c_2(b)$ and $c(b)$ are constants. In summary, as in the case $b = 1$, we conclude that $H_n \sim (b+1)\log_{Q_{b+1}} n$ (pr.), or more precisely for any $\varepsilon > 0$,

$$\Pr\{(1-\varepsilon)(b+1)\log_{Q_{b+1}} n \le H_n \le (1+\varepsilon)(b+1)\log_{Q_{b+1}} n\} = 1 - O\left(\frac{1}{n^\varepsilon}\right),$$

$$(4.13)$$

where $Q_{b+1}^{-1} = P_{b+1} = p^{b+1} + q^{b+1}$ (see Theorem 4.7 for a precise formulation of this result).

Markovian Source. Our last generalization deals with the Markovian model. We now assume that all strings $X^1, \ldots, X^n$ are generated independently by a *stationary Markovian source* over a finite alphabet $\mathcal{A}$ of size V. More precisely, the underlying Markov chain is stationary with the transition matrix P and the stationary distribution π satisfying $\pi = \pi\mathsf{P}$. For simplicity, we also assume $b = 1$. We shall follow the footsteps of the previous analysis once we find formulas on $\Pr\{C_{ij} \ge k\}$ and $\Pr\{C_{ij} \ge k, C_{lm} \ge k\}$. In principle, it is not that difficult. We need, however, to adopt a new approach that Arratia and Waterman [22] called the *analysis by pattern*; Jacquet and Szpankowski [216] named it the *string-ruler approach*; and it was already used in Pittel [337]. The idea is to choose a *given* pattern $w \in \mathcal{A}^k$ of length k, and measure relationships between strings $X^1, \ldots, X^n$ by comparing them to w. In particular, it is easy to see that

$$C_{ij} \ge k \;\Rightarrow\; \exists_{w \in \mathcal{A}^k} \; X_1^i X_2^i \ldots X_k^i = X_1^j X_2^j \ldots X_k^j = w,$$

that is, $C_{ij} \ge k$ implies that there exists a word w of length k such that prefixes of length k of X^i and X^j are the same and equal to w. Let $P(w) := \Pr\{w\}$ and $P^s(w) = [\Pr\{w\}]^s$. Then

$$\Pr\{C_{ij} \ge k\} = \sum_{w \in \mathcal{A}^k} P^2(w), \qquad (4.14)$$

$$\Pr\{C_{ij} \ge k, C_{lm} \ge k\} = \sum_{w \in \mathcal{A}^k} \sum_{u \in \mathcal{A}^k} P^2(w) P^2(u) \quad i \ne l,\; j \ne m, \qquad (4.15)$$

$$\Pr\{C_{ij} \ge k, C_{im} \ge k\} = \sum_{w \in \mathcal{A}^k} P^3(w) \quad i = l,\; j \ne m. \qquad (4.16)$$

To complete our analysis, we must compute the probabilities $\Pr\{C_{ij} \ge k\}$ and $\Pr\{C_{ij} \ge k, C_{lm} \ge k\}$ as $k \to \infty$. As shown in (4.14)–(4.16) these probabilities depend on $P(w)$ where $w \in \mathcal{A}^k$ is a word of length k. Let $w = j_1 j_2 \ldots j_k$ where $j_i \in \mathcal{A}$ for $1 \le i \le k$. Then

$$P(w) = \pi_{j_1} p_{j_1, j_2} p_{j_2, j_3} \cdots p_{j_{k-1}, j_k}.$$

Thus for any r we have

$$\sum_{w \in \mathcal{A}^k} P^r(w) = \sum_{1 \le j_1, j_2, \ldots, j_k \le V} \left(\pi_{j_1} p_{j_1, j_2} \cdots p_{j_{k-1}, j_k}\right)^r. \qquad (4.17)$$

TABLE 4.1. Spectral Properties of Nonnegative Matrices

Let $\mathbf{r}$ and $\mathbf{l}$ be, respectively, the right eigenvector and the left eigenvector of a matrix $\mathbf{A}$ associated with the eigenvalue λ, that is,

$$\mathbf{lA} = \lambda\mathbf{l}, \qquad \mathbf{Ar} = \lambda\mathbf{r}.$$

To avoid heavy notation, we do not specify whether vectors are column or row vectors since this should be clear from the context. Consider now a nonnegative matrix $\mathbf{A}$ (all elements of $\mathbf{A}$ are nonnegative). We also assume it is irreducible (cf. [200, 327]; the reader may think of $\mathbf{A}$ as a transition matrix of an irreducible Markov chain). Let $\lambda_1, \ldots, \lambda_m$ be eigenvalues of $\mathbf{A}$ associated with the eigenvectors $\mathbf{r}_1, \ldots, \mathbf{r}_m$. We assume that $|\lambda_1| \geq |\lambda_2| \geq \cdots \geq |\lambda_m|$.

Theorem 4.5   (Perron–Frobenius) *Let $\mathbf{A}$ be a $V \times V$ irreducible nonnegative matrix. Then*

- *$\mathbf{A}$ has a positive real eigenvalue λ_1 of the largest value, that is, $\lambda_1 \geq |\lambda_{i \neq 1}|$.*

- *The eigenvector $\mathbf{r}_1$ associated with λ_1 has all positive coordinates.*

- *λ_1 is of multiplicity one.*

- *$\lambda_1 > |\lambda_{i \neq 1}|$ if there exists k such that all entries of $\mathbf{A}^k$ are strictly positive or the main-diagonal entries are strictly positive.*

Assume first that all eigenvalues are of multiplicity one. Then the left eigenvectors $\mathbf{l}_1, \mathbf{l}_2, \ldots, \mathbf{l}_V$ are orthogonal with respect to the right eigenvectors $\mathbf{r}_1, \mathbf{r}_2, \ldots, \mathbf{r}_V$, that is, $\langle \mathbf{l}_i, \mathbf{r}_j \rangle = 0$ for $i \neq j$ where $\langle \mathbf{x}, \mathbf{y} \rangle$ denotes the inner (scalar) product of $\mathbf{x}$ and $\mathbf{y}$. (Indeed, $\langle \mathbf{l}_i, \mathbf{r}_j \rangle = \frac{\lambda_j}{\lambda_i} \langle \mathbf{l}_i, \mathbf{r}_j \rangle = 0$ since $\lambda_i \neq \lambda_j$.) Setting $\langle \mathbf{l}_i, \mathbf{r}_i \rangle = 1$ for all $1 \leq i \leq V$ we can write for any vector $\mathbf{x} = \langle \mathbf{l}_1, \mathbf{x} \rangle \mathbf{r}_1 + \sum_{i=2}^{V} \langle \mathbf{l}_i, \mathbf{x} \rangle \mathbf{r}_i$, which yields

$$\mathbf{Ax} = \langle \mathbf{l}_1, \mathbf{x} \rangle \lambda_1 \mathbf{r}_1 + \sum_{i=2}^{V} \langle \mathbf{l}_i, \mathbf{x} \rangle \lambda_i \mathbf{r}_i.$$

Since $\mathbf{A}^k$ has eigenvalues $\lambda_1^k, \lambda_2^k, \ldots, \lambda_V^k$, then—dropping the assumption about eigenvalues $\lambda_2, \ldots, \lambda_V$ being simple—we arrive at

$$\mathbf{A}^k \mathbf{x} = \langle \mathbf{l}_1, \mathbf{x} \rangle \mathbf{r}_1 \lambda^k + \sum_{i=2}^{V} q_i(k) \langle \mathbf{l}_i, \mathbf{x} \rangle \mathbf{r}_i \lambda_i^k$$

where $q_i(k)$ is a polynomial in k ($q_i(k) \equiv 1$ when the eigenvalues $\lambda_2, \ldots, \lambda_V$ are simple). In particular, for irreducible nonnegative matrices by the above and the Perron–Frobenius theorem $\mathbf{A}^k \mathbf{x} = \beta \lambda^k (1 + O(\rho^k))$ for some β and $\rho < 1$.

We can succinctly rewrite the above using the Schur's product of matrices. Let $P \circ P \circ \cdots \circ P := P_{[r]} := \{p_{ij}^r\}_{i,j=1}^V$ be the rth Schur product of P, that is, elementwise product of elements of P. We also write $\psi = (1, 1, \ldots, 1)$ for the unit vector, and $\pi_{[r]} = (\pi_1^r, \ldots, \pi_V^r)$ for the rth power of the stationary vector. Then (4.17) becomes in terms of matrices just introduced

$$\sum_{w \in \mathcal{A}^k} P^r(w) = \langle \pi_{[r]}, P_{[r]}^{k-1}\psi \rangle,$$

where $\langle x, y \rangle$ is the scalar product of vectors x and y. Assuming that the underlying Markov chain is aperiodic and irreducible, by the Perron-Frobenius Theorem 4.5 (Table 4.1) the above can be rewritten for some $\rho < 1$ as follows:

$$\langle \pi_{[r]}, A_{[r]}^{k-1}\psi \rangle = \lambda_{[r]}^{k-1} \langle \pi_{[r]}, r \rangle \langle l, \psi \rangle (1 + O(\rho^k)),$$

where l and r are the left and the right principal eigenvectors of $P_{[r]}$. In summary, there exists a constant $\beta > 0$ such that

$$\sum_{w \in \mathcal{A}^k} P^r(w) = \beta \lambda_{[r]}^k (1 + O(\rho^k)), \tag{4.18}$$

hence

$$\Pr\{C_{ij} \geq k\} = \beta_1 \lambda_{[2]}^k (1 + O(\rho^k)), \tag{4.19}$$

$$\Pr\{C_{ij} \geq k, C_{lm} \geq k\} = \beta_1^2 \lambda_{[2]}^{2k}(1 + O(\rho^k)) \quad i \neq l, \; j \neq m, \tag{4.20}$$

$$\Pr\{C_{ij} \geq k, C_{im} \geq k\} = \beta_2 \lambda_{[3]}^k (1 + O(\rho^k)) \quad i = l, \; j \neq m \tag{4.21}$$

for some constants $\beta_1 > 0$ and $\beta_2 > 0$.

We need the following inequality on $\lambda_{[r]}^{1/r}$ that extends Lemma 4.4 to Markov models and is interesting in its own right.

Lemma 4.6   (Karlin and Ost, 1985)   *Let $P_{[r]}$ be the rth order Schur product of a transition matrix P of an aperiodic irreducible Markov chain, and let $\lambda_{[r]}$ be its largest eigenvalue. The function*

$$F(r) = \lambda_{[r]}^{1/r}$$

is nonincreasing for $r > 0$.

Proof. We follow the arguments of Karlin and Ost [228]. From (4.18) we observe that

$$\lim_{k \to \infty} \left(\sum_{w \in A^k} P^r(w) \right)^{\frac{1}{k}} = \lambda_{[r]} \leq 1.$$

But (by either a probabilistic argument or an algebraic one)

$$\sum_{w \in A^k} P^{r+s}(w) \leq \sum_{w \in A^k} P^r(w) \sum_{w \in A^k} P^s(w),$$

hence together with the above we conclude that

$$\lambda_{[r+s]} \leq \lambda_{[r]}\lambda_{[s]}. \tag{4.22}$$

Furthermore, it is easy to see that $\log \lambda_{[r]}$ is convex as a function of r. Let now $f(s) = \log \lambda_{[s]}$ and set for $r < s < 2r$

$$s = \frac{2r - s}{r} r + \frac{s - r}{r} 2r.$$

By convexity

$$f(s) \leq \frac{2r - s}{r} f(r) + \frac{s - r}{r} f(2r).$$

But (4.22) implies $f(2r) \leq 2f(r)$; hence the above yields

$$f(s) \leq \frac{s}{r} f(r),$$

and therefore $f(r)/r$ is nonincreasing. This proves the lemma.   ∎

The rest is an imitation of our derivation from the above. The reader is asked in Exercise 2 to complete the proof of the following for any $\varepsilon > 0$

$$\Pr\{2(1 - \varepsilon) \log_{Q_2} n \leq H_n \leq 2(1 + \varepsilon) \log_{Q_2} n\} = 1 - O(n^{-\varepsilon}), \tag{4.23}$$

where $Q_2^{-1} = \lambda_{[2]}$ is the largest eigenvalue of $P \circ P = P_{[2]}$. Thus $H_n \sim 2 \log_{Q_2} n$ (pr.).

Almost Sure Convergence.   For simplicity of derivations, we again assume here $b = 1$. In (4.12), (4.13), and (4.23) we proved that **whp** the height H_n of a trie is asymptotically equal to $2 \log_{Q_2} n$ with the rate of convergence $O(n^{-\varepsilon})$. This rate does not yet justify an application of the Borel-Cantelli Lemma in order to improve the result to almost sure convergence. Nevertheless, we shall show in this section that $H_n/(2 \log_{Q_2} n) \to 1$ (a.s.), thanks to the fact that H_n is a nondecreasing sequence. We apply here a method suggested by Kesten and reported in Kingman [240].

First of all, observe that for any n

$$H_n \leq H_{n+1},$$

that is, H_n—though random—is a nondecreasing sequence. Furthermore, the rate of convergence $O(n^{-\varepsilon})$ and the Borel-Cantelli Lemma justify almost sure convergence of $H_n/(2\log_{Q_2} n)$ along the exponential skeleton $n = s2^r$ for some integers s and r. Indeed, we have

$$\sum_{r=0}^{\infty} \Pr\left\{\left|\frac{H_{s2^r}}{2\log_{Q_2}(s2^r)} - 1\right| \geq \varepsilon\right\} < \infty.$$

We must extend the above to *every* n. Fix s. For every n we find such r that

$$s2^r \leq n \leq (s+1)2^r.$$

Since logarithm is a slowly varying function and H_n is a nondecreasing sequence, we have

$$\limsup_{n\to\infty} \frac{H_n}{2\log_{Q_2} n} \leq \limsup_{r\to\infty} \frac{H_{(s+1)2^r}}{2\log_{Q_2}(s+1)2^r} \frac{2\log_{Q_2}(s+1)2^r}{2\log_{Q_2} s2^r} = 1 \qquad \text{(a.s.)}.$$

In a similar manner, we prove

$$\liminf_{n\to\infty} \frac{H_n}{2\log_{Q_2} n} \geq \liminf_{r\to\infty} \frac{H_{s2^r}}{2\log_{Q_2}(s+1)2^r} = 1 \qquad \text{(a.s.)}.$$

We summarize the main findings of this section in the following theorem.

Theorem 4.7   (Pittel, 1985; Szpankowski, 1991)   *Consider a b-trie built over n independent strings generated according to a stationary Markov chain with the transition matrix* P. *Then*

$$\lim_{n\to\infty} \frac{H_n}{\ln n} = \frac{(b+1)}{\ln \lambda_{[b+1]}^{-1}} \qquad \text{(a.s.)},$$

where $\lambda_{[b+1]}$ *is the largest eigenvalue of the* $(b+1)$*st order Schur product of* P. *In particular, in the Bernoulli model* $\lambda_{[b+1]} = \sum_{i=1}^{V} p_i^{b+1}$ *where* p_i *is the probability of generating the ith symbol from the alphabet* $\mathcal{A} = \{1, \ldots, V\}$.

4.2.4   Height in PATRICIA Tries

In Section 1.1 we described how to obtain a PATRICIA trie from a regular trie: In PATRICIA we compress a path from the root to a terminal node by avoiding (compressing) unary nodes (see Figure 1.1). In this section, we look at the height H_n of PATRICIA and derive its typical probabilistic behavior. To simplify our analysis, we again assume a binary alphabet $\mathcal{A} = \{0, 1\}$ with $p \leq q = 1 - p$. Actually, we shall write $p_{max} = \max\{p, q\}$ and $Q_{max} = 1/p_{max}$. We again assume the memoryless model.

We start with an upper bound for the PATRICIA height. Following Pittel [337] we argue that for fixed k and b the event $H_n \geq k + b - 1$ in the PATRICIA trie implies that there exist b strings, say $X^{i_1}, \ldots, X^{i_b}$ such that their common prefix is of length at least k. In other words,

$$H_n \geq k + b - 1 \quad \Rightarrow \quad \exists_{i_1,\ldots,i_b} \ C_{i_1,i_2,\ldots,i_b} \geq k.$$

This is true since in PATRICIA there are no unary nodes. Hence, if there is a path of length at least $k+b-1$, then there must be b strings sharing the common prefix of length k. But, as in the analysis of b tries, we know that

$$\Pr\{C_{i_1,i_2,\ldots,i_b} \geq k\} = P_b^k,$$

where $P_b = p^b + q^b$. Thus, for fixed b and $k(b) = b(1 + \varepsilon) \log_{P_b^{-1}} n$

$$\Pr\{H_n \geq k + b - 1\} \leq n^b \Pr\{C_{i_1,i_2,\ldots,i_b} \geq k\} = O(n^{-\varepsilon}).$$

The above is true for *all* values of b; hence we now allow b to increase arbitrary slowly to infinity. Observe then

$$\lim_{b \to \infty} k(b) = (1 + \varepsilon) \lim_{b \to \infty} \frac{b \log n}{\log P_b^{-1}} = (1 + \varepsilon) \lim_{b \to \infty} \frac{\log n}{\log P_b^{-1/b}}$$

$$\leq (1 + \varepsilon) \frac{\log n}{\log p_{max}^{-1}} = (1 + \varepsilon) \log_{Q_{max}} n,$$

since $\lim_{b \to \infty} P_b^{1/b} \leq \lim_{b \to \infty} p_{max}^{\frac{b-1}{b}} = p_{max}$. In summary, for any $\varepsilon > 0$ we just proved that

$$\Pr\{H_n \geq (1 + \varepsilon) \log_{Q_{max}} n\} = O(n^{-\varepsilon}),$$

which is the desired upper bound.

Not surprisingly, we shall use the second moment method to prove a lower bound. Let "1" occur with the probability p_{max}. Define the following k strings: $X^{i_1} = 10\ldots, X^{i_2} = 110\ldots, \cdots, X^{i_k} = 111 \cdots 10\ldots$, where string X^{i_j} has the first j symbols equal to 1 followed by a 0. We show that with high probability for $k = (1 - \varepsilon) \log_{Q_{max}} n$ the above k strings appear among all n strings $X^1, \ldots X^n$. Observe that if this is true, then $H_n \geq k + 1$ **whp**.

Let now $\mathbf{i} = (i_1, i_2, \ldots, i_k)$, and define $Z_{\mathbf{i}} = 1$ if $X^{i_1} = 10\ldots, X^{i_2} = 110\ldots, \cdots, X^{i_k} = 111\cdots 10\ldots$, and zero otherwise. Let also

$$Z = \sum_{\mathbf{i} \in \mathcal{D}_{\mathbf{i}}} Z_{\mathbf{i}},$$

where $\mathcal{D}_{\mathbf{i}} = \{\mathbf{i} = (i_1, \ldots, i_k) : i_j \neq i_l \text{ whenever } j \neq l\}$. Observe that

$$\Pr\{Z > 0\} \leq \Pr\{H_n > k\}.$$

We estimate the left-hand side of the above by the second moment method. First of all,

$$\mathbf{E}[Z] = \binom{n}{k} p_{\max}^{\frac{k(k+1)}{2}} (1 - p_{\max})^k,$$

since the probability of X^i is $p_{\max}^i (1 - p_{\max})$, and the strings are independent. In a similar manner we compute the variance

$$\mathbf{Var}[Z] \leq \mathbf{E}[Z] + \sum_{\mathcal{D}_{\mathbf{i}} \neq \mathcal{D}_{\mathbf{j}}} \mathbf{Cov}[Z_{\mathbf{i}} Z_{\mathbf{j}}].$$

The covariance $\mathbf{Cov}[Z_{\mathbf{i}} Z_{\mathbf{j}}]$ depends on how many indices are the same in $\mathbf{i}$ and $\mathbf{j}$. If, say $0 < l < k$, indices are the same, then

$$\mathbf{Cov}[Z_{\mathbf{i}} Z_{\mathbf{j}}] \leq \mathbf{E}[Z_{\mathbf{i}} Z_{\mathbf{j}}] \leq \binom{n}{2k - l} p_{\max}^{k(k+1)-l(l+1)/2} (1 - p_{\max})^{2k-l}.$$

Thus by the second moment method

$$\Pr\{Z = 0\} \leq \frac{\mathbf{Var}[Z]}{\mathbf{E}[Z]^2}$$

$$\leq \frac{1}{\mathbf{E}[Z]} + \frac{1}{\mathbf{E}[Z]^2} \sum_{l=1}^{k-1} \binom{n}{2k - l} p_{\max}^{k(k+1)-l(l+1)/2} (1 - p_{\max})^{2k-l}.$$

$$(4.24)$$

Let us now evaluate the above. First of all, for $k = (1 - \varepsilon) \ln n / \ln p_{\max}^{-1}$ we obtain for some constant c

$$\mathbf{E}[Z] \geq c \frac{n^k}{k!} p_{\max}^{k^2/2} (1 - p_{\max})^k$$

$$\geq \exp\left((1 - \varepsilon) \frac{\ln^2 n}{\ln p_{\max}^{-1}} - \frac{(1 - \varepsilon)^2}{2} \frac{\ln^2 n}{\ln p_{\max}^{-1}} - O(\ln n \ln \ln n) \right)$$

$$= \exp\left(\frac{1}{2}(1-\varepsilon)\frac{\ln^2 n}{\ln p_{\max}^{-1}}\right) \to \infty,$$

where ε is an arbitrary small positive number. The lth term of (4.24) can be estimated as follows

$$\frac{\binom{n}{2k-l}p_{\max}^{k(k+1)-l(l+1)/2}(1-p_{\max})^{2k-l}}{E[Z]^2} = \frac{\binom{n}{2k-l}p_{\max}^{-l(l+1)/2}(1-p_{\max})^{-l}}{\binom{n}{l}^2}$$

$$= O\left(n^{-l}p_{\max}^{-l^2/2}(1-p_{\max})^{-l}\right) \to 0,$$

where the convergence to zero is true for $l \le k = (1-\varepsilon)\ln n / \ln p_{\max}^{-1}$ by the same arguments as above (indeed, the function $f(l) = n^l p_{\max}^{l^2/2}$ is nondecreasing with respect to l for $l \le k = \ln n / \ln p_{\max}^{-1}$). Thus

$$\Pr\{H_n \ge (1-\varepsilon)\ln n / \ln p_{\max}^{-1}\} \ge 1 - O\left((\exp(-\alpha \ln^2 n + \ln \ln n))\right)$$

for some $\alpha > 0$.

In summary, we prove the following result (the almost sure convergence is established in the same fashion as in the previous section).

Theorem 4.8 (Pittel, 1985) *Consider a PATRICIA trie built over n independent strings generated by a memoryless source with p_i being the probability of generating the ith symbol from the alphabet $\mathcal{A} = \{1, \ldots, V\}$. Then*

$$\lim_{n\to\infty}\frac{H_n}{\log n} = \frac{1}{\log p_{\max}^{-1}} \quad (a.s.),$$

where $p_{\max} = \max\{p_1, p_2, \ldots, p_V\}$.

4.2.5 Height in Digital Search Trees

To complete our discussion of digital trees, we shall analyze the height of a digital search tree and prove a result similar to the one presented in Theorem 4.8.

We start with a lower bound, and use an approach similar to the one discussed for PATRICIA tries. Actually, the lower bound from Section 4.2.4 works fine for digital search trees. Nevertheless, we provide another derivation so the reader can again see the second moment method at work. Set $Z_0 = n$, and let Z_i for $i \ge 1$ be the number of strings whose prefix of length i consists of 1's. We recall that 1 occurs with the probability $p_{\max}$. Let also $k = (1-\varepsilon)\ln n / \ln p_{\max}^{-1}$. Observe that

$$Z_k > 0 \implies H_n > k,$$

where H_n is the height in a digital search tree. We prove that $\Pr\{Z_k > 0\}$ **whp** when $k = (1 - \varepsilon) \ln n / \ln p_{max}^{-1}$ using the second moment method. Thus, we must find $\mathbf{E}[Z_k]$ and $\mathbf{Var}[Z_k]$.

All strings $X^1, \ldots, X^n$ are independent, hence Z_1 has the binomial distribution with parameters n and p_{max}, that is, $Z_1 \sim Binomial(n, p_{max})$. In general, $Z_i \sim Binomial(Z_{i-1}, p_{max})$. Using conditional expectation we easily prove that (the reader is asked to provide details in Exercise 8)

$$\mathbf{E}[Z_k] = n p_{max}^k, \tag{4.25}$$

$$\mathbf{Var}[Z_k] = n p_{max}^k (1 - p_{max}^k). \tag{4.26}$$

Then for $Q_{max} = 1/p_{max}$

$$\Pr\{H_n \le (1 - \varepsilon) \log_{Q_{max}} n\} \le \Pr\{Z_k = 0\} \le \frac{\mathbf{Var}[Z_k]}{\mathbf{E}[Z_k]^2} \le \frac{1}{n p_{max}^k} = n^{-\varepsilon}.$$

This proves that $H_n / \log n \ge 1 / \log p_{max}^{-1}$ **whp**.

The upper bound is slightly more complicated. Let us consider a *given* word w (of possibly infinite length) whose prefix of length k is denoted as w_k. We shall write $P(w_k) := \Pr\{w_k\}$. Define $T_n(w)$ to be the length of a path in a digital search tree that follows the symbols of w until it reaches the last node of the tree on its path. For example, referring to Figure 1.1 we have $T_4(00000\ldots) = 3$. The following relationship is quite obvious:

$$\Pr\{T_n(w) \ge k\} = \Pr\{T_{n-1}(w) \ge k\} + \Pr\{T_{n-1}(w) = k - 1\} P(w_k),$$

since when inserting a new string to the tree we either do not follow the path of w (the first term above) or we follow the path leading to the second term of the recurrence. Observe now that for given w

$$T_n(w) \ge T_{n-1}(w),$$

so that $T_n(w)$ is a nondecreasing sequence with respect to n. Since, in addition, $P(w_k) \le p_{max}^k$ and $\Pr\{T_{n-1}(w) = k - 1\} \le \Pr\{T_{n-1}(w) \ge k - 1\}$, we can iterate the above recurrence to get

$$\Pr\{T_n(w) \ge k\} \le \Pr\{T_{n-2} \ge k\} + 2 p_{max}^k \Pr\{T_{n-1}(w) \ge k - 1\},$$

so that after iterations with respect to n the above yields

$$\Pr\{T_n(w) \ge k\} \le n p_{max}^k \Pr\{T_{n-1}(w) \ge k - 1\}.$$

Iterating now with respect to k, we find

$$\Pr\{T_n(w) \geq k\} \leq n^k p_{\max}^{k^2}.$$

Set now $k = (1 + \varepsilon) \log n / \log p_{\max}^{-1}$ to obtain

$$\Pr\{H_n > (1 + \varepsilon) \log_{p_{\max}} n^{-1}\} \leq \sum_{w \in \mathcal{A}^k} \Pr\{T_n(w) \geq k\}$$

$$\leq 2^k n^k p_{\max}^{k^2}$$

$$\leq \exp\left(k \ln 2 + k \ln n - k^2 \ln p_{\max}^{-1}\right)$$

$$\leq \exp\left(-k(1 + \varepsilon) \ln n + k \ln n + k \ln 2\right)$$

$$\leq \exp(-\alpha \ln^2 n) \to 0$$

for some $\alpha > 0$ as long as $\varepsilon > \ln 2 / \ln n$.

Theorem 4.9 (**Pittel, 1985**) *Consider a digital search tree built over n independent strings generated by a memoryless source with p_i being the probability of generating the ith symbol from the alphabet $\mathcal{A} = \{1, \ldots, V\}$. Then*

$$\lim_{n \to \infty} \frac{H_n}{\log n} = \frac{1}{\log p_{\max}^{-1}} \quad \text{(a.s.),}$$

where $p_{\max} = \max\{p_1, p_2, \ldots, p_V\}$.

4.2.6 Height in a Suffix Tree

Finally, we consider the height of a suffix tree built from the first n suffixes of a (possibly infinite) string X_1^∞ generated by a memoryless source. As before, we assume a binary alphabet with p and $q = 1 - p$ being the probability of generating the symbols. We denote by $X(i) = X_i^\infty$ the ith suffix, where $1 \leq i \leq n$. (We recall that a suffix tree is a trie built from suffixes $X(1), \ldots, X(n)$.) We restrict the analysis to the memoryless model. The reader is asked in Exercise 7 to extend this analysis to the mixing model.

The analysis of the height follows the same footsteps as in the case of tries (i.e., we use the first moment method to prove an upper bound and the second moment method to prove a lower bound) except that we must find the probability law governing the self-alignment C_{ij} that represents the length of the longest prefix common to $X(i)$ and $X(j)$. We recall that the height H_n is related to C_{ij} by $H_n = \max_{i \neq j}\{C_{ij}\} + 1$.

We need to know the probability of $A_{ij} = \{C_{ij} \geq k\}$. Let $d = |i - j|$ and consider suffixes $X(i)$ and $X(i + d)$. Below, we assume that $j = i + d$. When $d \geq k$, then $X(i)$ and $X(i + d)$ are independent in the memoryless model, hence as in tries we easily find that

$$\Pr\{C_{i,i+d} \geq k\} = P_2^k \quad d \geq k,$$

where $P_2 = p^2 + q^2$. The problem arises when $d < k$. We need to identify conditions under which k symbols starting at position i are the same as k symbols following position $i + d$; that is,

$$X_i^{i+k-1} = X_{i+d}^{i+d+k-1},$$

where $X_i^{i+k-1} = X_i, X_{i+1} \ldots X_{i+k-1}$ and $X_{i+d}^{i+d+k-1} = X_{i+d} \ldots X_{i+d+k-1}$ are k-length prefixes of suffixes $X(i)$ and $X(i+d)$, respectively. The following simple combinatorial lemma provides a solution.

Lemma 4.10 **(Lothaire, 1982)** *Let X_1^∞ be a string whose ith and jth suffixes are $X(i) = X_i^\infty$ and $X(j) = X_j^\infty$, respectively, where $j - i = d > 0$. Also let Z be the longest common prefix of length $k \geq d$ of $X(i)$ and $X(j)$. Then there exists a word $w \in \mathcal{A}^d$ of length d such that*

$$Z = w^{\lfloor k/d \rfloor} \bar{w}, \tag{4.27}$$

where $\bar{w}$ is a prefix of w, and w^l is the string resulting from the concatenation of $l = \lfloor k/d \rfloor$ copies of the word w.

We leave the formal proof of the above lemma to the reader (see Exercise 9). Its meaning, however, should be quite clear. To illustrate Lemma 4.10, let is assume that $X_1^{k+d} = X_1^{3d+2} = x_1 x_2 \ldots x_d x_1 x_2 \ldots x_d \ldots x_1 x_2 \ldots x_d x_1 x_2$, where $k = 2d + 2$ and $d > 2$. One easily identifies $w = x_1 x_2 \ldots x_d$ and $\bar{w} = x_1 x_2$. The common prefix Z of X_1^{k+d-1} and X_{d+1}^{k+2d} can be represented as $Z_1^{2d+2} = w^2 \bar{w}$, as stated in Lemma 4.10.

In view of the above, we can now derive a formula on the probability $\Pr\{C_{i,i+d} \geq k\}$, which is pivotal to our analysis of the height. An application of Lemma 4.10 leads to

$$\Pr\{C_{i,i+d} \geq k\} = \sum_{w \in \mathcal{A}^{k \wedge d}} P(w^{\lfloor \frac{k}{d} \rfloor + 1} \bar{w}) \tag{4.28}$$

$$= \left(p^{\lfloor \frac{k}{d} \rfloor + 1} + q^{\lfloor \frac{k}{d} \rfloor + 1} \right)^{d-r} \left(p^{\lfloor \frac{k}{d} \rfloor + 2} + q^{\lfloor \frac{k}{d} \rfloor + 2} \right)^r, \tag{4.29}$$

where $r = k - d\lfloor k/d \rfloor$ and $k \wedge d = \min\{d, k\}$. Observe that by restricting the length of w to $k \wedge d$, we also cover the case $k < d$, where $\Pr\{C_{i,i+d} \geq k\} = (p^2 + q^2)^k$.

To see why (4.28) implies (4.29), say for $d < k$, we observe that $d - r$ symbols of w must be repeated $l = \lfloor k/d \rfloor$ times in w^l, while the last r symbols of w must occur $l + 1$ times. The probability of having the same symbol on a given position in l strings is $p^{l+1} + q^{l+1}$. Thus (4.29) follows.

Before we proceed, we need an inequality on (4.29), which we formulate in the form of a lemma.

Lemma 4.11 *For $d < k$ we have*

$$\Pr\{C_{i,i+d} \geq k\} \leq (p^2 + q^2)^{(k+d)/2} = P_2^{\frac{k+d}{2}} \tag{4.30}$$

Proof. The result follows from Lemma 4.4 (cf. (4.11)) and (4.29). Indeed, let $P_l = (p^l + q^l)$. Then

$$\Pr\{C_{i,i+d} \geq k\} = P_{l+1}^{d-r} P_{l+2}^{r} \leq P_2^{\frac{1}{2}[(d-r)(l+1)+r(l+2)]} = P_2^{\frac{k+d}{2}},$$

since $k = dl + r$.   ∎

Now we are ready to prove an upper bound for the height. We use the Boole inequality and the above to obtain

$$\Pr\{\max_{i,d}\{C_{i,i+d}\} \geq k\} \leq n \left(\sum_{d=1}^{k} \Pr\{C_{i,i+d} \geq k\} + \sum_{d=k+1}^{n} \Pr\{C_{i,i+d} \geq k\} \right)$$

$$\leq n \left(\sum_{d=1}^{k} P_2^{\frac{k+d}{2}} + \sum_{d=k+1}^{n} P_2^{k} \right)$$

$$\leq n \left(P_2^{\frac{k+1}{2}} (1 - \sqrt{P_2})^{-1} + n P_2^{k} \right),$$

where the second inequality is a consequence of Lemma 4.11. From the above we conclude that

$$\Pr\{H_n > 2(1 + \varepsilon) \log_{Q_2} n\} = \Pr\{\max_{i,d}\{C_{i,i+d}\} \geq 2(1 + \varepsilon) \log_{Q_2} n\} \leq \frac{c}{n^{\varepsilon}},$$

where (as before) $Q_2 = P_2^{-1}$ and c is a constant. Thus **whp** we have $H_n \leq 2(1 + \varepsilon) \log_{Q_2} n$.

For the lower bound, we use the Chung and Erdős second moment method, as one can expect. We need, however, to overcome some difficulties. Let $\mathcal{D} = \{(i, j) : |i - j| > k\}$. Observe that

$$\Pr\{H_n > k\} = \Pr\{\bigcup_{i \neq j} A_{ij}\} \geq \Pr\{\bigcup_{i,j \in \mathcal{D}} A_{ij}\} \geq \frac{S_1^2}{S_1 + S_2},$$

where

$$S_1 = \sum_{i,j \in \mathcal{D}} \Pr\{A_{ij}\},$$

$$S_2 = \sum_{(i,j) \neq (l,m) \in \mathcal{D}} \Pr\{A_{ij} \cap A_{lm}\}.$$

The first sum is easy to compute from what we have learned so far. Since $|i - j| > k$ in $\mathcal{D}$ we immediately find that

$$S_1 \geq (n^2 - (2k + 1)n) P_2^k.$$

To compute S_2 we split it into three terms, S_{21}, S_{22} and S_{23}, such that the summation in S_{2i} ($i = 1, 2, 3$) is over the set $\mathcal{D}_i$ defined as follows:

$\mathcal{D}_1 = \{(i, j), (l, m) \in \mathcal{D} :$

$\quad \min\{|l - i|, |l - j|\} \geq k$ and $\min\{|m - i|, |m - j|\} \geq k\}$

$\mathcal{D}_2 = \{(i, j), (l, m) \in \mathcal{D} :$

$\quad \min\{|l - i|, |l - j|\} \geq k$ and $\min\{|m - i|, |m - j|\} < k$

$\quad$ or $\min\{|l - i|, |l - j|\} < k$ and $\min\{|m - i|, |m - j|\} \geq k\}$

$\mathcal{D}_3 = \{(i, j), (l, m) \in \mathcal{D} :$

$\quad \min\{|l - i|, , |l - j|\} < k$ and $\min\{|m - i|, |m - j|\} < k\}$

Using arguments as in the case of tries, we obtain

$$S_{21} = \sum_{(i,j),(l,m) \in \mathcal{D}_1} \Pr\{A_{ij} \cap A_{lm}\} \leq n^4 P_2^{2k},$$

$$S_{22} = \sum_{(i,j),(l,m) \in \mathcal{D}_2} \Pr\{A_{ij} \cap A_{lm}\}$$

$$\leq \sum_{(i,j),(i,m) \in \mathcal{D}_2} \Pr\{A_{ij} \cap A_{im}\} \leq 8kn^3 P_3^k \leq 8kn^2 P_2^{\frac{3}{2}k}$$

$$S_{23} = \sum_{(i,j),(l,m) \in \mathcal{D}_3} \Pr\{A_{ij} \cap A_{lm}\} \leq \sum_{(i,j),(i,j) \in \mathcal{D}_3} \Pr\{A_{ij} \cap A_{ij}\} \leq 16k^2 n^2 P_2^k,$$

where, as before, $P_3 = p^3 + q^3$. The last inequality for S_{22} follows from Lemma 4.4. The rest is a matter of algebra. Proceeding as in Section 4.2.3 we arrive at

$$\Pr\{H_n > 2(1 - \varepsilon) \log_{Q_2} n\} \geq \frac{1}{1 + c_1 n^{-\varepsilon} \log n + c_2 n^{-2\varepsilon}} = 1 - O\left(\frac{\log n}{n^{\varepsilon}}\right),$$

for some constants c_1, c_2. As before we can extend this convergence in probability to almost sure convergence by considering an exponential skeleton $n = s2^r$. Thus, we just proved the following interesting result.

Theorem 4.12 (Devroye, Rais, and Szpankowski, 1992) *Consider a suffix tree built from n suffixes of a string generated by a memoryless source with p_i being the probability of generating the ith symbol from the alphabet $\mathcal{A} = \{1, \ldots, V\}$. Then*

$$\lim_{n \to \infty} \frac{H_n}{\log n} = \frac{2}{\log P_2^{-1}} \quad \text{(a.s.)},$$

where $P_2 = \sum_{i=1}^{V} p_i^2$.

In passing we point out that results of Sections 4.2.3–4.2.6 can be extended to the mixing probabilistic model (cf. [337, 411, 412]). The reader is asked to try to prove such extensions in Exercises 6 and 7. Additional details can be found in Chapter 6.

4.3 EXTENSIONS AND EXERCISES

4.1 Prove the following alternative formulation of the second moment method. Let $\{A_i\}_{i=1}^{n}$ be a set of identically distributed events such that $n\Pr\{A_i\} \to \infty$. If

$$\sum_{i \neq j} \frac{\Pr\{A_i | A_j\}}{n\Pr\{A_i\}} \to 1$$

as $n \to \infty$, then $\Pr\{\bigcup_{i=1}^{n} A_i\} \to 1$.

4.2 Consider the height of a b-trie in a Markovian model and prove

$$\Pr\{(b+1)(1-\varepsilon) \log_{Q_{b+1}} n \leq H_n \leq (b+1)(1+\varepsilon) \log_{Q_{b+1}} n\} = 1 - O(n^{-\varepsilon}),$$

where $Q_{b+1}^{-1} = \lambda_{[b+1]}$ is the largest eigenvalue of $\mathsf{P}_{[b+1]} = \mathsf{P} \circ \cdots \mathsf{P}$.

4.3   Find another proof for the lower bound in the derivation of the height in PATRICIA tries.

4.4   ⚠ Extend the analysis of PATRICIA tries to a Markovian source. What is the equivalence of p_{max} in the Markovian model? (see Chapter 6.)

4.5   ⚠ Extend the analysis of digital search trees to a Markovian source.

4.6   ⚠ (Pittel 1985) Establish typical behaviors of heights in all three digital trees in the mixing model (see Chapter 6).

4.7   ⚠ (Szpankowski, 1993) Extend the analysis of the height in a suffix tree (see Section 4.2.6) to the mixing model.

4.8   Prove (4.25) and (4.26).

4.9   Prove Lemma 4.10.

4.10 ⚠ (Pittel, 1985) Consider the fill-up level F_n (see Section 1.1) in tries, PATRICIA tries and digital search trees. Prove the following theorem:

Theorem 4.13   (Pittel, 1985)   *Consider tries, PATRICIA tries, and digital search trees built over n independent strings generated by a memoryless source with p_i being the probability of generating the ith symbol from the alphabet $A = \{1, \ldots, V\}$. Then*

$$\lim_{n \to \infty} \frac{F_n}{\log n} = \frac{1}{\log p_{min}^{-1}} \quad (a.s.),$$

where $p_{min} = \min\{p_1, p_2, \ldots, p_V\}$.

Can this theorem be extended to suffix trees?

4.11 ▽ Consider the shortest path s_n in digital trees, including suffix trees. Prove or disprove the following result.

Theorem 4.14   *Consider digital trees built over n independent strings generated by a memoryless source with p_i being the probability of generating the ith symbol from the alphabet $A = \{1, \ldots, V\}$. Let s_n be the shortest path in such trees. Then*

$$\lim_{n \to \infty} \frac{s_n}{\log n} = \frac{1}{\log p_{min}^{-1}} \quad (a.s.),$$

where $p_{min} = \min\{p_1, p_2, \ldots, p_V\}$.

4.12 Consider the following "comparison" of two random strings. Let X_1^n and Y_1^m of length n and $m < n$, respectively, be generated by a binary memoryless source with probabilities p and $q = 1 - p$ of the two symbols. Define C_i to be the number matches between X_i^{i+m-1} and Y; that is,

$$C_i = \sum_{j=1}^{n-m+1} equal(X_{i+j-1}, Y_j),$$

where $equal(x, y)$ is one when $x = y$ and zero otherwise (i.e., the Hamming distance). Define $M_{mn} = \max_{1 \le i \le n-m+1}\{C_i\}$. Prove that *if* $\log n = o(m)$ *then*

$$\lim_{n \to \infty} \frac{M_{m,n}}{m} = P_2 \quad (a.s.),$$

where $P_2 = p^2 + q^2$.

4.13 ⚠(Atallah, Jacquet, and Szpankowski 1993) Consider the same problem as in Exercise 12 above. Let $n = \Theta(m^\alpha)$ and $P_3 = p^3 + q^3$. Define

$$\beta = \frac{(P_2 - P - 3)(P_2 - 3P_2^2 + 2P_3)}{6(P_3 - p_2^2)}.$$

Prove that if $P_2 - P_3 \le (1 - \alpha)\beta$ then $M_{m,n} - mP \sim \sqrt{2m(P_2 - P_3)\log n}$ (pr.) as $n, m \to \infty$.

4.14 Consider an n-dimensional unit cube $I_n = \{0, 1\}^n$ (i.e., a binary sequence $(x_1, \ldots, x_n)$ is regarded as a vertex of the unit cube in n dimensions). Let T_n be the cover time of a random walk; that is, the time required by a simple walk to visit all 2^n vertices of the cube. Prove that $T_n \sim (1 + n)2^n \log 2$ **whp**.

5

Subadditive Ergodic Theorem and Large Deviations

The celebrated *ergodic theorem* of Birkhoff finds many applications in computer science. However, for the probabilistic analysis of algorithms a generalization of it due to Kingman, called the *subadditive ergodic theorem*, is even more important. Kingman's result proves the existence of a limit, but neither says what the limit is nor what is the rate of convergence. Azuma's inequality tries to remedy the latter problem. Its easiest and most general derivation is based on Hoeffding's inequality for *martingale differences*. We also review *martingales* since they are useful probabilistic tools for the analysis of algorithms. Finally, we discuss large deviations for sums of i.i.d. random variables.

The ergodic theorem of Birkhoff asserts that a *sample mean* of a stationary ergodic process converges almost surely and in moments to the actual mean of the process. In the analysis of algorithms we often encounter a situation in which a quantity of interest, say X_0^n, satisfies either a *subadditivity* property (i.e., $X_0^n \leq X_0^m + X_m^n$ for $m \leq n$) or a *superadditivity* property (i.e., $X_0^n \geq X_0^m + X_m^n$). Does it suffice to conclude that X_0^n/n has a limit as $n \to \infty$? In 1976 Kingman [240] proposed conditions under which the answer is in the affirmative. We shall discuss in this chapter a generalization of Kingman's **subadditive ergodic theorem** and its applications to the analysis of algorithms.

The subaddtive ergodic theorem has its "cavity." It says that a limit exists but does not say what it is. Actually, finding this limit might be quite troublesome. But we can do the next best thing, namely, show that the process X_0^n/n is well concentrated around its limit. This can be accomplished through a powerful lemma initiated by Azuma, and further developed by others (cf. McDiarmid [312], Talagrand [418, 419]). We present here a proof of Azuma's inequality based on *martingale differences*. Martingales find many applications in the probabilistic analysis of algorithms, and we shall review here some of their properties. In recent years, Talagrand [418] developed a novel isoperimetric theory of concentration inequalities. We will not discuss it here. The reader is referred either to the original paper of Talagrand [418], or a more readable account of Steele [402], Talagrand [418] or McDiarmid [185]. Finally, we observe that Azuma's inequality is an example of a large deviations result. We discuss here in depth large deviations theory of i.i.d. random variables. We shall return to large deviations in Chapter 8.

In the applications section of this chapter we first discuss two pattern matching problems, namely, the edit distance problem and the Knuth-Morris-Pratt algorithm. The third application deals with a certain problem that arose in the analysis of the hashing with lazy deletion, in which martingales play a crucial role.

5.1   SUBADDITIVE SEQUENCE

Let us start with a deterministic situation. Assume a deterministic sequence $\{x_n\}_{n=0}^{\infty}$ satisfies the *subadditivity property*, that is,

$$x_{m+n} \leq x_n + x_m$$

for all integers $m, n \geq 0$. We will prove that

$$\lim_{n \to \infty} \frac{x_n}{n} = \inf_{m \geq 1} \frac{x_m}{m} = \alpha$$

for some $\alpha \in [-\infty, \infty)$. Indeed, it suffices to fix $m \geq 0$, write $n = km + l$ for some $0 \leq l < m$, and observe that by consecutive applications of the subadditivity property we arrive at

$$x_n \leq k x_m + x_l.$$

Dividing by n, and considering $n \to \infty$ with $n/k \to m$, we finally arrive at

$$\limsup_{n \to \infty} \frac{x_n}{n} \leq \inf_{m \geq 1} \frac{x_m}{m} = \alpha,$$

where the inequality follows from arbitrariness of m. This completes the derivation since

$$\liminf_{n \to \infty} \frac{x_n}{n} \geq \alpha$$

is automatic from the definition of $\liminf$ (cf. Table 5.1). We just derived the following theorem of Fekete.

Theorem 5.1   **(Fekete, 1923)**   *If a sequence of real numbers $\{x_n\}$ satisfies the subadditive condition*

$$x_{m+n} \leq x_n + x_m \tag{5.1}$$

for all integers $m, n \geq 0$, then

$$\lim_{n \to \infty} \frac{x_n}{n} = \inf_{m \geq 1} \frac{x_m}{m}. \tag{5.2}$$

If the subadditivity (5.1) is replaced by the superadditivity property

$$x_{m+n} \geq x_n + x_m \tag{5.3}$$

for all integers $m, n \geq 0$, then

$$\lim_{n \to \infty} \frac{x_n}{n} = \sup_{m \geq 1} \frac{x_m}{m}. \tag{5.4}$$

Example 5.1:   *Longest Common Subsequence*   In Section 1.5 we discussed the *longest common subsequence* (LCS) problem, which is a special case of the edit distance problem. To recall: Two ergodic stationary sequences $X = X_1, X_2, \ldots, X_n$ and $Y = Y_1, Y_2, \ldots, Y_n$ are given. Let

$$L_n = \max\{K : X_{i_k} = Y_{j_k} \text{ for } 1 \leq k \leq K, \text{ where } 1 \leq i_1 < i_2 < \cdots < i_K \leq n,$$

$$\text{and } 1 \leq j_1 < j_2 < \cdots < j_K \leq n\}$$

be the length of the longest common subsequence. Observe that

$$L_{1,n} \geq L_{1,m} + L_{m,n}, \quad m \geq n,$$

where $L_{m,n}$ is the longest common subsequence of X_m^n and Y_m^n. The above inequality follows from a simple observation that the longest subsequence may cross the boundary of X_1^m, Y_1^m and X_m^n, Y_m^n, hence it may be bigger than the sum of LCS in each subregion $(1, m)$ and (m, n). Thus, $a_n = \mathbf{E}[L_{1,n}]$ is superadditive. By Theorem 5.1

$$\lim_{n \to \infty} \frac{a_n}{n} = \alpha = \sup_{m \geq 1} \frac{\mathbf{E}[L_m]}{m}.$$

Interestingly enough, we still do not know the value of α even for the i.i.d case (memoryless source). Steele in 1982 conjectured that $\alpha = 2/(1 + \sqrt{2}) \approx 0.8284$, and some simulation studies support this guess. The best bounds were due to Deken [88], who proved that $0.7615 \leq \alpha \leq 0.8376$, until 1995, when Dančik and Paterson [80] improved them slightly.

Example 5.2: *Rate of Convergence in Theorem 5.1 Can Be Arbitrarily Slow*
Let $x_n = f(n) \geq 0$ with $f(n)/n$ decreasing where f is an arbitrary function. Then

$$x_{n+m} = f(n + m) = m\frac{f(n + m)}{n + m} + n\frac{f(n + m)}{n + m} \leq m\frac{f(m)}{m} + n\frac{f(n)}{n}$$

$$= x_m + x_n.$$

Thus

$$\lim_{n \to \infty} \frac{x_n}{n} = \inf_{m \geq 1} \frac{f(m)}{m}$$

and due to the arbitrariness of f the convergence can be as slow as desired. ∎

It should not be a surprise that the subadditivity property can be relaxed while still keeping unchanged the main thesis of Fekete's theorem. For example, if $x_{m+n} \leq x_n + x_m + c$, where c is a constant, then $y_n = x_n + c$ satisfies the subadditivity property, hence x_n/n has a limit that is equal to $\inf x_m/m$. Can we replace the constant c by a sequence $c_n = o(n)$ so that the thesis of Fekete's theorem still holds? The next result solves this problem, while exercises provide further generalizations. We shall follow the presentation of Steele [402].

TABLE 5.1. Basic Properties of Limsup and Liminf

We start with a formal definition.

Definition 5.2   *Let a_n be a sequence of real numbers. Limit superior and limit inferior are defined as*

$$\limsup_{n\to\infty} a_n = \inf_{n\geq 0}\{\sup_{k\geq n}\{a_k\}\} = \lim_{n\to\infty}\left(\sup_{m\geq n} a_m\right),$$

$$\liminf_{n\to\infty} a_n = \sup_{n\geq 0}\{\inf_{k\geq n}\{a_k\}\} = \lim_{n\to\infty}\left(\inf_{m\geq n} a_m\right).$$

The following theorem is useful.

Theorem 5.3   *We say that $L < \infty$ is the limit superior of a_n if the following two conditions hold:*

(i) *For every $\varepsilon > 0$ there exists an integer N such that $n > N$ implies*

$$a_n < L + \varepsilon,$$

that is, the above holds for all but finitely many n;

(ii) *For every $\varepsilon > 0$ and given m, there exists an integer $n > m$ such that*

$$a_n > L - \varepsilon,$$

that is, infinitely many a_n satisfy the above condition.

A similar statement is true for limit inferior.

In words, we can characterize the above limits as follows: Let us consider $\liminf a_n$. Take all possible convergent subsequences a_{n_k} of a_n, and let E be the set of all limits of such subsequences. Then $\liminf a_n = \inf E$. Finally, observe that $\liminf a_n$ and $\limsup a_n$ always exist, and $\lim a_n = a$ if and only if $\liminf a_n = a = \limsup a_n$. Finally, in applications we often use the following simple result.

Theorem 5.4   *Assume that $a_n \leq b_n$ for all $n = 1, 2, \ldots$ Then we have*

$$\liminf_{n\to\infty} a_n \leq \liminf_{n\to\infty} b_n, \qquad \limsup_{n\to\infty} a_n \leq \limsup_{n\to\infty} b_n.$$

Theorem 5.5 **(De Bruijn and Erdős, 1952)** *Let c_n be a positive and nondecreasing sequence fulfilling*

$$\sum_{k=1}^{\infty} \frac{c_k}{k^2} < \infty. \tag{5.5}$$

If x_n satisfies

$$x_{n+m} \leq x_n + x_m + c_{n+m}, \tag{5.6}$$

then

$$\lim_{n\to\infty} \frac{x_n}{n} = \inf_{m\geq 1} \frac{x_m}{m}.$$

Proof. The idea is to find a new sequence $b_n = o(n)$ such that $a_n = x_n + b_n$ is subadditive. Let

$$b_n = n \sum_{k=n}^{\infty} \frac{c_k}{k^2},$$

which is $o(n)$ due to (5.5). The fact that c_n is nondecreasing and positive implies

$$\sum_{k=\ell}^{h} \frac{c_k}{k^2} \geq c_\ell \int_\ell^{h+1} t^{-2} dt = c_\ell \left(\frac{1}{\ell} - \frac{1}{h+1} \right).$$

Define $a_n = x_n + b_n$, and assume without loss of generality that $m \geq n$. Then

$$a_{m+n} - a_n - a_m = x_{n+m} - x_n - x_m - m \sum_{k=m}^{m+n} \frac{c_k}{k^2} - n \sum_{k=n}^{m+n} \frac{c_k}{k^2}$$

$$\leq c_{n+m} - m c_m \left(\frac{1}{m} - \frac{1}{n+m+1} \right) - n c_n \left(\frac{1}{n} - \frac{1}{n+m+1} \right)$$

$$\leq c_m \left(\frac{m}{m+n+1} \right) - c_n \left(1 - \frac{n}{n+m+1} \right)$$

$$\leq c_n \left(\frac{m+n}{m+n+1} - 1 \right)$$

$$\leq 0,$$

where the inequalities in the third and fourth line follow from the monotonicity of c_n, and the last inequality is a consequence of the positivity of c_n. Thus we prove that $x_n + b_n$ is subadditive and its limit is equal to $\inf_{m\geq 1}(x_m/m + b_m/m)$. But, by (5.5) $b_m/m = o(1)$, which completes the proof. ∎

As the reader may expect, there are many other interesting embellishments of the subadditive theorem. We discuss some of them in the exercises section, but two stand out above all and we present a short account of them here following Steele's presentation from [402]. The first one deals with the rate of convergence (for a generalization see Exercise 2).

Theorem 5.6 (Pólya and Szegő, 1924) *If a real sequence x_n satisfies*

$$x_n + x_m - c \leq x_{n+m} \leq x_n + x_m + c \tag{5.7}$$

for all $n, m \geq 1$, then there is a finite constant α such that

$$\left| \frac{x_n}{n} - \alpha \right| \leq \frac{c}{n} \tag{5.8}$$

for all $n \geq 1$.

Proof. First of all, observe that setting $m = n$ in (5.7) we obtain

$$\left| \frac{x_{2m}}{2m} - \frac{x_m}{m} \right| \leq \frac{c}{2m}$$

for all $m \geq 1$. Also, since $x_n + c$ is subadditive, by Fekete's theorem $x_n/n \to \inf x_m/m = \alpha$. Using this and the above we obtain

$$\left| \alpha - \frac{x_m}{m} \right| < \left| \frac{x_{2m}}{2m} - \frac{x_m}{m} \right| + \left| \frac{x_{4m}}{4m} - \frac{x_{2m}}{2m} \right| + \cdots < \sum_{k \geq 1} \frac{c}{2^k m} = \frac{c}{m},$$

as desired.   ■

Finally, the next theorem due to De Bruijn and Erdős shows that we can relax the requirement that the subadditivity holds for all m and n. We ask the reader to prove it in Exercise 3 (cf. [402]).

Theorem 5.7 (De Bruijn and Erdős, 1952) *If a real sequence x_n satisfies the subadditivity condition for the restricted range of m and n*

$$x_{m+n} \leq x_n + x_m \quad \text{for} \quad \frac{1}{2}n \leq m \leq 2n,$$

then

$$\lim_{n \to \infty} \frac{x_n}{n} = \inf_{m \geq 1} \frac{x_m}{m} = \alpha,$$

where $-\infty \leq \alpha < \infty$.

5.2 SUBADDITIVE ERGODIC THEOREM

In the early seventies there was a resurgence of interest in generalizing Fekete's deterministic subadditivity result to a sequence of random variables. Such an extension would have an impact on many research problems of those days. For example, Chvatal and Sankoff [67] used ingenious tricks to establish the probabilistic behavior of the *Longest Common Superstring* problem (see Example 5.1), while it is a trivial consequence of a stochastic extension of the subadditivity result. In 1976 Kingman [240] presented the first proof of what later will be known as the *Subadditivity Ergodic Theorem*. Below, we present an extension of Kingman's result due to Liggett [288] and Derriennic [93].

To formulate it we must consider a sequence of doubly indexed random variables $X_m^n = (X_m, X_{m+1}, \ldots, X_n)$. However, traditionally, when dealing with subadditivity theorems, one denotes such a substring as $X_{m,n} = X_m^n$, and we adopt it here. Our goal is to say something about probabilistic behavior of $X_{0,n}$ as $n \to \infty$, when $X_{0,n}$ satisfies the subadditivity property. One expects that $X_{0,n} \sim n\alpha$ in a probabilistic sense, where α is a constant. This turns out to be true under certain assumptions specified below in the *subadditive ergodic theorem*.

Theorem 5.8 (Kingman, 1976; Liggett, 1985) *Let $X_{m,n}$ ($m < n$) be a sequence of random variables satisfying the following properties:*

(i) $X_{0,n} \leq X_{0,m} + X_{m,n}$ *(subadditivity);*

(ii) *For every k, $\{X_{nk,(n+1)k}, n \geq 1\}$ is a stationary sequence.*

(iii) *The distribution of $\{X_{m,m+k}, k \geq 1\}$ does not depend on m.*

(iv) $\mathbf{E}[X_{0,1}] < \infty$ *and for each n, $\mathbf{E}[X_{0,n}] \geq c_0 n$ where $c_0 > -\infty$.*

Then

$$\lim_{n \to \infty} \frac{\mathbf{E}[X_{0,n}]}{n} = \inf_m \frac{\mathbf{E}[X_{0,m}]}{m} := \alpha, \qquad (5.9)$$

and also

$$\lim_{n \to \infty} \frac{X_{0,n}}{n} = X \quad \text{a.s. and in } L^1, \qquad (5.10)$$

such that $\mathbf{E}[X] = \alpha$. Finally, if all stationary sequences in (ii) are ergodic, then

$$\lim_{n \to \infty} \frac{X_{0,n}}{n} = \alpha \quad \text{(a.s.)}. \qquad (5.11)$$

Clearly, (5.9) follows directly from Fekete's Theorem 5.1 while the proof of almost sure convergence and in mean (5.10) is beyond the scope of this book. An accessible proof can be found in Durrett [117]. Also, the original proof of Kingman [240] is readable. In fact, Kingman proved a weaker version of the theorem in which (ii) and (iii) are replaced by

(ii') $X_{m,n}$ is stationary (i.e., the joint distributions of $X_{m,n}$ are the same as $X_{m+1,n+1}$) and ergodic.

There are examples in which (ii') does not hold while Liggett's version works fine (cf. [117]).

Before we proceed with examples (see also Section 5.5) we mention a generalization of the subadditive ergodic theorem that is known as the *almost subadditive ergodic theorem*.

Theorem 5.9 (Derriennic, 1983) *Let (ii') and (iv) above be fulfilled and (i) be replaced by*

$$X_{0,n} \leq X_{0,m} + X_{m,n} + A_{m,n} \tag{5.12}$$

where $A_m \geq 0$ and $\lim_{n\to\infty} \mathbf{E}[A_{0,n}/n] = 0$, then (5.11) holds.

We must point out, however, that the above results establish only the existence of a constant α such that (5.11) holds. They say *nothing* of how to compute the limit, and in fact many ingenious methods have been devised in the past to bound the constant.

Example 5.3: *The First Birth Problem For an Age-Dependent Branching Process* Let us consider an *age-dependent branching process* in which each individual i lives for time t_i distributed as F before producing k offspring with probability p_k. The process starts with one individual in generation 0 who is born at time 0, and when dying its offspring start independent copies of the original process. Our interest lies in estimating the birth time B_n of the first member of generation n (cf. [46, 47, 241]). To estimate B_n, let $B_{0,m}$ be the birth of the first individual in generation m, and $B_{m,n}$ be the time needed for this individual to have an offspring in generation $n > m$. Little reflection is needed to find out that $B_{0,n} \leq B_{0,m} + B_{m,n}$; however, it is *not* true that for $l > 0$ we also have $B_{l,n} \leq B_{l,m} + B_{m,n}$. Since $B_n = B_{0,n}$ is a nonnegative process, all assumptions of Theorem 5.8 are fulfilled, and therefore there exists a constant α such that

$$\frac{B_n}{n} \to \alpha \quad \text{(a.s.)}.$$

Interestingly enough, in this case we can compute the constant α. In Exercise 16 we guide the reader through Biggins' derivation (cf. [46]).

The first birth problem is of prime importance to the analysis of algorithms since in 1986 Devroye [96] used it to prove the long-standing conjecture concerning the height of a binary search tree (cf. also Drmota [105, 106] and Reed [352]).

Example 5.4: *Increasing Sequences in Random Permutations* Let π be a permutation of $\{1, 2, \ldots, n\}$ and let $l_n(\pi)$ be the longest increasing sequence in π, that is, the largest k for which there are $i_1 < i_2 < \cdots < i_k$ such that $\pi(i_1) < \pi(i_2) < \cdots < \pi(i_k)$. This example is interesting since it is easy to see that $l_n(\pi)$ is *not* linear in n, hence the subadditive ergodic theorem cannot be applied. We have the following estimate by Boole's inequality

$$\Pr\{l_n(\pi) \geq 2e\sqrt{n}\} \leq \exp(-2e\sqrt{n}).$$

Indeed, there are $\binom{n}{k}$ subsequences of length k and each has probability $1/k!$, so that

$$\Pr\{l_n(\pi) \geq k\} \leq \binom{n}{k}\frac{1}{k!}.$$

Using Stirling's approximation with $k = 2e\sqrt{n}$, one proves the above estimate. But for any k

$$\mathbf{E}[l_n(\pi)] = \sum_{0 \leq i \leq k} i\Pr\{l_n = i\} + \sum_{k+1 \leq i \leq n} i\Pr\{l_n = i\}$$
$$\leq k + n\Pr\{l_n(\pi) > k\},$$

hence for sufficiently large n we have $\mathbf{E}[l_n(\pi)] \leq c\sqrt{n}$ for some $c > e$. The longest increasing sequence cannot grow linearly!

To circumvent this problem, we follow Hammersley and *poissonize*, that is, embed the process $l_n(\pi)$ into a Poisson stream of rate 1 in the plane. Poissonization is a powerful probabilistic technique to which we devote Chapter 10 of this book. We denote by $l(t, s)$ $(t < s)$ the length of the longest increasing sequence in the rectangle with corners (s, s), (s, t), (t, t) and (t, s). By the longest sequence in the plane we mean the largest k for which there are points (x_i, y_i) in the Poisson process with $s < x_1 < \cdots < x_k < t$ and $s < y_1 < \cdots < y_k < t$. Now, we are in the superadditive world, that is,

$$l(0, s + t) \geq l(0, s) + l(s, s + t),$$

since the longest increasing sequence can go through two other rectangles not included in $l(0, s)$ and $l(s, s + t)$. Then by the superadditive ergodic theorem

$$\lim_{t \to \infty} \frac{l(t)}{t} = \alpha \quad \text{(a.s.)}.$$

The ease with which we obtain this result has its price. We must *depoissonize* it to recover the probabilistic behavior of $l_n(\pi)$. Chapter 10 is devoted to *analytic depoissonization*. Here we apply a theorem from that chapter adopted to our situation. Let us concentrate on estimating the average $\mathbf{E}[l_n(\pi)]$, which we denote for simplicity as g_n. The idea is to choose such t that on a square of the area t^2 there are exactly n points. In other words, we shall condition on the Poisson process having n points in the square $(0, t)^2$. It is well known (cf. [402, 369]) that, under the conditioning on n, $l(0, t)$ has the same distribution as $l_n(\pi)$. But we just proved that

$$\mathbf{E}[l(t)] = \sum_{k=0}^{\infty} g_k \frac{(t^2)^k}{k!} e^{-t^2} \sim \alpha t \tag{5.13}$$

for large t. We need to recover $g_n = \mathbf{E}[l_n(\pi)]$ for large n from the above. This is exactly where the analytic depoissonization can help. Setting $t^2 = z$ in (5.13), we observe that the Poisson transform $\widetilde{G}(z) = \sum_{k=0}^{\infty} g_k \frac{z^k}{k!} e^{-z}$ of $g_n = \mathbf{E}[l_n(\pi)]$ grows like $\alpha \sqrt{z}$; hence by the depoissonization Theorem 10.3 in Chapter 10 we conclude that $\mathbf{E}[l_n(\pi)] = g_n = \alpha \sqrt{n} + o(\sqrt{n})$. ∎

5.3   MARTINGALES AND AZUMA'S INEQUALITY

The subadditive ergodic theorem is a very powerful tool that allows us to predict the growth rate of a quantity of interest. As we have seen before (cf. Example 5.4), the technique is not restricted to linear growth, but then one must also use other approaches. The drawback of the subadditivity method lies in its inability to evaluate the constant α. But we can try to do the next best thing, namely, assess the deviation from the most likely value αn. This is known as the *concentration of measure (probability)*. Surprisingly enough, there is a general technique for the concentration of measure even without any knowledge about α. Such results are known in literature under the name *Azuma's inequality* or the *method of bounded differences* (cf. [312]).

Azuma's inequality is an outcome of Hoeffding's large deviations lemma applied to martingales. Before we plunge into technical details, we would like to make two comments. Large deviations is discussed in depth in the next section of this chapter and we return to it in Chapter 8. Azuma's inequality could be viewed as a special case of large deviations, but its applicability and usefulness deserve a special attention. In fact, at the heart of Azuma's inequality are *martingales*. Martingale is another very useful tool in the probabilistic analysis of algorithms, so we recall their basic properties in Table 5.2. The applications Section 5.5.3 provides a more sophisticated example of martingales applications. In passing, we mention here that Régnier [355] (cf. Mahmoud [305]) used the *martingales convergence theorem* to establish the limiting distribution of the path length in a binary search tree or the quicksort algorithm (cf. [76, 110, 193, 258, 313, 370, 381, 420]).

Martingales are standard probabilistic tools, and the interested reader is referred to Durrett [117] for a more detailed discussion. Of supreme importance for martingale is its convergence theorem (see Theorem 5.12). However, in the analysis of algorithms a bound on the degree of fluctuation of a martingale is even more interesting. This bound will lead directly to Azuma's inequality, which finds an abundance of applications.

Let then $\{Y_n\}_{n=0}^{\infty}$ be a martingale with respect to the filtration $\mathcal{F}$, or simply Y_n is a martingale with respect to $X_0, X_1, \ldots$, as discussed in Definitions 5.10 and 5.11 of Table 5.2. We now define the **martingale difference** sequence $\{D_n\}_{n=1}^{\infty}$ as

$$D_n = Y_n - Y_{n-1}, \tag{5.14}$$

so that

$$Y_n = Y_0 + \sum_{i=1}^{n} D_i. \tag{5.15}$$

Clearly, D_n depends only on the knowledge contained in $X_0, \ldots, X_n$, or simply is $\mathcal{F}_n$-measurable where $\mathcal{F}_n = \sigma(X_0, \ldots, X_n)$ is the smallest σ-field generated by $(X_0, \ldots, X_n)$. Also, by the definition of martingales $\mathbf{E}[|D_n|] < \infty$. In fact, for all n,

$$\mathbf{E}[D_{n+1}| \mathcal{F}_n] = \mathbf{E}[Y_{n+1}| \mathcal{F}_n] - \mathbf{E}[Y_n|\mathcal{F}_n] = Y_n - Y_n$$
$$= 0, \tag{5.16}$$

hence $\mathbf{E}[D_n] = 0$. Actually, let us consider the product $D_{n+1}D_n$. Observe that

$$\mathbf{E}[D_{n+1}D_n| \mathcal{F}_n] = D_n\mathbf{E}[D_{n+1}| \mathcal{F}_n] = 0.$$

In general, we can conclude that for any $1 \leq i_1 < i_2 < \cdots < i_k \leq n$

$$\mathbf{E}[D_{i_1} D_{i_2} \cdots D_{i_k}] = 0. \tag{5.17}$$

TABLE 5.2. Basic Properties of Martingales

Martingales, besides Markov processes and stationary processes, are another broad class of processes that often arise in the analysis of algorithms.

Definition 5.10 *A sequence $Y_n = f(X_1, \ldots, X_n)$ $(n \geq 0)$ is a martingale with respect to the sequence $X_0, X_1, \ldots$ if for all $n \geq 0$*

(i) $\mathbf{E}[|\,Y_n\,|] < \infty$,

(ii) $\mathbf{E}[Y_{n+1} \mid X_0, X_1, \ldots, X_n] = Y_n$.

We should observe that $\mathbf{E}[Y_{n+1} \mid X_0, X_1, \ldots, X_n]$ defines a *random variable* depending on $X_0, \ldots, X_n$ (or more precisely, defined on the σ-field with respect to each $X_0, X_1, \ldots, X_n$). The best-known and simplest example of a martingale is constructed as follows: define $S_n = X_1 + \cdots + X_n$ with X_i being i.i.d. with mean $\mu = \mathbf{E}[X_i]$. Then $Y_n = S_n - n\mu$ is a martingale since

$$\mathbf{E}[S_{n+1} \mid X_0, X_1, \ldots, X_n] = \mathbf{E}[S_n + X_{n+1} \mid X_0, X_1, \ldots, X_n] = S_n + \mu.$$

Above we use $\mathbf{E}[S_n \mid X_1, \ldots, X_n] = S_n$ since S_n is completely defined on $X_1, \ldots, X_n$.

From the above, we may conclude that for the definition of martingale we require only the *knowledge* (i.e., information) obtained from $X_0, X_1, \ldots, X_n$. In fact, this vague idea can be made rigorous, if we agree to enter the realm of σ-fields. We do, since the advantage of such an approach will be immediately visible (e.g., shortening our notation). In general, let $\mathcal{F} = \{\mathcal{F}_0, \mathcal{F}_1, \ldots\}$ be a sequence of sub-σ-fields of $\mathcal{F}$. We call $\mathcal{F}$ a *filtration* if $\mathcal{F}_n \subset \mathcal{F}_{n+1}$ for all n (i.e., $\mathcal{F}_n$ is more coarser than $\mathcal{F}_{n+1}$). A sequence Y_n $(n \geq 0)$ is said to be *adapted* to the filtration $\mathcal{F}$, if Y_n is $\mathcal{F}_n$ measurable for all n (i.e., can be properly defined on $\mathcal{F}_n$). With this in mind, we can generalize the definition of martingales to the following one.

Definition 5.11 *We say that the sequence Y_n is a martingale with respect to the filtration $\mathcal{F}$ if for all $n \geq 0$ (i) above holds together with*

(ii) $\mathbf{E}[Y_{n+1} \mid \mathcal{F}_n] = Y_n$.

We recover the previous definition by choosing $\mathcal{F}_n = \sigma(X_0, X_1, \ldots, X_n)$, i.e., the smallest σ-fields with respect to which the variables X_i are defined. Finally, we discuss one result known as the *martingale convergence theorem*, which we present here in a simpler version.

Theorem 5.12 *If Y_n is a martingale with $\mathbf{E}[Y_n^2] \leq M < \infty$ for some M, then there exists a random variable Y such that Y_n converges to Y almost surely and in mean square.*

This property, as pointed out by Steele [402], is the only one required to prove Theorem 5.13 below (but we formulate it in a more traditional way and ask the reader to generalize it in Exercise 4).

Theorem 5.13   (Hoeffding's Inequality)   *Let $\{Y_n\}_{n=0}^{\infty}$ be a martingale with respect to the filtration $\mathcal{F}$, and let there exist constant c_n such that for all $n \geq 0$*

$$|Y_n - Y_{n-1}| \leq c_n. \tag{5.18}$$

Then for $x > 0$

$$\Pr\{|Y_n - Y_0| \geq x\} = \Pr\left\{\left|\sum_{i=1}^{n} D_i\right| \geq x\right\} \leq 2\exp\left(-\frac{x^2}{2\sum_{i=1}^{n} c_i^2}\right). \tag{5.19}$$

Proof. We start with a simple inequality for the exponential function. Observe that for $\beta > 0$ the exponential function $e^{\beta d}$ is convex for any d, and hence on the interval $d \in (-1, 1)$ we obtain

$$e^{\beta d} \leq \frac{1}{2}(1 - d)e^{-\beta} + \frac{1}{2}(1 + d)e^{\beta}.$$

If now d is replaced by a random variable D with mean $\mathbf{E}[D] = 0$, then the above becomes

$$\mathbf{E}[e^{\beta D}] \leq \frac{1}{2}\left(e^{-\beta} + e^{\beta}\right) < e^{\frac{1}{2}\beta^2}, \tag{5.20}$$

where the last inequality follows immediately from the Taylor expansions of both sides and a comparison of the coefficients at β^{2k}.

We shall only derive (5.19) for $Y_n - Y_0 \geq x$ since our arguments will be valid for the other side of the inequality. Observe now that by Markov's inequality (cf. (2.6)) we have

$$\Pr\{Y_n - Y_0 \geq x\} = \Pr\{e^{\theta(Y_n - Y_0)} \geq e^{\theta x}\} \leq e^{-\theta x}\mathbf{E}[e^{\theta(Y_n - Y_0)}] \tag{5.21}$$

for some $\theta > 0$. Let $D_n = Y_n - Y_{n-1}$ be the martingale difference. Observe that $Y_n - Y_0 = Y_{n-1} - Y_0 + D_n$. Conditioning now on $\mathcal{F}_{n-1}$ (the reader can think of conditioning on all knowledge extracted from $Y_0, Y_1, \ldots, Y_{n-1}$) we obtain

$$\mathbf{E}[e^{\theta(Y_n - Y_0)} \mid \mathcal{F}_{n-1}] = e^{\theta(Y_{n-1} - Y_0)}\mathbf{E}[e^{\theta D_n} \mid \mathcal{F}_{n-1}]$$

$$\leq e^{\theta(Y_{n-1} - Y_0)}e^{\frac{1}{2}\theta^2 c_n^2},$$

where the last inequality follows from (5.20) applied to $D_n/c_n \leq 1$ (cf. (5.18)). Now taking the expectation of the above with respect to $\mathcal{F}_n$ we have

$$\mathbf{E}[e^{\theta(Y_n-Y_0)}] \leq e^{\frac{1}{2}\theta^2 c_n^2}\mathbf{E}[e^{\theta(Y_{n-1}-Y_0)}],$$

and after n iterations we arrive at

$$\mathbf{E}[e^{\theta(Y_n-Y_0)}] \leq \exp\left(\frac{1}{2}\theta^2 \sum_{i=1}^n c_i^2\right).$$

Hence, by (5.21) we prove that

$$\Pr\{Y_n - Y_0 \geq x\} \leq \exp\left(-\theta x + \frac{1}{2}\theta^2 \sum_{i=1}^n c_i^2\right),$$

which becomes (5.19) after the substitution $\theta = x \left(\sum_{i=1}^n c_i^2\right)^{-1}$ (which maximizes the exponent of the above). This completes the proof. ∎

Our derivation of (5.19) is quite simple but we did not get the optimal constant in the exponent. The reader is asked in Exercise 5 to prove that (5.19) can be improved to

$$\Pr\{|Y_n - Y_0| \geq x\} = \Pr\left\{\left|\sum_{i=1}^n D_i\right| \geq x\right\} \leq 2\exp\left(-\frac{2x^2}{\sum_{i=1}^n c_i^2}\right). \tag{5.22}$$

Before we derive the Azuma concentration of measure (cf. Theorem 5.15), we illustrate it in an example. The reader should very carefully study this example since it explains the main idea behind the Azuma inequality.

Example 5.5: *Bin Packing*   Consider n objects with random sizes $X_1, X_2, \ldots,$ X_n having a common distribution on $[0, 1]$, and an unlimited collection of bins each of size 1. Let $B_n := B_n(X_1, \ldots, X_n)$ be the minimum number of bins required to pack the objects. First of all, it is not difficult to see that B_n is subadditive, that is, $B_{0,n} \leq B_{0,m} + B_{m,n}$, hence by Theorem 5.8 there exists a constant α such that

$$\lim_{n \to \infty} \frac{\mathbf{E}[B_n]}{n} = \alpha,$$

$$\lim_{n \to \infty} \frac{B_n}{n} = \alpha \quad \text{(a.s.)}.$$

The evaluation of α is quite complicated, if possible at all, but we can answer the next interesting question: how close is B_n to its mean value $E[B_n] \sim \alpha n$? As expected, we shall use Hoeffding's inequality. Thus, we must define a martingale and martingale differences. Here it comes. Let

$$Y_i = E[B_n(X_1, \ldots, X_n) | X_1, \ldots, X_i]$$
$$= E[B_n | \mathcal{F}_i],$$

where throughout we assume that $\mathcal{F}_i = \sigma(X_1, \ldots, X_i)$. In other words, Y_i is the average of the minimum number of bins required to pack n items, provided someone reveals to us sizes $X_1, \ldots, X_i$ of the first i objects. Is it a martingale? Yes, it is, and we prove it below. Our task is to show that $\{Y_i\}_{i=0}^n$ is a finite martingale; that is, for all $0 \le i \le n - 1$

$$E[Y_{i+1} | \mathcal{F}_i] = E[E[B_n | \mathcal{F}_{i+1}] | \mathcal{F}_i] = E[B_n | \mathcal{F}_i] = Y_i. \tag{5.23}$$

The second equality, which we shall prove in its whole generality in Lemma 5.14, is at the heart of many martingale applications. In this example, instead of being general and abstract we present a simple and boring proof that reveals all the details: We simply prove the above for $n = 3$; that is, we show

$$E[Y_2 | X_1] = E[E[B_n(X_1, X_2, X_3) | X_1, X_2] | X_1]$$
$$= E[B_n(X_1, X_2, X_3) | X_1] = Y_1. \tag{5.24}$$

Indeed, let us compute the above averages. Observe that the right-hand side of the above becomes

$$E[B_n(X_1, X_2, X_3) | X_1] = \int_{x_2} \int_{x_3} B(X_1, x_2, x_3) f(X_1, x_2, x_3 | X_1) dx_2 dx_3,$$
$$\tag{5.25}$$

where $f(X_1, x_2, x_3 | X_1) = f(X_1, x_2, x_3)/f(X_1)$ is the conditional density of (X_1, x_2, x_3) given X_1. Needless to say, the above average is a random variable (since X_1 is a random variable). Let us now compute the left-hand side of (5.24). Let $Z(X_1, X_2) = E[B_n(X_1, X_2, X_3) | X_1, X_2]$; that is, $Z(X_1, X_2)$ is a random function of X_1, X_2 whose value is

$$Z(X_1, X_2) = E[B_n(X_1, X_2, X_3) | X_1, X_2]$$
$$= \int_{x_3} B(X_1, X_2, x_3) \frac{f(X_1, X_2, x_3)}{f(X_1, X_2)} dx_3.$$

Its distribution is $f(x_1, x_2)$ when $X_1 = x_1$ and $X_2 = x_2$. Then,

$$E[Z(X_1, X_2)|\ X_1] = \int_{x_2} Z(X_1, x_2) \frac{f(X_1, x_2)}{f(X_1)} dx_2$$

$$= \int_{x_2} \int_{x_3} B(X_1, x_2, x_3) \frac{f(X_1, x_2, x_3)}{f(X_1)} dx_2 dx_3,$$

which is equal to $E[B_n(X_1, X_2, X_3)|\ X_1]$ as established in (5.25). This proves (5.24). The same analysis shows that for any $1 \le i \le n$ we have $E[Y_{i+1}|\ \mathcal{F}_i] = Y_i$.

Knowing that Y_i is a martingale, we define the martingale difference as $D_i = Y_i - Y_{i-1}$, and verify that

$$E[D_i|\mathcal{F}_{i-1}] = E[E[B_n|\ \mathcal{F}_i]|\ \mathcal{F}_{i-1}] - E[B_n|\mathcal{F}_{i-1}]$$

$$= E[B_n|\ \mathcal{F}_{i-1}] - E[B_n|\mathcal{F}_{i-1}] = 0, \qquad (5.26)$$

as desired. In fact, the equality (5.26) follows from (5.23) (see also Lemma 5.14 below).

It remains now to observe that $Y_n = E[B_n|X_1, \ldots, X_n] = B_n$ and $Y_0 = E[B_n]$; hence we are almost ready to apply Hoeffding's inequality, if we can prove that $D_i = Y_i - Y_{i-1}$ is bounded. The only difference between Y_i and Y_{i-1} is that the former knows one more size, namely X_i. But, in the worst case this cannot improve the packing by more than one bin! Thus $|D_i| = |Y_i - Y_{i-1}| \le 1$. In view of this, Theorem 5.13 allows us to write

$$Pr\{|B_n - E[B_n]| \ge x\} \le 2 \exp\left(-\frac{x^2}{2n}\right).$$

We do not know exactly $E[B_n]$, except that it is $O(n)$. Let then $x = \varepsilon n$ for any $\varepsilon > 0$. Thus

$$Pr\{|B_n - \alpha n| \ge \varepsilon n\} \le 2 \exp\left(-\frac{1}{2}\varepsilon^2 n(1 + o(1))\right)$$

which is a useful bound as long as $n\varepsilon^2 \to \infty$.   ∎

In the above example we used the following result, which is crucial for any martingale application, in particular, for the Azuma inequality.

Lemma 5.14 *If $\mathcal{F}_1 \subset \mathcal{F}_2$, then*

$$E[E[Y|\mathcal{F}_1]|\mathcal{F}_2] = E[Y|\mathcal{F}_1], \qquad (5.27)$$

$$E[E[Y|\mathcal{F}_2]|\mathcal{F}_1] = E[Y|\mathcal{F}_1]. \qquad (5.28)$$

In words, the smaller σ-field always wins. In particular,

$$\mathbf{E}[\mathbf{E}[Y| X_1, X_2]| X_1] = \mathbf{E}[Y| X_1] = \mathbf{E}[\mathbf{E}[Y| X_1]| X_1, X_2].$$

Proof. The proof of (5.27) is quite simple. Observe that $\mathbf{E}[f(X)| X] = f(X)$ since $f(X) \in \sigma(X)$ for a measurable function f. But $\mathbf{E}[Y|\mathcal{F}_1] \in \mathcal{F}_2$ since $\mathcal{F}_1 \subset \mathcal{F}_2$; hence (5.27) follows. The second part is a little more intricate, but we have already given a sketch of the proof above in Example 5.5. Here is another derivation: Notice that $\mathbf{E}[Y|\mathcal{F}_1] \in \mathcal{F}_1$ and for $A \in \mathcal{F}_1 \subset \mathcal{F}_2$ we have

$$\int_A \mathbf{E}[Y|\mathcal{F}_1]dP = \int_A YdP = \int_A \mathbf{E}[Y|\mathcal{F}_2]dP,$$

which completes the proof. ∎

We are now ready to derive the Azuma inequality. Let

$$Y_n = F_n(X_1, \ldots, X_n)$$

be a martingale with respect to $X_1, \ldots, X_n$. Define

$$
\begin{aligned}
D_i &= \mathbf{E}[Y_n|\mathcal{F}_i] - \mathbf{E}[Y_n|\mathcal{F}_{i-1}] \\
&= \mathbf{E}[Y_n| X_1, \ldots, X_i] - \mathbf{E}[Y_n| X_1, \ldots, X_{i-1}],
\end{aligned}
$$

which is a martingale difference since by Lemma 5.14

$$\mathbf{E}[D_i|\mathcal{F}_{i-1}] = \mathbf{E}[\mathbf{E}[Y_n|\mathcal{F}_i|\mathcal{F}_{i-1}] - \mathbf{E}[Y_n|\mathcal{F}_{i-1}] = \mathbf{E}[Y_n|\mathcal{F}_{i-1}] - \mathbf{E}[Y_n|\mathcal{F}_{i-1}] = 0.$$

Also, it is easy to see that

$$\mathbf{E}[Y_n|\mathcal{F}_n] = Y_n, \quad \text{and} \quad \mathbf{E}[Y_n|\mathcal{F}_0] = \mathbf{E}[Y_n],$$

since $\mathcal{F}_0$ does not have any information about $X_1, \ldots, X_n$ while $\mathcal{F}_n$ completely describes Y_n.

In view of the above, we can apply Hoeffding's inequality to D_i provided they are bounded, say $|D_i| \leq c_i$. However, the real trick of Azuma is to explore this generic representation to find a simple way of bounding D_i. In order to accomplish this, let $\hat{X}_i$ be a new random variable independent of X_i and with the same distribution as X_i. The point to observe is that $\mathcal{F}_i$ has no information about this new random variable, so that

$$\mathbf{E}[F_n(X_1, \ldots, X_i, \ldots, X_n)|\mathcal{F}_{i-1}] = \mathbf{E}[F_n(X_1, \ldots, \hat{X}_i, \ldots, X_n)|\mathcal{F}_i].$$

Hence

$$D_i = \mathbf{E}[F_n(X_1, \ldots, X_i, \ldots, X_n)|\mathcal{F}_i] - \mathbf{E}[F_n(X_1, \ldots, X_i, \ldots, X_n)|\mathcal{F}_{i-1}]$$
$$= \mathbf{E}[F_n(X_1, \ldots, X_i, \ldots, X_n)|\mathcal{F}_i] - \mathbf{E}[F_n(X_1, \ldots, \hat{X}_i, \ldots, X_n)|\mathcal{F}_i].$$

In view of this, if we postulate for every $1 \leq i \leq n$

$$|F_n(X_1, \ldots, X_i, \ldots, X_n)|\mathcal{F}_i] - F_n(X_1, \ldots, \hat{X}_i, \ldots, X_n)|\mathcal{F}_i]| \leq c_i,$$

then $|D_i| \leq c_i$, and Hoeffding's inequality immediately implies the following Azuma's result.

Theorem 5.15   (Azuma's Inequality)   *Let $Y_n = F_n(X_1, \ldots, X_n)$ be a martingale with respect to $X_1, \ldots, X_n$ such that for every $1 \leq i \leq n$ there exist constant c_i*

$$|F_n(X_1, \ldots, X_i, \ldots, X_n) - F_n(X_1, \ldots, \hat{X}_i, \ldots, X_n)| \leq c_i, \qquad (5.29)$$

where $\hat{X}_i$ is independent of X_i and has the same distribution as X_i. Then

$$\Pr\{|Y_n - \mathbf{E}[Y_n]| \geq x\} = \Pr\{|F_n(X_1, \ldots, X_n) - \mathbf{E}[F_n(X_1, \ldots, X_n)]| \geq x\}$$
$$\leq 2\exp\left(-\frac{x^2}{2\sum_{i=1}^n c_i^2}\right) \qquad (5.30)$$

for some $x > 0$.

In applications the following corollary is often used. Its proof is a direct consequence of our previous discussion.

Corollary 5.16   *If $X_1, \ldots, X_n$ are independent random variables satisfying (5.29) for a function F_n, then (5.30) holds.*

In the applications section (Section 5.5) we discuss some nontrivial applications of Azuma's inequality. The reader can also consult McDiarmid [312] for many interesting applications in graph theory. In passing, we point out that the concentration of mean for the bin packing Example 5.5 can be obtained directly from the Azuma inequality.

5.4 LARGE DEVIATIONS

Hoeffding's inequality presented in the previous section is an example of a large deviations bound when applied to the martingale $S_n - n\mathbf{E}[X_1]$ where $S_n = X_1 + \cdots + X_n$. We devote this section to a detailed discussion of large deviations of S_n. Before we plunge into technical details, we first offer an explanation of why one must investigate large deviations and why the central limit theorem studied so extensively in probability theory is not sufficient.

Let us consider a sequence $X_1, \ldots, X_n$ of i.i.d. random variables, and let $S_n = X_1 + \cdots + X_n$. Define $\mu := \mathbf{E}[X_1]$ and $\sigma^2 := \mathbf{Var}[X_1]$. To study limiting laws of S_n one must normalize it appropriately. For example, for the central limit theorem the following normalization is useful:

$$s_n := \frac{S_n - n\mu}{\sigma \sqrt{n}}.$$

Let $F_n(x) := \Pr\{s_n \le x\}$ and $\Phi(x)$ be the distribution function of the *standard normal distribution*, that is,

$$\Phi(x) := \frac{1}{\sqrt{2\pi}} \int_{-\infty}^{x} e^{-\frac{1}{2}t^2} dt.$$

The *central limit theorem* asserts that $F_n(x) \to \Phi(x)$ for continuity points of $\Phi(\cdot)$, provided $\sigma < \infty$ (cf. [117, 123]). A stronger version is due to Berry-Essén (see Lemma 8.28 of Chapter 8), who proved that

$$|F_n(x) - \Phi(x)| \le \frac{2\rho}{\sigma^2 \sqrt{n}}, \tag{5.31}$$

where $\rho = \mathbf{E}[|X - \mu|^3] < \infty$. Finally, Feller [123] showed that if the centralized moments $\mu_2, \ldots, \mu_r$ exist, then

$$F_n(x) = \Phi(x) - \frac{1}{\sqrt{2\pi}} e^{-\frac{1}{2}x^2} \sum_{k=3}^{r} n^{-\frac{1}{2}k+1} R_k(x) + O\left(n^{-\frac{1}{2}r+\frac{1}{2}}\right) \tag{5.32}$$

uniformly in x, where $R_k(x)$ is a polynomial of degree k depending only on $\mu_1, \ldots, \mu_r$ but not on n and r. We shall return to these problems in Chapter 8, where we study large deviations by analytic methods.

One should notice, however, the weakness of the central limit result that is able only to assess the probability of *small deviations* from the mean. Observe that (5.31) or (5.32) are useful only for $x = O(1)$ (i.e., for $S_n \in (\mu n - O(\sqrt{n}), \mu n + O(\sqrt{n}))$ due to a polynomial rate of convergence as shown in (5.31). To see this even more explicitly we quote here a result of Greene and Knuth [171], proved in

Section 8.4.3, that is valid for discrete random variables, namely,

$$\Pr\{S_n = \mu n + r\}$$

$$= \frac{1}{\sigma\sqrt{2\pi n}} \exp\left(\frac{-r^2}{2\sigma^2 n}\right)\left(1 - \frac{\kappa_3}{2\sigma^4}\left(\frac{r}{n}\right) + \frac{\kappa_3}{6\sigma^6}\left(\frac{r^3}{n^2}\right)\right) + O\left(n^{-\frac{3}{2}}\right),$$

$$(5.33)$$

where κ_3 is the third cumulant of X_1, and r is such that $\mu n + r$ is integer. Now, it should be clear that when $r = O(\sqrt{n})$ (i.e., $x = O(1)$ in our previous formulas) the error term is of the same order as the leading term of the asymptotic expression; thus the whole estimate becomes useless.

From the above discussion, one should conclude that the central limit theorem has a limited range of application, and one should expect another law for *large deviations* from the mean, that is, when $x \to \infty$. The most interesting from the application point of view is the case when $x = O(\sqrt{n})$ (or $r = O(n)$), that is, for $\Pr\{S_n = n(\mu + \delta)\}$ for $\delta \neq 0$. Hereafter, we shall discuss this large deviations behavior.

Actually, we already touched large deviations results when we derived Hoeffding's inequality. To see this, let us define $Y_n = S_n - n\mu$ where $X_1, \ldots, X_n$ are i.i.d. random variables. Notice that Y_n is a martingale with respect to $\mathcal{F}_n = \sigma(X_1, \ldots, X_n)$ since

$$\mathbf{E}[Y_{n+1}|\mathcal{F}_n] = \mathbf{E}[S_n + X_{n+1} - (n+1)\mu|\mathcal{F}_n] = S_n - (n+1)\mu - \mu = Y_n.$$

Defining the martingale difference $D_n = Y_n - Y_{n-1}$ and recognizing that $\mathbf{E}[Y_n|\mathcal{F}_n] = Y_n$ as well as $\mathbf{E}[Y_n|\mathcal{F}_0] = \mathbf{E}[Y_n]$, we derive from (5.22) Theorem 5.17 below (several generalizations and extensions of this results are discussed in the exercises section).

Theorem 5.17   (Hoeffding, 1963)   *Let $X_1, \ldots, X_n$ be i.i.d. random variables such that $|X_i| \leq 1$ for every i. Then*

$$\Pr\{S_n \geq n(\mu + \delta)\} \leq \exp\left(-\frac{1}{2}n\delta^2\right) \qquad (5.34)$$

for $\delta > 0$.

In passing we observe that directly from Theorem 5.13 we can obtain a weaker bound, namely, $\Pr\{S_n \geq n(\mu + \delta)\} \leq \exp(-\frac{1}{8}n\delta^2)$.

Let us now further explore the large deviations behavior of $S_n = X_1 + \cdots X_n$ for i.i.d. (not necessary bounded) random variables. We start with evaluating

$\Pr\{S_n \geq an\}$ as $n \to \infty$. Observe that

$$\Pr\{S_{n+m} \geq (n+m)a\} \geq \Pr\{S_m \geq ma, \ S_{n+m} - S_m \geq na\}$$
$$= \Pr\{S_n \geq na\}\Pr\{S_m \geq ma\},$$

since S_m and $S_{n+m} - S_m$ are independent. Taking logarithm of both sides, and recognizing that $\log \Pr\{S_n \geq an\}$ is a superadditive sequence, by Fekete's Theorem 5.1 we obtain

$$\lim_{n \to \infty} \frac{1}{n} \log \Pr\{S_n \geq na\} = -I(a),$$

where $I(a) \geq 0$. Thus S_n may decay *exponentially* (provided $I(a) > 0$) when far away from its mean, as we have already seen it for bounded random variables. Unfortunately, we obtain the above result from the subadditive property that allows us to conclude the existence of the above limit, but says nothing about $I(a)$. In particular, we must discover when $I(a) = 0$, that is, the decay of S_n is *subexponential*. We discuss this next.

We want to know when S_n has a subexponential or *heavy* tail, which will lead to $I(a) = 0$. Intuitively, when the tail of the distribution of a random variable X is not exponential, then the *moment generating function* defined as

$$M(\lambda) = \mathbf{E}[e^{\lambda X}]$$

cannot exist for $\lambda > 0$. (We shall discuss in depth generating functions in Chapter 7 for complex λ, but here we assume that λ is real and positive.) To make this claim precise, let us consider a sequence $X_1, \ldots, X_n$ of i.i.d. random variables such that for all $\lambda > 0$

$$M(\lambda) = \mathbf{E}[e^{\lambda X_1}] = \infty. \tag{5.35}$$

Then we claim

$$\liminf_{n \to \infty} \frac{1}{n} \log \Pr\{S_n \geq an\} = 0, \tag{5.36}$$

where, as always, $S_n = X_1 + \cdots + X_n$. It suffices to prove that

$$\liminf_{t \to \infty} \frac{1}{t} \log \Pr\{t \leq S_n \leq t + 1\} = 0. \tag{5.37}$$

Indeed, observe that

$$\liminf_{n \to \infty} \frac{1}{n} \log \Pr\{S_n \geq an\} \geq \liminf_{n \to \infty} \frac{1}{n} \log \Pr\{an \leq S_n < an + 1\}$$

$$= |a| \liminf_{n \to \infty} \frac{1}{|a|n} \log \Pr\{an \leq S_n < an + 1\}$$

$$\geq |a| \liminf_{t \to \infty} \frac{1}{t} \log \Pr\{t \leq S_n \leq t + 1\}.$$

We prove now that (5.35) implies (5.37). Let us assume the contrary and postulate the existence of $\delta > 0$ such that for $t > t_0$ the following holds: $\Pr\{t \leq S_n \leq t + 1\} \leq e^{-t\delta}$. Then for some constant $C_{t_0} > 0$, we obtain

$$\mathbf{E}[e^{\lambda S_n}] = \int_{-\infty}^{\infty} e^{\lambda s} dP(s) \leq C_{t_0} + \int_{t_0}^{\infty} e^{\lambda s} dP(s)$$

$$= C_{t_0} + \sum_{t=t_0}^{\infty} \int_t^{t+1} e^{\lambda s} dP(s)$$

$$\leq C_{t_0} + \sum_{t=t_0}^{\infty} e^{\lambda t} \Pr\{t \leq S_n \leq t + 1\}$$

$$< \infty \qquad \text{for} \qquad \lambda < \delta.$$

Thus the moment generating functions of S_n and X_1 exist at least for $\lambda < \delta$. This contradicts our basic assumption (5.35) and proves (5.36).

Hereafter, we adopt the hypothesis that the moment generating function of X_1 exists; that is, we assume that

$$M(\lambda) = \mathbf{E}[e^{\lambda X_1}] < \infty \text{ for some } \lambda > 0. \tag{5.38}$$

In addition, we define the *cumulant function* of X_1

$$\kappa(\lambda) = \log M(\lambda).$$

Certainly, $\kappa'(0) = \mathbf{E}[X_1] = \mu$ and $\kappa''(0) = \text{Var}[X_1]$. In Exercise 12 we ask the reader to prove that $\kappa(\lambda)$ is a convex function.

For a moment we drop the assumption about independence of $X_1, \ldots, X_n$. Then by Markov's inequality

$$e^{\lambda na} \Pr\{S_n \geq na\} = e^{\lambda na} \Pr\{e^{\lambda S_n} \geq e^{\lambda na}\} \leq \mathbf{E}[e^{\lambda S_n}]$$

for any $\lambda > 0$. Actually, due to arbitrariness of λ, subject to $\lambda > 0$, we just derived the *Chernoff's bound*.

Lemma 5.18 **(Chernoff's Bound)** *Let $X_1, \ldots, X_n$ be random variables for which the moment generating functions $M_i(\lambda)$ exist for some $\lambda > 0$. Then*

$$\Pr\{S_n \geq na\} \leq \min_{\lambda > 0} \left\{ e^{-\lambda na} \mathbf{E}[e^{\lambda S_n}] \right\} \tag{5.39}$$

for any $a > 0$.

Example 5.6: *Bounds on the Binomial Distribution* Let X_i be Bernoulli distributed with $\Pr\{X_1 = 1\} = p$ and $q = 1 - p$. Hence, $S_n = X_1 + \cdots + X_n$ has the *Binomial(n, p)* distribution. By Chernoff's bound we have for $\lambda > 0$

$$\Pr\{S_n \geq na\} \leq \exp\left(-na\lambda + n \ln(pe^\lambda + q)\right).$$

Set now $a = p(1 + \varepsilon)$. Instead of finding the optimal λ, we assume that $\lambda = \varepsilon\beta$, and we select $\beta > 0$ later. Since

$$\ln(pe^\lambda + q) \leq p\varepsilon\beta + p\varepsilon^2 \frac{\beta^2}{2} + p \sum_{i=3}^{\infty} \frac{(\varepsilon\beta)^i}{i!},$$

we find

$$\Pr\{S_n \geq na\} \leq \exp\left(-np\varepsilon^2 \left(\beta - \frac{1}{2}\beta^2 - O(\varepsilon)\right)\right).$$

Setting $\beta = 1 - (1 - O(\varepsilon))/\sqrt{3}$, for sufficiently small $\varepsilon > 0$ we finally arrive at

$$\Pr\{S_n \geq np(1 + \varepsilon)\} \leq \exp(-np\varepsilon^2/3). \tag{5.40}$$

The reader may follow the above derivation to prove also that

$$\Pr\{S_n \leq np(1 - \varepsilon)\} \leq \exp(-np\varepsilon^2/2) \tag{5.41}$$

for sufficiently small $\varepsilon > 0$. ∎

In the i.i.d. case the bound (5.39) becomes

$$\Pr\{S_n \geq na\} \leq \min_{\lambda > 0} \{\exp(-n(a\lambda - \kappa(\lambda)))\}$$

since

$$\mathbf{E}[e^{\lambda(X_1 + \cdots + X_n)}] = \mathbf{E}\left[\prod_{i=1}^{n} e^{\lambda X_i}\right] = \prod_{i=1}^{n} \mathbf{E}[e^{\lambda X_i}] = e^{n\kappa(\lambda)},$$

where the second equality follows from independence. We shall write the above in a different manner by introducing the Fenchel-Legendre transform:

$$I(a) = \sup_{\lambda > 0}\{a\lambda - \kappa(\lambda)\}.$$

Then

$$\Pr\{S_n \geq na\} \leq e^{-nI(a)}$$

and we shall prove below a matching lower bound.

We first need to study some properties of the Fenchel–Legendre transform. Let now

$$L(\lambda) = a\lambda - \kappa(\lambda)$$

$$I(a) = \sup_{\lambda > 0} L(\lambda).$$

To find an explicit formula for $I(a)$ we must solve a simple optimization problem, that is, to find such λ_a, if it exists, that $L(\lambda_a) = I(a)$. We shall look at λ_a among solutions of $L'(\lambda_a) = 0$, that is, such that

$$a = \kappa'(\lambda_a) = \frac{M'(\lambda_a)}{M(\lambda_a)}. \tag{5.42}$$

To study the existence of λ_a and to prove a lower bound for the large deviations result, we introduce the *twisted distribution* known also as the *shift of mean method* (cf. Section 8.4.3) and *exponential change of measure*. Let X^θ be a random variable with the distribution

$$F_\theta(x) = \frac{1}{M(\theta)} \int_{-\infty}^{x} e^{\theta y} dF(y).$$

It is easy to check that $F_\theta(x)$ is a distribution function, and

$$E[X^\theta] = \kappa'(\theta) = \frac{M'(\theta)}{M(\theta)}.$$

The exponential change of measure plays an important role in probability since the mean of X^θ can be shifted to any position by selecting proper θ (e.g., by choosing $\theta = \lambda_a$ the mean of X^θ becomes a). Then, we may apply several useful probability laws that are true only around the mean (e.g., weak law of numbers and central limit theorem). Moreover, since

$$\kappa''(\theta) = \mathrm{Var}[X^\theta] \geq 0,$$

$\kappa'(\theta) = \mathbf{E}[X^{\theta}]$ is strictly increasing as long as *the distribution F is not a point mass at* μ. In view of this, and the fact that $\kappa'(0) = \mu$, we conclude that for $a > \mu$ equation (5.42) has at most one solution $\lambda_a \geq 0$ such that it maximizes the concave function $L(\lambda)$.

We now know that λ_a maximizes $L(\lambda)$; however, to use this fact for the large deviations, we must relate it to the assumption (5.38) since it might happen that $M(\lambda)$ is defined only for $\lambda < \lambda_a$. To handle this properly, we introduce

$$x_0 = \sup\{x : F(x) < 1\},$$

$$\lambda_{\max} = \sup\{\lambda : M(\lambda) < \infty\}.$$

We are now ready to formulate the basic large deviations result.

Theorem 5.19   (Large Deviations I)   *Assume* $X_1, \ldots, X_n$ *are i.i.d. Let* $M(\lambda) = \mathbf{E}[e^{\lambda X_1}] < \infty$ *for* $0 < \lambda < \lambda_{\max}$, *and the distribution of* X_i *is not a point mass at* μ. *Set* $\mu < a < x_0$ *and assume there is* $\lambda_a \in (0, \lambda_{\max})$ *such that*

$$a = \frac{M'(\lambda_a)}{M(\lambda_a)}.$$

Then

$$\lim_{n \to \infty} \frac{1}{n} \log \Pr\{S_n \geq na\} = -(a\lambda_a - \log M(\lambda_a)) = -I(a).$$

Proof. The upper bound follows immediately from the Chernoff bound and the discussion above. Following Durrett [117], for the lower bound we introduce the twisted distribution F_λ such that $\lambda \in (\lambda_a, \lambda_{\max})$, and consider $X_1^\lambda, \ldots, X_n^\lambda$ with distribution F_λ. Observe that $S_n^\lambda = X_1^\lambda + \cdots + X_n^\lambda$ has mean $\mathbf{E}[S_n^\lambda] = n\kappa'(\lambda) > na$ where the inequality follows from the monotonicity of $\kappa'(\lambda)$ and $\lambda > \lambda_a$. It is also easy to compute the distribution F_n^λ of S_n^λ in terms of the distribution F_n of S_n. Indeed, since $M^n(\lambda)$ is the moment generating function of S_n, we realize that

$$\frac{dF_n(x)}{dF_n^\lambda(x)} = e^{-\lambda x} M^n(\lambda).$$

Now we obtain

$$\Pr\{S_n > an\} = \int_{na}^{\infty} dF_n(x) = \int_{na}^{\infty} e^{-\lambda x} M^n(\lambda) dF_n^\lambda(x)$$

$$\geq M^n(\lambda) e^{-\lambda na} (1 - \Pr\{S_n^\lambda \leq na\})$$

$$\to M^n(\lambda) e^{-\lambda na},$$

where the last line follows from the weak law of large numbers since $E[S_n^\lambda] > na$, as shown above. Thus

$$\liminf_{n\to\infty} \frac{1}{n} \log \Pr\{S_n > na\} \geq -(a\lambda - \log M(\lambda)),$$

and the theorem follows from the arbitrariness of $\lambda > \lambda_a$. ∎

In Theorem 5.19, we assumed that $\lambda_a < \lambda_{max}$; however, there is an abundance of examples (see below) when this condition is violated. In this case, we should replace Theorem 5.19 by the result below, which we ask the reader to prove in Exercise 14.

Theorem 5.20   (Large Deviations II)   *Suppose that*

$$\lim_{\lambda\to\lambda_{max}} \kappa'(\lambda) = x_1 \leq a.$$

Then

$$\lim_{n\to\infty} \frac{1}{n} \log \Pr\{S_n \geq na\} = -(a\lambda_{max} - \log M(\lambda_{max}))$$

for $a \geq x_1$.

Example 5.7:   *Large Deviations of Some Known Distributions*   We apply here the large deviations results to some distributions.

1. **Standard Normal Distribution:** In this case it is easy to check that

$$M(\lambda) = \frac{1}{\sqrt{2\pi}} \int_{-\infty}^{\infty} e^{\lambda y} e^{-\frac{1}{2}y^2} dy = e^{\frac{1}{2}\lambda^2}$$

by completing the square in the exponent. Then $\kappa(\lambda) = \frac{1}{2}\lambda^2$, $\kappa'(\lambda) = \lambda$, and hence $\lambda_a = a$, which implies $I(a) = \frac{1}{2}a^2$ so that for $a > 0$

$$\log \Pr\{S_n \geq an\} \sim -\frac{n}{2}a^2.$$

2. **Poisson Distribution:** Consider the Poisson distribution with mean θ. Then easy calculations show

$$M(\lambda) = \exp\left(\theta(e^\lambda - 1)\right),$$

and $\lambda_a = \log(a/\theta)$ so that

$$I(a) = a\left(\log(a/\theta) - 1\right) + \theta.$$

Theorem 5.19 implies for $a > \theta$

$$\Pr\{S_n \geq an\} = \left(\frac{a}{\theta}\right)^{-na} e^{-n(\theta - a) + o(n)}.$$

3. **Perverted Exponential Distribution:** Define the density f as follows: it is zero for $x < 1$ and for $x \geq 1$ we set $f(x) = Cx^{-3}e^{-x}$ with C chosen so that f is a probability density (cf. [117]). Observe that $M(\lambda)$ can exist only for $\lambda \leq 1$ and

$$\kappa'(\lambda) \leq \kappa'(1) = 2.$$

Thus $\lambda_{max} = 1$, $M(\lambda_{max}) = C/2$, and $\kappa'(\lambda) = a$ does not have a solution for $a > 2$. In other words, as $\lambda \to \lambda_{max}$ we only have $\kappa'(\lambda) \to 2 = x_1$. By Theorem 5.20 for $a > 2$ we have

$$n^{-1} \log \Pr\{S_n \geq an\} \to -((a - \log(C/2))$$

as $n \to \infty$. ∎

A major strengthening of the large deviation theorem is due to Gärtner [165] and Ellis [111], who extended it to weakly dependent random variables. Let us consider S_n as a sequence of random variables (e.g., $S_n = X_1 + \ldots + X_n$), and let $M_n(\lambda) = \mathbf{E}[e^{\lambda S_n}]$. The following is quite useful. Its proof can be found in Dembo and Zeitouni [90].

Theorem 5.21 **(Gärtner, 1977; Ellis, 1984)**   *For dependent random variable* $X_1, \ldots, X_n$, *let for real* λ

$$\lim_{n \to \infty} \frac{\log M_n(\lambda)}{n} = c(\lambda)$$

exist and it is finite in a subinterval of the real axis. If there exists λ_a *such that* $c'(\lambda_a)$ *is finite and* $c'(\lambda_a) = a$, *then*

$$\lim_{n \to \infty} \frac{1}{n} \log \Pr\{S_n \geq na\} = -(a\lambda_a - c(\lambda_a)).$$

We return to large deviations of weakly dependent random variables in Chapter 8 where we analyze the *quasi-power* case, for which $M_n(s) \sim a(s)c^n(s)$ where $a(s)$ and $c(s)$ are functions of complex s.

5.5   APPLICATIONS

This is a long chapter and we need a few good applications of the subadditive ergodic theorem, Azuma's inequality, martingales, Chernoff's bound, and large deviations to illustrate these important techniques. We hope this section presents some interesting problems. We shall continue using these techniques throughout the book.

Hereafter, we discuss three applications. First, we deal with the edit distance problem already described in Section 1.5 (cf. [296]). Then, we turn our attention to the Knuth-Morris-Pratt pattern matching algorithm and prove that its complexity grows linearly with the size of the text (cf. [357]). For this application, unlike other examples discussed so far, the hardest part is to establish the subadditive property. Finally, we investigate an interesting problem on a random sequence that arose in the analysis of hashing with lazy deletions (cf. [6]).

5.5.1   Edit Distance

We consider the edit distance problem presented in Section 1.5 and in Example 5.1 of this chapter. We want to estimate the minimum cost $C_{\min}$ or the maximum cost $C_{\max}$ of transforming Y_1^ℓ of length ℓ into X_1^s of length s. As mentioned before, this problem can be reduced to finding the longest (shortest) path in a special grid graph (see Figure 1.8). Let us assume that Y_1^ℓ and X_1^s are generated by a memoryless source and the weights of insertion W_I, deletion W_D and substitution W_Q are independent random variables. We further postulate that $\ell = \beta s$ for some $\beta > 0$; the reader may assume for simplicity that $\ell = s$; however, we express our results in terms of $n = \ell + s$. We start with the following result

$$\lim_{n \to \infty} \frac{C_{\max}}{n} = \lim_{n \to \infty} \frac{\mathbf{E}[C_{\max}]}{n} = \alpha \quad \text{(a.s.)}$$

where $\alpha > 0$ is a constant. Indeed, let us consider the $\ell \times s$ grid of Figure 1.8 with the starting point O and the ending point E. Call it Grid(O,E). We also choose an arbitrary point, say A, inside the grid so that we can consider two grids, Grid(O,A) and Grid(A,E). Actually, point A splits the edit distance problem into two subproblems with objective functions $C_{\max}(O, A)$ and $C_{\max}(A, E)$. Clearly, $C_{\max}(O, E) \geq C_{\max}(O, A) + C_{\max}(A, E)$. Thus under our assumption regarding the weights, the objective function $C_{\max}$ is superadditive, and a direct application of Theorem 5.8 proves the result.

As expected, the above calculation does not indicate how to compute α. From Example 5.1, we know that even in the simplest case of the longest common superstring problem, the constant α is unknown. But we can prove that $C_{\max}$ is concentrated around αn.

Theorem 5.22 (Louchard and Szpankowski, 1995) *Consider the (maximal) edit distance cost C_{max}.*

(i) *If all weights are bounded random variables such that* $\max\{W_I, W_D, W_Q\} \leq 1$, *then for arbitrary $\varepsilon > 0$ and large n*

$$\Pr\{|C_{max} - \mathbf{E}[C_{max}]| > \varepsilon n\alpha]\} \leq 2\exp(-\frac{1}{2}\varepsilon^2\alpha^2 n). \tag{5.43}$$

(ii) *If the weights are unbounded but such that $W_{max} = \max\{W_I, W_D, W_Q\}$ satisfies the following:*

$$n\Pr\{W_{max} \geq n^{1/2-\delta}\} \leq U(n) \tag{5.44}$$

for some $\delta > 0$ and a function $U(n) \to 0$ as $n \to \infty$, then

$$\Pr\{|C_{max} - \mathbf{E}[C_{max}]| > \varepsilon n\alpha]\} \leq 2\exp(-\beta n^\delta) + U(n) \tag{5.45}$$

for any $\varepsilon > 0$ and some $\beta > 0$.

Proof. Part (i) is a direct consequence of Azuma's inequality (cf. Theorem 5.15). We first define $Z_1^n = Y_1^\ell \cdot X_1^s$ as a concatenation of Y_1^ℓ and X_1^s of length $n = \ell + s$. Observe that $Z_1, \ldots, Z_n$ are independently distributed. Moreover,

$$|C_{max}(Z_1, \ldots, Z_i, \ldots, Z_n) - C_{max}(Z_1, \ldots, \hat{Z}_i, \ldots, Z_n)|$$
$$\leq \max_{1 \leq i \leq n}\{W_{max}(i)\} \leq 1, \tag{5.46}$$

where $W_{max}(i)$ is the ith independent version of W_{max} defined in the theorem. An application of Corollary 5.16 proves (5.43).

To prove part (ii), we proceed as follows for some c:

$$\Pr\{|C_{max} - \mathbf{E}[C_{max}]| \geq t\} = \Pr\{|C_{max} - \mathbf{E}[C_{max}]| \geq t, \ \max_{1 \leq i \leq n}\{W_{max}(i)\} \leq c\}$$

$$+ \Pr\{|C_{max} - \mathbf{E}[C_{max}]| \geq t, \ \max_{1 \leq i \leq n}\{W_{max}(i)\} > c\}$$

$$\leq 2\exp(-\frac{1}{2}t^2/nc^2) + n\Pr\{W_{max} > c\}.$$

Set now $t = \varepsilon n\alpha$ and $c = O(n^{1/2-\delta})$, then

$$\Pr\{|C_{max} - \mathbf{E}[C_{max}]| \geq \varepsilon n\alpha\} \leq 2\exp(-\beta n^\delta) + n\Pr\{W_{max} > n^{1/2-\delta}\},$$

for some constant $\beta > 0$. This implies (5.45) provided (5.44) holds. ∎

In passing, we add that the probabilistic analysis of the edit problem was initiated by Chvatal and Sankoff [67], who analyzed the longest common subsequence problem. After an initial success in obtaining some probabilistic results for this problem and its extensions by a rather straightforward application of the subadditive ergodic theorem, a deadlock was reached due to a strong statistical dependency. There is not much literature on the probabilistic analysis of the string edit problem and its variations (cf. [80]) with the notable recent exception of Arratia and Waterman [23], who proved their own conjecture concerning the phase transition in a sequence matching. The interested reader is advised to study this excellent paper.

5.5.2   Knuth-Morris-Pratt Pattern Matching Algorithms

Here, we shall analyze the number of comparisons C_n performed by a Knuth-Morris-Pratt like algorithm (described in detail in Section 1.3) when it searches for occurrences of a *given* pattern H of length m in a *random* text T of size n. In the derivation we shall use the subadditive ergodic Theorem 5.8. Unlike all other examples in this section, the main difficulty here lies in establishing the subadditivity property for C_n. We follow here the presentation of Régnier and Szpankowski [357], but we also discuss some new results.

We recall that Knuth-Morris-Pratt (KMP) algorithms scan the text from left to right. The main idea can be described as follows: Let (when searching for pattern occurrences) the algorithm find a mismatch at, say, position l of the text and position k of the pattern. Then the next comparison is made at position l of the text and position i of the pattern such that i is the largest integer satisfying

$$H_1^{i-1} = H_{k-i}^{k-1},$$

that is, the largest suffix of the already searched pattern that is a prefix of the pattern. The pattern is shifted by $k - i$ positions to be aligned at $l - i + 1$ position of the text (cf. Section 1.3). We shall write $M(l, k) = 1$ if the the algorithm makes a comparison at position l of the text with position k of the pattern; otherwise $M(l, k) = 0$.

Our goal is to prove the following result.

Theorem 5.23   (Régnier and Szpankowski, 1995)   *Consider the Knuth-Morris-Pratt algorithm for pattern matching. Let H be a given pattern of length m.*

(i) *If the text T_1^n is given, then*

$$\lim_{n\to\infty} \frac{\max_T C_n}{n} = \alpha_1(H),$$
(5.47)

where $\alpha_1(H) \geq 1$ is a constant.

(ii) *If the text is generated by a stationary and ergodic source, then*

$$\lim_{n\to\infty} \frac{C_n}{n} = \alpha_2(H) \quad \text{a.s.}$$

$$\lim_{n\to\infty} \frac{E[C_n]}{n} = \alpha_2(H),$$

where $\alpha_2(H) \geq 1$ is a constant.

We prove this theorem by showing that C_n is subadditive. For this, we denote by $C_{r,m}$ the number of comparisons performed on T_r^m ($r \leq m$). Actually, we prove a stronger result, namely, that the subadditivity holds for the ℓ-*convergent* pattern matching algorithms that are defined below together with some other useful notations.

Definition 5.24 *Consider a string matching algorithm.*

(i) *A position AP in the text T satisfying for some k ($1 \leq k \leq m$)*

$$M[AP + (k-1), k] = 1$$

*is said to be an **alignment position** (i.e., the first position of the pattern is aligned with position AP of the text).*

(ii) *For a given pattern H, a position i in the text T_1^n is an **unavoidable alignment position** for an algorithm if for any r, l such that $r \leq i$ and $l \geq i+m$, the position i is an alignment position when the algorithm is run on T_r^l.*

(iii) *An algorithm is said to be ℓ-**convergent** if, for any text T and pattern H, there exists an increasing sequence $\{U_i\}_{i=1}^n$ of unavoidable alignment positions satisfying $U_{i+1} - U_i \leq \ell$ where $U_0 = 0$ and $n - \max_i U_i \leq \ell$.*

Our main effort will concentrate on proving the following lemma.

Lemma 5.25 *An ℓ-convergent pattern matching algorithm satisfies the following inequality for all r such that $1 \leq r \leq n$:*

$$|C_{1,n} - (C_{1,r} + C_{r,n})| \leq m^2 + \ell m ,$$
(5.48)

provided any comparison is done only once.

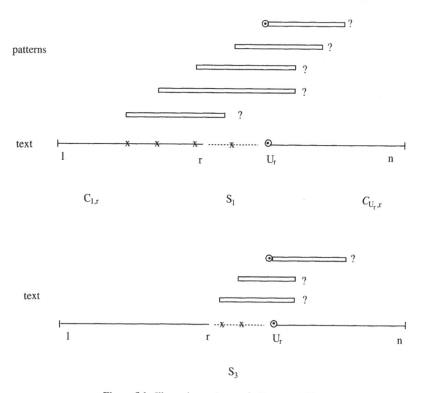

Figure 5.1. Illustration to the proof of Lemma 5.25

Proof. Let U_r be the smallest unavoidable position greater than r. We evaluate in turn $C_{1,n} - (C_{1,r} + C_{U_r,n})$ and $C_{r,n} - C_{U_r,n}$ (cf. Figure 5.1). We start our analysis by considering $C_{1,n} - (C_{1,r} + C_{U_r,n})$. This part involves the following contributions:

- Those comparisons that are performed after position r but with alignment positions before r. We call this contribution S_1. Observe that those comparisons contribute to $C_{1,n}$ but not to $C_{1,r}$. To avoid counting the last character r twice, we must subtract one comparison. Thus

$$S_1 = \sum_{AP < r} \sum_{i \geq r} M(i, i - AP + 1) - 1.$$

- The next contribution, called S_2, accounts for alignments AP satisfying $r \leq AP \leq U_r$ that only contribute to $C_{1,n}$, that is,

$$S_2 = \sum_{AP=r}^{U_r-1} \sum_{i \leq m} M(AP + (i-1), i).$$

- Finally, since the alignment positions after U_r on the text $T_{U_r}^n$ and T_1^n are the same, the only difference in contribution may come from the amount of information saved from previous comparisons done on T_1^r. This is clearly bound by

$$|C_{1,n} - (C_{1,r} + C_{U_r,n} + S_1 + S_2)| \leq m.$$

Now, we evaluate $C_{r,n} - C_{U_r,n}$ (see second part of Figure 5.1). We assume that the algorithm runs on T_r^n and let AP be any alignment position satisfying $r \leq AP < U_r$. The following contributions must be considered:

- The contribution S_3

$$S_3 = \sum_{AP=r}^{U_r-1} \sum_i M(AP + (i-1), i)$$

counts for the number of comparisons associated positions $r \leq AP < U_r$. This sum is the same as S_2 but the associated alignment positions and searched text positions $AP + k - 1$ may be different.

- Additional contribution may come from the alignment at position U_r. But no more than m comparisons can be saved from previous comparisons, hence

$$|C_{r,n} - C_{U_r,n} - S_3| \leq m.$$

To complete the proof, it remains to find upper bounds on S_1, S_2, and S_3. For $\ell \geq U_r - r$ we easily see that S_2 and S_3 are smaller than ℓm. So are their differences. With respect to S_1, for a given alignment position AP, we have $|i - AP| \leq m$; hence $|r - AP| \leq m$, and for any AP the index i has at most m different values. Thus $S_1 \leq m^2$, and this proves the lemma. ∎

To prove Theorem 5.23, it is enough to show that the KMP algorithm is an $\ell = m$-convergent algorithm. Before we establish this fact, we prove an interesting property of KMP-like algorithms.

Lemma 5.26 *Given a pattern H and a text T, KMP-like algorithms have the same set of unavoidable alignment positions $\mathcal{U} = \bigcup_{l=1}^n \{U_l\}$, where*

$$U_l = \min\{ \min_{1 \le k \le l} \{T_k^l = H_1^{l-k+1}\}, \ l+1 \}.$$

Proof. Let l be a text position such that $1 \le l \le n$, and r be any text position satisfying $r \le U_l$. Let $\{A_i\}$ be the set of alignment positions defined by a KMP-like algorithm that runs on T_r^n. As it contains r, we may define (see Figure 5.2)

$$A_J = \max\{A_i : \ A_i < U_l\}.$$

Hence, we have $A_{J+1} \ge U_l$. Using an adversary argument, we shall prove that $A_{J+1} > U_l$ cannot be true, thus showing that $A_{J+1} = U_l$. Let $y = \max\{k : M(A_J + (k-1), k) = 1\}$, that is, y the rightmost position where we can do a comparison starting from A_J. We observe that we have $y \le l$. Otherwise, according to the KMP algorithm rule, T_{A_J} would be a prefix of H, which contradicts the definition of U_l. Also, since KMP-like algorithms are sequential (i.e., the text-pattern comparisons define nondecreasing sequences), then $A_{J+1} \le y+1 \le l+1$. Hence $U_l = l+1$ contradicts the assumption $A_{J+1} > U_l$ and we may assume $U_l \le l$. In that case, $H_{U_l}^l$ is a prefix of H, and an occurrence of H at position U_l is consistent with the available information. Let the adversary assume that H does occur. Since a sequence $\{A_i\}$ is nondecreasing and A_{J+1} is greater than U_l, this occurrence will not be detected by the algorithm. This leads to the desired contradiction. Thus $A_{J+1} = U_l$, as needed. ∎

To complete the proof of Theorem 5.23, it is necessary to show that KMP is m-convergent. But this is easy. Let AP be an alignment position and define $l = AP + m$. As $|H| = m$, one has $l - (m-1) \le U_l \le l$. Hence, $U_l - AP \le m$ which establishes the m-convergence of the KMP algorithm, and the main Theorem 5.23.

As in other applications, since we obtained the result through the subadditive ergodic theorem, the constants α_1 and α_2 are unknown. In [72] some bounds for α_1 were found. But even if we cannot find the value of α_2, we still are able to show that C_n is concentrated around $\alpha_2 n$. As expected, we shall use the Azuma inequality. Let us now assume that T is generated by a memoryless source. It suffices to observe that C_n, as a function of random text $T = T_1, \ldots, T_n$, satisfies

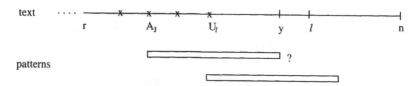

Figure 5.2. Illustration to the proof of Lemma 5.26.

$$|C_n(T_1, \ldots, T_i, \ldots, T_n) - C_n(T_1, \ldots, \hat{T}_i, \ldots, T_n)| \leq 2m^2$$

where $\hat{T}_i$ is an independent copy of the ith symbol T_i of the text. In view of this and Theorem 5.15 we derive the following.

Theorem 5.27 *Let T be generated by a memoryless source while the pattern H of length m is given. The number of comparisons C_n made by the Knuth-Morris-Pratt algorithm is concentrated around its mean $\mathbf{E}[C_n] = \alpha_2 n(1 + o(1))$. More precisely,*

$$\Pr\{|C_n - \mathbf{E}[C_n]| \geq \varepsilon n\alpha_2\} \leq 2\exp\left(-\frac{\varepsilon^2 \alpha_2^2}{8} \frac{n}{m^4}\right)$$

for any $\varepsilon > 0$.

5.5.3   Large Deviations of a Random Sequence

We consider here a large deviations result proved in Aldous, Hofri, and Sz-pankowski [6], who studied hashing with lazy deletion. We reformulate the problem as a coding problem.

Let $\{Y_i\}_{i=1}^{\infty}$ be a sequence of independent random variables with uniform distribution over $\mathcal{A} = \{1, \ldots, V\}$. Let also $f : \{1, 2, 3, \ldots\} \rightarrow \{1, 2, 3, \ldots\}$ be a $1 - 1$ function such that $f(i) > i$ for all $i \geq 1$. Our goal is to estimate the probability that

$$Y_{f(i)} \quad \text{is different than all of} \quad Y_1, \ldots, Y_i.$$

Let A_i be the event that the above holds, and let $B = \sum_{i=1}^{\infty} I(A_i)$. We want to estimate $\Pr\{B \geq b\}$ for large b.

A direct application of Chernoff's bound leads to

$$\Pr\{B \geq b\} \leq \min_{z>1}\left\{z^{-b}\mathbf{E}[z^B]\right\},$$

where we set $z = e^{\lambda}$ in Theorem 5.18. The problem is that we do not know how to assess the generating function of B since $I(A_i)$ are dependent. To overcome this difficulty, we adopt a different approach based on martingales. Let the events A_i be adapted to the increasing σ-fields $\mathcal{F}_i$ (i.e., $\mathcal{F}_i$ contains all the information about $A_1, \ldots, A_i$). We first prove that for some $w \geq 1$

$$M_n = \frac{w^{\sum_{i=1}^{n} I(A_i)}}{\prod_{i=1}^{n} \mathbf{E}[w^{I(A_i)}|\mathcal{F}_{i-1}]}$$

is a positive martingale. Indeed,

$$E[M_{n+1}|\mathcal{F}_n] = \frac{E[w^{\sum_{i=1}^{n} I(A_i)} w^{I(A_{n+1})}|\mathcal{F}_n]}{\prod_{i=1}^{n} E[w^{I(A_i)}|\mathcal{F}_{i-1}] E[w^{I(A_{n+1})}|\mathcal{F}_n]}$$

$$= \frac{w^{\sum_{i=1}^{n} I(A_i)}}{\prod_{i=1}^{n} E[w^{I(A_i)}|\mathcal{F}_{i-1}]} \cdot \frac{E[w^{I(A_{n+1})}|\mathcal{F}_n]}{E[w^{I(A_{n+1})}|\mathcal{F}_n]}$$

$$= M_n.$$

Observe that the above martingale is the "multiplicative" analogue of the additive martingale we have been working so far. But by the *martingale convergence theorem* (Theorem 5.12) M_n converges almost surely and in mean to the random variable

$$M_\infty = \frac{w^B}{D_w},$$

where

$$D_w = \prod_{i \geq 1} E[w^{I(A_i)}|\mathcal{F}_{i-1}] = \prod_{i \geq 1}(1 + (w - 1)\Pr\{A_i|\mathcal{F}_{i-1}\}).$$

In addition, we know that $E[M_\infty] \leq E[M_0] = 1$, hence

$$E[w^B/D_w] \leq 1. \tag{5.49}$$

Now we proceed as follows. Instead of using Chernoff's bound directly, we write for some $a > 0$

$$\Pr\{B \geq b\} \leq \Pr\{D_w > a\} + \Pr\{B \geq b,\ D_w \leq a\}$$

$$\leq E[D_w/a] + \Pr\{w^B/D_w \geq w^b/a\}$$

$$\leq E[D_w/a] + a/w^b$$

$$= 2\sqrt{w^{-b}E[D_w]}, \qquad a = \sqrt{w^b E[D_w]},$$

where the second inequality follows from Markov's inequality, and the third one is a consequence of (5.49). Since $w \geq 1$ is arbitrary, we finally arrive at

$$\Pr\{B \geq b\} \leq 2\sqrt{\inf_{w>1} w^{-b}E[D_w]}. \tag{5.50}$$

We now evaluate the required infimum in (5.50). First of all, observe that

$$\Pr\{A_i|\mathcal{F}_{i-1}\} = L_i/V,$$

where L_i is the number of values *not* taken by $Y_1, \ldots, Y_i$. Thus, for $w > 1$ the quantity D_w is

$$D_w = \prod_{i \geq 1} (1 + (w - 1)L_i/V) \leq \exp\left(\frac{w - 1}{V} \sum_{i \geq 1} L_i\right). \qquad (5.51)$$

But

$$\sum_{i \geq 1} L_i = \sum_{k=1}^{V} k W_k, \qquad (5.52)$$

where W_k is the waiting time for the process L_i to go from k to $k - 1$ (e.g., $L_0 = V$, $L_1 = V - 1$). A little thought reveals that the random variables W_k are independent with (different) geometric distributions

$$\Pr\{W_k = \ell\} = \left(1 - \frac{k}{V}\right)^{\ell-1} \frac{k}{V}, \quad \ell \geq 1.$$

(This is just the elementary argument for the classical coupon collector's problem with equally likely coupons.) Hence, the associated generating function $\mathbf{E}[z^{W_k}]$ may be written as

$$\mathbf{E}[z^{W_k}] = \frac{zk/V}{1 - z(1 - k/V)} = \frac{1}{1 - V/k(1 - z^{-1}))} \qquad (5.53)$$

for $|z| < V/(V - k)$. Combining (5.51)–(5.53) gives

$$\mathbf{E}[D_w] = \mathbf{E}\left[\prod_{i \geq 1}\left(1 + \frac{w - 1}{V}L_i\right)\right]$$

$$\leq \mathbf{E}\left[\exp\left(\frac{w - 1}{V}\sum_{i \geq 1} L_i\right)\right] = \mathbf{E}\left[\exp\left(\frac{w - 1}{V}\sum_{k=1}^{V} k W_k\right)\right]$$

$$= \prod_{k=1}^{V}\mathbf{E}\left[\exp\left(\frac{w - 1}{V} k W_k\right)\right] = \prod_{k=1}^{V}\left(1 - \frac{V}{k}\left(1 - e^{-k(w-1)/V}\right)\right)^{-1}$$

$$\leq \prod_{k=1}^{V}(1 - (w - 1))^{-1} = (2 - w)^{-V},$$

where the last inequality follows from $1 - y^{-1}(1 - e^{-ay}) \geq 1 - a$ for $a, y > 0$. By (5.50) we obtain

$$\Pr\{B \geq b\} \leq 2\sqrt{\inf_{1<w<2} w^{-b}(2-w)^{-V}}.$$

Elementary calculus gives the exact infimum at $w = 2b/(b+V)$. Since throughout the proof we require $w > 1$, our result holds only for $b > V$.

In summary, we prove the following lemma.

Lemma 5.28   *Let* $f : \{1, 2, 3, \ldots\} \rightarrow \{1, 2, 3, \ldots\}$ *be a* $1 - 1$ *function with* $f(i) > i$. *Let* $(Y_i; i \geq 1)$ *be independent, uniform on* $\{1, \ldots, V\}$. *Define* A_i *to be the event*

$$Y_{f(i)} \quad \text{is different than all of} \quad Y_1, \ldots, Y_i.$$

Let B *be the counting random variable* $B = \sum_{i \geq 1} I(A_i)$. *Then*

$$\Pr\{B \geq b\} \leq 2\left(\frac{V+b}{2b}\right)^{b/2}\left(\frac{V+b}{2V}\right)^{V/2}$$

for $b > V$.

5.6   EXTENSIONS AND EXERCISES

5.1   Prove the following generalization of the De Bruijn and Erdős result:

Theorem 5.29   *Let* x_n *and* c_n *be two sequences of real numbers such that*

$$x_{m+n} \leq x_m + x_n + c_n$$

for all $m, n \geq 1$. *Then* $\lim_{n \to \infty} x_n/n$ *exists provided* $\lim_{n \to \infty} c_n/n = 0$, *however, the limit may be greater than* $\inf x_m/m$.

5.2   Prove or disprove the following generalization of the Pólya and Szegő result:

Theorem 5.30   *If a real sequence* x_n *satisfies*

$$x_m + x_n - b(n + m) \leq x_{n+m} \leq x_m + x_n + b(n + m)$$

for all $m, n \geq 1$, *where* $b(n)$ *is a positive sequence, then there is a finite constant* ω *such that*

$$\left|\frac{x_n}{n} - \omega\right| \leq \frac{\beta_n}{n} \to 0,$$

provided that for every $m \geq 1$

$$\beta_m = \sum_{k \geq 1} b(2^k m) 2^{-k} < \infty$$

and $\beta_m = o(m)$.

5.3 Prove the De Bruijn and Erdős Theorem 5.7. Extend all other subadditive theorems so that the subadditivity property holds only in the range $\frac{1}{2}n \leq m \leq 2n$.

5.4 Prove Hoeffding's inequality in Theorem 5.13 using only (5.17) (cf. [402]).

5.5 Prove a refined Theorem 5.13 in which (5.19) is replaced by the optimal

$$\Pr\{|Y_n - Y_0| \geq x\} = \Pr\left\{|\sum_{i=1}^{n} D_i| \geq x\right\} \leq 2 \exp\left(-\frac{2x^2}{\sum_{i=1}^{n} c_i^2}\right).$$

5.6 *Doob-Levy's Martingale.* Let Y have a finite second moment, and let $X_1, \ldots, X_n$ be a sequence of variables. Prove that $Y_n = E[Y|X_1, \ldots, X_n]$ is a martingale with respect to $X_1, \ldots, X_n$.

5.7 Construct a counterexample to the following statement:

$$E[E[Y|\mathcal{F}_1]|\mathcal{F}_2] = E[E[Y|\mathcal{F}_2]|\mathcal{F}_1].$$

Observe that we drop the condition $\mathcal{F}_1 \subset \mathcal{F}_2$ from Lemma 5.14.

5.8 ⚠ **Knapsack Problem**. The objective of the knapsack problem is to pack a knapsack of a finite capacity with objects of maximum wealth. More precisely, let $V_1, \ldots, V_n$ and $W_1, \ldots, W_n$ be independent nonnegative random variables with finite means and $W_i \leq M$ for all i and some fixed M. Assume that the knapsack has finite capacity C and your goal is to pack as many objects as you can in a way that maximizes the wealth of the packed objects. In other words, we search for a zero–one vector $(z_1, z_2, \ldots, z_n)$ such that

$$\sum_{i=1}^{n} z_i V_i \leq C,$$

and

$$Z_{\max} = \max\left\{\sum_{i=1}^{n} z_i W_i\right\}$$

is maximized. Prove that

$$\Pr\{|Z_{\max} - \mathbf{E}[Z_{\max}]| \geq x\} \leq 2\exp\left(-\frac{x^2}{2nM^2}\right)$$

for $x > 0$.

5.9 ⚠ Prove the following generalization of the Efron-Stein inequality due to Steele [401]: Let $S(x_1, \ldots, x_n)$ be any function of n variables, and X_i $\hat{X}_i$ for $1 \leq i \leq n$ be $2n$ i.i.d. random variables. Define $\hat{S}_i = S(X_1, \ldots, X_{i-1}, \hat{X}_i, X_{i+1}, \ldots, X_n)$. Then

$$\mathbf{Var}[S(X_1, \ldots, X_n)] \leq \frac{1}{2}\mathbf{E}\left[\sum_{i=1}^{n}(S(X_1, \ldots, X_n) - \hat{S}_i)^2\right].$$

5.10 Prove that the following are equivalent: (i) $I(a) = \infty$, (ii) $\Pr\{X_1 \geq a\} = 0$, and (iii) $\Pr\{S_n \geq na\} = 0$ for all n.

5.11 Prove the following large deviations result due to Bernstein: *Let $X_1, \ldots, X_n$ be i.i.d. with mean μ and variance σ^2, and suppose that $|X_i - \mathbf{E}[X_i]| \leq M < \infty$. Then for all $x > 0$*

$$\Pr\{S_n \geq n\mu + x\sqrt{n}\} \leq \exp\left(-\frac{x^2}{2\sigma^2 + \frac{2Mx}{3\sqrt{n}}}\right),$$

where, as usual, $S_n = X_1 + \cdots + X_n$.

5.12 Prove that the cumulant function $\kappa(\lambda) = \log\mathbf{E}[e^{\lambda X}]$, if it exists, is a convex function.

Hint: Use the Hölder inequality.

5.13 Prove that the Fenchel-Legendre transform $I(a)$ attains its minimum at $a = \mathbf{E}[X] = \mu$ and $I(\mu) = 0$.

5.14 Prove Theorem 5.20.

5.15 (Feller, 1971) Let $F_n(x) = \Pr\{S_n \leq x\}$, where $S_n = X_1 + \cdots X_n$ and $X_1, \ldots X_n$ are i.i.d. Prove the following large deviations result for $x = o(\sqrt{n})$

$$\frac{1 - F_n(x)}{1 - \Phi(x)} = \left(1 + O\left(\frac{x}{\sqrt{n}}\right)\right)\exp\left(\frac{\mu_3}{6\sigma^3}\frac{x^3}{\sqrt{n}}\right),$$

where $\Phi(x)$ is the standard normal distribution function.

5.16 ⚠ (Biggins, 1976) Consider the age-dependent branching process from Example 5.3. Let t_i be the life-time for the ith individual, and let $T_n = t_1 + \cdots + t_n$. We recall that each individual generates k offspring with probability p_k when dying. Let $\mu = \sum_{k \geq 0} k p_k$. From large deviations theory, we know that for $a < \mathbf{E}[t_i]$

$$\lim_{n \to \infty} \frac{1}{n} \log \Pr\{T_n \leq an\} = -I(a),$$

where $I(a) = a\lambda_a - \log \mathbf{E}[e^{\lambda_a t_1}]$ and λ_a is a solution of $a\mathbf{E}[e^{\lambda t_1}] = \frac{d}{d\lambda}\mathbf{E}[e^{\lambda t_1}]$. In Example 5.3 we prove that the first birth of the nth generation, B_n, grows linearly with the constant α. Show that α can be computed as follows:

$$\alpha = \inf\{a : \log\mu - I(a) > 0\}.$$

Hint: Follow Biggins' footsteps from [46] by first considering a new random variable $Z_n(an)$ that denotes the number of individuals in generation n born by time an. Show:

(i) $\mathbf{E}[Z_n(an)] = \mu^n \Pr\{T_n \leq an\}$;

(ii) $\Pr\{Z_n(an) > 0\} > 0$ if $\mathbf{E}[Z_n(an)] > 1$, hence $B_n \leq an$ and $\alpha \leq a$.

5.17 ⚠ (Szpankowski, 1995) Consider the general combinatorial optimization problem described in Section 1.6. To recall, we define $Z_{\max}$ and $Z_{\min}$ as

$$Z_{\max}(\mathcal{S}_n) = \max_{\alpha \in \mathcal{B}_n}\left\{\sum_{i \in \mathcal{S}_n(\alpha)} w_i(\alpha)\right\}, \quad Z_{\min}(\mathcal{S}_n) = \min_{\alpha \in \mathcal{B}_n}\left\{\sum_{i \in \mathcal{S}_n(\alpha)} w_i(\alpha)\right\},$$

where $\mathcal{B}_n$ is a set of all feasible solutions, $\mathcal{S}_n(\alpha)$ is a set of objects from $\mathcal{S}_n$ belonging to the αth feasible solution, and $w_i(\alpha)$ is the weight assigned to the ith object in the αth feasible solution. Let us adopt the following assumptions:

(A) The cardinality $|\mathcal{B}_n|$ of $\mathcal{B}_n$ is fixed and equal to m. The cardinality $|\mathcal{S}_n(\alpha)|$ of the set $\mathcal{S}_n(\alpha)$ does *not* depend on $\alpha \in \mathcal{B}_n$ and for all α it is equal to N, i.e., $|\mathcal{S}_n(\alpha)| = N$.

(B) For all $\alpha \in \mathcal{B}_n$ and $i \in \mathcal{S}_n(\alpha)$ the weights $w_i(\alpha)$ are identically and independently distributed (i.i.d.) random variables with common distribution function $F(\cdot)$, and the mean value μ, the variance σ^2, and the third moment μ_3 are finite.

Prove the following result: *Under assumptions (A) and (B), as $N, m \to$ ∞ with $n \to \infty$*

$$Z_{\min} = N\mu - o(N) \qquad (pr.) \qquad Z_{\max} = N\mu + o(N),$$

provided $\log m = o(N)$.

6

Elements of Information Theory

Entropy and *mutual information* were introduced by Shannon, and this began a remarkable development of *information theory*. Over the last 50 years information theory underwent many changes and found new applications. The Shannon-McMillan-Breiman theorem and random coding technique nowadays are standard tools of the analysis of algorithms. In this chapter, we discuss elements of information theory and illustrate its applications to the analysis of algorithms. In particular, we prove three main results of Shannon, those of source coding, channel coding, and rate distortion. In the applications section, we discuss a variety of data compression schemes based on exact and approximate pattern matching.

Entropy of a random variable X with probability mass $P(a) = \Pr\{X = a\}$ is defined as

$$h(X) = -\sum_{a \in \mathcal{A}} P(a) \log P(a),$$

where by convention $0 \log 0 = 0$. We intentionally left the base of the logarithm unspecified. In this chapter it is convenient to use the logarithm to base 2. If the base of the logarithm is 2, we measure entropy in *bits*; with the natural logarithm we measure entropy in *nats*. As we shall see, entropy represents average uncertainty of X, and in fact it is the number of bits that on the average one requires to describe a random variable (see Exercise 1). For example, one needs 5 bits to identify a value of a random variable distributed uniformly over 32 outcomes, and entropy of X is $h(X) = 5$ bits.

The real virtue of entropy is best manifested when investigating entropy rates of stochastic processes. The **Asymptotic Equipartition Property** (AEP) that follows from the most celebrated result of information theory, that of the **Shannon-McMillan-Breiman** theorem, is the key result.

Let us start with a simple observation: Consider an i.i.d. binary sequence of symbols of length n, say $(X_1, \ldots, X_n)$, with p being the probability of generating one symbol and $q = 1 - p$ the probability of the other symbol. When $p = q = 1/2$, then $\Pr\{X_1, \ldots, X_n\} = 2^{-n}$ and it does not matter what are the actual values of $X_1, \ldots, X_n$. In general, $\Pr\{X_1, \ldots, X_n\}$ is not the same for all possible values of $X_1, \ldots, X_n$, however, we shall see that all **typical** sequences $(X_1, \ldots, X_n)$ have "asymptotically" the same probability. Consider $p \neq q$ in the example above. In a typical binary sequence of length n there are approximately $np + O(\sqrt{n})$ symbols of one kind and $nq + O(\sqrt{n})$ symbols of the second kind. We expect that the probability of such a sequence is well approximated by

$$p^{np+O(\sqrt{n})} q^{nq+O(\sqrt{n})} = 2^{-n(-p \log p - q \log q) + O(\sqrt{n})} \sim 2^{-nh},$$

where $h = -p \log p - q \log q$ is the *entropy* of the underlying Bernoulli random variable. We conclude that typical sequences have approximately the same probability equal to 2^{-nh}. We make this statement rigorous in this chapter.

Information theory was born in October 1948 when Shannon published his classic paper "A Mathematical Theory of Communication." He established three main results of information theory: source coding, channel coding, and rate distortion. In source coding one wants to find the shortest description of a message. It turns out that you cannot beat a certain limit, which is equal to the entropy times the length of the original message. In channel coding, the goal is the opposite to source coding, namely, we want to *reliably* send as much information as possible. Finally, there are sources of information that cannot be described by a finite number of bits (e.g., an arbitrary real number, a continuous function, etc.). In such cases, we can only *approximately* describe the source of information. How well we can do? To answer this question we enter the realm of *rate distortion theory*. We touch in this chapter some of these problems. The reader is referred to the excellent books of Cover and Thomas [75] and Csiszár and Körner [78] for further information. In fact, a large part of this chapter is based on [75].

In the analysis of algorithms, which is the main topic of this book, entropy appears quite often in disparate problems as we shall see in the applications section of this chapter. Actually, in the analysis of algorithms the Rényi's entropy occurs almost as frequently as the Shannon entropy. We study some properties of Rényi's entropy and illustrate its applications for problems on strings. In particular, we study the phrase length in the Lempel-Ziv scheme and its lossy extension, the shortest common superstring problem, and the bit rate of a lossy extension of the fixed database Lempel-Ziv scheme.

6.1 ENTROPY, RELATIVE ENTROPY, AND MUTUAL INFORMATION

In this section we present basic definitions of entropy, relative entropy, Rényi's entropy, and mutual information for single random variables. We shall follow the presentation of Cover and Thomas [75].

Let us consider a random variable X defined over a finite alphabet $\mathcal{A}$. We write $P(X)$ for a random function of X such that $P(a) = \Pr\{X = a\}$ for every $a \in \mathcal{A}$. The **Shannon entropy** is defined as

$$h(X) = -\mathbf{E}[\log P(X)] = -\sum_{a \in \mathcal{A}} P(a) \log P(a), \qquad (6.1)$$

where throughout this chapter we assume that the logarithm base is 2 and $0 \log 0 = 0$. We sometimes write $h(P)$ or simply h for the entropy. In some situations, we specifically show that the average in (6.1) is taken over the measure P, that is, we write $h(X) = \mathbf{E}_P[-\log P(X)]$.

As mentioned in the introduction, the entropy is a measure of uncertainty of X. It is also the minimum average number of bits required to describe X, as illustrated in the following example.

Example 6.1: *Entropy and Random Variable Representation*   Let X be a random variable taking eight values with the corresponding probabilities $(\frac{1}{2}, \frac{1}{4}, \frac{1}{8}, \frac{1}{16}, \frac{1}{64}, \frac{1}{64}, \frac{1}{64}, \frac{1}{64})$. The entropy of X is

$$h(X) = 2 \text{ bits.}$$

Suppose we want to relay to another person the outcome of X. We can use 3 bits to encode 8 messages; however, we can do much better by encoding more likely outcomes with shorter codes and less likely with longer. For example, if we assign the following codes to eight messages: 0, 10, 110, 1110, 111100, 111101, 111110, 111111, then we need on the average only 2 bits. Notice that the description length in this case is equal to the entropy. In Exercise 1 the reader is asked to extend it to general discrete random variables.   ∎

The entropy $h(X)$ leads to several new definitions that we discuss in the sequel.

Definition 6.1 *Let X and Y be random variables.*

(i) *Joint Entropy $h(X, Y)$ is defined as*

$$h(X, Y) = -\mathbf{E}[\log P(X, Y)] = -\sum_{a \in \mathcal{A}} \sum_{a' \in \mathcal{A}'} P(a, a') \log P(a, a'). \quad (6.2)$$

(ii) *Conditional Entropy $h(Y|X)$ is*

$$h(Y|X) = -\mathbf{E}_{P(X,Y)}[\log P(Y|X)] = \sum_{a \in \mathcal{A}} P(a) h(Y| X = a)$$

$$= -\sum_{a \in \mathcal{A}} \sum_{a' \in \mathcal{A}'} P(a, a') \log P(a'|a). \quad (6.3)$$

(iii) *Relative Entropy or Kullback Leibler Distance between two distributions P and Q defined on the same probability space is*

$$D(P \parallel Q) = \mathbf{E}_P \left[\log \frac{P(X)}{Q(X)} \right] = \sum_{a \in \mathcal{A}} P(a) \log \frac{P(a)}{Q(a)}, \quad (6.4)$$

where by convention $0 \log 0/Q = 0$ and $P \log P/0 = \infty$.

(iv) *Mutual Information of X and Y is the relative entropy between the joint distribution of X and Y and the product distribution $P(X)P(Y)$, that is,*

$$I(X; Y) = \mathbf{E}_{P(X,Y)} \left[\log \frac{P(X, Y)}{P(X)P(Y)} \right]$$

$$= \sum_{a \in \mathcal{A}} \sum_{a' \in \mathcal{A}'} P(a, a') \log \frac{P(a, a')}{P(a)P(a')}. \quad (6.5)$$

(v) *Rényi's Entropy of order b $(-\infty \leq b \leq \infty)$ is defined as*

$$h_b(X) = -\frac{\log \mathbf{E}[P^b(X)]}{b} = -\frac{1}{b} \log \sum_{a \in \mathcal{A}} P^{b+1}(a), \quad (6.6)$$

provided $b \neq 0$, where

$$h_{-\infty} = \min_{i \in \mathcal{A}} \{P(i)\}, \tag{6.7}$$

$$h_{\infty} = \max_{i \in \mathcal{A}} \{P(i)\} \tag{6.8}$$

by Hilbert's inequalities on means (2.13) and (2.14). Observe that $h(X) = \lim_{b \to 0} h_b(X)$.

Before we justify these definitions and provide an interpretation, let us explore some mathematical relationships between them. These properties are easy to prove by elementary algebra, and the reader is asked in Exercise 3 to verify them (see Exercise 4 for more properties of entropy).

Theorem 6.2 *The following holds:*

$$h(X, Y) = h(X) + h(Y|X), \tag{6.9}$$

$$h(X_1, X_2, \ldots, X_n) = \sum_{i=1}^{n} h(X_i | X_{i-1}, \ldots, X_1), \tag{6.10}$$

$$I(X; Y) = h(X) - h(X|Y) = h(Y) - h(Y|X), \tag{6.11}$$

$$I(X; X) = h(X). \tag{6.12}$$

It is intuitively clear that the conditional entropy $h(X|Y)$ should not be bigger than the entropy $h(X)$ since when Y is correlated with X we learn something about X knowing Y; hence its uncertainty decreases. We prove this fact and some others below.

Theorem 6.3 *The following inequalities hold*

$$D(P \parallel Q) \geq 0, \tag{6.13}$$

$$I(X; Y) \geq 0, \tag{6.14}$$

$$h(X) \geq h(X|Y), \tag{6.15}$$

$$h(X_1, \ldots, X_n) \leq \sum_{i=1}^{n} h(X_i), \tag{6.16}$$

with equality in (6.13) if and only if $P(a) = Q(a)$ for all $a \in \mathcal{A}$, equality in (6.14) and (6.15) if and only if X and Y are independent, and equality in (6.16) if and only if $X_1, \ldots, X_n$ are independent.

Proof. Since $\log x \le x - 1$ with equality if and only if $x = 1$, we obtain

$$-D(P \parallel Q) = \mathbf{E}_P \left[\log \frac{Q}{P} \right]$$

$$\le \sum_{a \in \mathcal{A}} Q(a) - \sum_{a \in \mathcal{A}} P(a) = 1 - 1 = 0,$$

which proves (6.13). Then (6.14) follows from $I(X; Y) = D(P(X, Y) \parallel P(X)P(Y))$, and (6.15) is a consequence of the above and (6.11). Finally, (6.16) follows from (6.10) and (6.15). ∎

6.2 ENTROPY RATE AND RÉNYI'S ENTROPY RATES

We first precisely define typical sequences and then prove the Shannon-McMillan-Breiman theorem for memoryless, mixing, and stationary and ergodic sources. In the second part of this section, we introduce Rényi's entropy rates that are widely used in the analysis of algorithms.

6.2.1 The Shannon-McMillan-Breiman Theorem

Let $\{X_k\}_{k=1}^{\infty}$ be a one-sided stationary sequence, and let $P(X_1^n)$ be a random variable equal to $\Pr\{X_1^n = (x_1, \ldots, x_n)\}$ whenever $X_1^n = (x_1, \ldots, x_n) \in \mathcal{A}^n$. We can define the *entropy rate* of X_1^n in two different ways, namely:

$$h := -\lim_{n \to \infty} \frac{\mathbf{E}[\log P(X_1^n)]}{n} = \lim_{n \to \infty} \frac{h(X_1, \ldots, X_n)}{n}, \qquad (6.17)$$

$$h' := \lim_{n \to \infty} h(X_n \mid X_{n-1}, X_{n-2}, \ldots, X_1), \qquad (6.18)$$

provided the limits above exist. We shall prove that for stationary processes the limits exist and are equal.

First of all, observe that the conditional entropy $h(X_n \mid X_{n-1}, X_{n-2}, \ldots, X_1)$ is nonincreasing with n and hence has a limit. Indeed, by (6.15) and stationarity

$$h(X_{n+1} \mid X_n, X_{n-1}, \ldots, X_1) \le h(X_{n+1} \mid X_n, X_{n-1}, \ldots, X_2)$$

$$= h(X_n \mid X_{n-1}, X_{n-2}, \ldots, X_1).$$

Now, we are in a position to prove the following statement.

Theorem 6.4 *For a stationary process, the limits in (6.17) and (6.18) exist and are equal, that is,*

$$h = h'.$$

Proof. By the chain rule (6.10),

$$\frac{h(X_1, \ldots, X_n)}{n} = \frac{1}{n} \sum_{i=1}^{n} h(X_i | X_{i-1}, \ldots, X_1).$$

But, we just proved that the conditional entropies $h(X_i | X_{i-1}, \ldots, X_1)$ tend to the limit h'; hence by the *Cesáro mean* theorem, the right-hand side converges to h'. We must then conclude that the limit of the left-hand side exists, and is equal to $h = h'$. ∎

Example 6.2: *Entropy Rates for Memoryless and Markovian Sources* Let us first consider a memoryless source. Then $P(X_1^n) = P^n(X_1)$ and we have $h(X_1^n) = nE[-\log P(X_1)] = nh(X_1)$.

For a Markovian source, we appeal to the definition (6.18). Let us consider a stationary Markov source of order one. Then

$$h = \lim_{n \to \infty} h(X_n | X_{n-1}, \ldots, X_1) = \lim_{n \to \infty} h(X_n | X_{n-1}) = h(X_2 | X_1).$$

Thus, by definition (6.3), we have

$$h = h(X_2 | X_1) = - \sum_{i,j \in A} \pi_i p_{ij} \log p_{ij}, \tag{6.19}$$

where π_i is the stationary distribution and $P = \{p_{ij}\}_{i,j \in A}$ the transition matrix of the stationary Markov chain. ∎

We are now well prepared to derive the most important theorem of this chapter, namely the Shannon-McMillan-Breiman theorem, which roughly says that the the average operator $E[\cdot]$ can be taken away from the definition of the entropy (6.17) provided the deterministic limit is replaced by almost sure convergence of the random variable $\log P(X_1^n)$. The main consequence of this theorem is the *Asymptotic Equipartition Property* discussed in the next section.

Let us put some rigor into the above rough description. We first show how to derive the Shannon-McMillan-Breiman theorem for memoryless sources and then for mixing sources. Finally, we shall formulate the theorem in its full generality and prove it for stationary ergodic sources using the "sandwich" approach of Algoet and Cover [8]. Consider first a memoryless source. Then

$$-\frac{\log P(X_1^n)}{n} = -\frac{1}{n} \sum_{i=1}^{n} \log P(X_i)$$

$$\to E[-\log P(X_1)] = h \quad \text{(a.s.),}$$

where the last implication follows from the *strong law of large numbers* applied to the sequence $(-\log P(X_1), \ldots, -\log P(X_n))$. One *should* notice a difference between the definition of the entropy (6.17) and the above result. In (6.17) we take the *average* of $\log P(X_1^n)$ while in the above we proved that *almost surely* the probability $P(X_1^n)$ can be well approximated by 2^{-nh}. We are aiming now at showing that the above conclusion is true for much more general sources.

As the next step, let us consider mixing sources that include as a special case Markovian sources (see Chapter 2). In fact, we consider a weak mixing since we only assume that there exists a constant c such that

$$P(X_1^{n+m}) \leq c P(X_1^n) P(X_{n+1}^{n+m})$$

for integers $n, m \geq 0$. Taking logarithm we obtain

$$\log P(X_1^{n+m}) \leq \log P(X_1^n) + \log P(X_{n+1}^{n+m}) + \log c.$$

Thus $\log P(X_1^n) + \log c$ is subadditive, hence by Theorem 5.5 we prove that

$$h = -\lim_{n \to \infty} \frac{\log P(X_1^n)}{n} \quad \text{(a.s.)}.$$

Again, the reader should notice the difference between this result and the definition of the entropy.

We are finally ready to state the Shannon-McMillan-Breiman theorem in its full generality. We prove it using the Markov approximation discussed in Lemma 4.2 (see also Theorem 6.33 in Exercise 11 below).

Theorem 6.5   (Shannon-McMillan-Breiman)   *For a stationary and ergodic sequence $\{X_k\}_{k \geq 1}$ the following holds*

$$h = -\lim_{n \to \infty} \frac{\log P(X_1^n)}{n} \quad \text{(a.s.)},$$

where h is the entropy rate of the process $\{X_k\}_{k \geq 1}$.

Proof. We present here the Algoet and Cover sandwich proof. The main idea is to use the kth order Markov approximation of a stationary distribution. Let P be the stationary distribution, and

$$P^k(X_1^n) := P(X_1^k) \prod_{i=k+1}^{n} P(X_i | X_{i-k}^{i-1})$$

be its kth order Markov approximation. Using (4.6) of Lemma 4.2, we obtain

$$\limsup_{n \to \infty} \frac{1}{n} \log \frac{1}{P(X_1^n)} \leq \limsup_{n \to \infty} \frac{1}{n} \log \frac{1}{P^k(X_1^n)} = h^k$$

for any $k \geq 1$. In the above, h^k is the entropy of the kth order Markov chain, which we know to exist (cf. Example 6.2).

Now, we consider a lower bound by using (4.7) of Lemma 4.2, that is,

$$\liminf_{n \to \infty} \frac{1}{n} \log \frac{1}{P(X_1^n)} \geq \lim_{n \to \infty} \frac{1}{n} \log \frac{1}{P(X_1^n | X_{-\infty}^0)} = h^\infty,$$

where h^∞ exists by the ergodic theorem since $\log P(X_1^n | X_{-\infty}^0)$, as a function of an ergodic process $\{X_i\}_{i=-\infty}^\infty$, is an ergodic process (i.e., $Y_n = f(X_{-\infty}^n)$ is ergodic if $\{X_i\}_{i=-\infty}^\infty$ is ergodic). Putting everything together, we have

$$h^\infty \leq \liminf_{n \to \infty} \frac{1}{n} \log \frac{1}{P(X_1^n)} \leq \limsup_{n \to \infty} \frac{1}{n} \log \frac{1}{P(X_1^n)} \leq h^k.$$

We shall now prove that $h^k \downarrow h^\infty$ completing the proof since by Theorem 6.4 $h = h^\infty$. By the Doob-Levy martingale construction (see Exercise 5.6 in Chapter 5), we know that $P(x_0 | X_{-k}^{-1})$ is a martingale. Therefore, by the martingale convergence Theorem 5.30

$$P(x_1 | X_{-k+1}^0) \to P(x_1 | X_{-\infty}^0) \quad \text{(a.s.)}$$

for all $x_1 \in \mathcal{A}$. Since the alphabet $\mathcal{A}$ is finite, $x \log x$ is bounded for $0 < x < 1$ and continuous, the *bounded convergence theorem* allows us to interchange expectation and limit, yielding

$$\lim_{k \to \infty} h^k = \lim_{k \to \infty} \mathbf{E} \left[- \sum_{x_1 \in \mathcal{A}} P(x_1 | X_{-k+1}^0) \log P(x_1 | X_{-k+1}^0) \right]$$

$$= \mathbf{E} \left[- \sum_{x_1 \in \mathcal{A}} P(x_1 | X_{-\infty}^0) \log P(x_1 | X_{-\infty}^0) \right] = h^\infty,$$

which completes the proof.   ∎

In passing we should mention that there is no universal convergence rate to the entropy for general stationary ergodic processes. However, for memoryless, Markovian, and mixing processes Marton and Shields [310] (cf. also Shields [388]) proved that the convergence rate to the entropy is exponential.

6.2.2 Rényi's Entropy Rates

In many problems on words **Rényi's entropy rates** are widely used (see Section 6.5 and [22, 23, 25, 411, 412]). For $-\infty \leq b \leq \infty$, the bth *order Rényi* *entropy* is defined as

$$h_b = \lim_{n \to \infty} \frac{-\log \mathbf{E}[P^b(X_1^n)]}{bn}$$

$$= \lim_{n \to \infty} \frac{-\log \left(\sum_{w \in \mathcal{A}^n} P^{b+1}(w)\right)^{1/b}}{n}, \tag{6.20}$$

provided the above limit exists. In the above, we write $P(w) = \Pr\{X_1^n = w\}$ for $w \in \mathcal{A}^n$. We will prove below that for mixing processes generalized Rényi's entropies exist. For now, we observe that

$$h = \lim_{b \to 0} h_b,$$

and by Hilbert's inequalities on means (2.13) and (2.14)

$$h_{-\infty} = \lim_{n \to \infty} \frac{-\log \left(\min\{P(X_1^n), P(X_1^n) > 0\}\right)}{n},$$

$$h_{\infty} = \lim_{n \to \infty} \frac{-\log \left(\max\{P(X_1^n), P(X_1^n) > 0\}\right)}{n}.$$

Example 6.3: *Rényi's Entropy for Memoryless and Markovian Sources*   We compute here the Rényi's entropies for memoryless and Markovian sources.

1. **Memoryless Source.** In this case, for $b > 0$ and a binary alphabet

$$\sum_{w \in \mathcal{A}^n} P^{b+1}(w) = \sum_{k=0}^{n} \binom{n}{k} p^{(b+1)k} q^{(b+1)(n-k)} = (p^{b+1} + q^{b+1})^n,$$

where, as always, $q = 1 - p$. For a general alphabet $\mathcal{A}$ of size V and probabilities $p_1, \ldots, p_V$ we obtain

$$h_b = -\frac{1}{b} \log(p_1^{b+1} + \cdots + p_V^{b+1}) \quad b > 0, \tag{6.21}$$

$$h_{-\infty} = -\log p_{\min}, \tag{6.22}$$

$$h_{\infty} = -\log p_{\max}, \tag{6.23}$$

where $p_{\max} = \max\{p_1, \ldots, p_V\}$ and $p_{\min} = \min\{p_1, \ldots, p_V, \ p_i > 0\}$.

2. **Markovian Source.** In Section 4.2.3 we proved that

$$\sum_{w \in \mathcal{A}^n} P^{b+1}(w) = \beta \lambda_{[b+1]}^n (1 + O(\rho^n))$$

for constants $\beta > 0$ and $\rho < 1$, where $\lambda_{[b]}$ is the largest eigenvalue of the Schur product (i.e., element-wise product) $\mathsf{P}_{[b]} = \mathsf{P} \circ \mathsf{P} \circ \cdots \circ \mathsf{P} = \{p_{ij}\}_{i,j=1}^V$. In view of this,

$$h_b = -\frac{1}{b} \log \lambda_{[b+1]}, \quad b > 0. \tag{6.24}$$

With respect to $h_{-\infty}$ and h_∞, we need a result from digraphs (cf. [234, 368]). Consider a digraph on $\mathcal{A}$ with weights equal to $-\log p_{ij}$ where $i, j \in \mathcal{A}$. Let $\mathcal{C} = \{i_1, i_2, \ldots, i_v, i_1\}$ be a cycle for some $v \leq V$ such that $i_j \in \mathcal{A}$, and let $\ell(\mathcal{C}) = -\sum_{k=1}^v \log(p_{i_k, i_{k+1}})$ (with $i_{v+1} = i_1$) be the total weight of the cycle $\mathcal{C}$. Then

$$h_{-\infty} = \min_{\mathcal{C}} \left\{ \frac{\ell(\mathcal{C})}{|\mathcal{C}|} \right\},$$

$$h_\infty = \max_{\mathcal{C}} \left\{ \frac{\ell(\mathcal{C})}{|\mathcal{C}|} \right\}.$$

Karp [234] showed how to compute these quantities efficiently.   ∎

We should point out that we have already encountered Rényi's entropies in this book. In Section 4.2.3 the entropies h_b and h_∞ appeared in the analysis of the height of digital trees, while $h_{-\infty}$ popped up in the study of the shortest path and fill-up level in a digital tree (see Exercises 4.10 and 4.11 in Chapter 4). The first Rényi's entropy h_1 plays a role in the analysis of molecular sequences as discussed in Arratia and Waterman [23, 25, 445].

6.3   ASYMPTOTIC EQUIPARTITION PROPERTY

An important conclusion from the Shannon-McMillan-Breiman theorem is the **Asymptotic Equipartition Property** (AEP), which asserts that asymptotically all *typical* sequences have the same probability approximately equal to 2^{-nh}. We first precisely define the basic AEP property, and then extend it to jointly typical sequences, formulate AEP for relative entropy (divergence), and finally discuss a lossy typical sequences, formulate AEP for relative entropy (divergence), and finally discuss a lossy generalization of AEP.

6.3.1   Typical Sequences

Let us start with the following definition:

Definition 6.6   *For given $\varepsilon > 0$ the typical set $\mathcal{G}_n^\varepsilon$ with respect to the probability measure $P = \mathsf{Pr}\{\cdot\}$ is the set of all sequences $(x_1, x_2, \ldots, x_n) \in \mathcal{A}^n$ such that*

$$2^{-n(h(P)+\varepsilon)} \leq P(x_1, \ldots, x_n) \leq 2^{-n(h(P)-\varepsilon)}, \tag{6.25}$$

where $P(x_1, \ldots, x_n) := \mathsf{Pr}\{X_1^n = (x_1, \ldots, x_n)\}.$

The Shannon-McMillan-Breiman theorem implies the following property that is at the heart of information theory.

Theorem 6.7 (**Asymptotic Equipartition Property**) *For a stationary and ergodic sequence* X_1^n, *for given* $\varepsilon > 0$ *the state space* $\mathcal{A}^n$ *can be partitioned into two subsets,* $\mathcal{B}_n^\varepsilon$ *("bad set") and* $\mathcal{G}_n^\varepsilon$ *("good set"), such that there is* N_ε *so that for* $n \geq N_\varepsilon$ *we have*

$$2^{-nh(1+\varepsilon)} \leq P(x_1^n) \leq 2^{-nh(1-\varepsilon)} \quad for \quad x_1^n \in \mathcal{G}_n^\varepsilon, \tag{6.26}$$

$$\mathsf{Pr}\{\mathcal{B}_n^\varepsilon\} \leq \varepsilon. \tag{6.27}$$

Moreover,

$$(1-\varepsilon)2^{nh(1-\varepsilon)} \leq |\mathcal{G}_n^\varepsilon| \leq 2^{nh(1+\varepsilon)} \tag{6.28}$$

for sufficiently large n.

Proof. By Shannon-McMillan-Breiman theorem we have for any $\varepsilon > 0$

$$\mathsf{Pr}\left\{ \left| -\frac{1}{n} \log P(X_1, \ldots, X_n) - h \right| < \varepsilon \right\} > 1 - \varepsilon$$

for, say $n > N_\varepsilon$. Denote by $\mathcal{B}_n^\varepsilon$ the set of $x_1^n = (x_1, \ldots, x_n)$ that does not satisfy the above. Its probability is smaller than ε. Call those x_1^n that fulfill the above inequality as $\mathcal{G}_n^\varepsilon$. Then for $x_1^n \in \mathcal{G}_n^\varepsilon$

$$h - \varepsilon \leq -\frac{1}{n} \log P(x_1, \ldots, x_n) \leq h + \varepsilon$$

which proves (6.26) and (6.27). Certainly, $\mathcal{G}_n^\varepsilon$ is the typical set according to Definition 6.6.

To prove the second part of the theorem, we proceed as follows:

$$1 = \sum_{x_1^n \in \mathcal{A}^n} P(x_1^n) \geq \sum_{x_1^n \in \mathcal{G}_n^\varepsilon} P(x_1^n) \geq \sum_{x_1^n \in \mathcal{G}_n^\varepsilon} 2^{-nh(1+\varepsilon)} = 2^{-nh(1+\varepsilon)} |\mathcal{G}_n^\varepsilon|,$$

and

$$1 - \varepsilon < \mathsf{Pr}\{\mathcal{G}_n^\varepsilon\} \leq \sum_{x_1^n \in \mathcal{G}_n^\varepsilon} 2^{-nh(1-\varepsilon)} = 2^{-nh(1-\varepsilon)} |\mathcal{G}_n^\varepsilon|.$$

This completes the proof. ∎

6.3.2 Jointly Typical Sequences

In some applications (see Section 6.4.2), we need a generalization of AEP to a pair of sequences (X_1^n, Y_1^n). Let us start with a definition of **jointly typical set**.

Definition 6.8   *The typical set $\mathcal{G}_n^\varepsilon$ of sequences (x_1^n, y_1^n) with respect to the probability measure $P = \Pr\{\cdot, \cdot\}$ is the set of all n-sequences such that for given $\varepsilon > 0$ the sequence x_1^n is $P(x)$ ε-typical, y_1^n is $P(y)$ ε-typical and*

$$2^{-n(h(X,Y)+\varepsilon)} \le P(x_1^n, y_1^n) \le 2^{-n(h(X,Y)-\varepsilon)}, \qquad (6.29)$$

where $P(x_1^n, y_1^n) := \Pr\{X_1^n = (x_1, \ldots, x_n), Y_1^n = (y_1, \ldots, y_n)\}$. In short,

$$\mathcal{G}_n^\varepsilon = \Big\{ (x_1^n, y_1^n) : \; 2^{-nh(X)(1+\varepsilon)} \le \; P(x_1^n) \; \le 2^{-nh(X)(1-\varepsilon)},$$
$$2^{-nh(Y)(1+\varepsilon)} \le \; P(y_1^n) \; \le 2^{-nh(Y)(1-\varepsilon)},$$
$$2^{-n(h(X,Y)+\varepsilon)} \le P(x_1^n, y_1^n) \le 2^{-n(h(X,Y)-\varepsilon)} \Big\}$$

for any $\varepsilon > 0$.

The following is a generalization of AEP to jointly typical sequences.

Theorem 6.9   (Joint AEP)   *Let $\{X_k\}_{k\ge1}$, $\{Y\}_{k\ge1}$ and $\{X_k, Y_k\}_{k\ge1}$ be mixing processes. Then for given $\varepsilon > 0$ there is N_ε so that for $n \ge N_\varepsilon$ we have*

$$\Pr\{(X_1^n, Y_1^n) \in \mathcal{G}_n^\varepsilon\} \ge 1 - \varepsilon, \qquad (6.30)$$

$$(1 - \varepsilon)2^{n(h(X,Y)-\varepsilon)} \le |\mathcal{G}_n^\varepsilon| \le 2^{n(h(X,Y)+\varepsilon)}. \qquad (6.31)$$

In addition, if $\widetilde{X}_1^n$ and $\widetilde{Y}_1^n$ are independent with the same marginals as $P(X_1^n, Y_1^n)$, then

$$(1 - \varepsilon)2^{-n(I(X,Y)+3\varepsilon)} \le \Pr\{(\widetilde{X}_1^n, \widetilde{Y}_1^n) \in \mathcal{G}_n^\varepsilon\} \le 2^{-n(I(X,Y)-3\varepsilon)} \qquad (6.32)$$

for sufficiently large n.

Proof. Since $\{X_k\}_{k\ge1}$ and $\{Y_k\}_{k\ge1}$ are mixing processes, by the Shannon-McMillan-Breiman theorem we observe that for given $\varepsilon > 0$ there exists n_1 such that for $n > n_1$

$$\Pr\left\{\left|-\frac{1}{n}\log P(X_1^n) - h(X)\right| > \varepsilon\right\} < \frac{\varepsilon}{3},$$

$$\Pr\left\{\left|-\frac{1}{n}\log P(Y_1^n) - h(Y)\right| > \varepsilon\right\} < \frac{\varepsilon}{3}.$$

Also, since the pair $\{X_k, Y_k\}$ is a mixing process, by the subadditive ergodic theorem (applied to $\log P(X_1^n, Y_1^n)$) we show that for given $\varepsilon > 0$ there exists n_2 such that for $n > n_2$

$$\Pr\left\{\left|-\frac{1}{n}\log P(X_1^n, Y_1^n) - h(X, Y)\right| > \varepsilon\right\} < \frac{\varepsilon}{3}.$$

In view of the above, for $n > \max\{n_1, n_2\}$ the probability of the union of the above sets is less than ε. Thus the intersection of the complements of these sets forms $\mathcal{G}_n^\varepsilon$, and therefore

$$\Pr\{\mathcal{G}_n^\varepsilon\} > 1 - \varepsilon.$$

To prove (6.31) we proceed exactly as in the proof of AEP of a single sequence, so we omit it here. Finally, for (6.32) we observe that

$$\Pr\{(\tilde{X}_1^n, \tilde{Y}_1^n) \in \mathcal{G}_n^\varepsilon\} = \sum_{(x_1^n, y_1^n) \in \mathcal{G}_n^\varepsilon} P(x_1^n) P(y_1^n)$$

$$\leq 2^{n(h(X,Y)+\varepsilon)} 2^{-n(h(X)-\varepsilon)} 2^{-n(h(Y)-\varepsilon)}$$

$$= 2^{-n(I(X,Y)-3\varepsilon)}.$$

Using the left-hand side of (6.31), we establish in a similar manner the lower bound of (6.32). This completes the proof. ∎

The above finding has profound implications for channel coding. We shall discuss it in depth in Section 6.4.2; however, a brief discussion is in order. Observe that there are about $2^{nh(X)}$ typical X sequences, about $2^{nh(Y)}$ typical Y sequences, but only $2^{nh(X,Y)}$ jointly typical sequences (recall, $h(X, Y) \leq h(X) + h(Y)$). The probability that any randomly chosen pair (X, Y) is jointly typical is approximately

$$\frac{2^{nh(X,Y)}}{2^{nh(X)} 2^{nh(Y)}} = 2^{-nI(X,Y)}.$$

In other words, for fixed Y_1^n, we may inspect about $2^{nI(X,Y)}$ pairs before we find a jointly typical pair.

6.3.3 AEP for Biased Distributions

Consider the following *classification problem*: A nineteenth-century manuscript written in Polish was found in Paris. One suspects that it is a lost manuscript of the famous Polish poet Adam Mickiewicz. To prove this assertion a classification algorithm is proposed that will compare (cf. [455] for possible algorithms) the found text with samples of Mickiewicz's writings.

From the information theory point of view, we may assume that the found text is generated by an unknown distribution $P = \Pr\{\cdot\}$ that must be compared with the known distribution Q obtained from Mickiewicz's samples. In other words, we estimate the probability of x_1^n generated by an unknown source P with respect to the known distribution Q. We call this problem **AEP for a biased distribution**.

To start, we must define the relative entropy or *divergence* $D(P \parallel Q)$ for random sequences. Let, as before, $X_1^n = (X_1, \ldots, X_n)$, be a random sequence generated by source P over a finite alphabet. We define the **divergence rate** as

$$D(P \parallel Q) := - \lim_{n \to \infty} \frac{\mathbf{E}_P[\log Q(X_1^n)/P(X_1^n)]}{n}, \qquad (6.33)$$

provided the limit exists. Unlike the entropy rate, the above limit may not exist for general stochastic processes, and in general one needs lim sup in the above definition. Below, we restrict our discussion to jointly mixing processes that will guarantee the existence of the limit. In such a case, we can also write

$$-\lim_{n \to \infty} \frac{\mathbf{E}_P[\log Q(X_1^n)]}{n} = -\lim_{n \to \infty} \frac{\mathbf{E}_P[\log Q(X_1^n)/P(X_1^n)] + \mathbf{E}_P[\log P]}{n}$$
$$= h(P) + D(P \parallel Q).$$

Let us now extend AEP to biased distributions. First, however, we prove the existence of the limit in (6.33) for the *jointly* mixing processes.

(JMX) Let the measures P and Q be defined on the same probability space and let $\mathbb{F}_m^n$ be the σ-algebra generated by X_m^n. The measures P and Q are *weakly mixing* if there exist constants c_1, C_1, c_2, C_2 such that for any $m, g \geq 1$ and any two events $A \in \mathbb{F}_1^m$ and $B \in \mathbb{F}_{m+g}^\infty$ the following holds

$$c_1 P(A)P(B) \leq P(AB) \leq c_2 P(A)P(B), \qquad (6.34)$$

$$C_1 Q(A)Q(B) \leq Q(AB) \leq C_2 Q(A)Q(B) \qquad (6.35)$$

and $c_1, C_1 > 0$.

In passing, we should observe that the condition $c_1, C_1 > 0$ is quite restrictive. For example, not all Markov chains belong to (JMX), but Markov chains with *positive* transition probabilities are jointly mixing (see Theorem 2.1).

It is easy to see that if both Q and P are mixing (cf. (2.2), then

$$\log Q(X_1^n)/P(X_1^n) \leq \log C_2/c_1 + \log Q(X_1^m)/P(X_1^m) + \log Q(X_m^n)/P(X_m^n),$$

and $\log Q/P$ is subadditive. By Theorem 5.5 we have

$$-\lim_{n \to \infty} \frac{\log Q(X_1^n)/P(X_1^n)}{n} = D(P \parallel Q) \quad \text{(a.s)},$$

and hence

$$-\lim_{n \to \infty} \frac{\log Q(X_1^n)]}{n} = h(P) + D(P \parallel Q) \quad \text{(a.s.)}. \tag{6.36}$$

This leads to the following generalization of AEP (details of the proof are left as an exercise).

Theorem 6.10   (AEP for Biased Distributions)   *Let $\{X_k\}_{k \geq 1}$ be generated according to P, which, together with another distribution Q, is jointly mixing as defined in (JMX). Then the state space $\mathcal{A}^n$ can be partitioned into two subsets, $\mathcal{B}_n^\varepsilon$ ("bad set") and $\mathcal{G}_n^\varepsilon(Q)$ ("good set"), such that for given $\varepsilon > 0$ there is N_ε so that for $n \geq N_\varepsilon$ we have*

$$2^{-n(h(P)+D(P\|Q))(1+\varepsilon)} \leq Q(x_1^n)$$

$$\leq 2^{-n(h(P)+D(P\|Q))(1-\varepsilon)}, \quad x_1^n \in \mathcal{G}_n^\varepsilon(Q), \tag{6.37}$$

$$Q(\mathcal{B}_n^\varepsilon) \leq \varepsilon. \tag{6.38}$$

Moreover,

$$(1 - \varepsilon)2^{n(h(P)+D(P\|Q))(1-\varepsilon)} \leq |\mathcal{G}_n^\varepsilon(Q)| \leq 2^{n(h(P)+D(P\|Q))(1+\varepsilon)} \tag{6.39}$$

for sufficiently large n.

We should point out that the above AEP for biased distributions cannot be extended to general stationary ergodic processes (cf. [179, 180]). In fact, as we mentioned above even the limit in the definition (6.33) may not exist in a general case. However, the following general result is known (the reader is asked to prove it in Exercise 10; cf. also Gray [75, 181]).

Lemma 6.11 *If P is a stationary process and P^k represents the measure of a k order stationary Markov process, then the limit in (6.33) exists and*

$$D(P \parallel P^k) = -h_P(X) - \mathbf{E}_P[\ln P^k(X_k|X_0^{k-1})] = -h_P(X) - h_{P^k}(X_k|X_0^{k-1}).$$

6.3.4 Lossy Generalizations of AEP

Let $w \in \mathcal{A}^k$ be a word of length k. Often one asks for the probability that a prefix of a sequence generated by a stationary source is equal to w. Using AEP, we can asymptotically estimate this probability (e.g., by checking whether w is a typical sequence). But in some situations (e.g., for a continuous signal) another question is more appropriate, namely, how well w approximates a k-prefix of a sequence. To answer such a question a lossy or approximate extension of AEP is needed. We discuss this next.

In order to more precisely define the notion of *approximate*, we need to introduce a *distortion measure* or a *distance* (cf. Berger [42] for a detailed discussion). In information theory, one defines the *distortion* function between symbols as

$$d : \mathcal{A} \times \hat{\mathcal{A}} \to \mathbb{R}^+.$$

Then the distortion-per-symbol becomes

$$d_n(x_1^n, \hat{x}_1^n) = \frac{1}{n} \sum_{i=1}^n d(x_i, \hat{x}_i). \tag{6.40}$$

For example, for the Hamming distance one sets $d(x_i, \hat{x}_i) = 0$ if $x_i = \hat{x}_i$ and $d(x_i, \hat{x}_i) = 1$ otherwise. Observe that in this case, the distortion $d_n(x_1^n, \hat{x}_1^n)$ represents the fraction of mismatches between x_1^n and $\hat{x}_1^n$. Another distortion measure is useful for image compression, for example, the *square error* distortion defined as $d(x_i, \hat{x}_i) = (x_i - \hat{x}_i)^2$.

Let us now fix $D > 0$ and define a *ball* of radius D and center $c_1^k \in \mathcal{A}^k$ as

$$B_D(c_1^k) = \{x_1^k : d_k(x_1^k, c_1^k) \le D\}.$$

In particular, $B_0(x_1^k) = x_1^k$. If a random sequence X_1^k is generated by source P, we shall write $P(B_D(X_1^k)) = \sum_{\hat{x}_1^k \in B_D(X_1^k)} \Pr\{d_k(X_1^k, \hat{x}_1^k)\}$. With these definitions in mind, we can now generalize Rényi's entropy.

Definition 6.12 (Generalized Rènyi's Entropy of order b) *For any* $-\infty \le b \le \infty$

$$r_b(D) = -\lim_{k \to \infty} \frac{\log \mathbf{E}[P^b(B_D(X_1^k))]}{bk} \tag{6.41}$$

$$= -\lim_{k \to \infty} \frac{\log \left(\sum_{w_1^k \in \mathcal{A}^k} P^b(B_D(w_1^k)) P(w_1^k) \right)}{bk},$$

where for $b = 0$ we understand $r_0(D) = \lim_{b \to 0} r_b(D)$, that is,

$$r_0(D) = -\lim_{k \to \infty} \frac{\mathbf{E}[\log P(B_D(X_1^k))]}{k}, \tag{6.42}$$

provided the above limits exist.

In Theorem 6.13 below, we first establish the existence of the generalized Rényi's entropies for mixing processes. Then we prove a generalization of the Shannon-McMillian-Breiman theorem, which will imply AEP for approximate/lossy case.

Theorem 6.13 (AEP for Approximate/Lossy Case) *Let $\{X_k\}_{k \geq 1}$ be a weakly mixing process.*

(i) *If $c_1 > 0$ (cf. (6.34)), then the generalized bth order entropy $r_b(D)$ is well defined (i.e., the limit in (6.41) exists) for any $-\infty \leq b \leq \infty$.*

(ii) *The generalized Shannon entropy $r_0(D)$ satisfies*

$$r_0(D) = -\lim_{k \to \infty} \frac{\log P(B_D(X_1^k))}{k} \quad (a.s.). \tag{6.43}$$

(iii) *Given $\varepsilon > 0$, the state space $\mathcal{A}^n$ can be partitioned into two subsets, $\mathcal{B}_n^\varepsilon$ ("bad set") and $\mathcal{G}_n^\varepsilon$ ("good set"), such that there exists N_ε so that for all $n > N_\varepsilon$ we have*

$$2^{-n(r_0(D)+\varepsilon)} \leq P(B_D(x_1^n)) \leq 2^{-n(r_0(D)-\varepsilon)}, \tag{6.44}$$

$$P(\mathcal{B}_n^\varepsilon) \leq \varepsilon, \tag{6.45}$$

$$(1-\varepsilon)2^{n(r_0(D)-\varepsilon)} \leq |\mathcal{G}_n^\varepsilon|. \tag{6.46}$$

Proof. We consider only $0 \leq b < \infty$, while the $b < 0$ case is left as an exercise. For (i) it suffices to show that for some constant $c > 0$

$$\mathbf{E}[P^b(B_D(X_1^{n+m}))] \geq c\mathbf{E}[P^b(B_D(X_1^n))]\mathbf{E}[P^b(B_D(X_1^m))]. \qquad (6.47)$$

Provided (6.47) is true we simply use Fekete's Theorem 5.1 to establish our claim. In the course of proving (6.47) we shall see that $P^b(B_D(X_1^{n+m})) \geq cP^b(B_D(X_1^n))P^b(B_D(X_1^m))$, which by the *subadditive ergodic* Theorem 5.5 applied to $\log P(B_D(X_1^n))$ will imply (6.43) of part (ii).

We now wrestle with (6.47). Observe that for any string $w_1^{n+m} \in \mathcal{A}^{n+m}$ of length $n+m$ we have

$$
\begin{aligned}
P^b(B_D(w_1^{n+m})) &= \left(\sum_{z_1^{n+m} \in B_D(w_1^{n+m})} P(z_1^{n+m}) \right)^b \\
&\geq c \left(\sum_{z_1^{n+m} \in B_D(w_1^{n+m})} P(z_1^n)P(z_{n+1}^{n+m}) \right)^b \\
&\geq c \left(\sum_{z_1^n \in B_D(w_1^n)} P(z_1^n) \right)^b \left(\sum_{z_{n+1}^{n+m} \in B_D(w_{n+1}^{n+m})} P(z_{n+1}^{n+m}) \right)^b \\
&= cP^b(B_D(w_1^n))P^b(B_D(w_{n+1}^{n+m})),
\end{aligned}
$$

where $c > 0$ is a positive constant. The first inequality of the above follows from the weak mixing condition (6.34), and the second one is a simple consequence of the following easy-to- establish property of the distortion-per-symbol

$$d_{n+m}(x_1^{n+m}, \hat{x}_1^{n+m}) = \frac{n}{n+m}d_n(x_1^n, \hat{x}_1^n) + \frac{m}{n+m}d_m(x_{n+1}^{n+m}, \hat{x}_{n+1}^{m+n}).$$

To complete the proof we again use the mixing condition to get

$$
\begin{aligned}
\mathbf{E}[P^b(B_D(X_1^{n+m}))] &= \sum_{w_1^{n+m} \in \mathcal{A}^{n+m}} P^b(B_D(w_1^{n+m}))P(w_1^{n+m}) \\
&\geq c \sum_{w_1^n \in \mathcal{A}^n} P^b(B_D(w_1^n))P(w_1^n) \sum_{w_{n+1}^{m+n} \in \mathcal{A}^m} P^b(B_D(w_{n+1}^{n+m}))P(w_{n+1}^{m+n}) \\
&= c\mathbf{E}[P^b(B_D(X_1^n))]\mathbf{E}[P^b(B_D(X_{n+1}^{n+m}))].
\end{aligned}
$$

Parts (i) and (ii) are proved by Fekete's theorem and the superadditive ergodic theorem (see Chapter 5).

Part (iii) is a direct consequence of Part (ii). For example, to prove (6.46) we proceed as follows

$$1 - \varepsilon \le \sum_{x_1^n \in \mathcal{G}_n^\varepsilon} P(x_1^n) \le \sum_{x_1^n \in \mathcal{G}_n^\varepsilon} P(B_D(x_1^n)) \le |\mathcal{G}_n^\varepsilon| 2^{-n(r_0(D)-\varepsilon)}.$$

This completes the proof.   ■

The evaluation of the generalized Rényi's entropies is harder than for the lossless case. Rényi's entropies for memoryless sources are computed in the lemma below. There are no simple formulas for the generalized Rényi's entropies for Markovian sources; however, some useful relationships may be found in Yang and Kieffer [439].

Lemma 6.14 (Łuczak and Szpankowski, 1997) *Assume a binary alphabet* $\mathcal{A} = \{0, 1\}$ *with symbol occurrence* p *and* $q = 1 - p$. *Define* $h(D, x) = (1 - D) \log((1 - D)/x) + D \log(D/(1 - x))$ *for* $0 < D, x < 1$.

(i) *Let* $p_{\min} = \min\{p, q\}$ *and* $p_{\max} = \max\{p, q\}$, *then*

$$r_{-\infty}(D) = \begin{cases} h(D, p_{\min}) & \text{for } D \le p_{\max} \\ 0 & \text{for } D > p_{\max}. \end{cases}$$

and

$$r_{\infty}(D) = \begin{cases} h(D, p_{\max}) & \text{for } D \le p_{\min} \\ 0 & \text{for } D > p_{\min}. \end{cases}$$

In addition, $r_{-\infty}(D)$ *and* $r_{\infty}(D)$ *are convex functions of* D.

(ii) *If* $p = q = 1/2$ *then, for every* $-\infty \le b \le \infty$ *and* $D \le p_{\min}$, *we have* $r_b(D) = h(D, 1/2)$.

(iii) *Let* $p \ne q$ *and* $-\infty < b < \infty$. *Then* $r_b(D) = 0$ *whenever* $D > 2pq$, *while for* $0 \le D \le 2pq$ *and* $b \ne 0$

$$r_b(D) = (1/b) \min_{0 \le x \le 1} \Big\{ x \log(x/p) + (1 - x) \log((1 - x)/q) - b \Big(D \log(p/q)$$

$$+ x \log(px) + (1 - x) \log(q(1 - x)) - x \log(x - F(x)) \tag{6.48}$$

$$- D \log(D - F(x)) - (1 - x - D) \log(1 - x - D + F(x)) \Big) \Big\},$$

where $F(x)$ *is defined as*

$$F(x) = \frac{x + D}{2} + \frac{\sqrt{(p^2 + (x + D)(q - p))^2 + 4xq^2 D(p - q)} - p^2}{2(p - q)}.$$
$$(6.49)$$

In particular, we have

$$r_1(D) = \begin{cases} h(D, P) & \text{for } D \leq 1 - P = 2pq \\ 0 & \text{for } D > 1 - P = 2pq, \end{cases}$$

where $P = p^2 + q^2$. The function $r_1(D)$ is convex with respect to D.

(iv) *If $p \neq q$, then $r_0(D) = 0$ for $D > 2pq$, and for $0 \leq D \leq 2pq$*

$$r_0(D) = -\Big(D \log(p/q) + 2p \log p + 2q \log q - p \log(p - F(p))$$

$$- D \log(D - F(p)) - (q - D) \log(q - D + F(p))\Big), \quad (6.50)$$

where F is the function defined by (6.49). In addition, $r_0(D)$ is convex with respect to D.

Proof. We only prove part (iii). The reader is referred to [303] for the whole proof, or even better, is encouraged to provide the missing parts.

Let $p \neq q$ and $-\infty < b < \infty$, $b \neq 0$. From the definition of the expectation, for $E[P^b(B_D(X_1^k))]$ we have

$$E[P^b(B_D(X_1^k))]$$

$$= \sum_{i=0}^{k} \binom{k}{i} p^i q^{k-i} \left(\sum_{j=0}^{\lfloor Dk \rfloor} \sum_{\ell=\max\{0, i+j-k\}}^{\min\{i, j\}} \binom{i}{\ell} \binom{k - i}{j - \ell} p^{i+j-2\ell} q^{k-i-j+2\ell} \right)^b,$$

where i counts the number of ones in X_1^k, j stays for the overall number of mismatches, and ℓ is the number of disagreements among ones. Let us look first at the sum

$$s(k, i) = \sum_{j=0}^{\lfloor Dk \rfloor} \sum_{\ell=\max\{0, i+j-k\}}^{\min\{i, j\}} \binom{i}{\ell} \binom{k - i}{j - \ell} p^{i+j-2\ell} q^{k-i-j+2\ell}$$

$$= \sum_{j=0}^{\lfloor Dk \rfloor} \sum_{\ell=\max\{0, i+j-k\}}^{\min\{i, j\}} r(k, i, j, \ell).$$

Since there are at most k^2 terms in the sum, certainly we have

$$\max_{j, \ell} r(k, i, j, \ell) \leq s(k, i) \leq k^2 \max_{j, \ell} r(k, i, j, \ell).$$

(Note that all ratios that grow polynomially with k will disappear if we divide the logarithm of $s(k, i)$ by k; thus they will not affect the value of $r_b(D)$.) Similarly,

$$\max_i \binom{k}{i} p^i q^{k-i} s^b(k, i) \leq E[P^b(B_D(X_1^k))]$$

$$\leq k \max_i \binom{k}{i} p^i q^{k-i} s^b(k, i).$$

By Stirling's formula (cf. (8.21)) for every $x \in (0, 1)$ we have

$$\binom{k}{xk} \sim \left(\frac{1}{(1-x)^{1-x} x^x}\right)^k. \tag{6.51}$$

We use the above to estimate the binomial coefficients and set $i = xk$, $j = yk$, $\ell = zk$ to arrive at the following asymptotic formula for $E[P^b(B_D(X_1^k))]$:

$$\max_{0 \leq x \leq 1} \left\{ \left(\frac{p^x q^{1-x}}{x^x (1-x)^{(1-x)}} \right. \right.$$

$$\left. \left(\max_{A(x, D)} \left\{ \frac{p^{x+y-2z} q^{1-x-y+2z} x^x (1-x)^{1-x}}{z^z (x-z)^{x-z} (y-z)^{y-z} (1-x-y+z)^{1-x-y+z}} \right\} \right)^b \right)^k \right\},$$

where $A(x, D) \subset \mathbb{R}^2$ is defined as

$$A(x, D) = \{(y, z) \in \mathbb{R}^2 : 0 \leq y \leq D, \ \max\{0, x + y - 1\} \leq z \leq \min\{x, y\}\}.$$

Consequently, we get the following formula

$$r_b(D) = (1/b) \min_{0 \leq x \leq 1} \left\{ x \log(x/p) + (1-x) \log((1-x)/q) \right.$$

$$- b \max_{A(x, D)} \left\{ (x + y - 2z) \log(p/q) + \log q + x \log x \right. \tag{6.52}$$

$$+ (1-x) \log(1-x) - z \log z - (x-z) \log(x-z)$$

$$\left. \left. - (y-z) \log(y-z) - (1-x-y+z) \log(1-x-y+z) \right\} \right\}.$$

Simple algebra reveals that (6.48) follows from (6.52). Indeed, let us assume first that $D > 2pq$. Then, the value of the maximum in (6.52) is 0 and is achieved for $y - z = (1-x)p$ and $z = qx$. Furthermore, the first two terms vanish for $x = p$. Hence, for such a D, we have $r_b(D) = 0$ for every $-\infty < b < \infty$.

If $D \leq 2pq$, then the function which appears under the maximum in (6.52) grows with y, so we must put $y = D$. Furthermore, easy calculations show that to choose an optimal value of z one must solve the equation

$$(p - q)z^2 + (p^2 + (q - p)(x + D))z - xDq^2 = 0.$$

Thus we should set $z = F(x)$, where F is defined by (6.49). Then (6.48) follows.

Finally, let us notice that the case $b = 0$ can be easily deduced from (6.48). Indeed, for $b \to 0$ the sum of the first two terms $x \log(x/p)$ and $(1 - x) \log((1 - x)/q)$ must vanish, which is possible only for $x = p$. Thus (6.50) follows. ∎

We must point out, however, that one can envision another AEP for the lossy case that is actually quite useful in the rate distortion theory discussed in Section 6.4.3. We briefly discuss it here.

We start with a definition of **distortion typical set** denoted as $\mathcal{G}_{d,n}^\varepsilon$.

Definition 6.15 *The distortion ε-typical set $\mathcal{G}_{d,n}^\varepsilon$ of sequences $(x_1^n, \hat{x}_1^n)$ with respect to the probability measure $P = \Pr\{\cdot, \cdot\}$ is the set of all n-sequences such that for given $\varepsilon > 0$*

$$\mathcal{G}_{d,n}^\varepsilon = \Big\{ (x_1^n, \hat{x}_1^n) : \; 2^{-nh(X)(1+\varepsilon)} \leq P(x_1^n) \leq 2^{-nh(X)(1-\varepsilon)},$$

$$2^{-nh(\hat{X})(1+\varepsilon)} \leq P(\hat{x}_1^n) \leq 2^{-nh(\hat{X})(1-\varepsilon)},$$

$$2^{-n(h(X,Y)+\varepsilon)} \leq P(x_1^n, \hat{x}_1^n) \leq 2^{-n(h(X,Y)-\varepsilon)}$$

$$\mathbf{E}_{X_1^n}[d(X_1^n, \hat{X}_1^n)] - \varepsilon \leq d(x_1^n, \hat{x}_1^n) \leq \mathbf{E}_{X_1^n}[d(X_1^n, \hat{X}_1^n)] + \varepsilon \Big\}.$$

The following extension of joint AEP can be stated.

Theorem 6.16 (AEP with Distortion) *Let $\{X_k\}_{k\geq 1}$, $\{\hat{X}_k\}_{k\geq 1}$ and $\{X_k, \hat{X}_k\}_{k\geq 1}$ be mixing processes. Then for given $\varepsilon > 0$ there is N_ε so that for $n \geq N_\varepsilon$ we have*

$$\Pr\{(X_1^n, \hat{X}_1^n) \in \mathcal{G}_{d,n}^\varepsilon\} \geq 1 - \varepsilon, \tag{6.53}$$

while for $(x_1^n, \hat{x}_1^n) \in \mathcal{G}_{n,d}^\varepsilon$ we obtain

$$P(\hat{x}_1^n | x_1^n) 2^{-n(I(X;\hat{X})-3\varepsilon)} \leq P(x_1^n) \leq P(\hat{x}_1^n | x_1^n) 2^{-n(I(X;\hat{X})+3\varepsilon)} \tag{6.54}$$

for sufficiently large n.

Proof. We prove only (6.54) since (6.53) follows from the subadditivity of $\log P(X_1^n)$, $\log P(\hat{X}_1^n)$, $\log P(X_1^n, \hat{X}_1^n)$ and $d(X_1^n, \hat{X}_1^n)$ for mixing processes.

Observe that for $(x_1^n, \hat{x}_1^n) \in \mathcal{G}_{d,n}^{\varepsilon}$

$$P(\hat{x}_1^n | x_1^n) = P(\hat{x}_1^n) \frac{P(x_1^n, \hat{x}_1^n)}{P(x_1^n)P(\hat{x}_1^n)}$$

$$\leq P(\hat{x}_1^n) \frac{2^{-n(h(X,\hat{X})-\varepsilon)}}{2^{-n(h(X)+\varepsilon)}2^{-n(h(\hat{X})+\varepsilon)}} = P(\hat{x}_1^n)2^{n(I(X,\hat{X})-3\varepsilon)}.$$

The lower bound is proved in a similar fashion.   ∎

6.4 THREE THEOREMS OF SHANNON

In 1948 Claude Shannon published his seminal paper "*A Mathematical Theory of Communication*", where he formulated the three fundamental results of information theory: the *universal source coding* theorem, the *channel coding* theorem, and the *rate distortion* theorem. In this section, we discuss them and provide simple proofs. We shall follow the presentation of Cover and Thomas [75]. In Shannon's spirit, we shall deal mostly with memoryless sources to succinctly present the main ideas. Throughout this section, we use a new technique, not yet seen in this book, called *random coding*. This method selects codes at random to establish the existence of a good code that achieves the best possible code rate.

6.4.1 Source Coding Theorem

A *source code* $\mathscr{C}$ for a source (random variable) X taking values in the set $\mathcal{A}$ is a mapping

$$\phi : \mathcal{A} \to \{0, 1\}^*,$$

that we also write as $\mathscr{C} = \phi(X)$. Such a code is one-to-one if

$$x_i \neq x_j \in \mathcal{A} \quad \Rightarrow \quad \mathscr{C}(x_i) \neq \mathscr{C}(x_j).$$

The code $\mathscr{C}_n$ for a sequence $X_1, \ldots, X_n$ is an *extension* of $\mathscr{C}$ and it is defined by a mapping

$$\phi_n : \mathcal{A}^n \to \{0, 1\}^*.$$

Such a code is:

- *Uniquely decodable* if it is one-to-one;

- *Prefix code* or *instantaneous code* if no codeword is a prefix of another codeword.

It turns out that lengths of prefix codewords cannot be completely arbitrary and must satisfy a certain constraint known as *Kraft's inequality*. We prove it below for the simplest case. Its extensions can be found in [75].

Theorem 6.17 (Kraft's Inequality) *For any prefix code (over a binary alphabet), the codeword lengths $\ell_1, \ell_2, \ldots, \ell_m$ satisfy the inequality*

$$\sum_{i=1}^{m} 2^{-\ell_i} \leq 1. \tag{6.55}$$

Conversely, if codeword lengths satisfy this inequality, then one can build a prefix code.

Proof. This is an easy exercise on trees. Observe that one can associate a binary tree with a code by assigning, say, "0" to left branch and "1" to the right branch. A path from the root to a node creates a code. Prefix codes guarantee that *none* of the codewords are assigned to an *internal node*; that is, all codewords are terminal nodes (or *leaves* of the tree). Let $\ell_{\max}$ be the maximum codeword length. Observe that at level $\ell_{\max}$ some nodes are codewords, some are descendants of codewords, and some are neither. Since the number of descendants at level $\ell_{\max}$ of a codeword located at level ℓ_i is $2^{\ell_{\max}-\ell_i}$ (if root is at level 0 of the tree, then at level ℓ there are no more than 2^ℓ nodes), we obtain

$$\sum_{i=1}^{m} 2^{\ell_{\max}-\ell_i} \leq 2^{\ell_{\max}},$$

which is the desired inequality. The converse part is easy to prove, and we omit it here. ∎

The average length $\bar{\ell}_n(\mathscr{C}_n)$ of the source code $\mathscr{C}_n$ is defined as

$$\bar{\ell}_n(\mathscr{C}_n) = \sum_{x_1^n \in \mathcal{A}^n} \ell(x_1^n) P(x_1^n),$$

where P is the source distribution; that is, $P(x_1^n) = \Pr\{X_1^n = x_1^n\}$. We cannot make $\bar{\ell}(\mathscr{C}_n)$ arbitrarily small, as Theorem 6.18 below indicates. In other words, there is a limit how much we can compress on the average.

Theorem 6.18   *For any source code $\mathscr{C}_n$ satisfying the Kraft inequality, the average code length $\bar{\ell}(\mathscr{C}_n)$ must be bigger than or equal to the entropy $h(X_1^n)$ of the source, that is, $\bar{\ell}(\mathscr{C}_n) \geq h_n(X_1^n)$.*

Proof. Let $K = \sum_{x_1^n} 2^{-\ell(x_1^n)} \leq 1$. Then

$$\bar{\ell}(\mathscr{C}_n) - h_n(X_1^n) = \sum_{x_1^n \in \mathcal{A}^n} P(x_1^n)\ell(x_1^n) + \sum_{x_1^n \in \mathcal{A}^n} P(x_1^n) \log P(x_1^n)$$

$$= \sum_{x_1^n \in \mathcal{A}^n} P(x_1^n) \log \frac{P(x_1^n)}{2^{-\ell(x_1^n)}/K} - \log K$$

$$\geq 0,$$

since the first term is a divergence and cannot be negative by Theorem 6.3, while $K \leq 1$ by Kraft's inequality.   ∎

The quantity

$$\bar{r}_n(\mathscr{C}_n) = \bar{\ell}(\mathscr{C}_n) - h_n(X_1^n)$$

is known as the average *code redundancy*. We just proved that for prefix codes it cannot be negative. We also observe that the average bits per symbol $\bar{\ell}(C_n)/n$ is bounded from the below by the entropy rate of the source. In Theorem 6.19 below we show that this is true even if we allow errors on the decoding side. We shall return to the code redundancy in Sections 8.7.2 and 9.4.2, where we use analytic tools to evaluate precisely the code redundancy for some classes of processes.

To formulate a *universal source coding problem*, we start by introducing the bit rate R of a code $\mathscr{C}_n$. Hereafter, we deal only with extended codes for $X_1, \ldots, X_n$ generated by a source with the underlying probability measure P. Let M be the number of codewords for $\mathscr{C}_n$. The **bit rate** of such a code is defined as

$$R = \frac{\log M}{n}.$$

We often write $M = 2^{nR}$ to mean $M = \lceil 2^{nR} \rceil$, and the corresponding code we denote as $(2^{nR}, n)$. Observe that R represents the number of bits per source symbol, and it can be viewed as the reciprocal of the compression ratio. We expect $R < 1$ for good compression codes.

The *fixed rate block code* $\mathscr{C}_n$ of rate R for $X_1, \ldots, X_n$ is defined by two mappings:

1. Encoder

$$\phi_n : \mathcal{A}^n \rightarrow \{1, 2, \ldots, 2^{nR}\};$$

2. Decoder

$$\psi_n : \{1, 2, \ldots, 2^{nR}\} \rightarrow \mathcal{A}^n;$$

where we explicitly set $M = \lceil 2^{nR} \rceil$.

Let $\hat{X}_1^n = \psi_n(\phi_n(X_1^n)) = \psi_n(\mathscr{C}_n)$, and we allow that $\hat{X}_1^n \neq X_1^n$. In this case, we define the probability of error P_E with respect to the source probability P as

$$P_E := \Pr\{X_1^n \neq \hat{X}_1^n\}.$$

The problem of **universal source coding** is to find the smallest bit rate R (or biggest compression ratio) such that $P_E < \varepsilon$ for given $\varepsilon > 0$ irrespectively of the (usually unknown) distribution P.

The first Shannon theorem states that one can build a universal source code as long as the bit rate R is bigger than the entropy $h(P)$ of the source. In other words, on average the best bit rate is equal to the entropy of the source.

Theorem 6.19   (Shannon, 1948)   *Let $(2^{nR}, n)$ denote a source code with bit rate R.*

   (i) *There exists a sequence of $(2^{nR}, n)$ universal source codes with rate $R > h(P)$ such that for a given $\varepsilon > 0$ the probability of error $P_E < \varepsilon$ for large n.*
  (ii) *Conversely, for all universal codes with $P_E \rightarrow 0$ the bit rate $R > h(P)$.*

Proof. We prove the achievability part of the theorem by constructing *a random code* that satisfies the conditions of the theorem. For a given $\varepsilon > 0$, let $\mathcal{G}_n^\varepsilon$ be a set of $\mathcal{A}^n$ such that $P(\mathcal{G}_n^\varepsilon) \geq 1 - \varepsilon$. We assign codes that are correctly reproduced to elements of the set $\mathcal{G}_n^\varepsilon$. Clearly, then, the error P_E is bounded by ε. By AEP, we know that $\mathcal{G}_n^\varepsilon$ is the set of P-typical sequences and its cardinality can be bounded as $(1 - \varepsilon)2^{n(h(P)-\varepsilon)} \leq |\mathcal{G}_n^\varepsilon| \leq 2^{n(h(P)+\varepsilon)}$. Thus, $R = \log |\mathcal{G}_n^\varepsilon|/n$ is bounded between $h(P) - \varepsilon$ and $h(P) + \varepsilon$, for sufficiently large n. This proves the first part of the theorem.

To prove the converse, we establish that if $R < h(P)$, then $P_E > 1 - \varepsilon$ for a given $0 < \varepsilon < 1$. Let $\mathscr{C}'$ be the subset of all codewords $\mathscr{C}$ that are encoded without an error. Observe that $|\mathscr{C}'| = 2^{nR}$. Since we assume that $R < h(P)$ there exists $\delta > 0$ such that

$$|\mathscr{C}'| < 2^{n(h-2\delta)}.$$

We now use AEP with $2\delta = \varepsilon$ to obtain (we write below $\mathcal{B}_n^\varepsilon = \mathcal{A}^n - \mathcal{G}_n^\varepsilon$)

$$1 - P_E = \sum_{x_1^n \in \mathscr{C}'} P(x_1^n)$$

$$= \sum_{x_1^n \in \mathscr{C}' \cap \mathcal{B}_n^{\varepsilon/2}} P(x_1^n) + \sum_{x_1^n \in \mathscr{C}' \cap \mathcal{G}_n^{\varepsilon/2}} P(x_1^n)$$

$$\le \varepsilon/2 + 2^{n(h-2\delta)} 2^{-n(h-\delta)} < \varepsilon,$$

for sufficiently large n. This proves the theorem.   ∎

6.4.2   Channel Coding Theorem

Let A relay a message W drawn from the index set of size M to B, who receives message $\hat{W}$ and must guess W. Such a communication is performed by sending the signal $X(W)$ that is received as a signal Y depending on X, hence on the message W. This communication process is illustrated in Figure 6.1, in which the communication medium is called the *channel* and is characterized by the conditional probability $P(Y|X)$. As before, we shall write P for the probability measure characterizing the source and the channel. In fact, we should point out that to send M messages one needs a code of length $n \ge \log M$, so from now on we shall deal with X_1^n and Y_1^n for some $n \ge \log M$.

Throughout this section, we consider the channel $P(Y_1^n|X_1^n)$ satisfying the following three properties:

$$P(y_1^n|x_1^n) = \prod_{i=1}^{n} P(y_i|x_i), \tag{6.56}$$

$$P(y_k|x_1^k, y_1^{k-1}) = P(y_k|x_k), \qquad k = 1, 2, \ldots, n, \tag{6.57}$$

$$P(x_k|x_1^{k-1}, y_1^{k-1}) = P(x_k|x_1^{k-1}), \quad k = 1, 2, \ldots, n. \tag{6.58}$$

Such a channel is called a *discrete memoryless channel* (cf. (6.56) and (6.57)) *without feedback* (cf. (6.58)).

For the channel $P(Y_1^n|X_1^n)$, we define an (M, n) *channel code* with rate R and $M = \lceil 2^{nR} \rceil$ as follows:

Figure 6.1. A communication system.

- Encoding function

$$X_1^n : \{1, 2, \ldots, M\} \to \mathcal{A}^n,$$

yielding *codewords* $X_1^n(1), \ldots, X_1^n(M)$ that constitute the *codebook*;

- Decoding function

$$\chi : \mathcal{B}^n \to \{1, 2, \ldots, M\},$$

where $\mathcal{B}$ is the alphabet of the output signal Y. Observe that the received message is $\hat{W} = \chi(Y_1^n)$.

The point to observe is that we consider a *noisy* channel that can alter the signal, so that the received signal is not necessary equal to the original one. To capture this situation, we introduce the (average) probability of error P_E defined as

$$P_E = \Pr\{W \ne \hat{W}\} \tag{6.59}$$

$$= \frac{1}{M} \sum_{i=1}^{M} \Pr\{\chi(Y_1^n) \ne i | W = i\} = \frac{1}{M} \sum_{i=1}^{M} \sum_{y_1^n : \chi(y_1^n) \ne i} P(y_1^n | x_1^n(i)).$$

This definition implies that a message to be sent is selected uniformly from the index set $\{1, 2, \ldots, M\}$ where $M = \lceil 2^{nR} \rceil$.

The **channel coding problem** is to find the largest possible R (hence M) for which there exists a channel code $(2^{nR}, n)$ such that the probability of error $P_E \to 0$ as $n \to \infty$. In a sense, the channel coding is an opposing problem to the source coding. In the latter, we try to find the best compression, thus the smallest possible bit rate R, while in the former we want to reliably send as many messages as possible, hence making the code rate R as big as possible.

To formulate the main result of this section, we introduce the **channel capacity** defined as

$$C = \max_{P(X)} \{I(X; Y)\}. \tag{6.60}$$

(In Exercises 17 and 18 we ask the reader to compute the channel capacity for the binary symmetric channel and the binary erasure channel.) We shall prove that the channel capacity is the achievability rate of channel codes. To see this intuitively, we observe that there are approximately $2^{nH(Y)}$ typical Y-sequences. We can divide this set into $2^{n(H(Y)-H(Y|X))} = 2^{nI(X;Y)}$ distinguishable sets of size $2^{nH(Y|X)}$ corresponding to different input sequences (Theorem 6.9). Hence, we can send at most $2^{nI(X;Y)}$ distinguishable sequences of length n. Below we make this statement more rigorous. In our presentation, we shall closely follow the excellent exposition of Cover and Thomas [75].

To follow this idea, we first derive the *converse* channel coding theorem by showing that if $R > C$ then $P_E > 1 - \varepsilon$ for sufficiently large n. Then we formulate and complete the proof of the Shannon coding theorem.

Our goal now is to show that

$$P_E \geq 1 - \frac{C}{R} - \frac{1}{nR}, \tag{6.61}$$

which would imply that if $R > C$, then the probability of error is bounded away from zero for sufficiently large n. To derive it, we first observe that

$$h(W|Y_1^n) = h(W) - I(W; Y_1^n)$$
$$\geq h(W) - I(X_1^n; Y_1^n) \geq n(R - C), \tag{6.62}$$

since $I(X_1^n; Y_1^n) \leq I(W; Y_1^n)$, $nR \leq h(W)$ (indeed, $2^{nR} \leq \lceil 2^{nR} \rceil = M$), and $I(X_1^n; Y_1^n) \leq nC$. To see that the latter is true, we write the following chain of relationships that follow directly from the definition of the discrete memoryless channel (6.57) and (6.16):

$$I(X_1^n; Y_1^n) = h(Y_1^n) - h(Y_1^n|X_1^n) = h(Y_1^n) - \sum_{i=1}^{n} h(Y_i|Y_1, \ldots, Y_{i-1}, X_1^n)$$

$$= h(Y_1^n) - \sum_{i=1}^{n} h(Y_i|X_i) \leq \sum_{i=1}^{n} h(Y_i) - \sum_{i=1}^{n} h(Y_i|X_i)$$

$$= nI(X_i, Y_i) \leq nC.$$

We now derive an inequality on $h(W|Y_1^n)$. Let $E = I(\hat{W} \neq W)$, where, as always, I is the indicator function. Using the chain rule (6.10)

$$h(X, Y|Z) = h(X|Z) + h(Y|X, Z) = h(Y|Z) + h(X|Y, Z),$$

and noting that $h(E|W, Y_1^n) = 0$, we have

$$h(E, W|Y_1^n) = h(W|Y_1^n) = h(E|Y_1^n) + h(W|E, Y_1^n) \leq 1 + h(W|E, Y_1^n),$$

where the inequality follows from $h(E|Y_1^n) \leq h(E) \leq 1$. We estimate $h(W|E, Y_1^n)$ as follows:

$$h(W|E, Y_1^n) = \Pr\{E = 0\}h(W|Y_1^n, E = 0) + \Pr\{E = 1\}h(W|Y_1^n, E = 1)$$
$$\leq (1 - P_E)0 + P_E nR \leq nRP_E,$$

since $h(W|Y_1^n, E = 0) = 0$ due to the fact that in this case W is completely determined by Y. Combining everything we obtain the Fano inequality

$$h(W|Y_1^n) \leq 1 + nRP_E,$$ (6.63)

which, after substituting in (6.62), proves (6.61).

In view of the above, we know that the rate of a reliable transmission (i.e., $P_E < \varepsilon$) cannot exceed the capacity of the channel C. But, can we achieve this capacity? In 1948 Shannon was the first to use a number of new ideas (e.g., random coding technique) to prove this rather counterintuitive fact. We formulate it in the following channel coding theorem. Below, we deal with the average probability of error P_E, but the theorem holds also for the maximal probability of error $P_{\max} = \max_i \Pr\{\chi(Y_1^n) \neq i | W = i\}$ (the fact that the reader is asked to prove in Exercise 13).

Theorem 6.20 (Shannon's Channel Coding Theorem) *Let $(2^{nR}, n)$ denote a channel code with bit rate R.*

(i) *For every $\varepsilon > 0$ and rate $R < C$, there exists a sequence of $(2^{nR}, n)$ codes with $P_E < \varepsilon$ for sufficiently large n.*

(ii) *Conversely, any sequence of $(2^{nR}, n)$ codes with $P_E \to 0$ must have $R < C$.*

Proof. The converse part was already proved above, so we now concentrate on the achievability part. The proof of this part is a classic example of the *random coding technique*. For $R < C$ we construct a random code that achieves P_E as small as desired for sufficiently large n. We independently generate a $(2^{nR}, n)$ code according to the distribution $P(x_1^n) = \prod_{i=1}^n P(x_i)$. The wth codeword is denoted as $x_1^n(w)$ for $1 \leq w \leq 2^{nR}$. Clearly, the probability of generating a particular code $\mathscr{C}$ is

$$P(\mathscr{C}) = \prod_{w=1}^{2^{nR}} \prod_{i=1}^n P(x_i(w)).$$

We postulate that:

- The code $\mathscr{C}$ is revealed to both sender and receiver.
- A message W is chosen uniformly, that is,

$$\Pr\{W = w\} = 2^{-nR}$$

for $w = 1, 2, \ldots, 2^{nR}$.

- The codeword of $x_1^n(w)$ sent over the channel is received as y_1^n, where

$$P(y_1^n|x_1^n(w)) = \prod_{i=1}^n P(y_i|x_i(w)).$$

- *Typical Decoding Procedure.* The receiver declares that $\hat{W}$ was sent if: (i) $(X_1^n(\hat{W}), Y_1^n)$ is *jointly typical*; (ii) there is no other index $K \neq W$ such that $(X_1^n(K), Y_1^n)$ is jointly typical (cf. joint AEP Theorem 6.9).

Under the above decoding rule, we now estimate the probability of error $P_E = \Pr\{W \neq \hat{W}\}$. Let $P_E^w(\mathscr{C}) = \Pr\{\hat{W} \neq w | W = w\}$ when $\mathscr{C}$ code was used. Clearly

$$P_E = \sum_{\mathscr{C}} P(\mathscr{C})\Pr\{\hat{W} \neq W|\mathscr{C}\} = \frac{1}{2^{nR}} \sum_{w=1}^{2^{nR}} \sum_{\mathscr{C}} P(\mathscr{C})P_E^w(\mathscr{C}).$$

But by symmetry of the code construction the average probability of error averaged over all codes does not depend on the particular index w; hence we further assume $w = 1$. The above can be written as

$$P_E = \sum_{\mathscr{C}} P(\mathscr{C})P_E^1(\mathscr{C}).$$

Let now

$$A_i = \{(X_1^n(i), Y_1^n) \in \mathcal{G}_n^\varepsilon, \ i \in \{1, 2, \dots, 2^{nR}\}\},$$

where $\mathcal{G}_n^\varepsilon$ is the set of jointly (X, Y)–typical sequences. Clearly,

$$P_E = \Pr\{\bar{A}_1 \cup A_2 \cup \cdots \cup A_{2^{nR}}\} \leq \Pr\{\bar{A}_1\} + \sum_{i=2}^{2^{nR}} \Pr\{A_i\},$$

where $\bar{A}_1$ is the complement of A_i and represents the event that $X_1^n(1)$ and Y_1^n are not jointly typical. Observe that $X_1^n(1)$ and $X_1^n(i)$ are independent, as well as $X_1^n(i)$ and Y_1^n for $i \neq 1$ due to the way the code is constructed. Then by Theorem 6.9,

$$P_E \leq \Pr\{\bar{A}_1\} + \sum_{i=2}^{2^{nR}} \Pr\{A_i\}$$

$$\leq \varepsilon + \sum_{i=2}^{2^{nR}} 2^{-n(I(X,Y)-3\varepsilon)}$$

$$\leq \varepsilon + 2^{3n\varepsilon} 2^{-n(I(X,Y)-R)}$$

$$\leq 2\varepsilon$$

for sufficiently large n and $R < I(X; Y) - 3\varepsilon$. To complete the proof, we choose the source distribution P to be the distribution P that achieves the capacity C. Hence, the condition $R < I(X, Y)$ in the above can be replaced by $R < C$. ∎

6.4.3 Rate Distortion Theorem

In some situations, we cannot or do not want to recover exactly the message sent by the encoder. The former situation arises when dealing with continuous signals while the latter occurs in a lossy data compression when one agrees to lose some information in order to get better compression. Rate distortion theory deals with such problems and can be succinctly described as follows: Given a source distribution and a distortion measure, what is the minimum rate required to achieve a particular distortion?

To state our problem in precise mathematical terms, we introduce some definitions. Let $M = \lceil 2^{nR} \rceil$. A rate distortion code consists of an encoding mapping

$$\phi_n : \mathcal{A}^n \rightarrow \{1, 2, \ldots, 2^{nR}\},$$

and a decoding (reproduction) function

$$\psi_n : \{1, 2, \ldots, 2^{nR}\} \rightarrow \hat{\mathcal{A}}^n,$$

where $\hat{\mathcal{A}}$ is the reproduction alphabet. We often write

$$\hat{X}_1^n = \psi_n(\phi_n(X_1^n))$$

and refer to $\hat{X}_1^n$ as the *reproduction* vector, and $\hat{\mathcal{A}}^n$ as the reproduction space. The point to observe is that $X_1^n \neq \hat{X}_1^n$ and $\hat{X}_1^n$ approximates the source code X_1^n. We now introduce a distortion measure that was already mentioned in Section 6.3.4. Throughout this section, we consider only the bounded distortion function, that is,

$$\max_{a \in \mathcal{A}, \hat{a} \in \hat{\mathcal{A}}} \{d(a, \hat{a})\} = d_{\max} < \infty.$$

Roughly speaking, given $D > 0$, we call a rate distortion code $(2^{nR}, n, D)$ D-semifaithful if either $d(X_1^n, \hat{X}_1^n) \leq D$ with high probability or $\mathbf{E}_{X_1^n}[d(X_1^n, \hat{X}_1^n)] \leq D$. We shall show below that these two definitions are asymptotically equivalent, at least for memoryless sources, which are assumed throughout this section. We shall follow here Csiszár and Körner [78] and Cover and Thomas [75].

Definition 6.21   (i) *A pair (R, D) is called ε-achievable rate at distortion level D for source $\{X_k\}_{k\geq 1}$, if for every $\delta > 0$ and sufficiently large n there exists a sequence of rate distortion codes (ϕ_n, ψ_n, D) of rate less than $R + \delta$ such that*

$$\Pr\{d_n(X_1^n, \psi(\phi(X_1^n)) \leq D + \delta\} \geq 1 - \varepsilon. \tag{6.64}$$

(ii) *The real number R is an achievable rate at distortion D if it is ε-achievable for every $0 < \varepsilon < 1$.*

(iii) *The pair (R, D) is called an achievable rate-distortion pair if it is achievable.*

(iv) *The infimum of achievable (resp. ε-achievable) rates at distortion D is called the rate distortion function and will be denoted $R(D)$ (resp. $R_\varepsilon(D)$). Clearly*

$$R(D) = \lim_{\varepsilon \to 0} R_\varepsilon(D). \tag{6.65}$$

Before we go any further, we would like to observe that condition (6.64) can be replaced by a more global one:

$$\lim_{n \to \infty} \mathbf{E}_{X_1^n}[d(X_1^n, \psi_n(\phi_n(X_1^n)))] \leq D. \tag{6.66}$$

without any harm to the final results. To see this, let us first assume (6.64) holds. We show that (6.66) follows. Indeed, let $\mathcal{S}_n$ be the set of $\hat{x}_1^n$ for which (6.64) holds. Then

$$\mathbf{E}[d(X_1^n, \hat{X}_1^n)] = \sum_{x_1^n} P(x_1^n) d(x_1^n, \hat{x}_1^n)$$

$$= \sum_{x_1^n \in \mathcal{S}_n} P(x_1^n) d(x_1^n, \hat{x}_1^n) + \sum_{x_1^n \notin \mathcal{S}_n} P(x_1^n) d(x_1^n, \hat{x}_1^n)$$

$$\leq D + \delta + \varepsilon d_{\max}.$$

Now assume (6.66) holds, and let $N(a, \hat{a}|x_1^n, \hat{x}_1^n)$ be the number of $a \in \mathcal{A}$ and $\hat{a} \in \hat{\mathcal{A}}$ occurring in x_1^n and $\hat{x}_1^n$, respectively. If $(X_k, \hat{X}_k)$ are i.i.d., then from the law of large numbers we infer for any $\varepsilon > 0$

$$\lim_{n \to \infty} \Pr\{|N(a, \hat{a}|X_1^n, \hat{X}_1^n)/n - P(a, \hat{a})| > \varepsilon\} = 0, \tag{6.67}$$

where, as always in this section, $P(\cdot, \cdot)$ is the probability measure of $(X_k, \hat{X}_k)$. (Observe that (6.67) holds for a much larger class of processes.) In view of this, we have:

$$d(X_1^n, \psi_n(\phi_n(X_1^n)) = \frac{1}{n} \sum_{a \in \mathcal{A}} \sum_{\hat{a} \in \hat{\mathcal{A}}} N(a, \hat{a}|X_1^n, \hat{X}_1^n) d(a, \hat{a})$$

$$\overset{(6.67)}{\leq} \sum_{a \in \mathcal{A}} \sum_{\hat{a} \in \hat{\mathcal{A}}} P(a, \hat{a}) d(a, \hat{a}) + 2\varepsilon |\mathcal{A}||\hat{\mathcal{A}}| d_{\max} \text{ with prob. } 1 - \varepsilon$$

$$\overset{(6.66)}{\leq} D + \delta \qquad \text{with prob. } 1 - \varepsilon,$$

for $\delta > 0$. This implies (6.64). Therefore, we shall freely use either (6.66) or (6.64) in the discussions below.

We now define the *information rate distortion function* $R_I(D)$, and later prove that for memoryless sources it is equal to the rate distortion function $R(D)$; that is, $R_I(D)$ represents the infimum of rates that achieve a particular distortion. This is the main result of this section, which was proved by Shannon in 1959.

Definition 6.22 *The information rate distortion function $R_I(D)$ for source $\{X_k\}_{k \geq 1}$ with distortion measure $d(x, \hat{x})$ is defined as*

$$R_I(D) = \min_{P(\hat{x}|x) : \mathbf{E}[d(X, \hat{X})] \leq D} I(X; \hat{X}), \qquad (6.68)$$

where the minimization is over all conditional distributions $P(\hat{x}|x)$, for which the joint distribution $P(x, \hat{x}) = P(x) P(\hat{x}|x)$ satisfies

$$\mathbf{E}[d(X, \hat{X})] = \sum_{(x, \hat{x})} P(x) P(\hat{x}|x) d(x, \hat{x}) \leq D.$$

In Exercise 21 we ask the reader to prove the following property of $R_I(D)$.

Lemma 6.23 *The information rate distortion function $R_I(D)$ is a non-increasing convex and continuous function of D.*

The main result of this section can be stated as follows.

Theorem 6.24 (Shannon Rate Distortion Theorem) *The rate distortion function $R(D)$ for memoryless sources X with distribution $P(\cdot)$ and bounded distortion function $d(x, \hat{x})$ is equal to the information rate distortion function*

$R_I(D)$, that is,

$$R(D) = R_I(D) = \min_{P(\hat{x}|x):\mathrm{E}[d(X,\hat{X})]\le D} I(X;\hat{X}) \tag{6.69}$$

is the minimum achievable rate at distortion level D.

Proof. We start with the converse part, and later deal with the achievability part.

Converse. We need to show that for any $(2^{nR}, n, D)$ rate distortion code, the code rate R satisfies $R > R_I(D)$. We proceed now as in Cover and Thomas [75], taking into account the fact that $X_1, \ldots, X_n$ are independent:

$$nR \ge h(\hat{X}_1^n) \ge h(\hat{X}_1^n) - h(\hat{X}_1^n|X_1^n) = I(\hat{X}_1^n; X_1^n)$$

$$\overset{\text{Theorem 6.2}}{=} h(X_1^n) - h(X_1^n|\hat{X}_1^n) = \sum_{i=1}^{n} h(X_i)$$

$$- \sum_{i=1}^{n} h(X_i|\hat{X}_1^n, X_1, \ldots, X_{i-1})$$

$$\overset{(6.15)}{\ge} \sum_{i=1}^{n} h(X_i) - \sum_{i=1}^{n} h(X_i|\hat{X}_i) = \sum_{i=1}^{n} I(X_i; \hat{X}_i)$$

$$\overset{(6.68)}{\ge} \sum_{i=1}^{n} R_I(\mathrm{E}[d(X_i, \hat{X}_i)])$$

$$\overset{\text{Lemma 6.23}}{\ge} nR_I\left(\frac{1}{n}\sum_{i=1}^{n}(\mathrm{E}[d(X_i, \hat{X}_i)])\right) = nR_I(\mathrm{E}[d(X_1^n, \hat{X}_1^n)])$$

$$\overset{\text{Lemma 6.23}}{\ge} nR_I(D).$$

Achievability. Let $P(\hat{x}|x)$ be the conditional probability that achieves equality in the definition of $R_I(D)$, that is, $I(X; \hat{X}) = R_I(D)$. We will prove that for any $\delta > 0$ there exists a rate distortion code $(2^{nR}, n, D)$ with rate $R > R_I(D)$ and distortion smaller than $D + \delta$. As in the channel coding theorem, we construct a random code as follows:

- *Generation.* Generate randomly a rate distortion codebook $\mathscr{C}$ consisting of 2^{nR} sequences $\hat{X}_1^n$ drawn i.i.d. according to $\prod_{i=1}^{n} P(\hat{x}_i)$. Index this codewords by $w \in \{1, 2, \ldots, 2^{nR}\}$.

- *Encoding.* Encode x_1^n by such w that $(X_1^n, \hat{X}_1^n(w)) \in \mathcal{G}_{n,d}^\varepsilon$, where $\mathcal{G}_{n,d}^\varepsilon$ is the set of distortion typical sequences, as defined in Definition 6.15. If there is no such w, set $w = 1$. Observe that nR bits suffice to encode such an index w.
- *Decoding.* The reproduction sequence is $\hat{X}_1^n(w)$.

We need to estimate the expected distortion $\mathbf{E}_{X_1^n, \mathscr{C}}[d(X_1^n, \hat{X}_1^n)]$ over *the random choice of codebooks* $\mathscr{C}$. Observe that

$$\mathbf{E}_{X_1^n, \mathscr{C}}[d(X_1^n, \hat{X}_1^n)] \le D + \varepsilon + P_E d_{\max},$$

where P_E is the total probability of all sequences x_1^n for which there is no codeword $\hat{X}_1^n(w)$ which is distortion typical with x_1^n. It suffices now to show that $P_E \to 0$ for sufficiently large n.

We estimate now P_E. Let $\mathcal{S}_n(\mathscr{C})$ be the set of source sequences x_1^n for which there is at least one codeword in $\mathscr{C}$ that is distortion typical with x_1^n. We obtain

$$P_E = \sum_{\mathscr{C}} P(\mathscr{C}) \sum_{x_1^n \notin \mathcal{S}_n(\mathscr{C})} P(x_1^n) = \sum_{x_1^n} P(x_1^n) \sum_{\mathscr{C}: x_1^n \notin \mathcal{S}_n(\mathscr{C})} P(\mathscr{C})$$

$$= \sum_{x_1^n} P(x_1^n) \left(1 - \Pr\{(x_1^n, \hat{X}_1^n) \in \mathcal{G}_{d,n}^\varepsilon\}\right)^{2^{nR}}, \tag{6.70}$$

where the last equation is a consequence of a random generation of 2^{nR} independent codewords. Using distortion AEP Theorem 6.16, we obtain

$$\Pr\{(x_1^n, \hat{X}_1^n) \in \mathcal{G}_{d,n}^\varepsilon\} = \sum_{\hat{x}_1^n : (x_1^n, \hat{x}_1^n) \in \mathcal{G}_{d,n}^\varepsilon} P(\hat{x}_1^n)$$

$$\ge 2^{-n(I(X;\hat{X})+3\varepsilon)} \sum_{\hat{x}_1^n : (x_1^n, \hat{x}_1^n) \in \mathcal{G}_{d,n}^\varepsilon} P(\hat{x}_1^n | x_1^n). \tag{6.71}$$

Let $A(x_1^n) = \sum_{\hat{x}_1^n : (x_1^n, \hat{x}_1^n) \in \mathcal{G}_{d,n}^\varepsilon} P(\hat{x}_1^n | x_1^n)$. We continue now with (6.70), and use the following simple inequality

$$(1 - xy)^n \le 1 - x + e^{-ny}, \quad 0 \le x, y \le 1 \tag{6.72}$$

to obtain

$$P_E \overset{(6.71)}{\le} \sum_{x_1^n} P(x_1^n) \left(1 - 2^{-n(I(X;\hat{X})+3\varepsilon)} A(x_1^n)\right)^{2^{nR}}$$

$$\overset{(6.72)}{\leq} \sum_{x_1^n} P(x_1^n)(1 - A(x_1^n)) + \exp\left(-2^{n(R-I(X;\hat{X})-3\varepsilon)}\right)$$

$$= \sum_{(x_1^n,\hat{x}_1^n)\notin \mathcal{G}_{d,n}^\varepsilon} P(x_1^n,\hat{x}_1^n) + \exp\left(-2^{n(R-I(X;\hat{X})-3\varepsilon)}\right)$$

$$= \Pr\{(X_1^n, \hat{X}_1^n) \notin \mathcal{G}_{d,n}^\varepsilon\} + \exp\left(-2^{n(R-I(X;\hat{X})-3\varepsilon)}\right)$$

$$\rightarrow 0 \quad \text{if} \quad R > I(X; \hat{X}) + 3\varepsilon,$$

since $\Pr\{(X_1^n, \hat{X}_1^n \notin \mathcal{G}_{d,n}^\varepsilon\} < \varepsilon$ due to (6.53) of Theorem 6.16. In summary, if we choose $P(\hat{x}|x)$ to be the conditional distribution that achieves the minimum in the rate information function $R_I(D)$, then $R > R_I(D)$ implies $R > I(X; \hat{X})$, and we can choose ε small enough so that $E[d(X_1^n, \hat{X}_1^n)] \leq D + \delta$. This completes the proof. ∎

Before we leave this section, we formulate yet another representation for the rate distortion function $R(D)$ that captures quite well the main thrust of $R(D)$ and is easy to understand intuitively.

Definition 6.25   (Operational Rate Distortion)   *Let $A = \hat{A}$, and define a D-ball in A^n with center y_1^n as*

$$B_D(y_1^n) = \{x_1^n : d(x_1^n, y_1^n) \leq D\}.$$

Let $N_\varepsilon(n, D)$ be the minimum number of D-balls covering A^n up to a set of probability ε, that is, the smallest N for which there exist $y_1^n(1), \ldots, y_1^n(N)$ such that the set $S_n \subseteq \bigcup_{i=1}^N B_D(y_1^n(i))$ satisfies

$$\Pr\{S_n\} \geq 1 - \varepsilon.$$

Then define

$$R_\varepsilon(n, D) = \min_{S_n:\, P(S_n)\geq 1-\varepsilon} \frac{\log N_\varepsilon(n, D)}{n}.$$

The following

$$R_O(D) = \lim_{\varepsilon\to 0} \lim_{n\to\infty} R_\varepsilon(n, D) \tag{6.73}$$

*defines the so–called **operational rate-distortion**.*

We shall now argue that $R(D) = R_O(D)$. As for the converse, let us consider ε-achievable codes $(2^{nR}, n, D)$, and let $\mathcal{S}_n$ be the set of x_1^n for which (6.64) holds. We sequentially encode $x_1^n \in \mathcal{S}_n$ by assigning to it the index of the closest $\hat{x}_1^n \in \mathcal{S}_n$ that is within distance D. The other x_1^n are encoded arbitrary (e.g., we assign to them index $w = 1$). Clearly, the set $\mathcal{S}_n$ can be covered by 2^{nR} balls of radius $D + \delta$. But, $N_\varepsilon(n, D)$ is the smallest number of balls covering $1 - \varepsilon$ set, hence $N_\varepsilon(n, D) \leq 2^{nR_\varepsilon(D+\delta)}$, and therefore

$$\limsup_{n \to \infty} \frac{\log N_\varepsilon(n, D)}{n} \leq R_\varepsilon(D),$$

since $R(D)$ is a continuous function of D. To obtain the achievability part, we construct a code $(2^{nR}, n, D)$ with any $R > R_O(D)$ that achieves asymptotically the distortion D. Actually, it suffices to set $\hat{x}_1^n(i) = y_1^n(i)$ (center of D-balls) for $i = 1, \ldots, 2^{nR}$. The encoder assigns x_1^n to the closest $y_1^n(i) = \hat{x}_1^n(i)$ such that $d(x_1^n, \hat{x}_1^n) \leq D$. Clearly, $\Pr\{d(x_1^n, \hat{x}_1^n) \leq D\} \geq 1 - \varepsilon$.

We just sketched the derivation of the following important result (cf. [78]).

Theorem 6.26 *For memoryless sources and bounded distortion function*

$$R(D) = R_O(D) = \lim_{\varepsilon \to 0} \lim_{n \to \infty} R_\varepsilon(n, D). \tag{6.74}$$

6.5 APPLICATIONS

In this applications section, we discuss three problems involving entropy and AEP. At first, we look at the typical depth in a suffix tree (see Section 6.5.1). This turns out to be equivalent to the computation of the phrase length in the Wyner-Ziv [452] compression scheme, as discussed in Section 1.2. We follow this path and extend the analysis to a lossy Lempel-Ziv scheme. The next problem we tackle is more of interest to DNA recombination than to data compression. We compare greedy and optimal algorithms that find the shortest common superstring described Section 1.4. We prove that these algorithms are asymptotically optimal (Section 6.5.2). Finally, we turn our attention to the *fixed database* model of a lossy data compression and estimate the compression ratio (Section 6.5.3).

6.5.1 Phrase Length in the Lempel-Ziv Scheme and Depth in a Suffix Tree

We start with a more general case, namely, a lossy extension of the Lempel-Ziv scheme, and conclude with a lossless scheme. Let us assume that a mixing source MX (see Section 2.1) emits a stationary mixing sequence $\{X_k\}_{k \geq 1}$. We also assume that the first n symbols, X_1^n, called the *database* or the *training sequence*,

are revealed to the decoder and encoder. To make the analysis a little more interesting, we actually postulate that the first n symbols are coming from a distorted
source, and we denote them as $\hat{X}_1^n$. Let $\hat{P}$ be the underlying probability measure
of $\hat{X}_1^n$, and we shall postulate that $\{\hat{X}_k, X_k\}_{k \geq 1}$ is a ψ-strongly mixing process.

We consider now a string X_{n+1}^∞ generated by a ψ-mixing source with the underlying probability P. For simplicity, we only consider binary alphabets $\mathcal{A} =
\hat{A} = \{0, 1\}$ and the Hamming distance as the distortion measure. Let $D > 0$ be
fixed. Define two important parameters:

Let D_n be the largest K such that a prefix of X_{n+1}^∞ of length K is within distance D
from $\hat{X}_i^{i-1+K}$ for some $1 \leq i \leq n - K + 1$, that is, $d(\hat{X}_i^{i-1+K}, X_{n+1}^{n+K}) \leq D$;

and

For *fixed* $M \leq n$, let $D_n(M)$ be the length K of the longest prefix of X_M^∞ for which
there exists $M + K \leq i \leq n + 1$ such that $d(\hat{X}_i^{i+K}, X_M^{M+K}) \leq D$.

We observe that for $D = 0$ the phrase lengths D_n and $D_n(M)$ become the *depth
of insertion* and the Mth depth of a suffix tree built from $\hat{X}_1^n$ (see Definition 1.1
of Section 1.1).

To formulate the main result of this subsection, we recall the definition of
the generalized Shannon entropy $\hat{r}_0(D)$ with respect to measure $\hat{P}$ (see Definition 6.12)

$$\hat{r}_0(D) = - \lim_{k \to \infty} \frac{\mathbf{E}_P[\log \hat{P}(B_D(X_1^k))]}{k}, \tag{6.75}$$

where $\mathbf{E}_P$ is the average operator with respect to P.

Theorem 6.27 (Łuczak and Szpankowski, 1997) *Let $\{\hat{X}_k, X_k\}_{k \geq 1}$ be a ψ-
strongly mixing process. Let also $\hat{r}_0(D) > 0$. Then*

$$\lim_{n \to \infty} \frac{D_n(M)}{\log n} = \lim_{n \to \infty} \frac{D_n}{\log n} = \frac{1}{\hat{r}_0(D)} \quad (pr.) - P \times \hat{P} \tag{6.76}$$

provided

$$\sum_{\ell \geq k} \Pr\{\mathcal{B}_\ell^\varepsilon\} = \delta(k) \to 0, \tag{6.77}$$

*where $\mathcal{B}_k^\varepsilon$ is the set of bad states in the lossy AEP with respect to $\hat{r}_0(D)$ (see
Theorem 6.13). More precisely, for any $\varepsilon > 0$*

$$\Pr\left\{\left|\frac{D_n}{\log n} - \frac{1}{\hat{r}_0(D)}\right| > \varepsilon\right\} = O\left(\max\{\psi(n^{\varepsilon/4}), \delta(\log n), \frac{\log n}{n^{\varepsilon/4}}\}\right). \quad (6.78)$$

Proof. We establish separately an upper bound and a lower bound for D_n. The quantity $D_n(M)$ can be treated in a similar manner. Throughout, we use the lossy AEP formulated in Theorem 6.13 from which we know that the space $\mathcal{A}^n$ can be divided into sets $\mathcal{G}_n^\varepsilon$ and $\mathcal{B}_n^\varepsilon$ such that for $\hat{x}_1^n \in \mathcal{G}_n^\varepsilon$

$$2^{-n\hat{r}_0(D)(1+\varepsilon)} \leq \hat{P}(B_D(x_1^n)) \leq 2^{-n\hat{r}_0(D)(1-\varepsilon)}, \quad (6.79)$$

and $P(\mathcal{B}_n^\varepsilon) < \varepsilon$. We start with the upper bound and use the first moment method (see Chapter 3). Let Z_n be the number of positions $1 \leq i \leq n - k + 1$ such that the prefix of X_{n+1}^{n+k} is within distance D from $\hat{X}_i^{i+k-1}$, that is,

$$Z_n = |\{1 \leq i \leq n - k + 1 : d(\hat{X}_i^{i+k-1}, X_{n+1}^{n+k}) \leq D\}|.$$

To find an upper bound we need to estimate the probability $\Pr\{(D_n \geq k\}$ for $k = \lfloor(1 + \varepsilon)\hat{r}_0^{-1}(D)\log n\rfloor$. Surprisingly, there is a certain detail we have to take care of to get it right. It turns out that $\{D_n \geq k\}$ does not imply that there exists a position i in the database such that $X_{n+1}^{n+k} \in B_D(\hat{X}_i^{i+k-1})$. To see it, let us assume that $d(\hat{X}_1^n, w_1^n) \leq D$ for a word $w_1^n \in \mathcal{A}^n$. When we increase n, we might have $d(\hat{X}_1^{n+l}, w_1^{n+l}) \leq D$ as well as $d(\hat{X}_1^{n+l}, w_1^{n+l}) \geq D$ for $l \geq 1$. Roughly speaking, the set $\{n : d(\hat{X}_1^n, w_1^n) \leq D\}$ does not consist of consecutive integers, that is, it has *gaps*. The correct implication is as follows:

$$\{D_n \geq k\} \implies \exists_{\ell \geq k} \exists_{1 \leq i \leq n - \ell + 1} \quad d(X_{i+1}^{i+\ell}, X_{n+1}^{n+\ell}) \leq D. \quad (6.80)$$

Then

$$\Pr\{D_n \geq k\} \leq \sum_{\ell \geq k} \sum_{i=1}^{n-\ell} \Pr\{\hat{X}_i^{i+\ell-1} \in B_D(X_{n+1}^{n+\ell}), \ X_{n+1}^{n+\ell} \in \mathcal{G}_n^{\varepsilon/2}\} + \sum_{\ell \geq k} \Pr\{\mathcal{B}_\ell^{\varepsilon/2}\}$$

$$= \sum_{\ell \geq k} \sum_{i=1}^{n-\ell} \sum_{w_1^\ell \in \mathcal{G}_\ell^{\varepsilon/2}} \Pr\{\hat{X}_i^{i+\ell-1} \in B_D(X_{n+1}^{n+\ell}), \ w_1^\ell = X_{n+1}^{n+\ell}\}$$

$$+ \sum_{\ell \geq k} \Pr\{\mathcal{B}_\ell^{\varepsilon/2}\}$$

$$\leq \sum_{\ell \geq k} \sum_{i=1}^{n-\ell} (1 + \psi(n + 2 - \ell - i)) \, 2^{-\ell\hat{r}_0(D)(1-\varepsilon/2)} + \sum_{\ell \geq k} \Pr\{\mathcal{B}_\ell^{\varepsilon/2}\}$$

$$\leq nC2^{-k\hat{r}_0(D)(1-\varepsilon/2)} + \sum_{\ell \geq k} \Pr\{\mathcal{B}_\ell^{\varepsilon/2}\},$$

where $C > 0$ is a constant. Set $k = \lfloor (1+\varepsilon)\hat{r}_0^{-1}(D) \log n \rfloor$. But $\sum_{\ell \geq k} \Pr\{\mathcal{B}_\ell^{\varepsilon/2}\} = \delta(k) \to 0$ (e.g., $\Pr\{\mathcal{B}_k^\varepsilon\} = O(1/k^{1+\delta})$ for some $\delta > 0$ suffices). Finally, we arrive at

$$\Pr\{D_n \geq \lfloor (1+\varepsilon)\hat{r}_0^{-1}(D) \log n \rfloor\} \leq \frac{c}{n^{\varepsilon/2(1-\varepsilon)}} + \delta(\log n)$$

for some constant $c > 0$. This completes the proof of the upper bound.

For the lower bound, we use the second moment method (see Chapter 3). Let $k = \lfloor (1-\varepsilon)\hat{r}_0^{-1}(D) \log n \rfloor$, and define

$$Z'_n = |\{1 \leq i \leq n/(k+g) : d(\hat{X}_{i(k+g)+1}^{(i+1)k+ig}, X_{n+1}^{n+k}) \leq D\}|$$

and $g = \Theta(\log n)$ is a *gap* between $\lfloor n/(k+g) \rfloor = \Theta(n/\log n)$ *nonoverlapping* substrings of length k. In words, instead of looking at all strings of length k we consider only $m = \lfloor n/(k+g) \rfloor$ strings with gaps of length g among them. These gaps are used to "weaken" dependency between the substrings of length k. Observe now that

$$\Pr\{D_n < k\} \leq \Pr\{Z'_n = 0\},$$

as one can expect. Indeed, if $Z'_n > 0$ then by the definition $D_n \geq k$. Note also that

$$\begin{aligned}
\Pr\{Z'_n = 0\} &= \Pr\{Z'_n = 0, X_{n+1}^{n+k} \in \mathcal{G}_k^{\varepsilon/2}\} + \Pr\{Z'_n = 0, X_{n+1}^{n+k} \in \mathcal{B}_k^{\varepsilon/2})\\
&\leq \sum_{w_1^k \in \mathcal{G}_k^{\varepsilon/2}} \Pr\{Z'_n = 0 | X_{n+1}^{n+k} = w_1^k\}\Pr\{w_1^k \in \mathcal{G}_k^{\varepsilon/2}\} + \Pr\{\mathcal{B}_k^{\varepsilon/2}\}\\
&\leq \sum_{w_1^k \in \mathcal{G}_k^{\varepsilon/2}} \Pr\{Z'_n(w_1^k) = 0\}\Pr\{w_1^k \in \mathcal{G}_k^{\varepsilon/2}\} + \Pr\{\mathcal{B}_k^{\varepsilon/2}\},
\end{aligned}$$

where

$$Z'_n(w_1^k) = |\{1 \leq i \leq n/(k+g) : d(\hat{X}_{i(k+g)+1}^{(i+1)k+ig}, w_1^k) \leq D\}|,$$

and $\Pr\{Z'_n(w_1^k) = 0\} = \Pr\{Z'_n = 0 | X_{n+1}^{n+k} = w_1^k\}$. Thus, it is suffices to show that $\Pr\{Z'_n(w_1^k) = 0\} \to 0$ *uniformly* for all $w_1^k \in \mathcal{G}_k^{\varepsilon/2}$. Hereafter, we assume that $w_1^k \in \mathcal{G}_k^{\varepsilon/2}$.

Let now $m = n/k = \Theta(n/\log n)$. From the definition of the set $\mathcal{G}_k^{\varepsilon/2}$ for every $w_1^k \in \mathcal{G}_k^{\varepsilon/2}$ we have

$$E[Z_n'(w_1^k)] = m\Pr\{B_D(w_1^k)\} \geq m2^{-k\hat{r}_0(D)(1+\varepsilon/2)} \geq c\frac{n^{\varepsilon/2(1+\varepsilon)}}{\log n}$$

for a constant c. We now compute the variance $\mathbf{Var}[Z_n'(w_1^n)]$ for $w_1^k \in \mathcal{G}_k^{\varepsilon/2}$. Let $Z_n^i = 1$ if w_1^k occurs approximately at position $i(k + g)$, otherwise $Z_n^i = 0$. Certainly, $Z_n'(w_1^n) = \sum_{i=1}^m Z_n^i$, and $\mathbf{Var}[Z_n'(w_1^n)] = \sum_{i=1}^m \mathbf{Var}[Z_n^i] + \sum_{i=1}^m \sum_{i,j=1}^m \mathbf{Cov}[Z_n^i, Z_n^j]$. Simple algebra reveals that

$$\sum_{i=1}^m \mathbf{Var}[Z_n^i] \leq m\mathbf{E}[Z_n^i] = m\Pr\{B_D(w_1^k)\} = \mathbf{E}[Z_n'(w_1^k)].$$

To compute the second term in the sum above, we split it as

$$\sum_{i,j=1}^n \mathbf{Cov}[Z_n^i, Z_n^j] = S_1 + S_2,$$

where

$$S_1 = \sum_{i=1}^m \sum_{|i-j| \leq n^{\varepsilon/4}} \mathbf{Cov}[Z_n^i, Z_n^j],$$

$$S_2 = \sum_{i=1}^m \sum_{|i-j| \geq n^{\varepsilon/4}} \mathbf{Cov}[Z_n^i, Z_n^j].$$

Observe that

$$\mathbf{Cov}[Z_n^i, Z_n^j] = \Pr\{Z_n^i Z_n^j = 1\} - \Pr\{Z_n^i = 1\}\Pr\{Z_n^j = 1\}$$

$$\leq \Pr\{Z_n^i = 1\} = \mathbf{E}[Z_n^i].$$

Hence $S_1 \leq 2n^{\varepsilon/4}\mathbf{E}[Z_n'(w_1^k)]$.

On the other hand, proceeding as in the above and using the mixing condition we also have $\mathbf{Cov}[Z_n^i, Z_n^j] \leq \psi(g)\Pr\{Z_n^i = 1\}\Pr\{Z_n^j = 1\}$ where $g \geq n^{\varepsilon/4}$. Thus, $S_2 \leq 2\psi(g)(\mathbf{E}^2[Z_n'(w_1^k)]$. Consequently, for every $w_1^k \in \mathcal{G}_k^{\varepsilon/2}$ we have ($\varepsilon < 1$)

$$\Pr\{Z_n'(w_1^k) = 0\} \le \frac{\mathrm{Var}[Z_n'(w_1^k)]}{(\mathbf{E}^2[Z_n'(w_1^k)]} \le 2\psi(g) + O\left(\frac{n^{\varepsilon/4}}{\mathbf{E}[Z_n'(w_1^k)]}\right)$$

$$\le 2\psi(g) + O\left(\frac{\log n}{n^{\varepsilon/4}}\right),$$

and finally we obtain

$$\Pr\{D_n < \lfloor (1 - \varepsilon)\hat{r}_0^{-1}(D) \log n \rfloor\} \to 0$$

as $n \to \infty$, which completes the lower bound.   ■

Here is another interesting question: Can we extend convergence in probability of $D_n/\log n$ in Theorem 6.27 to almost sure convergence? Observe that the rate of convergence in Theorem 6.27 does not justify such an extension. We recall, however, that in Section 4.2.3 we obtained such a generalization for the height H_n in a trie even when the rate of convergence was not good enough to apply the Borel-Cantelli lemma. But the height H_n is a nondecreasing function of n, and this allows us to extend convergence in probability to almost sure convergence. Clearly, D_n does not possess this property, and surprisingly enough we shall prove that $D_n/\log n$ does *not* converge almost surely. Actually, this was an open problem posed by Wyner and Ziv in [452] for the lossless case. It was answered in the negative by the author in [411] (cf. also Ornstein and Weiss [333] for another formulation of this problem).

To further study the above problem, we need a definition of the generalized height H_n and the generalized fill-up level F_n.

The height H_n is equal to the largest K for which there exist $1 \le i < j \le n+1-K$ such that $d(\hat{X}_i^{i-1+K}, X_j^{j-1+K}) \le D$;

and

The fill-up level F_n is the largest k such that for *every* $w_1^k \in \mathcal{A}^k$ there exists $1 \le i \le n+1-k$ such that $d(\hat{X}_i^{i-1+k}, w_1^k) \le D$.

Observe now that $F_n \le D_n \le H_n$. If we show that $F_n/\log n$ and $H_n/\log n$ converge (a.s.) to *different*(!) constants, then we should expect that $D_n/\log n$ oscillates between these two constants. We shall prove this fact below.

Regarding F_n, we establish the following generalization of the result presented in Exercise 4.10 of Chapter 4.

Theorem 6.28 **(Łuczak and Szpankowski, 1997)** *Let $\{X_k\}_{k \geq 1}$ be a mixing process with ψ-mixing coefficients such that for every $\kappa \geq 0$*

$$\lim_{g \to \infty} g^\kappa \alpha(g) = 0. \tag{6.81}$$

Then

$$\lim_{n \to \infty} \frac{F_n}{\log n} = \frac{1}{r_{-\infty}(D)} \quad (a.s.) \tag{6.82}$$

for $D \geq 0$.

Proof. We start with an upper bound which is quite simple in this case. Let us define $p_{\min}(k) = \min_{w_1^k \in \mathcal{A}^k} \{P(B_D(w_1^k))\}$, where we assume that the underlying probability measure of the source is P. By Theorem 6.13, $p_{\min}(k) \leq 2^{-kr_{-\infty}(D)(1-\varepsilon)}$. Observe that—unlike the lossless case—by definition of F_n we have

$$\{F_n > \ell\} \quad \Longrightarrow \quad \exists_{k > \ell} \forall_{w_1^k \in \mathcal{A}^k} \exists_{1 \leq i \leq n+1-k} \ d(X_i^{i-1+k}, w_1^k) \leq D.$$

Thus, in particular, $\Pr\{F_n > \ell\} \leq (n+1) \sum_{k > \ell} P(B_D(w_k^{\min}))$, where $w_k^{\min}$ is a word from $\mathcal{A}^k$ for which $\log(P(B_D(w_k^{\min})) \sim -kr_{-\infty}(D)$. Hence, for $\ell = \lfloor (1 + \varepsilon) r_{-\infty}^{-1}(D) \log n \rfloor$ we have

$$\Pr\{F_n > \ell\} \leq (n+1) \sum_{k > \ell} p_{\min}(k) = O(1/n^\varepsilon).$$

The lower bound requires a bit more work. Let us set $k = \lfloor (1-\varepsilon) r_{-\infty}^{-1}(D) \log n \rfloor$ and consider a set of *nonoverlapping* substrings of X_1^n of length $k = O(\log n)$ between which one inserts *gaps* of length $g = O(\log n)$. Thus, there are $m = \lfloor (n+1)/(k+g) \rfloor = O(n/\log n)$ substrings $\{X_{i(k+g)+1}^{(i+1)k+ig}\}_{i=1}^m$. We show that with probability tending to 1 as $n \to \infty$ for every $w_1^k \in \mathcal{A}^k$ one can find among these m substrings at least one which is within distance D from w_1^k and consequently $F_n \geq k$. Indeed, using mixing conditions we have

$$\Pr\{F_n < k\} \leq \Pr\{ \bigcup_{w_1^k \in \mathcal{A}^k} \bigcap_{i=1}^m \left(X_{i(k+g)+1}^{(i+1)k+ig} \neq w_1^k \right) \}$$

$$\leq \sum_{w_1^k \in \mathcal{A}^k} (1 + \psi(g))^m (1 - P(B_D(w_1^k)))^m$$

$$\leq 2^k (1 + \psi(g))^m (1 - p_{\min}(k))^m.$$

Taking into account (6.81) we immediately prove that

$$\Pr\{F_n < \lfloor (1-\varepsilon) r_{-\infty}^{-1}(D) \log n \rfloor\} \leq O(\exp(-n^{\varepsilon/2}/\log n)),$$

which completes the proof of the convergence in probability of F_n.

The rate of convergence for the upper bound does not yet justify applying the Borel-Cantelli lemma. But, F_n is a nondecreasing sequence of n; hence taking an exponentially increasing skeleton such as $n_l = 2^l$, as we explained in Section 4.2.6, we obtain almost sure convergence for the fill-up level.

∎

The height H_n is harder to handle. First of all, we formulate it for the lossless case. Its proof follows the idea already presented in Section 4.2.6 so we ask the reader to establish it in Exercise 23.

Theorem 6.29 **(Szpankowski, 1993)**   *Let $\{X_k\}_{k \geq 1}$ be a mixing process with ψ-mixing coefficients such that*

$$\sum_{g \geq 1} \psi(g) < \infty.$$

Then the height H_n in the lossless case ($D = 0$) satisfies

$$\lim_{n \to \infty} \frac{H_n}{\log n} = \frac{2}{h_1} \quad (a.s.) \tag{6.83}$$

where h_1 is the first Rényi entropy.

A generalization of Theorem 6.29 to the lossy situation is not easy. One expects that (6.83) becomes in the lossy case

$$\lim_{n \to \infty} \frac{H_n}{\log n} = \frac{2}{r_1(D)} \quad (a.s.)$$

where $r_1(D)$ is the first generalized Rényi entropy. This fact was proved only for memoryless and Markovian sources by Arratia and Waterman [22]. The difficulty arises when analyzing overlapping substrings. The reader is asked to give a try and prove it in Exercise 24 for memoryless sources.

Now, we are able to answer the question posed just after the proof of Theorem 6.27. Does $D_n/\log n$ converge (a.s.) to a limit? We consider only the lossless case to simplify the presentation. We shall show that

$$\liminf_{n \to \infty} \frac{D_n}{\log n} = \frac{1}{h_{-\infty}} \quad (a.s.) \qquad \limsup_{n \to \infty} \frac{D_n}{\log n} = \frac{2}{h_1} \tag{6.84}$$

provided (6.81) holds. Indeed, $D_n \leq H_n$, hence $D_n/\log n \leq H_n/\log n$, and obviously

$$\limsup_{n\to\infty} \frac{D_n}{\log n} \leq \lim_{n\to\infty} \frac{H_n}{\log n} \quad \text{(a.s.)}.$$

We now show that the above holds with $\leq$ replaced by $\geq$, which will complete the proof. Note that almost surely $D_n = H_n$ whenever $H_{n+1} > H_n$, which happens infinitely often (i.o.) since $H_n \to \infty$ (a.s.), and $\{X_k\}_{k\geq 1}$ is an ergodic sequence. Therefore, $\Pr\{D_n = H_n \text{ i.o.}\} = 1$ implies that almost surely there exists a subsequence, say $n_k \to \infty$, such that $D_{n_k} = H_{n_k}$. Thus

$$\lim_{n_k\to\infty} D_{n_k}/\log n_k = \lim_{n_k\to\infty} H_{n_k}/\log n_k \quad \text{(a.s.)}.$$

This implies that

$$\limsup_{n\to\infty} \frac{D_n}{\log n} \geq \lim_{n\to\infty} \frac{H_n}{\log n} \quad \text{(a.s.)},$$

that is, $\limsup_{n\to\infty} D_n/\log n = \lim_{n\to\infty} H_n/\log n$ (a.s.), and by (6.83) this proves the $\limsup$ part of (6.84). In a similar manner we can prove the $\liminf$ part by using F_n and Theorem 6.28. In Exercise 25 we ask if (6.84) can be extended to the lossy case.

6.5.2 Shortest Common Superstring Problem

We now turn our attention to the *shortest common superstring* problem that we described in Section 1.4. To recall and to reestablish notation, there are n strings, say $X_1^\ell(1), \ldots, X_1^\ell(n)$ of equal length $\ell \to \infty$. These strings are generated by n independent memoryless sources. Our goal is to construct the shortest superstring, where by a superstring we mean a string that contains given strings $X_1^\ell(1), \ldots, X_1^\ell(n)$ as substrings. In fact, instead of minimizing the superstring length we will maximize the *overlap* O_n^{opt} defined as the difference between $n\ell$ (the total length of all strings) and the shortest superstring (see Section 1.4). We compare O_n^{opt} to an overlap O_n^{gr} that is obtained by an application of a greedy algorithm to construct a superstring. A class of greedy algorithms was also discussed in Section 1.4. Here we concentrate on **RGREEDY**, which builds a superstring Z by finding the maximum overlap between (suffix of) Z and (prefixes of) those original strings that have not yet been used in Z. We prove a surprising result showing that both O_n^{opt} and O_n^{gr} asymptotically grow like $\frac{1}{h}n\log n$ where h is the entropy rate of the source provided ℓ is long enough.

Before we formulate our findings, let us get some sense of the problem. As explained in Section 1.4, to construct a superstring one takes a string, say $X_1^\ell(i_1)$, and attaches to its end another string, say $X_1^\ell(i_2)$, trying to overlap the longest suffix of $X_1^\ell(i_1)$ with the longest common prefix of $X_1^\ell(i_2)$. We continue this process until we exhaust the set of available strings. This, of course, does not construct an optimal superstring, which would require us to examine all $n!$ possible permutations of $X_1^\ell(1), \ldots, X_1^\ell(n)$. In fact, the problem is NP-hard (see Section 1.4 for a reduction to a Hamiltonian path problem). In the **RGREEDY** algorithm we choose the first string, say $X_1^\ell(i_1)$ arbitrary, but our next choice, say $X_1^\ell(i_2)$, is such that its overlap with $X_1^\ell(i_1)$ is the largest among all strings excluding $X_1^\ell(i_1)$. And so on, we choose in every step a remaining string that has the largest overlap with the superstring built so far.

Let us define C'_{ij} to be the length of the longest suffix of $X_1^\ell(i)$ that is equal to the prefix of $X_1^\ell(j)$ for $1 \le i \ne j \le n$. We introduce

$$D'_n(i) = \max_{1 \le j \le n, j \ne i} \{C'_{ij}\},$$

$$H_n = \max_{1 \le i \le n} \{D'_n(i)\}.$$

We should observe that C'_{ij} is distributed exactly as the alignment C_{ij} introduced in Definition 1.2 (i.e., the longest common prefix of $X(i)$ and $X^\ell(j)$). But then, $D'_n(i)$ is distributed as the ith depth $D_n(i)$ in a trie built from $X_1^\ell(1), \ldots X_1^\ell(n)$. Therefore, we write $D_n(i)$ for $D'_n(i)$ for $1 \le i \le n$. By Theorems 6.27 and 4.6 (see Section 4.2.3), we know that for any $\varepsilon > 0$

$$\lim_{n \to \infty} \Pr\left\{(1 - \varepsilon)\frac{1}{h} \ln n \le D_n(i) \le (1 + \varepsilon)\frac{1}{h} \ln n\right\} = 1 - O(1/n^\varepsilon) \quad 1 \le i \le n,$$

(6.85)

$$\lim_{n \to \infty} \Pr\left\{(1 - \varepsilon)\frac{2}{h_1} \ln n \le H_n \le (1 + \varepsilon)\frac{2}{h_1} \ln n\right\} = 1 - O(1/n^\varepsilon).$$

(6.86)

To get an upper bound on O_n^{opt}, we observe that

$$O_n^{\text{opt}} \le \sum_{i=1}^{n} D_n(i).$$

But with high probability (i.e., $1 - O(n^{-\varepsilon})$) all but $n\varepsilon$ depths $D_n(i)$ are bounded by $(1+\varepsilon)\frac{1}{h} \ln n$. Those $n\varepsilon$ longer strings cannot have depths bigger than H_n, hence their total contribution to the above sum is bounded by $O(n\varepsilon \ln n)$. In summary, with high probability (**whp**) for any $\varepsilon > 0$

$$\Pr\{O_n^{\text{opt}} \le (1 + \varepsilon)\frac{1}{h} n \ln n\} = 1 - O(1/n^\varepsilon)$$

(6.87)

Can we establish a matching lower bound for O_n^{opt}? Definitely *not*, unless the lengths ℓ of strings are unbounded (indeed, take $\ell = 1$ to see that $O_n^{\text{opt}} \leq n$). If the original strings are "long" enough, as specified below, then we prove the following main result.

Theorem 6.30   (Alexander, 1996; Frieze and Szpankowski, 1998)   *Consider the shortest common superstring problem, in which n memoryless sources generate independent strings $X_1^\ell(1), \ldots, X_1^\ell(n)$ over the alphabet $\mathcal{A} = \{\omega_1, \ldots, \omega_V\}$ of size $V = |\mathcal{A}|$. Then*

$$\lim_{n \to \infty} \frac{O_n^{\text{opt}}}{n \log n} = \frac{1}{h}, \qquad (pr.) \quad \lim_{n \to \infty} \frac{O_n^{\text{gr}}}{n \log n} = \frac{1}{h} \qquad (6.88)$$

provided

$$\ell > \frac{4}{h_1} \ln n \qquad (6.89)$$

where $h_1 = -\ln(p_1^2 + \cdots + p_V^2)$ is the first order Rényi's entropy and $p_t = \Pr\{X_k(i) = \omega_t\}$ for $1 \leq i \leq n$.

Proof. Since $O_n^{\text{gr}} \leq O_n^{\text{opt}}$, we need only establish a lower bound for the greedy algorithm. We first justify (6.89). Let $\mathcal{E}$ denote the event that there is no such pair, say i, j, where the overlap between $X_1^\ell(i)$ and $X_1^\ell(j)$ is bigger than $\ell/2$. Since $\mathcal{E}$ occurs if and only if $H_n \leq \ell/2$, we conclude from (6.83) that **whp** (i.e., $1 - O(n^{-\varepsilon})$) all overlaps are smaller than $\ell/2$ provided (6.89) holds. In view of this, we split every string $X_1^\ell(i)$ into a prefix $Y_1^{\ell/2}(i)$ and suffix $W_{\ell/2+1}^\ell(i)$ such that $X_1^\ell(i) = Y_1^{\ell/2}(i) W_{\ell/2+1}^\ell(i)$.

From the above, we conclude that a suffix of the superstring Z being constructed is random and independent of the previous history of the algorithm. Having this in mind, let us define for $1 \leq k \leq \ell/2$

$$N_t(y_1^k) = |\{1 \leq i \leq k : y_i = \omega_t \in \mathcal{A}, \ 1 \leq t \leq V\}|.$$

In words, $N_t(y_1^k)$ denotes the number of positions in y_1^k that are equal to the tth symbol of the alphabet. When k is large, we expect that $N_t(y_1^k) \sim kp_t$ since $N_t(y_1^k)$ is distributed as the *Binomial*$(k, p_t) := B(k, p_t)$. Indeed, let us define the set of typical sequences as

$$\mathcal{G}_k^\varepsilon = \{y_1^k : N_t(y_1^k) \leq (1 + \varepsilon)kp_t, \ 1 \leq t \leq V\}.$$

From AEP we know that $\Pr\{\mathcal{G}_k^\varepsilon\} \geq 1 - \varepsilon$, but we need a stronger version of it. In particular, we need to control the probability of $\Pr\{\mathcal{B}_k^\varepsilon\}$ where $\mathcal{B}_k^\varepsilon = \mathcal{A}^k - \mathcal{G}_k^\varepsilon$.

Recall that in Example 5.6 of Chapter 5 we proved the following bound on the tails of the binomial distribution $B = B(n, p)$ (cf. (5.40))

$$\text{Pr}\{B \le (1 - \varepsilon)np\} \le e^{-\varepsilon^2 np/2},$$

$$\text{Pr}\{B \ge (1 + \varepsilon)np\} \le e^{-\varepsilon^2 np/3}.$$

Thus

$$\text{Pr}\{\mathcal{B}_k^\varepsilon\} = \text{Pr}\{y_1^k(i) \notin \mathcal{G}_k^\varepsilon\} \le \sum_{t=1}^{V} e^{-\varepsilon^2 k p_t/3} = \theta, \tag{6.90}$$

and θ decays to zero for the following choice of ε and k

$$\varepsilon = (\log n)^{-1/3} \quad \text{and} \quad k = \left\lfloor (1 - 2\varepsilon)\frac{1}{h} \log n \right\rfloor$$

that we assume from now on. In other words, $\varepsilon^2 k \to \infty$ with n and

$$e^{-k(1-\varepsilon)h} \le \text{Pr}\{Y_1^k(i) = y_1^k\} = P(y_1^k) \le e^{-k(1+\varepsilon)h}, \quad y_1^k \in \mathcal{G}_k^\varepsilon. \tag{6.91}$$

As a side effect, we observe that **whp**

$$|\{i : y_1^k(i) \notin \mathcal{G}_k^\varepsilon\}| = o(\varepsilon n). \tag{6.92}$$

Finally, for $y_1^k \in \mathcal{G}_k^\varepsilon$ define

$$R(y_1^k) = |\{i : Y_1^k(i) = y_1^k\}|.$$

Then

$$\text{Pr}\{R(y_1^k) \le (1 - \varepsilon)n P(y_1^k)\} \le |\mathcal{G}_k^\varepsilon| e^{-\varepsilon^2 n P(y_1^k)/3}$$

$$\le V^k e^{-\varepsilon^2 n^\varepsilon/3} = o(1) \tag{6.93}$$

since $R(y_1^k)$ is distributed as $B(n, P(y_1^k))$ and $nP(y_1^k) \ge n^\varepsilon$ due to (6.91).

To model the progress of **RGREEDY**, we build a trie $\mathcal{T}_n$ from $Y_1^{\ell/2}(1)$, ..., $Y_1^{\ell/2}(n)$ and label every internal node v by $v(v)$ that denotes the size of the subtree rooted at v. In other words, if node v is at depth d and the path to v can be labeled as $v = y_1 y_2 \ldots y_d$, then $v(v)$ is the number of i such that $y_1^d(i) = y_1 y_2 \ldots y_d$. We sometimes write $v(y_1^d)$ instead of $v(v)$.

We now make the trie $\mathcal{T}_n$ a dynamic tree, by allowing random deletions leading to tries $\mathcal{T}_{n-1}, \ldots \mathcal{T}_1$. Let $W_1^{\ell/2}$ be an independent string. We delete from the trie $\mathcal{T}_n$ a string $Y_1^{\ell/2}(i)$ that has the largest overlap with a suffix of $W_1^{\ell/2}$. After the deletion, we modify labels $v(v)$ along the path of $W_1^{\ell/2}$. The reader should observe that $W_1^{\ell/2}(i)$ is a suffix of Z (i.e., it represents the second half of the string just added to the superstring Z at step i of **RGREEDY**).

Let κ_i be the length of the longest path along $W_1^{\ell/2}(i)$ in the trie $\mathcal{T}_{n-i+1}$ when deleting string $Y_1^{\ell/2}(i)$ for $1 \leq i \leq n$. We will show that **whp**

$$\kappa_1 + \kappa_2 + \cdots + \kappa_n \geq (1 - 5\varepsilon)\frac{1}{h}n \ln n. \tag{6.94}$$

The final argument goes as follows. We want to show that **whp** we will have $\kappa_t \geq k$ for $1 \leq t \leq n_0 = \lceil (1 - 3\varepsilon)n \rceil$, where $k = (1 - 2\varepsilon)\frac{1}{h}\log n$. Now, most of the time the k-suffix of Z lies in $\mathcal{G}_k^\varepsilon$. Indeed, the probability it doesn't is at most θ, which we prove in (6.90) to be small. If the k-suffix of Z belongs to $\mathcal{G}_k^\varepsilon$ and

$$v(y_1^k) \neq 0 \quad \text{for all} \quad y_1^k \in \mathcal{G}_k^\varepsilon, \tag{6.95}$$

then $\kappa \geq k$, where $v(y_1^k)$ is the label of the node reached by y_1^k. We argue next that **whp** (6.95) holds up to $n_0 = \lceil (1 - 3\varepsilon)n \rceil$. If we consider a fixed $y_1^k \in \mathcal{G}_k^\varepsilon$, then at this point the number of decrements $D(y_1^k)$ in $v(y_1^k)$ (when deleting n_0 strings) is distributed as $B(n_0, P(y_1^k))$. Hence, using $n_0 P(y_1^k) \geq (1 - 3\varepsilon)n^\varepsilon$, for $y_1^k \in \mathcal{G}_k^\varepsilon$ we have

$$\Pr\{D(y_1^k) \geq (1 + \varepsilon)n_0 P(y_1^k)\} \leq 2|\mathcal{G}_k^\varepsilon|e^{-(1-3\varepsilon)\varepsilon^2 n^\varepsilon/3} = o(1).$$

So **whp** at this point $v(y_1^k) \geq n(1 - \varepsilon)P(y_1^k) - n_0(1 + \varepsilon)P(y_1^k) > 0$ for every $y_1^k \in \mathcal{G}_k^\varepsilon$. Then (6.94) follows and this completes the proof. ∎

In Exercise 26 we ask the reader to extend the above result to mixing models and in Exercise 27 to other greedy algorithms discussed in Section 1.4.

In passing we should mention that the shortest common superstring is in general not a good compression algorithm. We start with a description of the compression code based on the shortest common superstring. Observe that instead of storing all n strings of total length $n\ell$ we can store the shortest common superstring and n pointers indicating the beginning of the original strings (plus lengths of all strings). The compression ratio c_n (understood as the ratio of the number of bits needed to transmit the compression code to the length of the original set of strings) is

$$c_n = \frac{n\ell - \frac{1}{H}n \log n + n \log_2(n\ell - \frac{1}{H}n \log n)}{n\ell}$$

where the first term of the numerator represents the length of the shortest super-string and the second term corresponds to the number of bits needed to encode the pointers. When the length of a string ℓ grows faster than $\log n$, then $c_n \to 1$ (i.e., no compression). When $\ell = O(\log n)$ some compression might take place. The fact that SCS does not compress well is hardly surprising: In the construction of SCS we do not use all available redundancy of all strings (as in the Lempel-Ziv schemes) but only that contained in suffixes/prefixes of the original strings.

6.5.3   Fixed-Database Lossy Lempel-Ziv Algorithm

We wrap up this long chapter with a computation of the compression ratio for a lossy extension of the Lempel-Ziv algorithm in the *fixed database model*. The reader is referred to Section 1.2.1 for a more detailed description. Below we only sketch the model.

Let us assume that the encoder and the decoder have access to a common *fixed database* $\hat{X}_1^n$ generated according to a Markovian source with distribution $\hat{P}$. The source sequence X_1^M is emitted also by a Markovian source with distribution P. We assume that the database sequence and the source sequence are independent, but one can extend this analysis to mixing dependent sequences.

For fixed $D > 0$, the source sequence X_1^M is partitioned into $\Pi_n = \Pi_n(X_1^M)$ variable length phrases $Z^1, Z^2, \ldots, Z^{\Pi_n}$ of respective lengths $D_n^1, \ldots, D_n^{\Pi_n}$. The first phrase Z^1 and its length D_n^1, are computed as

$$D_n^1 = \max\{k : d(\hat{X}_i^{i+k-1}, X_1^k) \leq D, \ 1 \leq i \leq n - k + 1\}$$

$$Z^1 = X_1^{D_n^1},$$

where $d(\cdot, \cdot)$ is an additive distortion measure. This implies that the first phrase comprises of the longest prefix of the input X_1^M that matches a substring in the database $\hat{X}_1^n$ within distortion D or less. The string $\hat{Z}^1$ recovered by the decoder is therefore given by

$$\hat{Z}^1 = \hat{X}_i^{i+D_n^1-1}. \tag{6.96}$$

Let now $K_n(m) = \sum_{i=1}^m D_n^i$ for some $m \geq 1$. Subsequent phrases are computed in a similar manner, that is, the $(m + 1)$st phrase is defined as

$$D_n^{m+1} = \max\{k : d\left(\hat{X}_i^{i+k-1}, X_{K_n(m)+1}^{K_n(m)+k}\right) \leq D, \ 1 \leq i \leq n - k + 1\},$$

$$Z^{m+1} = X_{K_n(m)}^{K_n(m)+D_n^m-1},$$

$$\hat{Z}^{m+1} = \hat{X}_i^{i+D_n^m-1}.$$

Observe that the source sequence is partitioned into Π_n as $X_1^M = Z^1 Z^2 \dots Z^{\Pi_n}$ while the decoder recovers the string $\hat{Z}^1 \hat{Z}^2 \dots \hat{Z}^{\Pi_n}$ that is within distortion D from X_1^M. We represent each $\hat{Z}^i$ by a pointer `ptr` to the database and its length; hence its description cost is $\log n + \Theta(\log L_n^i)$ bits. The total data compression code length is

$$\ell_n(X_1^n) = \sum_{i=1}^{\Pi_n} \log n + \Theta(\log L_n^i).$$

Thus the bit rate becomes

$$r_n(X_1^n) = \frac{1}{M} \sum_{i=1}^{\Pi_n} \left(\log n + \Theta(\log L_n^i) \right). \tag{6.97}$$

Our goal is to evaluate asymptotically $r_n(X_1^M)$ as $n, M \to \infty$. We shall prove that the asymptotic average bit rate is equal to the generalized Shannon entropy $\hat{r}_0(D)$ defined in (6.75). For convenience we repeat the definition of $\hat{r}_0(D)$

$$\hat{r}_0(D) = \lim_{n \to \infty} \frac{E_P[-\log \hat{P}(B_D(X_1^n))]}{n},$$

where $B_D(x_1^n) = \{y_1^n : d(x_1^n, y_1^n) \le D\}$ is the ball of radius D and center x_1^n. To see why $\hat{r}_0(D)$ appears as the limit of the bit rate, we recall that by Theorem 6.27 (in fact, by exactly the same arguments that led to Theorem 6.27) for any ε

$$\Pr\left\{ (1 - \varepsilon) \frac{1}{\hat{r}_0(D)} \log n \le D_n^1 \le (1 + \varepsilon) \frac{1}{\hat{r}_0(D)} \log n \right\} = 1 - O\left(\frac{\log n}{n^\varepsilon} \right) \tag{6.98}$$

for Markovian sources. (Indeed, for Markovian sources the $\psi(n)$ coefficients and $\delta(n)$ defined in (6.77) decay exponentially with n.)

Theorem 6.31 *Let us consider the fixed database model with the database $\hat{X}_1^n$ generated by a Markovian source $\hat{P}$ and the source sequence X_1^M emitted by a Markovian source P where all transition probabilities of both Markovian sources are positive. The average bit rate attains asymptotically*

$$\lim_{n \to \infty} \lim_{M \to \infty} E[r_n(X_1^M)] = \hat{r}_0(D), \tag{6.99}$$

where E denotes the expectation with respect to $P \times \hat{P}$.

Proof. We start with an upper bound. Here we adopt the approach of Wyner and Ziv [454] and its simplification proposed by Kontoyiannis [278]. Let us partition all phrases into "long" phrases and "short" phrases. A phrase k is long if $D_n^i \geq \frac{1}{\hat{r}_0(D)}(1 - \varepsilon) \log n$; otherwise it is a short phrase. Let $\mathcal{L}_M$ and $\mathcal{S}_M$ be the sets of long and short phrases, respectively. Observe that

$$N = |\mathcal{L}_M| \leq \frac{M\hat{r}_0(D)}{(1 - \varepsilon) \log n}. \tag{6.100}$$

Now we proceed as follows (c is a constant and $\varepsilon > 0$ is arbitrary small, which can change from line to line):

$$r_n(X_1^M) \leq \frac{1}{M} \sum_{i \in \mathcal{L}_M} (\log n + c \log D_n^i) + \frac{1}{M} \sum_{i \in \mathcal{S}_M} (\log n + c \log D_n^i)$$

$$\overset{(6.100)}{\leq} (1 + \varepsilon)\hat{r}_0(D) + c\frac{N}{M} \sum_{i \in \mathcal{L}_M} \frac{1}{N} \log D_n^i$$

$$+ \frac{1}{M} \sum_{i \in \mathcal{S}_M} (\log n + c \log D_n^i)$$

$$\overset{\text{Jensen}}{\leq} (1 + \varepsilon)\hat{r}_0(D) + c\frac{N}{M} \log \left(\frac{1}{N} \sum_{i \in \mathcal{L}_M} D_n^i \right)$$

$$+ \frac{1}{M} \sum_{i \in \mathcal{S}_M} (\log n + c \log D_n^i)$$

$$\leq (1 + \varepsilon)\hat{r}_0(D) + c\frac{N}{M} \log \left(\frac{M}{N} \right) + \frac{1}{M} \sum_{i \in \mathcal{S}_M} (\log n + c \log D_n^i)$$

$$\leq (1 + \varepsilon)\hat{r}_0(D) + \frac{c \log \log n}{\log n} + \frac{1}{M} \sum_{i \in \mathcal{S}_M} (\log n + c \log D_n^i).$$

We now evaluate the expected value of the third term above. We shall write below $\widetilde{D}_n^1$ for a match (phrase) that may occur in any position. Observe that D_n^1 and $\widetilde{D}_n^1$ have the same distribution, that is, $D_n^1 \overset{d}{=} \widetilde{D}_n^1$.

$$\mathbf{E} \left[\frac{1}{M} \sum_{i \in \mathcal{S}_M} (\log n + c \log D_n^i) \right] \leq \frac{c}{M} \log n \, \mathbf{E} \left[\sum_{i=1}^{|\Pi_n|} I(D_n^i \text{ is short}) \right]$$

$$\leq \frac{c}{M} \log n \, \mathbf{E} \left[\sum_{i=1}^{M} I(\widetilde{D}_n^1 \text{ is short}) \right]$$

$$\leq \ c \log n \ \Pr \left\{ D_n^1 \leq (1-\varepsilon) \frac{1}{\hat{r}_0(D)} \log n \right\}$$

$$\overset{(6.98)}{\leq} \ c \frac{\log^2 n}{n^\varepsilon},$$

where the third inequality above follows by considering not just all i's but all possible positions on X_1^M where a short match (i.e., $\widetilde{D}_n^1$) can occur. Thus

$$\limsup_{n\to\infty} \limsup_{M\to\infty} \mathbf{E}[r_n(X_1^M)] \leq \hat{r}_0(D),$$

and this establishes the desired upper bound.

We now deal with the lower bound. Let $\mathcal{H}$ be the set of those phrases whose length is not bigger than $(1+\varepsilon)\frac{1}{\hat{r}_0(D)} \log n$. Clearly

$$\mathbf{E}[r_n(X_1^M)] \geq \frac{1}{M} \mathbf{E}\left[\sum_{\mathcal{H}} \log n \right] = \frac{1}{M} \mathbf{E}[|\mathcal{H}|] \log n.$$

Thus it suffices to prove that

$$\mathbf{E}[|\mathcal{H}|] \geq \hat{r}_0(D) \frac{M(1 - o(1))}{(1+\varepsilon) \log n} \tag{6.101}$$

to establish the desired lower bound. We start with the following fact: For any $\delta > 0$ and all n, we have

$$\Pr\{D_n^1 > n^\delta\} \leq n \exp(-An^\delta) \tag{6.102}$$

where $A > 0$ is a positive constant. The above is a simple consequence of our previous findings (see Sections 4.2.6 and 6.5.1). Nevertheless, the reader is asked to prove (6.102) independently in Exercise 32. Let now $\bar{\mathcal{H}}$ be the complementary set to $\mathcal{H}$. We call phrases in $\bar{\mathcal{H}}$ as "very long." Observe that

$$M \leq |\mathcal{H}| \frac{1}{\hat{r}_0(D)} (1+\varepsilon) \log n + \sum_{i \in \bar{\mathcal{H}}} D_n^i$$

Taking the expectation of both sides of the above yields

$$M \ \leq \ \mathbf{E}[|\mathcal{H}|] \frac{1}{\hat{r}_0(D)} (1+\varepsilon) \log n$$

$$+ \mathbf{E}\left[\sum_{\{i:\ i \in \bar{\mathcal{H}} \& D_n^i \leq n^{\varepsilon/2}\}} D_n^i + \sum_{\{i:\ D_n^i > n^{\varepsilon/2}\}} D_n^i \right]$$

$$\overset{(6.102),\delta=\varepsilon/2}{\leq} \quad \mathbf{E}[|\mathcal{H}|]\frac{1}{\hat{r}_0(D)}(1+\varepsilon)\log n + n^{\varepsilon/2}\cdot \mathbf{E}\left[\sum_{i=1}^{M} I(\tilde{D}_n^1 \text{ is very long})\right]$$

$$+ \quad n\mathbf{E}\left[\sum_{i=1}^{M} I(\tilde{D}_n^1 > n^{\varepsilon/2})\right]$$

$$\leq \quad \mathbf{E}[|\mathcal{H}|]\frac{1}{\hat{r}_0(D)}(1+\varepsilon)\log n$$

$$+ \quad n^{\varepsilon/2}M\Pr\left\{D_n^1 > (1+\varepsilon)\frac{1}{\hat{r}_0(D)}\log n\right\} + nM\Pr\{D_n^1 > n^{\varepsilon/2}\}$$

$$\overset{(6.98)\ \&\ (6.102)}{\leq} \quad \mathbf{E}[|\mathcal{H}|]\frac{1}{\hat{r}_0(D)}(1+\varepsilon)\log n + M\frac{\log n}{n^{\varepsilon/2}} + n^2 M e^{-An^{\varepsilon/2}}.$$

Hence (6.101) follows and Theorem 6.31 is proved. ∎

6.6 EXTENSIONS AND EXERCISES

6.1 Consider a discrete random variable, X, defined over a finite set $\mathcal{A}$. Prove that the average number of bits, L, required to represent such a random variable satisfies the following inequality:

$$h(X) \leq L \leq h(X) + 1.$$

6.2 Prove or disprove that $h(X|Y) = h(Y|X)$. What about $h(X) - h(X|Y) = h(Y) - h(Y|X)$?

6.3 Prove Theorem 6.2.

6.4 Let $Y = g(X)$ where g is a measurable function. Prove

- $h(g(X)) \leq h(X)$;
- $h(g(X)|X) = 0$.

6.5 Random variables X, Y, Z form a Markov chain in that order (denoted $X \to Y \to Z$) if the conditional distribution of Z depends on Y and is independent of X, that is,

$$\Pr\{X = x, Y = y, Z = z\} = \Pr\{X = x\}\Pr\{Y = y|X = x\}\Pr\{Z = z|Y = y\}$$

for all possible x, y, z. Prove the *data processing inequality* that states

$$I(X; Y) \geq I(X; Z)$$

if $X \to Y \to Z$.

6.6 Consider a probability vector $\mathbf{p} = (p_1, \ldots, p_n)$ such that $\sum_{i=1}^{n} p_i = 1$. What $\mathbf{p}$ minimizes the entropy $h(\mathbf{p})$?

6.7 (Log Sum Inequality) Establish the following facts:

(i) Prove that for non-negative numbers $a_1, \ldots, a_n$ and $b_1, \ldots, b_n$

$$\sum_{i=1}^{n} a_i \log \frac{a_i}{b_i} \geq \left(\sum_{i=1}^{n} a_i \right) \log \frac{\sum_{i=1}^{n} a_i}{\sum_{i=1}^{n} b_i}$$

with equality if and only if $\frac{a_i}{b_i} = const.$

(ii) Deduce form (i) that for $p_1, \ldots, p_n$ and $q_1, \ldots, q_n$ such that $\sum_{i=1}^{n} p_i = \sum_{i=1}^{n} q_i = 1$ we have

$$\sum_{i=1}^{n} p_i \log \frac{1}{q_i} \geq \sum_{i=1}^{n} p_i \log \frac{1}{p_i},$$

that is,

$$\min_{q_i} \sum_{i=1}^{n} p_i \log \frac{1}{q_i} = \sum_{i=1}^{n} p_i \log \frac{1}{p_i}.$$

(iii) Show that the following extension of (ii) is **not true**

$$\sum_{i=1}^{n} p_i \left\lceil \log \frac{1}{q_i} \right\rceil \geq \sum_{i=1}^{n} p_i \left\lceil \log \frac{1}{p_i} \right\rceil$$

where $\lceil x \rceil$ is the smallest integer greater equal than x.

6.8 This is the problem considered by Alon and Orlitsky [10]. Let X be a discrete random variable over a finite set $\mathcal{A}$. A binary encoding $\phi : \mathcal{A} \to \{0, 1\}^*$ is an injection to the set of finite binary strings. Let $\ell(X)$ be the minimum expected length of the encoding ϕ of X (the encoding ϕ is *not* necessarily prefix free!). Prove that

$$\ell(X) \leq h(X) \qquad \text{(Wyner)}$$

$$h(X) - \log \log(|\mathcal{A}| + 1) \leq \ell(X) \qquad \text{(Dunham)}.$$

6.9 Provide details of the proof of Theorem 6.10.

6.10 Prove Lemma 6.11. Furthermore, prove the following representation for the divergence rate $D(P \parallel P^k)$ (cf. [180]).

Lemma 6.32   *Let $\{X_k\}_{k\geq 1}$ be a source described by process distributions P and P^k, where P is stationary and P^k is a kth order stationary Markov process. Then for $n \geq k$*

$$D(P \parallel P^k) = \lim_{n\to\infty} D_{P\parallel P^k}(X_0|X_{-1}, \ldots, X_{-n})$$

$$= -h_P(X) - \mathbf{E}_P[\log P^k(X_k|X_0^{k-1})].$$

6.11 ⚠ Using Lemma 6.32 established in Exercise 10 above, prove the following Markov approximation of a stationary distribution (cf. [75, 180] and Section 4.2.1).

Theorem 6.33   **(Markov Approximation)**   *Let $\{X_k\}_{k\geq 1}$ be a stationary process with distribution P. Define the kth order Markov approximation as*

$$P^k(x_1^n) := P(x_1^k) \prod_{i=k}^{n} P(x_i|X_{i-k}^{i-1} = x_{i-k}^{i-1}).$$

Then

$$\lim_{k\to\infty} D(P \parallel P^k) = 0.$$

6.12 Prove Part (i) of Theorem 6.13 for $b < 0$.

6.13 Prove the stronger version of Theorem 6.20 in which the average probability of error P_E is replaced by the maximal probability of error $P_{\max} = \max_i \Pr\{\chi(Y_1^n) \neq i|W = i\}$.

6.14 ⚠ (Strassen, 1962) Let X_1^n be generated by a memoryless source P. Define $M_\varepsilon(n)$ as the minimum cardinality of sets $A \in \mathcal{A}^n$ such that $P(A) \geq 1 - \varepsilon$. Prove that (cf. Strassen [400])

$$\log M_\varepsilon(n) = nh(P) + \sqrt{n\text{Var}[-\log P(X_1)]}\lambda - \frac{1}{2}\log n + O(1)$$

where λ is a solution of $\Phi(\lambda) = 1 - \varepsilon$ with Φ being the distribution of the standard normal distribution.

6.15 (Barron, 1985) Let $L(X_1^n)$ be the length of a fixed-to-variable codeword satisfying the Kraft inequality, where X_1^n is generated by a stationary ergodic source. Prove that *for any sequence c_n of positive constants with $\sum_n 2^{-c_n} < \infty$ the following holds*

$$\Pr\{L(X_1^n) \leq -\log P(X_1^n) - c_n\} \leq 2^{-c_n},$$

and conclude that

$$L(X_1^n) \geq -\log P(X_1^n) - c_n \quad \text{(a.s.)}.$$

6.16 (Kontoyiannis, 1997) Under the assumption of the previous exercise with X_1^n generated by a Markov source, show that there is a sequence of random variables Z_n such that

$$\frac{L(X_1^n) - nh}{\sqrt{n}} \geq Z_n \quad \text{(a.s.)}$$

where $Z_n \xrightarrow{d} N(0, \sigma^2)$ and $\sigma^2 = \mathbf{Var}[-\log P(X_1)]$ (also known as the "minimal coding variance").

6.17 Consider a *binary symmetric channel* in which the binary input $\{0, 1\}$ is transformed into the binary output $\{0, 1\}$ such that $P(Y = 0|X = 0) = P(Y = 1|X = 1) = 1-p$ and $P(Y = 0|X = 1) = P(Y = 1|X = 0) = p$. Prove that the capacity C of such a channel is

$$C = 1 - h(p) = 1 + p \log p + (1 - p) \log(1 - p).$$

6.18 Consider a *binary erasure channel* that transmits the binary input $\{0, 1\}$ into a ternary alphabet $\{0, 1, e\}$ where e means that a symbol is erased. Let $P(Y = 0|X = 0) = P(Y = 1|X = 1) = 1 - \alpha$ and $P(Y = e|X = 1) = P(Y = e|X = 0) = \alpha$. Prove that

$$C = 1 - \alpha.$$

6.19 ⚠ (Wolfowitz, 1961) Establish the following facts:
   (i) Consider the channel coding problem, and let $N_\varepsilon(n)$ be the maximum number of messages that can be reliably sent (i.e., the probability of error $P_E < \varepsilon$). Prove that (cf. [400, 449])

$$\exp(nC - K\sqrt{n}) \leq N_\varepsilon(n) \leq \exp(nC + K\sqrt{n})$$

   for a constant $K > 0$ and c is the channel capacity.
   (ii) Find an asymptotic expansion of $N_\varepsilon(n)$.

6.20 Consider a *Bernoulli*(p) source. For the Hamming distance, prove that the rate distortion function is

$$R(D) = \begin{cases} h(p) - h(D) & 0 \leq D \leq \min\{p, 1 - p\} \\ 0 & \text{otherwise} \end{cases}$$

where $h(x) = -x \log x - (1 - x) \log(1 - x)$ for $0 \leq x \leq 1$.

6.21 Prove Theorem 6.23.

6.22 ▽ Prove that $R_I(D) = R_O(D)$ with*out* using the random coding technique.

6.23 ⚠ (Szpankowski, 1993) Using the derivations from Section 4.2.6 prove Theorem 6.29 (cf. [411, 412]).

6.24 ⚠ (Arratia and Waterman, 1989) Consider the generalized height H_n in a lossy situation as defined in Section 6.5.1. For a memoryless source prove that

$$\lim_{n \to \infty} \frac{H_n}{\log n} = \frac{2}{r_1(D)} \qquad \text{(a.s.)}$$

where $r_1(D)$ is the first generalized Rényi's entropy. This is a difficult problem.

6.25 ▽ Consider a lossy extension of the Lempel-Ziv scheme as discussed in Section 6.5.1. Can we generalize (6.84) to the lossy case, that is, is the following true:

$$\liminf_{n \to \infty} \frac{D_n(D)}{\log n} = \frac{1}{r_{-\infty}(D)} \qquad \text{(a.s.)} \qquad \limsup_{n \to \infty} \frac{D_n(D)}{\log n} = \frac{2}{r_1(D)}.$$

6.26 ⚠ Consider the shortest common superstring problem as in Section 6.5.2 but with mixing sources, that is, the strings $X(1), \ldots, X(n)$ are generated by n independent mixing sources. Extend Theorem 6.30 to this situation; prove the following.

Theorem 6.34 (Frieze and Szpankowski, 1998) *Consider the shortest common superstring problem under the mixing model. Then* (**whp**)

$$\lim_{n \to \infty} \frac{O_n^{\text{opt}}}{n \log n} = \frac{1}{h} \qquad (pr.) \qquad \lim_{n \to \infty} \frac{O_n^{\text{gr}}}{n \log n} = \frac{1}{h}$$

provided

$$|X(i)| > \frac{4}{h_1} \log n$$

for all $1 \le i \le n$ where h_1 is the Rényi's entropy of order one.

6.27 ⚠ (Frieze and Szpankowski, 1998) Extend Theorem 6.30 for the shortest common superstring problem to other greedy algorithms discussed in Section 1.4, namely, **GREEDY** and **MGREEDY**.

6.28 ⚠ (Kontoyiannis, 1999) Let P^* be the probability mass function that achieves infimum in Definition 6.22 (cf. (6.68)) of the information rate distortion function $R(D)$. As before, we write $B_D(X_1^n)$ for the distortion-ball of all strings of length n that are within distortion D from the center X_1^n. Using the Gärtner-Ellis Theorem 5.15 prove that

$$\lim_{n \to \infty} \frac{-\log P^*(B_D(X_1^n))}{n} = R(D) \quad \text{(a.s.)}.$$

6.29 ⚠ Consider the *lossless* fixed database model where Y_1^n is database and X_1^M the source string. Both sequences are generated independently by a mixing source P. Let, as before, $Z^1 = X_1^{D_n^1}$ be the first phrase. Define $\mathcal{C}(X_1^n)$ to be the set of all possible Z^1. It is also called the *complete prefix set*. Observe that it can be easily generated from the suffix tree of Y_1^n. Indeed, $\mathcal{C}(X_1^n)$ is the set of all depths of insertion. The following lemma of Yang and Kieffer [438] is very useful in the analysis of lossless data compression. The reader is asked to prove it.

Lemma 6.35 (Yang and Kieffer, 1997) *Assume that the source sequence* $\{X_k\}_{k \geq 1}$ *is strongly mixing with summable mixing coefficients. Then for any string* $x_1^k \in \mathcal{A}^k$ *with* $k \leq n$

$$\Pr\{\exists_{i \leq k} : x_1^i \in \mathcal{C}(X_1^n)\} \leq \frac{c}{(n - k + 1)P(x)}$$

where c is a constant.

6.30 ⚠ Consider the *waiting time* N_ℓ defined as follows:

The waiting time N_ℓ is the smallest $N \geq 2\ell$ such that $d(X_1^\ell, X_{N-\ell+1}^N) \leq D$.

Alternatively, in the fixed database model, let $\{X_k\}_{k \geq 1}$ be a source sequence and $(\dots, Y_{-2}, Y_{-1})$ be an independent database sequence. Then N_ℓ is defined as the smallest $i \geq \ell$ such that $d(Y_{-i}^{-i+\ell-1}, X_0^{\ell-1}) \leq D$. Prove the following results.

Theorem 6.36 (Łuczak and Szpankowski, 1997; Yang and Kieffer, 1998) *Let* $\{X_k\}_{k \geq 1}$ *and* $\{Y_k\}_{k \geq 1}$ *be a strongly mixing sequence with* *summable mixing coefficients. The following holds*

$$\lim_{\ell \to \infty} \frac{\log N_\ell}{\ell} = r_0(D) \quad (a.s.)$$

where $r_0(D)$ is the generalized Shannon entropy.

6.31 ▽ Consider the kth phrase length D_n^k in the fixed database model of Section 6.5.3. Prove that for all $k \le n$

$$\lim_{n \to \infty} \frac{D_n^k}{\log n} = \frac{1}{\hat{r}_0(D)} \quad (pr.),$$

but $D_n^k / \log n$ does *not* converge almost surely when $k \to \infty$.

6.32 ⚠ Prove (6.102) from Section 6.5.3; that is, for any $\delta > 0$ and for all $n \ge 1$

$$\Pr\{D_n^1 > n^\delta\} \le n \exp(-An^\delta)$$

where D_n^1 is the length of the first phrase and A is a positive constant.

6.33 ▽ Consider the longest phrase $\max_{1 \le i \le M} D_n^i$ in the fixed database lossy Lempel-Ziv model. Is it true that **whp**

$$\max_{1 \le i \le M} D_n^i \to \frac{2}{\hat{r}(D)} \log n$$

where $\hat{r}(D)$ is related to the generalized Rényi's entropy of order one? (For a related problem see Ziv and Merhav [462].)

6.34 ⚠ Extend Theorem 6.31 to almost sure convergence; that is,

$$\lim_{n \to \infty} \lim_{M \to \infty} r_n(X_1^M) = \hat{r}_0(D) \quad (a.s.)$$

6.35 ▽ Consider the lossy extension of the Lempel-Ziv'78 algorithm and the lossy growing database model defined in Section 1.2. What is the asymptotic bit rate $r_n(X_1^n)$ in these cases? Observe that, when finding the next phrase, the comparison is made to a distorted database whose distribution is unknown.

Part III

ANALYTIC TECHNIQUES

7

Generating Functions

$$\int z\, p(a^n)\, k\omega \frac{1}{ski} =$$

Generating functions are one of the most popular analytic tools in the analysis of algorithms and combinatorics. They are helpful in establishing exact and recurrence formulas, deriving averages, variances, and other statistical properties, finding asymptotic expansions, showing unimodality and convexity, and proving combinatorial identities. In this chapter, we present the formal power series approach and analytic theory of ordinary generating functions, exponential generating functions, probability generating functions, and Dirichlet series. In particular, we construct generating functions for some popular combinatorial structures. In the applications section, we present a fairly general approach to certain recurrence equations arising in the analysis of digital trees, and derive generating functions of the number of pattern occurrences in a random text.

The generating function of a sequence $\{a_n\}_{n\geq 0}$ (e.g., representing the size of objects belonging to a certain class) is defined as

$$A(z) = \sum_{n\geq 0} a_n z^n,$$

where the meaning of z is explained below. In the *formal power series* we assume that $A(z)$ is an algebraic object; more precisely, the set of such formal power series forms a *ring*. In this case, z does not have any value, but one can identify the coefficient at z^n. Moreover, we can manipulate formal power series to discover new identities, and establish recurrences and exact formulas for the coefficients. Convergence of $A(z)$ is *not* an issue.

In *analytic theory* of generating functions, we assume that z is a complex number, and the issue of convergence is a pivotal one. In fact, singularity points of $A(z)$ (i.e., points where $A(z)$ is not defined) determine asymptotics of the coefficients. We study asymptotic methods in the next chapter.

We can define other generating functions. For example, the *exponential* generating function

$$A(z) = \sum_{n\geq 0} \frac{a_n}{n!} z^n$$

is very popular (e.g., when studying labeled combinatorial structures as we do in Section 7.2). An exotic one is the *tree-like* generating function defined as

$$A(z) = \sum_{n\geq 0} a_n \frac{n^{n-1}}{n!} z^n$$

that finds myriad applications in analysis of algorithms, coding theory, combinatorics, and information theory. We discuss its applications in this and the forthcoming chapters.

In some situations (e.g., enumeration problems), a combinatorial view on generating functions is more convenient. Let S be a set of objects (e.g., words, apples, numbers, graphs), and let $\alpha \in S$. We write $w(\alpha)$ for a weight function that we interpret here as the size of α; that is, $w(\alpha) = |\alpha|$. We define the counting generating function $A(z)$ as

$$A(z) = \sum_{\alpha\in S} z^{w(\alpha)}.$$

Certainly, this definition is equivalent to the previous one if one sets a_n to be the number of objects α such that $w(\alpha) = n$. This approach is quite useful in studying combinatorial properties of objects (e.g., the number of connected components in a graph, the number of rooted trees).

This chapter is the first one in the final part of this book, and it is devoted to analytic methods. We shall discuss formal power series and analytic theory of ordinary generating functions, exponential generating functions, probability generating functions, and Dirichlet series. In particular, we construct generating functions for a class of combinatorial structures, explain the Lagrange formula, and discuss the Borel transform and the Perron-Mellin summation formula. We illustrate our discussion with several examples and three major applications. First, we show how exponential generating functions can be used to solve exactly a certain class of recurrences arising in the analysis of digital trees. Then we consider a given string, called a *pattern*, and ask how many times it can occur (overlapping allowed) in a random string called the text. Finally, we use the theory of Dirichlet series to derive the total number of 1-digits in the binary representations of $1, 2, \ldots, n$.

There are a number of books dealing with generating functions. Generating functions in combinatorics are thoroughly discussed in Comtet [71], Goulden and Jackson [178], Stanley [397], and Wilf [447]. Applications of generating functions to discrete mathematics and analysis of algorithms can be found in Sedgewick and Flajolet [383] and Graham, Knuth and Patashnik [169].

7.1 ORDINARY GENERATING FUNCTIONS

We shall discuss below elementary properties of **ordinary generating functions** (in short: OGF). In particular, we show how to manipulate formal power series, and how to apply them to study combinatorial objects, and present elements of analytic theory of generating functions.

7.1.1 Formal Power Series

Define for a sequence $\{a_n\}_{n=0}^{\infty}$ an algebraic object

$$A(z) = \sum_{n \geq 0} a_n z^n$$

that we also write as

$$\{a_n\}_0^{\infty} \quad \longleftrightarrow \quad A(z).$$

It is convenient to introduce another notation, namely,

$$[z^n]A(z) := a_n,$$

which extracts the coefficient of $A(z)$ at z^n. In the *formal power series* approach there is no restriction on the domain of variation of z, and the series is not required to be convergent. In fact, we look at $A(z)$ (or simply A) as an algebraic object. The set of formal power series with standard addition and multiplication operations forms a *ring* $K[z]$. We recall that multiplication of A and B is defined as

$$\left\{ \sum_{k=0}^{n} a_k b_{n-k} \right\}_0^{\infty} \quad \longleftrightarrow \quad AB$$

With respect to multiplication, the ring $K[z]$ is a commutative semigroup with neutral element $1 + \sum_{n \geq 1} 0 \cdot z^n = 1$. The *reciprocal* of series A is $1/A$ understood as

$$A \cdot (1/A) = 1$$

provided $a_0 \neq 0$. For example, $(1 - z)$ is reciprocal to $A = \sum_{n \geq 0} z^n$ since

$$(1 - z)(1 + z + z^2 + \cdots) = 1.$$

We denote by $1/(1-z)$ the series $A = \sum_{n \geq 0} z^n$. We must underline that $1/(1-z)$ does *not* have any analytical meaning in the formal power series theory; it is just a name of the power series generated by $\{1\}_0^{\infty}$. We know, however, that $\frac{1}{1-z} = \sum_{n \geq 0} z^i$ for $|z| < 1$, but for this to make sense we need the analytic theory of generating functions that is discussed in the sequel.

We can define other operations and manipulation rules on the ring $K[z]$. The *derivative* $DA(z)$ of $A(z)$, also denoted as $A'(z)$, is

$$DA(z) = \sum_{n \geq 0} n a_n z^{n-1}.$$

The integration of $A(z)$ is

$$\int_0^z A(t) dt = \sum_{n \geq 1} \frac{a_{n-1}}{n} z^n.$$

Some manipulation rules are presented in Table 7.1. Rules (7.1)–(7.6) are easy to derive and left as a warmup exercise. For example, (7.6) is a consequence of our definition (7.4) and the multiplication rule.

Example 7.1: *Power Series of Harmonic Numbers*   Let $H_n = 1 + \frac{1}{2} + \cdots + \frac{1}{n}$ be the nth harmonic number. What is the formal power series of H_n? By Rule (7.6) it is $A(z)/(1 - z)$, where

$$A(z) = \sum_{n \geq 1} \frac{z^n}{n}.$$

TABLE 7.1. Basic Properties of Ordinary Generating Functions

Let $\{a_n\}_{n=0}^{\infty}$ be a sequence and $A(z)$ (or simply A), its ordinary generating functions understood as a formal power series, that is, $\{a_n\}_0^{\infty} \longleftrightarrow A$. Let zD be a formal derivative operator multiplied by z, that is,

$$(zD)A \longleftrightarrow \{na_n\}_0^{\infty}.$$

The following relationships hold:

$$\{a_n\}_0^{\infty} + \{b_n\}_0^{\infty} \longleftrightarrow A + B, \tag{7.1}$$

$$\{a_{n+k}\}_0^{\infty} \longleftrightarrow \frac{A(z) - a_0 - \cdots - a_{k-1}z^{k-1}}{z^k}, \tag{7.2}$$

$$\{n^k a_n\}_0^{\infty} \longleftrightarrow (zD)^k A, \tag{7.3}$$

$$\{1\}_0^{\infty} \longleftrightarrow A := \frac{1}{1-z}, \tag{7.4}$$

$$\left\{ \sum_{n_1 + \cdots + n_k = n} a_{n_1} \cdots a_{n_k} \right\}_0^{\infty} \longleftrightarrow A^k, \tag{7.5}$$

$$\left\{ \sum_{j=0}^{n} a_j \right\}_0^{\infty} \longleftrightarrow \frac{A}{1-z}. \tag{7.6}$$

Observe that the derivative $DA = 1/(1-z)$, so that

$$A = D^{-1}(1/(1-z)),$$

where D^{-1} is the inverse operator to D. It is convenient to give a name for such a power series. We shall see later that a good name is

$$D^{-1}(1/(1-z)) := \log \frac{1}{1-z}.$$

Thus

$$\sum_{n=1}^{\infty} H_n z^n = \frac{1}{(1-z)} \log \frac{1}{1-z}.$$

In fact, the reader may verify below identities that follow directly from the above and Rule (7.6)

$$1/(1-z)^2 \log(1/(1-z)) = \sum_{n=1}^{\infty} z^n \sum_{k=1}^{n} H_k,$$

$$1/(1-z)^2 \log(1/(1-z)) + 1/(1-z)^2 = \sum_{n=0}^{\infty} (n+1) H_{n+1} z^n.$$

Comparing these two expressions, we can obtain the following

$$\sum_{k=1}^{n} H_k = (n+1) H_{n+1} - n - 1.$$

Example 7.2: *Ordered Partition of an Integer*   Let $f_k(n)$ be the number of ways that the nonnegative number n can be partitioned as an *ordered* sum of k nonnegative integers. For example: $f_2(3) = 4$ since $3 = 3 + 0 = 2 + 1 = 1 + 2 = 0 + 3$. Observe that

$$f_k(n) = \sum_{n_1 + \cdots + n_k = n} 1,$$

hence by Rule (7.5) we have

$$\sum_{n=0}^{\infty} f_k(n) z^n = 1/(1-z)^k.$$

We shall need analytic theory of generating functions to recover $f_k(n)$ from the above.   ∎

7.1.2   Combinatorial Calculus

As we mentioned before, generating functions are often used in the enumeration of combinatorial structures. This is known as *combinatorial calculus*. In such enumerations, it is desirable to introduce classes of combinatorial objects. There are excellent books dealing with such problems (e.g., Flajolet and Sedgewick [149], Goulden and Jackson [178] and Stanley [397]), so we restrict our discussion to the two simplest cases and a few examples. We shall follow here the presentation of Flajolet and Sedgewick [149].   Let $\mathcal{A}$ be a class of combinatorial objects (e.g., set of binary trees, strings, graphs). (If $\mathcal{A}$ is a set of words, then we often call it a *language*.) The associated generating function is defined as

$$A(z) = \sum_{\alpha \in \mathcal{A}} z^{|\alpha|}, \tag{7.7}$$

where $|\alpha|$ is the size of the object α. Let now $\mathcal{B}$ be another class. We can combine $\mathcal{A}$ and $\mathcal{B}$ into a new class. We define a "disjoint union" denoted as $\mathcal{A}+\mathcal{B}$ that gives the class consisting of disjoint copies of the members of $\mathcal{A}$ and $\mathcal{B}$. We can also define a "Cartesian product" denoted as $\mathcal{A} \times \mathcal{B}$ that produces a class of ordered pairs of objects, one from $\mathcal{A}$ and the other from $\mathcal{B}$. It is easy to see that

$$\mathcal{C} = \mathcal{A} + \mathcal{B} \longleftrightarrow C(z) = A(z) + B(z), \tag{7.8}$$

$$\mathcal{C} = \mathcal{A} \times \mathcal{B} \longleftrightarrow C(z) = A(z)B(z). \tag{7.9}$$

For example, to prove (7.9) we argue as follows

$$C(z) = \sum_{\gamma \in \mathcal{A} \times \mathcal{B}} z^{|\gamma|} = \sum_{\alpha \in \mathcal{A}} \sum_{\beta \in \mathcal{B}} z^{|\alpha|+|\beta|} = A(z)B(z).$$

The reader may wonder why we complicate the matter and use generating functions in the enumeration problems. The reason is that generating functions are very successful in the enumeration. And, they are successful because in (7.7) we deal with *all* objects belonging to a class $\mathcal{A}$ and do not need to worry about boundary effects of the size n. For example, consider all binary strings without two consecutive 1's (see Example 7.4). If we restrict the analysis to strings of length n, then we must worry about the bits at the end of the string. (This becomes more complicated when the avoidable pattern is more complicated.) By considering the *whole* set of *all* binary strings without consecutive 1's, the enumeration is as simple as in the example below.

Example 7.3: *Enumeration of Strings* Let $\mathcal{A}$ be the set of all binary strings with no two consecutive 1 bits. Such strings are either an object of size zero that we denote as ϵ, or a single 1 or a recursive (composed) object consisting of 0 or 10 followed by a string with no two consecutive 1 bits (hence belonging to $\mathcal{A}$). In other words, the class $\mathcal{A}$ can be written formally as

$$\mathcal{A} = \epsilon + \{1\} + \{0, 10\} \times \mathcal{A}.$$

By (7.8) and (7.9), the above immediately translates into

$$A(z) = 1 + z + (z + z^2)A(z).$$

Example 7.4: *Enumeration of Binary Trees* Let $\mathcal{B}$ be the class of (unlabeled) binary trees that are either empty or consist of a root and a left binary tree and a right binary tree. In symbolic calculus

$$\mathcal{B} = \epsilon + \{node\} \times \mathcal{B} \times \mathcal{B}$$

which translates into

$$B(z) = 1 + z B^2(z).$$

Example 7.5: *Partition of Integers* We consider now a somewhat more so-phisticated problem. As in Example 7.2, we partition the integer n into an un-limited number of terms, but this time we ignore the order. In other words, we assume

$$n = n_1 + n_2 + \cdots, \quad \text{subject to} \quad 1 \le n_1 \le n_2 \le \cdots$$

Let p_n be the number of such solutions. For example, $p_4 = 5$ since $4 = 1 + 1 + 1 + 1 = 1 + 1 + 2 = 1 + 3 = 2 + 2 = 4$. To find p_n, we represent the class $\mathcal{P}$ of all partitions for all n as a Cartesian product of simpler sets. Define $\{i^*\}$ as

$$\{i^*\} = \{\epsilon; \ i; \ i, i; \ i, i, i; \ \cdots\},$$

that is, $\{i^*\}$ is a set consisting of all finite repetitions of i (including the zeroth repetition). Clearly,

$$\mathcal{P} = \{1^*\} \times \{2^*\} \times \cdots$$

For example, the partition $1 + 1 + 2 = 4$ takes the third element from $\{1^*\}$, the second element from $\{2^*\}$, and the first element from all others. If $I_i(z)$ denotes the generating function of $\{i^*\}$, then

$$I_i(z) = \sum_{n=0}^{\infty} z^{in} = 1/(1 - z^i),$$

and therefore

$$P(z) = \sum_{n=0}^{\infty} p_n z^n = \prod_{i=1}^{\infty} \frac{1}{(1 - z^i)}.$$

Now, once we know the formula we can find an even easier derivation. Observe that n can also be represented as

$$n = 1 n_1 + 2 n_2 + \cdots + n n_n$$

where n_i is the number of i's in the partitions. Then

$$\sum_{n=0}^{\infty} p_n z^n = \sum_{n=0}^{\infty} \sum_{1 n_1 + 2 n_2 + \cdots + n n_n = n} z^{1 n_1 + 2 n_2 + \cdots + n n_n}$$

$$= (1 + z + z^2 + \cdots) \cdot (1 + z^2 + z^6 + \cdots) \cdots (1 + z^i + z^{2i} + \cdots) \cdots$$

$$= \prod_{i=1}^{\infty} \frac{1}{(1 - z^i)}.$$

To see the above, we just apply the rule of formal power series multiplication that yields

$$\left\{ \sum_{k_1+k_2+..=n} \prod_{i=1}^{\infty} a_{i,k_i} \right\}_0^{\infty} \longleftrightarrow \prod_{i=1}^{\infty} A_i(z),$$

where

$$\{a_{i,k}\}_{k=0}^{\infty} \longleftrightarrow A_i(z).$$

Setting $A_i(z) = 1/(1 - z^i)$, we prove the above identity.

This latter approach is quite useful in some situations. For example, in Section 3.3.3 we applied (3.24), which we repeat here

$$\sum_{m=0}^{\infty} \sum_{n_1+2n_2+\cdots mn_m=m} \frac{x^m}{n_1! n_2! \cdots n_m!} \prod_{i=1}^{m} c_i^{n_i} = \exp\left(\sum_{i=1}^{\infty} c_i x^i\right),$$

to prove the result of Karlin and Ost (see Theorem 3.8). We delayed the proof of (3.24), and now are are ready to derive it. We use the same approach as above. Starting from the right-hand side we have

$$\exp\left(\sum_{i=1}^{\infty} c_i x^i\right) = \left(1 + c_1 x + \frac{c_1^2 x^2}{2!} + \frac{c_1^3 x^3}{3!} + \cdots\right)$$

$$\cdots \left(1 + c_i x^i + \frac{c_1^2 x^{2i}}{2!} + \frac{c_1^3 x^{3i}}{3!} + \cdots\right) \cdots$$

$$= \sum_{m=0}^{\infty} \sum_{n_1+2n_2+\cdots mn_m=m} \frac{x^{n_1+2n_2+\cdots mn_m}}{n_1! n_2! \cdots n_m!} \prod_{i=1}^{\infty} c_i^{n_i}$$

$$= \sum_{m=0}^{\infty} \sum_{n_1+2n_2+\cdots mn_m=m} \frac{x^m}{n_1! n_2! \cdots n_m!} \prod_{i=1}^{m} c_i^{n_i}.$$

This proves the desired identity. ∎

7.1.3   Elements of Analytic Theory

In the formal theory of power series, we viewed the ordinary generating function $A(z)$ as an algebraic element and z is added for a convenience. Convergence of the series was not an issue. If, however, the power series converges to an analytic function, then we may use a whole machinery of the theory of analytic functions to find analytic information about $A(z)$. In particular, such information can be used to obtain asymptotic expansion of the coefficients (see Chapter 8).

The first problem we face is to see under what conditions the series

$$A(z) = \sum_{n=0}^{\infty} a_n z^n$$

converges for z complex. Fortunately, following Hadamard we can give a fairly complete answer to this question.

Theorem 7.1   (Hadamard)   *There exists a number $0 \leq R \leq \infty$ such that the series $A(z)$ converges for $|z| < R$ and diverges for $|z| > R$. The radius of convergence R can be expressed as*

$$R = \frac{1}{\limsup_{n \to \infty} |a_n|^{\frac{1}{n}}}, \tag{7.10}$$

where by convention $1/0 = \infty$ and $1/\infty = 0$. The function $A(z)$ is analytic for $|z| < R$.

Proof. The cases $R = 0$ and $R = \infty$ are left to the reader. Let $0 < R < \infty$ and $|z| < R$. We prove that the series $A(z)$ converges. Choose $\varepsilon > 0$ such that

$$|z| < \frac{R}{1 + \varepsilon R}.$$

By definition of $\limsup$, there exists N such that for all $n > N$ we have from (7.10)

$$|a_n|^{\frac{1}{n}} < R^{-1} + \varepsilon.$$

Thus

$$|a_n||z|^n < \left(|z|(R^{-1} + \varepsilon) \right)^n \leq \alpha^n$$

for $\alpha < 1$. The series converges absolutely for $|z| < R$.

To prove the second part of the theorem, let now $|z| > R$. We show that then $a_n z^n \nrightarrow 0$, hence the series cannot converge. Let for $\varepsilon > 0$

$$\theta = |z/R - \varepsilon z|.$$

Since $|z| > R$ we can choose $\varepsilon > 0$ such that $\theta > 1$. But by (7.10) and the definition of lim sup for *infinitely* many n we have

$$|a_n|^{\frac{1}{n}} > R^{-1} - \varepsilon,$$

hence for infinitely many n

$$|a_n||z|^n > \left|(R^{-1} - \varepsilon)z\right|^n = \theta^n,$$

which increases exponentially. This completes the proof. ■

Hadamard's theorem sheds light on asymptotics of the coefficients a_n. We will devote Chapter 8 to this problem. Here we only observe that Hadamard's theorem tells us what is happening with $A(z)$ for $|z| < R$ and $|z| > R$, but nothing about $|z| = R$. Really? In fact, from the theorem we should conclude that the function $A(z)$ has at least one *singularity* (a point where the function ceases to be well defined) on the circle $|z| = R$. To see this, we argue by contradiction. If there is no singularity on $|z| = R$, then we can cover it with a finite number of disks (by the Heine-Borel theorem) on which $A(z)$ is analytic. This would imply that $A(z)$ converges in a larger disk $|z| < R + \varepsilon$, which is the desired contradiction.

So far we were concerned with the calculation of generating functions from their coefficients. But often we need to find the coefficients knowing the generating function. One way to solve this problem is to have a list of generating functions together with their corresponding coefficients and find $[z^n]A(z)$ by inspection. In Table 7.2 we list some classic generating functions. We illustrate their usage on several examples.

Example 7.6: *A Simple Recurrence*

We start with a simple recurrence. Let

$$a_{n+1} = 2a_n + n, \quad n \geq 0,$$

with $a_0 = 1$. Multiplying both sides by z^n and summing over all $n \geq 0$, we obtain by (7.2) and (7.3)

$$\frac{A(z) - 1}{z} = 2A(z) + \frac{z}{(1 - z)^2},$$

TABLE 7.2. Classic Generating Functions

1. Binomial coefficients

$$\frac{1}{1-z-uz} = \sum_{n,k \geq 0} \binom{n}{k} u^k z^n$$

$$\frac{z^k}{(1-z)^{k+1}} = \sum_{n \geq k} \binom{n}{k} z^n$$

$$\frac{1}{(1-z)^{k+1}} = \sum_{n \geq 0} \binom{n+k}{n} z^n$$

$$(1+z)^\alpha = \sum_{n \geq 0} \binom{\alpha}{n} z^n$$

2. Bernoulli numbers

$$\frac{z}{(e^z-1)} = \sum_{n=0}^{\infty} B_n \frac{z^n}{n!}$$

3. Catalan numbers

$$\frac{1-\sqrt{1-4z}}{2z} = \sum_{n \geq 0} \frac{1}{n+1} \binom{2n}{n} z^n$$

4. Reciprocals

$$\ln \frac{1}{1-z} = \sum_{n \geq 1} \frac{z^n}{n}$$

5. Harmonic numbers

$$\frac{1}{1-z} \ln \frac{1}{1-z} = \sum_{n \geq 1} H_n z^n$$

6. Fibonacci numbers

$$\frac{z}{1-z-z^2} = \sum_{n \geq 0} F_n z^n$$

7. Tree function

$$z e^{T(z)} = T(z) = \sum_{n \geq 1} \frac{n^{n-1}}{n!} z^n$$

8. Stirling numbers of the first kind

$$\frac{1}{(1-z)^u} = \sum_{n,k \geq 0} \begin{bmatrix} n \\ k \end{bmatrix} u^k \frac{z^n}{n!}$$

$$u(u+1)\cdots(u+n-1) = \sum_{k \geq 0} \begin{bmatrix} n \\ k \end{bmatrix} u^k$$

9. Stirling numbers of the second kind

$$e^{u(e^z-1)} = \sum_{n,k \geq 0} \begin{Bmatrix} n \\ k \end{Bmatrix} u^k \frac{z^n}{n!}$$

$$(e^z - 1)^n = n! \sum_{k \geq n} \begin{Bmatrix} k \\ n \end{Bmatrix} \frac{z^k}{k!}$$

$$\frac{z^k}{(1-z)(1-2z)\cdots(1-kz)} = \sum_{n \geq 0} \begin{Bmatrix} n \\ k \end{Bmatrix} z^n$$

which can be solved to yield

$$A(z) = \frac{1 - 2z + 2z^2}{(1 - z)^2(1 - 2z)}$$

$$= \frac{2}{1 - 2z} - \frac{1}{(1 - z)^2}.$$

The last line is a partial fraction expansion of the first line. By Entry 1 of Table 7.2 we immediately obtain

$$a_n = 2^{n+1} - n - 1$$

for all $n \geq 0$.

Example 7.7: *Enumeration of Unlabeled Trees—Catalan Numbers* In Example 7.4 we proved that the ordinary generating function, $B(z)$, of the number of *unlabeled binary trees* satisfies $B(z) - 1 = zB^2(z)$. It can be solved to yield

$$B(z) = \frac{1 - \sqrt{1 - 4z}}{2z}.$$

Let us now find $[z^n]B(z)$, that is, the nth Catalan number. We shall prove that

$$[z^n]B(z) = \frac{1}{n+1}\binom{2n}{n}.$$

Observe that by the binomial series (see Table 7.2) we have

$$zB(z) = -\frac{1}{2}\sum_{n \geq 1}\binom{\frac{1}{2}}{n}(-4z)^n.$$

Extracting coefficients of both sides of the above, we obtain

$$[z^n]B(z) = -\frac{1}{2}\binom{\frac{1}{2}}{n+1}(-4)^{n+1}$$

$$= -\frac{1}{2}\frac{\frac{1}{2}(\frac{1}{2} - 1)\cdots(\frac{1}{2} - n)(-4)^n}{(n+1)!}$$

$$= \frac{1 \cdot 3 \cdot 5 \cdots (2n - 1)2^n}{(n+1)!}$$

$$= \frac{1}{n+1}\frac{1 \cdot 3 \cdot 5 \cdots (2n-1)}{n!}\frac{2 \cdot 4 \cdot \cdot 6 \cdots 2n}{1 \cdot 2 \cdot 3 \cdots n} = \frac{1}{n+1}\binom{2n}{n}.$$

Example 7.8: *Stirling Numbers of the Second Kind* Let us enumerate the number of set partitions (i.e., a collection of nonempty pairwise disjoint subsets whose union is the original set) of the set $[n] := \{1, 2, \ldots, n\}$ into k subsets known also as classes. For example, the set $[3]$ is partitioned into $k = 2$ classes as follows:

$$[3] = \{\{1\}\{2, 3\}; \{1, 3\}\{2\}; \{1, 2\}\{3\}\}.$$

This number is known as the *Stirling number of the second kind*, and it is denoted as $\left\{{n \atop k}\right\}$. We first find a recurrence on it, and then derive an explicit formula.

Let us start with a recurrence on $\left\{{n \atop k}\right\}$. The set of partitions of $[n]$ into k classes can be split into two groups: In the first group, you will have all partitions in which n is in a separate class itself, that is, we have a subset of the form $\{n\}$. The second group contains all other partitions. It is easy to notice that the number of partitions in the first group is $\left\{{n-1 \atop k-1}\right\}$. So it remains to enumerate the second group.

Observe that the second group is empty for $k = n$ and for $k < n$ it contains all partitions of $[n]$ in which the element n is not listed separately but together with other elements. Let us now delete n from all such partitions. We definitely obtain a partition of $[n - 1]$ into k classes, but some classes will be repeated. Which ones and how many of them? Since we have k classes, and n could appear in any of them, after the deletion of n we will end up with k copies of the same class. Thus, the second group contains $k\left\{{n-1 \atop k}\right\}$ partitions. This proves the following recurrence

$$\left\{{n \atop k}\right\} = \left\{{n-1 \atop k-1}\right\} + k\left\{{n-1 \atop k}\right\}, \qquad (n, k) \neq (0, 0). \qquad (7.11)$$

We also postulate that $\left\{{0 \atop 0}\right\} = 1$ and $\left\{{n \atop 0}\right\} = 0$ for $n \neq 0$.

To find the generating function

$$S_k(z) = \sum_{n \geq k} \left\{{n \atop k}\right\} z^n$$

and an explicit formula for $\left\{{n \atop k}\right\}$, we proceed as in the previous examples. The recurrence (7.11) translates into

$$S_k(z) = z S_{k-1}(z) + kz S_k(z)$$

for $k \geq 1$ and $S_0(z) = 1$. Thus,

$$S_k(z) = \frac{z}{1 - kz} S_{k-1}(z),$$

and finally

$$S_k(z) = \sum_{n \geq k} \left\{ {n \atop k} \right\} z^n = \frac{z^k}{(1-z)(1-2z)\cdots(1-kz)}$$

which is Entry 9 of Table 7.2.

Now we find an explicit formula for $\left\{ {n \atop k} \right\} = [z^n] S_k(z)$. As before, we need a partial fractional expansion that looks like this:

$$\frac{1}{(1-z)(1-2z)\cdots(1-kz)} = \sum_{j=1}^{k} \frac{a_j}{(1-jz)}.$$

After multiplying both sides by $(1 - rz)$ and evaluating at $z = 1/r$ we obtain

$$a_r = (-1)^{k-r} \frac{r^{k-1}}{(r-1)!(k-r)!}.$$

In view of this, we immediately find

$$\left\{ {n \atop k} \right\} = [z^n] \frac{z^k}{(1-z)(1-2z)\cdots(1-kz)}$$

$$= [z^{n-k}] \sum_{r=1}^{k} \frac{a_r}{(1-rz)} = \sum_{r=1}^{k} a_k r^{n-k}$$

$$= \frac{1}{k!} \sum_{r=1}^{k} (-1)^{k-r} \binom{k}{r} r^n$$

which is not much simpler than the recurrence itself. ∎

7.1.4 Generating Functions Over an Arithmetic Progression

Finally, we deal with the following problem: How to compute sums over an arithmetic progression (e.g., $\sum_{k \geq 0} \binom{n}{3k} x^{3k}$). In general, let $A(z)$ be the ordinary generating function of a sequence a_n, and for fixed r define

$$A_r(z) = \sum_{n=rk,\ k \geq 0} a_n z^n = \sum_{k=0}^{\infty} a_{rk} z^{rk}.$$

Can we estimate $A_r(z)$ knowing $A(z)$? We can, and here is how it works. Let $\omega^r = 1$ be the rth root of unity. There are r such roots that we denote as ω_k for $k = 0, 1, \ldots, r-1$. In fact, $\omega_k = e^{(2\pi i k)/r}$. Observe that for r not dividing n (i.e.,

$r \nmid n)$ we have

$$\sum_{k=0}^{r-1} \omega_k^n = \sum_{k=0}^{r-1} e^{(2\pi ikn)/r} = \frac{1 - e^{2\pi in}}{1 - e^{(2\pi in)/r}} = 0.$$

If $r|n$, then $\omega_k^n = 1$ for all $k = 0, 1, \ldots, r - 1$. In summary,

$$\frac{1}{r} \sum_{\omega^r = 1} \omega^n = \begin{cases} 1 & \text{if } r|n \\ 0 & \text{otherwise.} \end{cases} \tag{7.12}$$

This immediately implies

$$A_r(z) = \frac{1}{r} \sum_{k=0}^{r-1} A(z\omega_k). \tag{7.13}$$

Example 7.9:   *A Combinatorial Sum*   We will find

$$R_n = \sum_k (-1)^k \binom{n}{3k}$$

applying (7.13) with $r = 3$. We first observe that

$$A(z) = \sum_k \binom{n}{k} z^k = (1 + z)^n.$$

Since $R_n = A_3(-1)$, $\omega_1 = e^{2\pi i/3}$, and $\omega_2 = e^{4\pi i/3}$ we obtain

$$A_3(-1) = \frac{1}{3} \left(A(z) + A(z\omega_1) + A(z\omega_2) \right)|_{z=-1}$$

$$= \frac{1}{3} \left((1 - \omega_1)^n + (1 - \omega_2)^n \right)$$

$$= \frac{1}{3} \left(\left(\frac{3 - \sqrt{3}i}{2} \right)^n + \left(\frac{3 + \sqrt{3}i}{2} \right)^n \right)$$

$$= \frac{2}{3} 3^{\frac{n}{2}} \cos\left(\frac{n\pi}{3} \right),$$

which would be hard to find without (7.13).   ■

7.2 EXPONENTIAL GENERATING FUNCTIONS

It is easy to construct a sequence a_n for which the ordinary generating function (OGF) either does not exist or cannot be computed in a closed form. Think of $a_n = n!$ or $a_n = n^n/n!$. Often a very simple trick solves the problem. Namely, one defines a weighted generating function such that

$$A(z) = \sum_{n=0}^{\infty} a_n \frac{z^n}{w_n}$$

where w_n is the weight. The most popular and widely used weight is $w_n = n!$, which produces the *exponential* generating function. But other weights work remarkably well in a variety of applications (e.g., $w_n = n!/n^n$ leads to a *tree-like* generating function that we shall discuss in the sequel).

7.2.1 Elementary Properties

Let us first concentrate on exponential generating functions. We sometimes must distinguish between ordinary and exponential generating functions; therefore, we often write (lowercase letter) $a(z)$ for the exponential generating function of $\{a_n\}_{n=0}^{\infty}$, that is,

$$a(z) = \sum_{n=0}^{\infty} a_n \frac{z^n}{n!}.$$

Elementary properties of exponential generating functions are listed in Table 7.3. As in the case of ordinary generating functions, we can view exponential generating functions as a formal power series, but we skip this formalism and assume that $a(z)$ is an analytic function. Clearly, what we said above about analytic theory of OGF applies here with a_n replaced by $a_n/n!$.

The properties (7.14)–(7.17) presented in Table 7.3 are easy to prove, and left for the reader. We only show here how to derive (7.17) from its ordinary generating function counterpart (7.5). If $a(z)$ and $b(z)$ are exponential generating functions, then

$$\left[\frac{z^n}{n!}\right] a(z)b(z) = n![z^n]a(z)b(z) = n! \sum_{k=0}^{n} \frac{a_k}{k!} \frac{b_{n-k}}{(n-k)!} = \sum_{k=0}^{n} \binom{n}{k} a_k b_{n-k},$$

which yields (7.17).

Example 7.10: *Bell Numbers*   In Example 7.8 we computed the Stirling numbers of the second kind $\left\{{n \atop k}\right\}$ that represent the number of ways the set $[n] =$

TABLE 7.3. Basic Properties of Exponential Generating Functions

Let $\{a_n\}_{n=0}^{\infty}$ be a sequence and $a(z)$ its exponential generating function. We write

$$\{a_n\}_0^{\infty} \quad \longleftrightarrow \quad a(z).$$

We shall also write $Da(z)$ or $a'(z)$ for the derivative of $a(z)$.
The following relationships hold:

$$\{a_n\}_0^{\infty} + \{b_n\}_0^{\infty} \longleftrightarrow a(z) + b(z), \tag{7.14}$$

$$\{a_{n+k}\}_0^{\infty} \longleftrightarrow D^k a(z) = \frac{d^k}{dz^k} a(z), \tag{7.15}$$

$$\{n^k a_n\}_0^{\infty} \longleftrightarrow (zD)^k a(z), \tag{7.16}$$

$$\left\{\sum_{k=0}^{n} \binom{n}{k} a_k b_{n-k}\right\}_{n=0}^{\infty} \longleftrightarrow a(z)b(z). \tag{7.17}$$

$\{1, \ldots, n\}$ can be partitioned into k classes. What if we are interested in the total number of partitions of $[n]$ irrespective of how many classes are there? These numbers are called the *Bell numbers* and are denoted as b_n. We prove that

$$b_{n+1} = \sum_{k=0}^{n} \binom{n}{k} b_k = \sum_{k=0}^{n} \binom{n}{k} b_{n-k}$$

with $b_0 = 1$ by convention. Indeed, let us consider the collection of $[n + 1]$ partitions. Its cardinality is b_{n+1}. We split this collection into $k = 0, 1, \ldots, n$ groups such that the kth group consists of all partitions, in which the class that includes symbol $n + 1$ contains exactly k others symbols. Certainly, the above follows.

To obtain the exponential generating function $b(z)$ of the Bell numbers, we use (7.15) and (7.17), which yields

$$b'(z) = e^z b(z), \quad b(0) = 1.$$

Solving it one arrives at

$$b(z) = \exp\left(e^z - 1\right).$$

A closed-form formula for b_k can be derived, but this is left as an exercise.

Example 7.11: *A Binomial Sum* In some computations, one needs to evaluate the generating function of the binomial sum s_n defined as

$$s_n = \sum_{k=0}^{n} \binom{n}{k} a_k$$

where a_k is a sequence. Clearly by (7.17) we have

$$\sum_{n \geq 0} s_n \frac{z^n}{n!} = e^z a(z)$$

where $a(z) = \sum_{n \geq 0} a_n \frac{z^n}{n!}$. The situation is a little more complex if one wants to compute the ordinary generating function of s_n. But

$$s_n = \sum_{k=0}^{n} \binom{n}{k} a_k \quad \longleftrightarrow \quad S(z) = \frac{1}{1-z} A\left(\frac{z}{1-z}\right). \qquad (7.18)$$

Indeed, by Entry 1 of Table 7.2 we have

$$\frac{1}{1-z} A\left(\frac{z}{1-z}\right) = \sum_{m=0}^{\infty} a_m z^m \frac{1}{(1-z)^{m+1}} = \sum_{m=0}^{\infty} a_m z^m \sum_{j=0}^{\infty} \binom{m+j}{j} z^j$$

$$= \sum_{n=0}^{\infty} z^n \sum_{k=0}^{n} \binom{n}{k} a_k$$

as desired. ∎

7.2.2 Labeled Combinatorial Structures

As in the case of ordinary generating functions, the exponential generating functions have a combinatorial interpretation. We consider here *labeled* structures (e.g., permutations, cycles in a permutation, set partitions, labeled trees, labeled graphs) where each "atom" of a structure (e.g., a node in graphs or trees) has a distinctive integer label. Let $\mathcal{A}$ be a class of labeled combinatorial objects, and let a_n be the counting sequence (i.e., the number of objects of size n). Then the exponential generating function (EGF) of $\mathcal{A}$ is

$$a(z) = \sum_{\alpha \in \mathcal{A}} \frac{z^{|\alpha|}}{|\alpha|!} = \sum_{n \geq 0} a_n \frac{z^n}{n!},$$

since there are $n!$ orderings of labels.

Foata presented in [158] a detailed formalization of labeled constructions. There are many excellent surveys and books that treat exponential combinatorial families in depth (cf. Flajolet [126], Flajolet and Sedgewick [149], Hofri [197],

Sedgewick and Flajolet [383], and Wilf [447]), so here we restrict our attention to only a few examples. We shall follow the presentation of Flajolet and Sedgewick [149].

In *labeled combinatorial structures* each object consists of atoms that are marked by distinct integers. The *size* of a structure is the number of its atoms (e.g., number of nodes in a graph or a tree). We assume, however, that such structures are *well labeled* by insisting that the labels are consecutive integers starting from 1. We allow *relabeling* that *preserves* the order relations between labels. For example, a structure with labels $\{5, 9, 11\}$ can be relabeled as $\{1, 2, 3\}$, but not $\{1, 3, 2\}$. Knowing this, we can introduce two new *labeled constructions*, namely, *disjoint unions* and *products*.

Disjoint union of structures $\mathcal{A}$ and $\mathcal{B}$ denoted as $\mathcal{A} + \mathcal{B}$ is defined as in the unlabeled case except that we need to relabel the composite structure to get well labeled structures. For example, $\{1\} + \{1, 2\} = \{1, 2, 3\}$. In a similar fashion, we define the product $\mathcal{C} = \mathcal{A} \star \mathcal{B}$ of $\mathcal{A}$ and $\mathcal{B}$ by forming ordered pairs from $\mathcal{A} \times \mathcal{B}$ and performing all possible *order consistent relabeling*, ensuring that the resulting pairs are well labeled. Using the same argument as above for the un-labeled structures (ordinary generating functions), we easily prove the following translations

$$\mathcal{C} = \mathcal{A} + \mathcal{B} \longleftrightarrow c(z) = a(z) + b(z), \tag{7.19}$$

$$\mathcal{C} = \mathcal{A} \star \mathcal{B} \longleftrightarrow c(z) = a(z)b(z). \tag{7.20}$$

The last relation also can be proved by observing that the counting number c_n of $\mathcal{C}$ is related to the counting numbers a_n and b_n of $\mathcal{A}$ and $\mathcal{B}$ by

$$c_n = \sum_{k=0}^{n} \binom{n}{k} a_k b_{n-k},$$

which leads directly to the product of the associated exponential generating functions.

We illustrate the above constructions in two interesting examples, namely, counting labeled trees (i.e., Cayley's trees) and the number of cycles in a permutation (i.e., Stirling numbers of the first kind).

Example 7.12: *Labeled or Cayley's Trees*   Let $\mathcal{T}$ be the class of all labeled rooted trees, also called the Cayley trees. Think of such trees as labeled connected graphs without cycles and with one node marked as the root (e.g., subtrees are not ordered; there are no left and right subtrees). Our goal is to find the number t_n of all labeled rooted trees of size n. In Figure 7.1 we show all $t_3 = 9$ labeled trees with $n = 3$ nodes. Let $T(z)$ be the EGF of t_n (we write $T(z)$ instead of $t(z)$ for historical reasons). Observe that a rooted labeled tree consists of the root

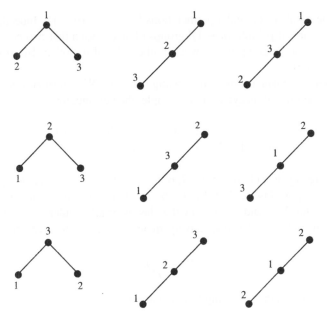

Figure 7.1. Nine rooted labeled trees built on $n = 3$ nodes with roots at the top.

and either one rooted labeled subtree or two rooted labeled subtrees or k rooted labeled subtrees, and so on. But k rooted labeled subtrees can be symbolically represented as a product of k rooted labeled trees $\mathcal{T}$, that is, $\mathcal{T}^k$. Finally, since the subtrees are *not* ordered we must divide the product $\mathcal{T}^k$ by $k!$ to construct the structure $\mathcal{T}$ (we recall that $\mathcal{T}^k$ is an order product). Having this in mind, and writing root for the root node, we can represent $\mathcal{T}$ as

$$\mathcal{T} = \text{root} \star \left(\epsilon + \mathcal{T} + \frac{1}{2!}\mathcal{T}^2 + \frac{1}{3!}\mathcal{T}^3 + \cdots + \frac{1}{k!}\mathcal{T}^k + \cdots \right),$$

where ϵ represents an empty tree. This translates into

$$T(z) = ze^{T(z)}. \tag{7.21}$$

Thus $T(z)$ is given only implicitly. We will return to this problem in Section 7.3.2, where we show that $t_n = [\frac{z^n}{n!}]T(z) = n^{n-1}$.

Example 7.13: *Cycles in Permutations or Stirling Numbers of the First Kind*
Let $\mathcal{K}_n$ represent the subclass of circular permutations of size n. A permutation is circular if a circular shift leads to the same permutation (e.g., $(1, 3, 2)$ is the same circular permutation as $(3, 2, 1)$ but not the same as $(1, 2, 3)$). It is easy to

see that the number of circular permutations k_n is $k_n = (n-1)!$. Indeed, we split $n!$ permutations of $[n]$ into $(n-1)!$ groups of size n such that two permutations belong to the same group if one is a circular shift of the other. But, $nk_n = n!$, hence the claim.

Now, let us consider a more interesting question. We consider a permutation of $[n]$ and identify all its cycles. For example, the permutation

$$
\begin{array}{cccccccccc}
1 & 2 & 3 & 4 & 5 & 6 & 7 & 8 & 9 & 10 \\
10 & 1 & 7 & 9 & 3 & 5 & 6 & 4 & 8 & 2
\end{array}
$$

contains three cycles, $(1, 10, 2)$, $(3, 7, 6, 5)$ and $(4, 9, 8)$, that can be represented as cycle digraphs (Figure 7.2). Let $\mathcal{C}_n^r$ be a subclass of all permutations of $[n]$ having r cycles. If $\mathcal{K}$ and $\mathcal{C}^{(r)}$ denote the classes of all circular permutations and permutations with r cycles, respectively, then a little thought reveals that

$$
\mathcal{C}^{(r)} = \frac{1}{r!}\mathcal{K}^r.
$$

This translates into the following EGF equation

$$
C^{(r)}(z) = \frac{1}{r!}K^r(z).
$$

But

$$
K(z) = \sum_{n \ge 1}(n-1)!\frac{z^n}{n!} = \log\frac{1}{1-z}.
$$

The numbers

$$
\begin{bmatrix} n \\ r \end{bmatrix} := \frac{n!}{r!}[z^n]\left(\log\frac{1}{1-z}\right)^r \tag{7.22}
$$

are called the *Stirling number of the first kind* and represent the number of permutations with r cycles.

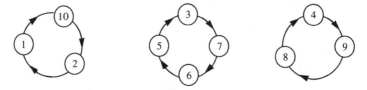

Figure 7.2. Three Cycles of the Permutation of Example 7.13.

Let us see if we can get a more explicit formula for the generating function $C^{(r)}(z)$. We first introduce a *bivariate* generating function

$$C(z, u) = \sum_{r \geq 0} C^{(r)}(z) u^r.$$

Observe that

$$C(z, u) = \sum_{r \geq 0} \left(\log \frac{1}{1-z} \right)^r \frac{u^r}{r!} = \exp\left(-u \log(1-z) \right) = \frac{1}{(1-z)^u}.$$

Using Entry 1 of Table 7.2, we obtain

$$\left[\frac{z^n}{n!} \right] C(z, u) = \sum_{r=0}^{n} \begin{bmatrix} n \\ r \end{bmatrix} u^r$$

$$= \left[\frac{z^n}{n!} \right] (1-z)^{-u} = n! \binom{u+n-1}{n}$$

$$= u(u+1) \cdots (u+n-1).$$

This agrees with Entry 8 of Table 7.2.   ■

7.3 CAUCHY, LAGRANGE AND BOREL FORMULAS

We have been lucky so far, because in order to find $[z^n] A(z)$ we need only to look up an entry in Table 7.2. But this is not always the case as already seen in Example 7.12. We discuss below two formulas, one by Cauchy and the other one by Lagrange, that provide guidance for finding $[z^n] A(z)$. We shall learn how to extract coefficients from generating functions given either explicitly (cf. Cauchy's formula) or implicitly (cf. Lagrange's inversion formula). We shall also discuss the Borel transform, which allows us to transfer an exponential generating function into an ordinary generating function.

7.3.1 Cauchy Coefficient Formula

We start with the Cauchy formula that we already discussed in Chapter 2 (see Theorem 2.6). Observe that if $A(z)$ is analytic in the vicinity of $z = 0$, then

$$a_n = [z^n] A(z) = \frac{1}{2\pi i} \oint \frac{A(z)}{z^{n+1}} dz \tag{7.23}$$

where the integral is around any simple curve encircling $z = 0$. Indeed, since $A(z)$ is analytic at $z = 0$, it has a convergent series representation, say

$$A(z) = \sum_{k \geq 0} a_k z^k.$$

Thus

$$\oint \frac{A(z)}{z^{n+1}} dz = \sum_{k \geq 0} a_k \oint z^{-(n+1-k)} dz = 2\pi i a_n,$$

where the last equality follows from (2.23) which we repeat here:

$$\oint z^{-n} dz = \int_0^1 e^{-2\pi i(n-1)} dt = \begin{cases} 2\pi i & \text{for } n = 1 \\ 0 & \text{otherwise.} \end{cases}$$

The interchange of the integral and sum above is justified since the series converges uniformly. Using Cauchy's formula and Cauchy's residue theorem we can evaluate exactly or asymptotically the coefficient a_n. Since the Cauchy formula (7.23) is mostly used in the asymptotic evaluation of the coefficients, we delay our discussion until the next chapter.

Finally, we should mention that Cauchy's formula is often used to establish a bound on the coefficients a_n. Let

$$M(R) = \max_{|z| \leq R} |A(z)|.$$

Then for all $n \geq 0$

$$|a_n| \leq \frac{M(R)}{R^n}, \tag{7.24}$$

which follows immediately from (7.23) and a trivial majorization under the integral.

7.3.2 Lagrange Inversion Formula

Cauchy's formula can also be used to prove Lagrange's formula, which allows us to extract the coefficient of *implicitly* defined generating functions. We already saw examples of generating functions that are defined by functional equations (see Example 7.12). We were lucky so far to be able to explicitly solve such functional equations. We cannot count on this forever. For example, while enumerating *unlabeled* trees we found an explicit formula discussed in Example 7.7, but this is not true for enumerating *labeled trees*. In the latter case, the generating function

$T(z)$ of the number of *labeled trees* satisfies $T(z) = ze^{T(z)}$ (see Example 7.12). This function does not have a simple, explicit solution, but with the help of the Lagrange formula we shall easily find the coefficients.

We first digress and examine the inversion problem that is at the heart of the Lagrange formula. Define explicitly a function $x(z)$ as a solution of

$$x(z) = z\Phi(x(z)) \tag{7.25}$$

provided it exists. We assume that $\Phi(u)$ is analytic at some point x_0, and $\Phi(0) \neq 0$ for the problem to be well posed. To study the existence issue, let us rephrase (7.25) as

$$\Theta(x(z)) = z, \quad \text{where} \quad \Theta(u) = \frac{u}{\Phi(u)}. \tag{7.26}$$

In other words, we are looking for conditions under which the inverse function $x(z) = \Theta^{-1}(z)$ exists. We shall see that for the existence of the inverse function in a vicinity of x_0 we must require that $\Theta'(x_0) \neq 0$.

Since $\Theta(x)$ is analytic at x_0, we have

$$\Theta(x) = \Theta(x_0) + (x - x_0)\Theta'(x_0) + \frac{1}{2}(x - x_0)^2\Theta''(x_0) + \cdots.$$

As a first order approximation, we find

$$x - x_0 \sim \frac{1}{\Theta'(x_0)}(z - z_0)$$

provided $\Theta'(x_0) \neq 0$ where $z = \Theta(x)$ and $z_0 = \Theta(x_0)$. We can further iterate the above to derive a better approximation:

$$x - x_0 = \frac{1}{\Theta'(x_0)}(z - z_0) - \frac{1}{2}\frac{\Theta''(x_0)}{[\Theta'(x_0)]^3}(z - z_0)^2 + O((z - z_0)^3). \tag{7.27}$$

Iterating many times we can get a convergent series representation of the inverse $\Theta^{-1}(x)$ around x_0 provided $\Theta'(x_0) \neq 0$. We should mention that the problem of finding the inverse function is known in the literature as the **implicit-function theorem** or **inverse mapping theorem**. The reader is referred to any standard book on analysis or complex analysis (e.g., [363, 448]) for further reading.

We can continue the above iterative process, but unfortunately it becomes more and more complicated. We would like to extract coefficients of the inverse function $\Theta^{-1}(x)$ or $x(z)$ in a more simpler and direct way. Is it possible? The answer is emphatically *yes*, and it is known as the **Lagrange inversion formula**.

Theorem 7.2 **(Lagrange Inversion Formula)** *Let $\Phi(u)$ be analytic in some neighborhood of $u = 0$ with $[u^0]\Phi(u) \neq 0$, and let $X(z)$ be a solution of*

$$X = z\Phi(X). \tag{7.28}$$

The coefficients of $X(z)$ (or in general $\Psi(X(z))$), where Ψ is an analytic function in a neighborhood of the origin, satisfy

$$[z^n]X(z) = \frac{1}{n}[u^{n-1}](\Phi(u))^n, \tag{7.29}$$

$$[z^n]\Psi(X(z)) = \frac{1}{n}[u^{n-1}](\Phi(u)^n\Psi'(u)). \tag{7.30}$$

Proof. We prove only (7.30) since (7.29) is a special case. Formally we can proceed as follows, taking $u = X(z)$ as an independent variable:

$$[u^{n-1}](\Phi(u)^n\Psi'(u)) = [u^{n-1}](\Psi'(u)(u/z)^n) = [u^{-1}](\Psi'(u)/z^n)$$

$$= \frac{1}{2\pi i}\oint \frac{\Psi'(u)}{z(u)^n}du$$

$$= \frac{1}{2\pi i}\oint \frac{\Psi'(X(z))X'(z)}{z^n}dz \quad u = X(z)$$

$$= [z^n]\left(z\frac{d}{dz}\Psi(X(z))\right)$$

$$= n[z^n]\Psi(X(z)).$$

Most of the above equalities are trivial to justify (the second one is just the residue theorem) except for the third equality. But this one is a consequence of the inverse function theorem discussed above ($[u^0]\Phi(u) \neq 0$ is the equivalent of the requirement that the first derivative of the inverse function is nonzero, as we shall see below), and the change of variables. ∎

Before we proceed to discuss more sophisticated examples, let us check if (7.27) agrees with the Lagrange formula.

Example 7.14: *Checking (7.27).* As in (7.26), we define $\Theta(u) = u/\Phi(u)$ (i.e., $X(z) = \Theta^{-1}(z)$), hence for $u = 0$ we have $\Theta'(0) = 1/\Phi(0)$. From Lagrange's formula (7.29) we obtain for $n = 1$

$$[z]X(z) = [u^0]\Phi(u) = 1/\Theta'(0),$$

which coincides with (7.27). To find $[z^2]X(z) = [z^2]\Theta^{-1}(z)$ we need a little bit of algebra. First of all, observe that by (7.29)

$$[z^2]X(z) = \frac{1}{2}[u]\Phi^2(u).$$

But

$$\Phi^2(u) = \frac{u^2}{\Theta(u)^2} = \frac{u^2}{(u\Theta'(0) + \frac{1}{2}u^2\Theta''(0) + O(u^3))^2}$$

$$= \frac{1}{[\Theta'(0)]^2} - u\frac{\Theta''(0)}{[\Theta'(0)]^3} + O(u^2).$$

Thus

$$[z^2]X(z) = -\frac{1}{2}\frac{\Theta''(0)}{[\Theta'(0)]^3},$$

which again agrees with (7.27).   ∎

The most popular application of the Lagrange formula is to the tree function $T(z)$ defined in Example 7.12, which we repeat here:

$$T(z) = ze^{T(z)}. \tag{7.31}$$

In fact, the tree function is closely related to the Lambert W-function $W(x)$ $\exp(W(x)) = x$ (i.e., $T(z) = -W(-z)$). The interested reader is referred to [73] for a comprehensive survey on the Lambert function and its myriad applications.

Example 7.15: *Tree Function $T(z)$*   We compute here the coefficients $t_n = [z^n]T(z)$ that represent the number of labeled rooted trees. Setting $\Phi(u) = e^u$ and $\Psi(u) = u$ in the Lagrange formula, we obtain

$$[z^n]T(z) = \frac{1}{n}[u^{n-1}]e^{nu} = \frac{1}{n}\frac{n^{n-1}}{(n-1)!}$$

$$= \frac{n^{n-1}}{n!} \tag{7.32}$$

which agrees with Entry 7 in Table 7.2.

We still need to know the radius of convergence of $T(z)$. We will show that $T(z)$ is analytic for all $|z| < e^{-1}$. But this follows directly from the Hadamard Theorem 7.1. We present now an indirect proof of the same that is more general and may apply to a larger class of implicitly defined functions. Let us define, as in

(7.26), the function $\Theta(u) = u/\Phi(u) = ue^{-u}$, and observe that $T(z) = \Theta^{-1}(z)$. From our discussion preceding the Lagrange formula (cf. (7.27)), we conclude that $T(z)$, or the inverse of $\Theta(u)$, exists as long as $\Theta'(u) \neq 0$. Let τ be the smallest real number such that

$$0 = \Theta'(\tau) = e^{-\tau}(1 - \tau)$$

or equivalently

$$\Phi(\tau) - \tau\Phi'(\tau) = e^{\tau}(1 - \tau).$$

We find $\tau = 1$ and $z = \Theta(\tau) = e^{-1}$. Since $\Theta(u)$ increases from $u = 0$ to $u = \tau$, the tree function $T(z)$ has it first singularity at $z = e^{-1}$. Thus it is analytic for $|z| < e^{-1}$.
                                                                  ■

The tree function is so important that it deserves more discussion. One defines the **tree-like** generating function of a_n as

$$A(z) = \sum_{n=0}^{\infty} a_n \frac{n^n}{n!} z^n. \tag{7.33}$$

In particular, for $|z| < e^{-1}$

$$T(z) = \sum_{n=1}^{\infty} \frac{1}{n} \frac{n^n}{n!} z^n, \tag{7.34}$$

$$B(z) = \sum_{n=0}^{\infty} \frac{n^n}{n!} z^n = \frac{1}{1 - T(z)}. \tag{7.35}$$

To see how (7.35) follows from (7.34), observe that after differentiating (7.34) one finds $B(z) = 1 + zT'(z)$. On the other hand, differentiating the functional equation $T(z) = ze^{T(z)}$ we obtain

$$zT'(z) = \frac{T(z)}{1 - T(z)},$$

and the proof of (7.35) follows. In Exercise 3 the reader is asked to rederive (7.35) from Lagrange's formula.

Example 7.16: *Sum and Recurrence Arising in Coding Theory*   In informa-tion theory (e.g., channel coding, minimax redundancy of universal coding, and universal portfolios) the following sum is of interest:

$$S_n = \sum_{i=0}^{n} \binom{n}{i} \left(\frac{i}{n}\right)^i \left(1 - \frac{i}{n}\right)^{n-i}.$$

We can analyze it through the tree-like generating function $s(z) = \sum_{n \geq 0} S_n \frac{n^n}{n!} z^n$. Indeed, observe that $n^n S_n / n!$ is a convolution of two identical sequences $\frac{n^n}{n!}$, hence

$$s(z) = [B(z)]^2$$

where $B(z)$ is defined in (7.35).

A more interesting example is the following. When analyzing the minimax redundancy of source coding for memoryless sources (see also Section 8.7.2), Shtarkov [390] came up with the following summation problem

$$D_n(m) = \sum_{i=1}^{m} \binom{m}{i} D_n^*(i), \qquad (7.36)$$

where $D_0^*(1) = 0$, $D_n^*(1) = 1$ for $n \geq 1$, and $D_n^*(i)$ for $i > 1$ satisfies the following recurrence:

$$D_n^*(i) = \sum_{k=1}^{n} \binom{n}{k} \left(\frac{k}{n}\right)^k \left(1 - \frac{k}{n}\right)^{n-k} D_{n-k}^*(i-1) \qquad (7.37)$$

provided that $D_n^*(i) = 0$ for $i > n$. We express $D_n(m)$ and $D_n^*(m)$ as tree-like generating functions. Let us introduce a new sequence $\widehat{D}_n^*(m)$ defined as

$$\widehat{D}_n^*(m) = \sum_{k=0}^{n} \binom{n}{k} \left(\frac{k}{n}\right)^k \left(1 - \frac{k}{n}\right)^{n-k} D_{n-k}^*(m-1) = D_n^*(m) + D_n^*(m-1).$$

Its tree-like generating functions is

$$\widehat{D}_m^*(z) = \sum_{k=0}^{\infty} \frac{k^k}{k!} z^k \widehat{D}_k^*(m).$$

Observe that it can be re-written as

$$\frac{n^n}{n!} \widehat{D}_n^*(m) = \sum_{k=0}^{n} \frac{k^k}{k!} \cdot \frac{(n-k)^{n-k}}{(n-k)!} D_{n-k}^*(m-1).$$

We now multiply both sides of the above by z^n, sum it up, and then by the convolution formula of generating functions one arrives at

$$\widehat{D}_m^*(z) = B(z) D_{m-1}^*(z).$$

Thus

$$D_m^*(z) = (B(z) - 1)^{m-1} D_1^*(z) = (B(z) - 1)^m, \qquad (7.38)$$

since $D_1^*(z) = B(z) - 1$. By (7.38) we obtain

$$D_m(z) = \sum_{i=1}^{m} \binom{m}{i} D_i^*(z) = \sum_{i=1}^{m} \binom{m}{i} (B(z) - 1)^i = B^m(z) - 1. \qquad (7.39)$$

We return to this problem in Section 8.7.2.

Finally, we are in a position to formulate an even more general recurrence. Under some initial conditions, define a sequence x_n^m satisfying for $n \geq 1$ and $m \geq 1$

$$x_n^m = a_n + \sum_{i=0}^{n} \binom{n}{i} \left(\frac{i}{n}\right)^i \left(1 - \frac{i}{n}\right)^{n-i} (x_i^{m-1} + x_{n-i}^{m-1}), \qquad (7.40)$$

where a_n is a given sequence (called the *additive* term), and m is an additional parameter. Using the same arguments as above, one can solve this recurrence in terms of tree-like generating functions:

$$X_m(z) = A(z) + 2B(z)X_{m-1}(z) \qquad (7.41)$$

where

$$X_m(z) = \sum_{k=0}^{\infty} \frac{k^k}{k!} z^k x_k^m, \qquad A(z) = \sum_{k=0}^{\infty} \frac{k^k}{k!} z^k a_k.$$

The recurrence (7.41) can be solved by telescoping in terms of m. ∎

7.3.3 Borel Transform

In some applications, it is better to work with ordinary generating functions and in others with exponential generating functions. Therefore, one would like to have an easy transfer formula from one generating function to another. In Example 7.11, we show how to obtain the ordinary generating function $S(z)$ for the binomial sum from its exponential generating function $s(z) = e^z a(z)$ (cf. (7.18)). This is an example of a transfer formula. A question arises whether one can find a transform that allows us to recover the ordinary generating function from the exponential generating function. The answer is given by the **Borel transform** (cf. [424, 448]), which we discuss next.

Let a_n be a sequence such that its ordinary generating function $A(z)$ exists in $|z| < R$. Let also $a(z)$ be its exponential generating function. Define

$$f(z) = \int_0^\infty e^{-t} a(zt) dt$$

for $|z| < R$. We prove that this integral exists and $f(z) = A(z)$ for $|z| < R$. Indeed, observe that by Cauchy's bound (7.24) $|a_n R^n| < M$ for some $M > 0$, hence

$$|a(z)| < M \sum_{n \geq 0} \frac{|z^n|}{R^n n!} = e^{|z|/R}.$$

Therefore, for $|z| < R$

$$\left| \int_0^\infty e^{-t} a(zt) dt \right| < M \int_0^\infty e^{-t(1-|z|/R)} dt,$$

and the integral exists for $|z| < R$. Thus by Theorem 2.4(ii) we conclude that $f(z)$ is analytic, and the interchange of the integral and the sum below is justifiable for $|z| < R$:

$$\int_0^\infty e^{-t} \sum_{n \geq 0} a_n \frac{z^n t^n}{n!} dt = \sum_{n \geq 0} a_n z^n \frac{1}{n!} \int_0^\infty e^{-t} t^n dt = A(z)$$

where the last equation follows from $n! = \int_0^\infty e^{-t} t^n dt = \Gamma(n+1)$. In summary, we prove the following theorem.

Theorem 7.3   (Borel Transform)   *Let $A(z) = \sum_{n \geq 0} a_n z^n$ converge for $|z| < R$. Then for $|z| < R$*

$$A(z) = \int_0^\infty e^{-t} a(zt) dt, \qquad (7.42)$$

where $a(z) = \sum_{n \geq 0} a_n \frac{z^n}{n!}$ is the EGF of $\{a_n\}_{n=0}^\infty$.

7.4   PROBABILITY GENERATING FUNCTIONS

Generating functions and other transforms are important and useful tools in probability theory. They are particularly handy in the analysis of algorithms and combinatorics, that is, *discrete* structures. If X is a discrete random variable with $p_k = \Pr\{X = k\}$ $(k = 0, 1, \ldots)$, then the *probability generating function*, denoted as $G_X(z)$ or $G(z)$), is defined with respect to the sequence $\{p_k\}_{k \geq 0}$ as

$$G_X(z) = \sum_{k \geq 0} p_k z^k = \mathbf{E}[z^X].$$

We already saw moment generating functions in action when large deviations theory was discussed in Chapter 5. Here, we first briefly review terminology and elementary facts in order to later present more interesting applications of probability generating functions.

The usefulness of probability generating functions stems from the fact that some quantities of interest for X are easily computable through the generating functions. For example, *moments* of X are related to the derivatives of $G_X(z)$ at $z = 1$. Indeed, if X possesses all moments, then

$$\mathbf{E}[X^r] = \left(z \frac{d}{dz} \right)^r G_X(z) \bigg|_{z=1},$$

as is easy to check. Often, *factorial moments* $\mathbf{E}[(X)_r] := \mathbf{E}[X(X-1)\cdots(X-r+1)]$ are easier to compute since

$$\mathbf{E}[X(X-1)\cdots(X-r+1)] = \frac{d^r}{dz^r} G_X(z) \bigg|_{z=1}.$$

In particular,

$$\mathbf{E}[X] = G'_X(1),$$

$$\mathbf{Var}[X] = G''_X(1) + G'_X(1) - [G'_X(1)]^2.$$

Of course, one can recover the probability distribution p_k from the generating function since $p_k = [z^k] G_X(z)$; hence Cauchy and Lagrange formulas work fine. But also

$$p_k = \frac{d^k}{dz^k} G_X(z) \bigg|_{z=0}.$$

Probability generating functions are commonly used in probability theory since they replace complicated operations like convolutions by algebraic equations. For example, if X and Y are independent, then

$$G_{X+Y}(z) = \mathbf{E}[z^{X+Y}] = \mathbf{E}[z^X]\mathbf{E}[z^Y] = G_X(z)G_Y(z).$$

Also, the distribution function $F(k) = \Pr\{X \leq k\} = \sum_{i=0}^{k} p_i$ can be represented in PGF terms, since by (7.6) we have

$$\sum_{k \geq 0} F(k) z^k = \frac{G_X(z)}{1-z}.$$

In Chapter 5 we used another transform for probability distributions, namely, the *moment generating function* defined as

$$M_X(s) = G_X(e^s) = \mathbf{E}[e^{sX}]$$

where s is assumed to be complex. The moment generating function $M_X(s)$ is well defined at least for $\Re(s) \leq 0$. We saw in Section 5.4 that if the definition of $M_X(s)$ can be extended to a vicinity of $s = 0$, say for $\Re(s) \in [-c, c]$ for some $c > 0$, then X has an exponential tail.

Finally, it is also convenient to deal with the *cumulant generating function* $\kappa_X(s)$ defined as

$$\kappa_X(s) = \log M_X(s) = \log G_X(e^s).$$

Observe that

$$\mathbf{E}[X] = \kappa'(0),$$

$$\mathbf{Var}[X] = \kappa''(0),$$

provided $\kappa(s)$ exists in the vicinity of $s = 0$. Then

$$\kappa(s) = \mu s + \sigma^2 \frac{s^2}{2} + O(s^3), \tag{7.43}$$

where $\mu = \mathbf{E}[X]$ and $\sigma^2 = \mathbf{Var}[X]$.

Example 7.17: *The Average Complexity of Quicksort* Hoare's *Quicksort* algorithm is the most popular sorting algorithm due to its good performance in practice. The basic algorithm can be briefly described as follows [269, 305]: A partitioning key is selected at random from the unsorted list of keys, and used to partition the keys into two sublists to which the same algorithm is called recursively until the sublists have size one or zero. Let now X_n denote the number of comparisons needed to sort a random list of length n. It is known that after randomly selecting a key, the two sublists are still "random"; hence assuming any key is chosen with equal probability $1/n$ we obtain a stochastic equation

$$X_n = n - 1 + X_K + X_{n-1-K}$$

where K is the random variable denoting the selected key. It can be translated into the generating function $G_n(z) = \mathbf{E}[z^{X_n}]$ as

$$G_n(z) = z^{n-1} \sum_{k=0}^{n-1} \frac{1}{n} \mathbf{E}[z^{X_k}] \mathbf{E}[z^{X_{n-1-k}}] = \frac{z^{n-1}}{n} \sum_{k=0}^{n-1} G_k(z) G_{n-1-k}(z).$$

Then $c_n := \mathbf{E}[X_n] = G'_n(1)$ satisfies

$$c_n = n - 1 + \frac{2}{n} \sum_{i=0}^{n-1} c_i$$

with $c_0 = 0$. We solve this recurrence using the generating function $C(z) = \sum_{n \geq 0} c_n z^n$. By (7.6) and Entry 1 in Table 7.2, we derive

$$C(z) = \frac{z^2}{(1-z)^2} + 2 \int_0^z \frac{C(x)}{1-x} dx,$$

which becomes, after differentiation,

$$C'(z) = \frac{2z}{(1-z)^2} + \frac{2z^2}{(1-z)^3} + \frac{2C(z)}{1-z}.$$

This is a linear differential equation that is simple to solve. Using, for example, the method of variation of parameters, we arrive at

$$C(z) = \frac{2}{(1-z)^2} \left(\log \frac{1}{1-z} - z \right).$$

To find $C_n = [z^n] C(z)$ we proceed as in Example 7.1, and finally prove

$$C_n = 2(n+1) H_{n+1} - 4n - 2 = 2(n+1) H_n - 4n,$$

where H_n is the nth harmonic number.   ∎

7.5   DIRICHLET SERIES

Finally, we study one more generating function of $\{a_n\}_1^\infty$ defined as

$$A(s) = \sum_{n=1}^{\infty} \frac{a_n}{n^s},$$

and known as the *Dirichlet series*. When dealing with ordinary and exponential generating functions, we observed that the choice of a particular generating function largely depends on the convolution formula. This is also the case here. We shall demonstrate that the Dirichlet series is a very useful tool in number-theoretical problems.

7.5.1 Basic Properties

Let us start with analytic theory of the Dirichlet series. For what complex s does the series

$$A(s) = \sum_{n=1}^{\infty} \frac{a_n}{n^s}$$

converge? Observe that

$$\left| \frac{a_n}{n^s} \right| = \frac{|a_n|}{n^{\Re(s)}}.$$

Therefore, if the series converges absolutely for $s = r + it$, then it must also converge for $\Re(s) \geq r$. Let r_a be a lower bound of such r. The series converges for $\Re(s) > r_a$ and diverges for $\Re(s) < r_a$. The real number r_a is called the *abscissa* of absolute convergence. This observation leads to the following known result (for a formal proof see [15, 195]).

Theorem 7.4 *Suppose the series $\sum_{n=1}^{\infty} |a_n/n^s|$ does not converge or diverge for all s. Then there exists a real number r_a called the abscissa of absolute convergence, such that the Dirichlet series $\sum_{n=1}^{\infty} \frac{a_n}{n^s}$ converges absolutely for $\Re(s) > r_a$ and does not converge absolutely for $\Re(s) < r_a$.*

As for an example, let us look at the Riemann zeta function (discussed in Section 2.4.2), which is the Dirichlet series of $\{1\}_1^{\infty}$, that is,

$$\zeta(s) = \sum_{n=1}^{\infty} \frac{1}{n^s}.$$

It converges absolutely for $\Re(s) > 1$, and for $s = 1$ it diverges, so it diverges for $\Re(s) < 1$. In fact, Dirichlet series converges absolutely for all $\Re(s) > 1$ as long as $|a_n|$ is bounded for all $n \geq 1$.

Let us now turn our attention to some important properties of Dirichlet series. How does the product of two Dirichlet series relate to the Dirichlet convolution? Observe that

$$A(s)B(s) = \sum_{n=1}^{\infty} a_n n^{-s} \sum_{m=1}^{\infty} b_m m^{-s}$$

$$= \sum_{n=1}^{\infty} \sum_{m=1}^{\infty} a_n b_m (nm)^{-s}$$

$$= \sum_{k=1}^{\infty} k^{-s} \sum_{mn=k} a_n b_m.$$

Therefore, if $d|n$ reads "d divides n," then we just proved the following relationship:

$$A(s)B(s) \quad \longleftrightarrow \quad \left\{ \sum_{d|n} a_d b_{\frac{n}{d}} \right\}_{n=1}^{\infty}. \qquad (7.44)$$

In other words, Dirichlet convolution is a sum of products of the series coefficients whose product of the subscripts is fixed.

Example 7.18: *Some Dirichlet series* Here are two applications of the Dirichlet convolution formula.

1. *Divisor Function.* Let $d(n)$ be the number of divisors of the integer n. Clearly

$$d(n) = \sum_{d|n} 1 \cdot 1,$$

hence

$$D(s) = \sum_{n=1}^{\infty} \frac{d(n)}{n^s} = \zeta^2(s).$$

2. *Function $v_2(n)$.* Let $v_2(n)$ be the exponent of 2 in the prime decomposition of n. Since

$$v_2(2n) = v_2(n) + 1,$$

we find, after easy algebra,

$$V(s) = \frac{\zeta(s)}{2^s - 1}$$

which is valid for $\Re(s) > 1$.   ∎

7.5.2 Euler Products

For the purpose of this section, it is convenient to write $\{f(n)\}_1^\infty := \{f_n\}_1^\infty$. A function $f(n)$ whose domain is the set of positive integers is called *multiplicative* if

$$f(mn) = f(m)f(n)$$

for all pairs of *relatively prime* positive integers m and n (i.e., $\gcd(m, n) = 1$). Observe that since every integer n has a unique prime factorization

$$n = p_1^{e_1} p_2^{e_2} \cdots p_r^{e_r}, \tag{7.45}$$

where $p_1, \ldots, p_r$ are distinct primes and $e_1, \ldots, e_r$ are nonnegative integers, then for a multiplicative function f

$$f(n) = f(p_1^{e_1}) \cdots f(p_r^{e_r}).$$

Let us now consider the product

$$P(s) = \prod_p (1 + f(p)p^{-s} + f(p^2)p^{-2s} + \cdots)$$

where f is multiplicative and the product is over all prime numbers p. Let us look for a moment at $P(s)$ as a formal power series and multiply it out. What is the coefficient at n^{-s}? Assuming the factorization (7.45), we have

$$(p_1^{e_1} p_2^{e_2} \cdots p_r^{e_r})^{-s} f(p_1^{e_1}) f(p_2^{e_2}) \cdots f(p_r^{e_r}) = n^{-s} f(n).$$

Therefore, formally

$$\prod_{p>1} \left(1 + f(p)p^{-s} + f(p^2)p^{-2s} + \cdots \right) = \sum_{n=1}^\infty \frac{f(n)}{n^s}.$$

For $s = 0$ the above becomes the Euler product. In fact, it is not difficult to embrace analytic theory into our considerations. This leads to the following very important result (cf. [15] for a formal proof).

Theorem 7.5 **(Euler, 1737)** *Let f be a multiplicative function such that the series $\sum_{n=1}^\infty f(n)n^{-s}$ is absolutely convergent. Then*

$$\sum_{n=1}^\infty \frac{f(n)}{n^s} = \prod_p \left(1 + f(p)p^{-s} + f(p^2)p^{-2s} + \cdots \right) \tag{7.46}$$

and the infinite product is convergent, too.

Euler's theorem is a source of many interesting identities for the Riemann zeta function. We consider some of them below.

Example 7.19: *Another Representation for* $\zeta(s)$ Let us first assume $f(n) = 1$ for all n. Then

$$\zeta(s) = \prod_p \left(1 + p^{-s} + p^{-2s} + \cdots\right) = \prod_p \frac{1}{1 - p^{-s}}. \tag{7.47}$$

We now recall the definition of the Möbius function $\mu(n)$, which we discussed in Section 3.2. For prime p we define $\mu(p^e)$ as

$$\mu(p^e) = \begin{cases} +1 & \text{if } e = 0, \\ -1 & \text{if } e = 1, \\ 0 & \text{if } e \geq 2. \end{cases}$$

Then by Euler's theorem

$$\sum_{n=1}^{\infty} \frac{\mu(n)}{n^s} = \prod_p (1 - p^{-s}) = \frac{1}{\zeta(s)} \tag{7.48}$$

for $\Re(s) > 1$.   ∎

The last example shows how to invert relationships like

$$a_n = \sum_{d \mid n} b_d,$$

that is, to find b_n as a function of a_d. Let us again consider Example 3.4 of Chapter 3 where we counted the number of primitive sequences. If $f(n)$ is this number, then in Chapter 3 we already proved that

$$2^n = \sum_{d \mid n} f(d).$$

How can we recover $f(n)$? Let $A(s) = \sum_{n \geq 1} 2^n n^{-s}$. Then

$$A(s) = F(s)\zeta(s),$$

where $F(s) = \sum_{n \geq 1} f(n)n^{-s}$. Hence,

$$F(s) = \frac{A(s)}{\zeta(s)} = A(s) \sum_{n \geq 1} \frac{\mu(n)}{n^s}.$$

This translates into

$$f(n) = \sum_{d|n} \mu\left(\frac{n}{d}\right) 2^d,$$

as already shown in Example 3.4.   ∎

7.5.3 Perron-Mellin Formula

In this section, we present a surprising connection between *discrete* sums and complex integrals. We concentrate here on *exact* formulas for such sums. This opens a venue for a much deeper analysis presented in subsequent chapters (see Sections 8.2.1, 8.5, and 10.1), where complex integrals are used to derive asymptotic expansions of discrete sums. Needless to say, evaluation of discrete quantities by complex integrals is at the core of an analytic approach to analysis of algorithms and analytic combinatorics. We shall return to this issue several times in the sequel.

Let $\{a_k\}_{k=1}^n$ be a sequence of reals. We are interested in the evaluation of the following discrete sum

$$S_n = \sum_{k=1}^n a_k.$$

From elementary calculus we know that such sums are well approximated by $\int_1^n a(x)dx$, where $a(x)|_{x=n} = a_n$ and $a(x)$ is a monotone function (see also Section 8.2.1). Here, we aim at *exact* evaluation of such sums through the Cauchy residue Theorem 2.5 (see Section 2.3). We recall that Cauchy's residue theorem states that

$$\frac{1}{2\pi i} \oint_C f(z)dz = \sum_{j=1}^N \text{Res}[f(z); \; z = a_j],$$

as long as $f(z)$ is analytic inside C except at N poles $a_1, \ldots, a_N$. Thus, if one selects a "good" function $f(z)$, then the right-hand side of the above can represent a discrete sum. In many applications, it is often convenient to take an infinite loop C as shown in Figure 7.3, where $T, b \to \infty$. In such a case, we write

$$\int_{c-i\infty}^{c+i\infty} f(s)ds := \lim_{T\to\infty} \int_{c-iT}^{c+iT} f(s)ds$$

assuming the integral exists.

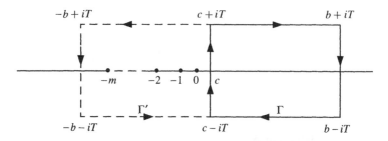

Figure 7.3. Illustration of the "closing-the-box" method.

We illustrate the above idea in the Perron-Mellin formula. We shall follow the presentation of Apostol [15] and Flajolet et al. [134]. Let $\{a_n\}_1^\infty$ be a sequence whose Dirichlet series is $A(s)$. We prove that for $m \geq 1$

$$\frac{1}{m!} \sum_{k=1}^{n-1} a_k \left(1 - \frac{k}{n}\right)^m = \frac{1}{2\pi i} \int_{c-i\infty}^{c+i\infty} A(s) n^s \frac{ds}{s(s+1)\cdots(s+m)} \qquad (7.49)$$

where $c > 0$ lies in the half-plane of absolute convergence of $A(s)$. When deriving the Perron-Mellin formula, we introduce the reader to a new technique, sometimes called the **closing-the-box** method, that allows us to evaluate integrals like the one above.

Consider the integral of the right-hand side of (7.49). We have

$$\frac{m!}{2\pi i} \int_{c-i\infty}^{c+i\infty} A(s) n^s \frac{ds}{s(s+1)\cdots(s+m)}$$

$$= \frac{m!}{2\pi i} \int_{c-i\infty}^{c+i\infty} \left(\sum_{k=1}^\infty \frac{a_k}{k^s}\right) \frac{n^s}{s(s+1)\cdots(s+m)} ds$$

$$= \sum_{k=1}^\infty a_k \frac{m!}{2\pi i} \int_{c-i\infty}^{c+i\infty} \frac{(n/k)^s}{s(s+1)\cdots(s+m)} ds,$$

where the interchange of the sum and the integral is justified since both converge absolutely. We need now to evaluate the integral

$$J_m(x) = \frac{m!}{2\pi i} \int_{c-i\infty}^{c+i\infty} \frac{x^s}{s(s+1)\cdots(s+m)} ds$$

for $x > 0$. We shall prove below that for all $m \geq 1$

$$J_m(x) = \begin{cases} 0 & \text{for } 0 < x < 1 \\ (1 - x^{-1})^m & \text{for } x > 1. \end{cases} \qquad (7.50)$$

Provided the above is true, the proof of (7.49) follows (with the exception of $x = k/n = 1$ and $m = 0$, which we treat below).

In order to prove (7.50), we apply the closing-the-box method. We consider two cases.

A. Case $0 < x < 1$ To evaluate the "infinite" integral in the definition of $J_m(x)$, we first consider a finite one, that is, $\int_{c-iT}^{c+iT}$ and later allow T to grow. The latter integral is calculated by considering the rectangular contour Γ shown in Figure 7.3 (solid line). First, observe that there is no singularity inside the contour Γ since the integrand function appearing in the definition of $J_m(x)$ is analytic. Hence

$$\int_\Gamma \frac{x^s}{s(s+1)\cdots(s+m)} = 0,$$

and therefore

$$\int_{c-iT}^{c+iT} = \int_{b+iT}^{c+iT} + \int_{b-iT}^{b+iT} + \int_{c-iT}^{b-iT}.$$

We prove now that the right-hand side of the above tends to zero as $T, b \to \infty$ proving (7.50) for $x < 1$. Let $s = r + iy$. We have the following estimates (here we use the bound (2.22) from Section 2.3):

$$\left| \int_{c-iT}^{c+iT} \frac{x^s}{s(s+1)\cdots(s+m)} ds \right| \leq 2 \int_c^b \frac{x^{\Re(s)}}{T(T+1)\cdots(T+m)} ds + 2T \frac{x^b}{b^{m+1}}$$

$$\leq \frac{2}{T^{m+1}} \int_c^\infty x^r dr + 2T \frac{x^b}{b^{m+1}}$$

$$= \frac{2}{T^{m+1}} \left(\frac{-x^c}{\ln x} \right) + 2T \frac{x^b}{b^{m+1}}$$

$$\to 0,$$

since x^b decays exponentially to 0 as $b \to \infty$ for $0 < x < 1$. In the above, we first take the limit with respect to b and then allow $T \to \infty$.

B: Case $x > 1$ In this case, we close the loop (rectangle) to the right of the line $(c - i\infty, c + i\infty)$, creating a new contour Γ' as shown in Figure 7.3 (dashed line). But, inside the loop Γ' there are $m + 1$ poles at $a_i = -i$ for $i = 0, 1, \ldots, m$. By Cauchy

$$\frac{1}{2\pi i} \int_{\Gamma'} \frac{x^s}{s(s+1)\cdots(s+m)} ds = \sum_{i=0}^m \text{Res}[f(s), s = -i],$$

where $f(s) = x^s/(s(s+1)\cdots(s+m))$. The above residues are easy to calculate. For example, using (2.24), we obtain

$$\text{Res}[f(s), s = -i] = (-1)^i \frac{x^{-i}}{i!(m-i)!},$$

thus

$$\frac{m!}{2\pi i} \int_{\Gamma'} \frac{x^s}{s(s+1)\cdots(s+m)} ds = (1 - x^{-1})^m.$$

On the other hand, we can evaluate the integral over the contour Γ' as

$$\int_{\Gamma'} f(s) ds = \int_{c-iT}^{c+iT} f(s) ds + \int_{-b-iT}^{c-iT} f(s) ds$$

$$+ \int_{-b+iT}^{-b-iT} f(s) ds + \int_{c+iT}^{-b+iT} f(s) ds.$$

We now show that the last three integrals vanish as $T, b \to \infty$, proving (7.50). Using the same argument as above, we have

$$\left| \int_{-b-iT}^{c-iT} \frac{x^s}{s(s+1)\cdots(s+m)} ds \right| \le \frac{1}{T^{m+1}} \int_{-\infty}^{c} x^r dr = \frac{x^c}{T^{m+1} \ln x},$$

$$\left| \int_{-b+iT}^{-b-iT} \frac{x^s}{s(s+1)\cdots(s+m)} ds \right| \le 2T \frac{x^{-b}}{b^{m+1}},$$

$$\left| \int_{-b+iT}^{c+iT} \frac{x^s}{s(s+1)\cdots(s+m)} ds \right| \le \frac{1}{T^{m+1}} \int_{-b}^{c} x^r dr \le \frac{1}{T^{m+1}} \frac{x^c}{\ln x}.$$

Since $x > 1$, we see that all three integrals tend to zero as $b \to \infty$ and then $T \to \infty$.

Finally, the case $x = 1$ and $m = 0$ must be considered separately. Direct calculations, that are left as an exercise for the reader, show that

$$\frac{1}{2\pi i} \int_{c-i\infty}^{c+i\infty} \frac{ds}{s} = \frac{1}{2}$$

for $c > 0$.

In summary, we have just proved the Perron-Mellin formula.

Theorem 7.6 (Perron-Mellin Formula) *Let $A(s)$ be the Dirichlet series of $\{a_n\}_1^\infty$. Let $c > 0$ lie inside the half-plane of absolute convergence of $A(s)$. Then*

$$\sum_{k=1}^{n-1} a_k + \frac{1}{2}a_n = \frac{1}{2\pi i} \int_{c-i\infty}^{c+i\infty} A(s)n^s \frac{ds}{s}, \tag{7.51}$$

and for all $m \geq 1$

$$\frac{1}{m!} \sum_{k=1}^{n-1} a_k \left(1 - \frac{k}{n}\right)^m = \frac{1}{2\pi i} \int_{c-i\infty}^{c+i\infty} A(s)n^s \frac{ds}{s(s+1)\cdots(s+m)}. \tag{7.52}$$

In Section 7.6.3 we shall use the Perron-Mellin formula to evaluate the total number of 1-digits in the binary representation of $1, 2, \ldots, n$. Here, we illustrate Theorem 7.6 on a simpler, but important example.

Example 7.20: *An Identity for* $\zeta(s)$ *Function* Let us apply the Perron-Mellin formula for $m = 1$ and $a_k \equiv 1$. Then

$$\frac{n-1}{2} = \frac{1}{2\pi i} \int_{2-i\infty}^{2+i\infty} \zeta(s)n^s \frac{ds}{s(s+1)}. \tag{7.53}$$

Now we evaluate the integral on the right-hand side by the closing-the-box approach. We consider a rectangle box with the line of integration shifted to the left up to $\Re(s) = -\frac{1}{4}$. Let us call the lines in the box East (equal to the original line of integration), West (i.e., $\Re(s) = -\frac{1}{4}$), North (i.e., $\Im(s) = T$ as $T \to \infty$), and South (i.e., $\Im(s) = -T$). Call this contour (traversed counterclockwise) C. Inside C there are two poles at $s = 0$ and $s = 1$ with residues $\mathrm{Res}[f(s); s = 0] = -1/2$ (since $\zeta(0) = -1/2$) and $\mathrm{Res}[f(s); s = 1] = n/2$, where $f(s) = \zeta(s)n^s/(s(s+1))$. Thus

$$\frac{1}{2\pi i} \int_C \zeta(s)n^s \frac{ds}{s(s+1)} = \frac{n-1}{2},$$

that is, the same value as the integral (7.53). Therefore,

$$\int_{-\frac{1}{4}-iT}^{-\frac{1}{4}+iT} + \int_{\text{North}} + \int_{\text{South}} = 0.$$

It easy to see (by a similar argument as above) that the integral over the South/North lines tends to zero as $T \to \infty$. We conclude that

$$\lim_{T\to\infty} \int_{-\frac{1}{4}-iT}^{-\frac{1}{4}+iT} \zeta(s)n^s \frac{ds}{s(s+1)} = \int_{-\frac{1}{4}-i\infty}^{-\frac{1}{4}+i\infty} \zeta(s)n^s \frac{ds}{s(s+1)} = 0, \tag{7.54}$$

which is a source of several *exact* instead of *asymptotic* formulas (e.g., see Section 7.6.3).   ∎

7.6 APPLICATIONS

We discuss three different applications of generating functions. In the first one, we consider a class of recurrences arising in the analysis of digital trees, and show how to solve them exactly. Then we study an old problem of counting the number of pattern occurrences in a random string. We use the combinatorial structure approach (i.e., language approach) to derive the generating function for the frequency of pattern occurrence when the text string is generated by a Markovian source. Finally, we use the Perron-Mellin formula to derive Delange's formula for the total number of 1 digits in the binary representation of $1, 2, \ldots, n$.

7.6.1 Recurrences Arising in the Analysis of Digital Trees

We consider certain recurrences that arise in problems on words, in particular, in digital trees discussed in Section 1.1. Let x_n be a generic notation for a quantity of interest (e.g., depth, size, or path length in a digital tree built over n strings). Given x_0 and x_1, the following three recurrences originate from problems on tries, PATRICIA tries, and digital search trees, respectively,

$$x_n = a_n + \beta \sum_{k=0}^{n} \binom{n}{k} p^k q^{n-k} (x_k + x_{n-k}), \quad n \geq 2, \tag{7.55}$$

$$x_n = a_n + \beta \sum_{k=1}^{n-1} \binom{n}{k} p^k q^{n-k} (x_k + x_{n-k}) - \alpha (p^n + q^n) x_n, \quad n \geq 2, \tag{7.56}$$

$$x_{n+1} = a_n + \beta \sum_{k=0}^{n} \binom{n}{k} p^k q^{n-k} (x_k + x_{n-k}) \quad n \geq 0, \tag{7.57}$$

where a_n is a known sequence (also called the additive term), α and β are some constants, and $p + q = 1$.

To solve these recurrences we apply exponential generating functions. We start with (7.55), multiplying it by z^n, summing up, and using the convolution formula for exponential generating function (7.17) from Table 7.3. Taking into account the initial conditions, we arrive at

$$x(z) = a(z) + \beta x(zp) e^{zq} + \beta x(zq) e^{zp} + d(z), \tag{7.58}$$

where $d(z) = d_0 + d_1 z$ and d_0 and d_1 depend on the initial condition for $n = 0, 1$. This functional equation is still hard to solve; hence we introduce the Poisson transform

$$\widetilde{X}(z) = x(z)e^{-z} = e^{-z} \sum_{n=0}^{\infty} x_n \frac{z^n}{n!}.$$

The Poisson transform is a very useful tool in the analysis of algorithms. We devote to it all of Chapter 10. Observe now that (7.58) reduces to the following functional equation

$$\widetilde{X}(z) = \widetilde{A}(z) + \beta \widetilde{X}(zp) + \beta \widetilde{X}(z) + d(z)e^{-z}, \qquad (7.59)$$

where $\widetilde{A}(z) = a(z)e^{-z} = e^{-z} \sum_{n=0}^{\infty} a_n \frac{z^n}{n!}$. (Throughout this book we write $\widetilde{A}(z)$ for the Poisson transform of the sequence $\{a_n\}_{n=0}^{\infty}$). Since $\widetilde{x}_n = n![z^n]\widetilde{X}(z)$ the quantities $\widetilde{x}_n$ and x_n are related by

$$x_n = \sum_{k=0}^{n} \binom{n}{k} \widetilde{x}_k.$$

After comparing coefficients of $\widetilde{X}(z)$ at z^n, we finally obtain

$$x_n = x_0 + n(x_1 - x_0) + \sum_{k=2}^{n} (-1)^k \binom{n}{k} \frac{\hat{a}_k + kd_1 - d_0}{1 - \beta(p^k + q^k)}, \qquad (7.60)$$

where $n![z^n]\widetilde{A}(z) = \widetilde{a}_n := (-1)^n \hat{a}_n$. In passing, we point out that $\hat{a}_n$ and a_n form the *binomial inverse relations*, and

$$\hat{a}_n = \sum_{k=0}^{n} \binom{n}{k} (-1)^k a_k, \qquad a_n = \sum_{k=0}^{n} \binom{n}{k} (-1)^k \hat{a}_k, \qquad (7.61)$$

that is, $\hat{\hat{a}}_n = a_n$.

Example 7.21: *The Average External Path Length in a Trie* To illustrate our discussion so far, let us consider a trie with a memoryless source and estimate the average ℓ_n of the external path length, i.e., $\ell_n = \mathbf{E}[L_n]$ (see Section 1.1). Clearly, $\ell_0 = \ell_1 = 0$ and for $n \geq 2$

$$\ell_n = n + \sum_{k=0}^{n} \binom{n}{k} p^k q^{n-k} (\ell_k + \ell_{n-k}).$$

Thus by (7.60)

$$\ell_n = \sum_{k=2}^{n} (-1)^k \binom{n}{k} \frac{k}{1 - p^k - q^k}$$

for $n \geq 2$.   ∎

Let us now consider recurrence (7.56), which is more intricate. It has an exact solution only for some special cases that we discuss below. We consider only a simplified version of (7.56), namely,

$$x_n(2^n - 2) = 2^n a_n + \sum_{k=1}^{n-1} \binom{n}{k} x_k$$

with $x_0 = x_1 = 0$ (for a more general recurrence of this type see [404]). The EGF $x(z)$ of x_n is

$$x(z) = (e^{z/2} + 1)x(z/2) + a(z) - a_0, \tag{7.62}$$

where $a(z)$ is the exponential generating function of a_n. To solve this recurrence we observe that after multiplying both sides by $z/(e^z - 1)$ and defining

$$\check{X}(z) = x(z) \frac{z}{e^z - 1},$$

we obtain a new, simpler functional equation, namely:

$$\check{X}(z) = \check{X}(z/2) + \check{A}(z). \tag{7.63}$$

In the above we assume for simplicity $a_0 = 0$. This is due to the identity $e^z - 1 = (e^{z/2} - 1)(e^{z/2} + 1)$ that luckily translates (7.62) into (7.63). But (7.63) is of the same type as the functional equation (7.59), and this allows us to extract the coefficient $\check{x}_n = n![z^n]\check{X}(z)$. One must, however, translate the coefficients $\check{x}_n$ into the original sequence x_n. We use the Bernoulli polynomials $B_n(x)$ and the Bernoulli numbers $B_n = B_n(0)$ (see Table 8.1 for detailed discussion) defined as

$$\frac{ze^{tz}}{e^z - 1} = \sum_{k=0}^{\infty} B_k(t) \frac{z^k}{k!}.$$

Furthermore, we introduce the *Bernoulli inverse relations* for a sequence a_n as

$$\check{a}_n = \sum_{k=0}^{n} \binom{n}{k} B_k a_{n-k} \quad \longleftrightarrow \quad a_n = \sum_{k=0}^{n} \binom{n}{k} \frac{\check{a}_k}{k+1}. \tag{7.64}$$

In Exercise 10 we ask the reader to prove

$$a_n = \binom{n}{r}q^n \quad \longleftrightarrow \quad \breve{a}_n = \binom{n}{r}q^r B_{n-r}(q) \tag{7.65}$$

for $0 < q < 1$. Using now (7.63), and comparing coefficients we show that for $a_n = \binom{n}{r}q^n$ the above recurrence has a particularly simply solution:

$$x_n = \sum_{k=1}^{n}(-1)^k \binom{n}{k}\frac{B_{k+1}(1-q)}{k+1}\frac{1}{2^{k+1}-1}.$$

A general solution to the above recurrence can be found in Szpankowski [404] (cf. also [167]), and it involves $\breve{a}_n$.

Example 7.22:   *The Average Unsuccessful Search*   As an example, consider the number of trials u_n in an unsuccessful search of a string in a PATRICIA trie in which strings are generated by an unbiased memoryless source (i.e., $p = q = 1/2$). Knuth [269] proved that

$$u_n(2^n - 2) = 2^n(1 - 2^{1-n}) + \sum_{k=1}^{n-1}\binom{n}{k}u_k$$

and $u_0 = u_1 = 0$. A simple application of the above leads, after some algebra, to

$$u_n = 2 - \frac{4}{n+1} + 2\delta_{n0} + \frac{2}{n+1}\sum_{k=2}^{n}\binom{n+1}{k}\frac{B_k}{2^{k-1}-1},$$

where $\delta_{n,k}$ is the Kronecker delta, that is, $\delta_{n,k} = 1$ for $n = k$ and zero otherwise.   ∎

In summary, so far we were able to exactly solve functional equations (7.55) and (7.56) since they can be reduced to a simple functional equation of the form (7.59). In particular, equation (7.56) became (7.59) since luckily $e^z - 1 = (e^{z/2} - 1)(e^{z/2} + 1)$, as already pointed out by Knuth [269], but one cannot expect that much luck with other functional equations. Nevertheless, there is a large class of recurrences that can be reduced to the following general functional equation

$$F(z) = a(z) + b(z)F(\sigma(z)), \tag{7.66}$$

where $a(z), b(z), \sigma(z)$ are known function. Formally, iterating this equation we obtain its solution as

$$F(z) = \sum_{k=0}^{\infty} a(\sigma^{(k)}(z)) \prod_{j=0}^{k-1} b(\sigma^{(j)}(z)), \tag{7.67}$$

where $\sigma^{(k)}(z)$ is the kth iterate of $\sigma(\cdot)$. When applying the above to solve real problems, one must assure the existence of the infinite series involved. In some cases we can provide asymptotic solutions to such formulas by appealing to the Mellin transform that we shall discuss in Chapter 9.

Example 7.23: *A Simple Functional Equation* Consider the following functional equation

$$f(z) = \beta a(z/2) f(z/2) + b(z),$$

where $|\beta| \leq 1$. By (7.67) we obtain

$$f(z) = \sum_{n=0}^{\infty} \beta^n b(z2^{-n}) \prod_{k=1}^{n} a(z2^{-k}),$$

provided the series converges. Define

$$\varphi(z) = \prod_{j=0}^{\infty} a(z2^j),$$

provided the infinite product converges. Then

$$f(z)\varphi(z) = \sum_{n=0}^{\infty} \beta^n b(z2^{-n}) \varphi(z2^{-n}). \tag{7.68}$$

This form is quite useful in some computations, as we shall in see subsequent chapters. ∎

Finally, we deal with recurrence (7.57). The EGF $x(z)$ of x_n satisfies

$$x'(z) = a(z) + x(zp)e^{zq} + x(zq)e^{zp},$$

which becomes after the substitution $\widetilde{X}(z) = x(z)e^{-z}$

$$\widetilde{X}'(z) + \widetilde{X}(z) = \widetilde{A}(z) + \widetilde{X}(zp) + \widetilde{X}(zq). \tag{7.69}$$

The above is a *differential*-functional equation that we did not discuss so far. It can be solved since a direct translation of coefficients gives: $\widetilde{x}_{n+1} + \widetilde{x}_n = \widetilde{a}_n +$

$\tilde{x}_n(p^n + q^n)$. Fortunately, this is a simple linear recurrence that has an explicit solution. Since

$$x_n = \sum_{k=0}^{n} \binom{n}{k} \tilde{x}_k,$$

we finally obtain

$$x_n = x_0 - \sum_{k=1}^{n} (-1)^k \binom{n}{k} \sum_{i=1}^{k-1} \hat{a}_i \prod_{j=i+1}^{k-1} (1 - p^j - q^j)$$

$$= x_0 - \sum_{k=1}^{n} (-1)^k \binom{n}{k} \sum_{i=1}^{k-1} \hat{a}_i \frac{Q_k}{Q_i}, \tag{7.70}$$

where $Q_k = \prod_{j=2}^{k}(1 - p^j - q^j)$, and $\hat{a}_n$ is the binomial inverse of a_n as defined in (7.61).

Example 7.24: *The Average External Path Length in a Digital Search Tree*
To illustrate this, let ℓ_n be the expected path length in a digital search tree. Then for all $n \geq 0$

$$\ell_{n+1} = n + \sum_{k=1}^{n} \binom{n}{k} p^k q^{n-k} (\ell_k + \ell_{n-k})$$

with $\ell_0 = 0$. By (7.57) it has the following solution

$$\ell_n = \sum_{k=2}^{n} (-1)^k \binom{n}{k} Q_{k-1},$$

where Q_k is defined above. ∎

In passing, we should observe that solutions of the recurrences (7.55)-(7.57) have a form of an alternating sum, that is,

$$x_n = \sum_{k=1}^{n} (-1)^k \binom{n}{k} f_k,$$

where f_n is a given sequence. In Section 8.5 we shall discuss in depth such sums and provide a tool to find asymptotic expansions for them.

We were quite lucky so far since we dealt mostly with linear recurrences of the *first* order. However, this is not longer true when we consider b-digital search trees (discussed in Section 1.1) in which one assumes that a node of such a tree can store up to b strings. Then the general recurrence (7.57) becomes

$$x_{n+b} = a_n + \beta \sum_{k=0}^{n} \binom{n}{k} p^k q^{n-k} (x_k + x_{n-k}) \quad n \geq 0, \tag{7.71}$$

provided $x_0, \ldots, x_{b-1}$ are given. Our previous approach would lead to a linear recurrence of order b that does not possess a nice explicit solution. The culprit lies in the fact that the exponential generating function of $\{x_{n+b}\}_{n=0}^{\infty}$ is the bth derivative of the exponential generating function of $\{x_n\}_{n=0}^{\infty}$. We know, however, that the *ordinary* generating functions of $\{x_{n+b}\}_{n=0}^{\infty}$ translate into $z^{-b}(X(z) - x_0 - \cdots - x_{b-1} z^{b-1})$, where $X(z)$ is OGF of $\{x_n\}_0^{\infty}$. This simple observation led Flajolet and Richmond [145] to reconsider the standard approach to the above binomial recurrences, and to introduce the ordinary generating function into the play. A careful reader observes, however, that then one must translate into ordinary generating functions sequences such as $s_n = \sum_{k=0}^{n} \binom{n}{k} a_k$ (which was easy under exponential generating functions since it becomes $a(z)e^z$). In Example 7.11 we prove that

$$s_n = \sum_{k=0}^{n} \binom{n}{k} a_k \quad \longleftrightarrow \quad S(z) = \frac{1}{1-z} A\left(\frac{z}{1-z}\right).$$

We shall return to this problem in Sections 9.4.1 and 10.5.2.

Example 7.25: *A Simple Differential-Functional Equation*   For example, recurrence (7.71) for $p = q = 1/2$ and any $b \geq 1$ can be translated into ordinary generating functions as

$$X(z) = \frac{1}{1-z} G\left(\frac{z}{1-z}\right),$$

$$G(z)(1+z)^b = 2z^b G(z/2) + P(z),$$

where $P(z)$ is a function of a_n and initial conditions. The latter functional equation falls under (7.66), and thus can be solved by the method presented above. However, for $p \neq 1/2$ an explicit solution is harder to find, and we must resort to an asymptotic solution. We shall discuss it again in Section 10.5.2.   ∎

7.6.2   Pattern Occurrences in a Random Text

Let us consider two strings, a pattern string $H = h_1 h_2 \ldots h_m$ and a text string $T = t_1 t_2 \ldots t_n$ of respective lengths equal to m and n over an alphabet S of size V. (In this section, we drop our standard notation $\mathcal{A}$ for the alphabet since it is traditionally reserved in the pattern matching problems for the autocorrelation language defined below.) We shall write $S = \{1, 2, \ldots, V\}$ to simplify the presentation. We assume that the pattern string is *fixed* and given, while the text string is random and generated by a stationary Markovian source. We also present results for memoryless sources. For the stationary Markov model, as always, we denote the transition matrix as $P = \{p_{i,j}\}_{i,j \in S}$, the stationary distribution as π, and we write Π for the stationary matrix that consists of V identical rows equal to π.

A word about notation: To extract a particular element, say with index (i, j), from a matrix, say P, we shall write $[P]_{i,j} = p_{i,j}$. Finally, we recall that $(I - P)^{-1} = \sum_{k \geq 0} P^k$ provided the inverse matrix exists (i.e., $\det(I - P) \neq 0$ or $||P|| < 1$ for any matrix norm $|| \cdot ||$). Below, we write $P(H_i^j) = \Pr\{T_{i+k}^{j+k} = H_i^j\}$ for the probability of the substring $H_i^j = h_i \ldots h_j$ occurring in the random text T_{i+k}^{j+k} between symbols $i + k$ and $j + k$ for any k (in particular, $P(H)$ denotes the probability of H appearing in the text).

Our goal is to estimate the frequency of pattern occurrences in a text generated by a Markov source. We allow patterns to overlap when counting occurrences (e.g., if $H = abab$, then it occurs twice in $T = abababb$ when overlapping is allowed; it occurs only once if overlapping is not allowed). We write O_n or $O_n(H)$ for the number of pattern H occurrences in the text of size n. We derive formulas for the following two generating functions:

$$T^{(r)}(z) = \sum_{n \geq 0} \Pr\{O_n(H) = r\}z^n,$$

$$T(z, u) = \sum_{r=1}^{\infty} T^{(r)}(z)u^r = \sum_{r=1}^{\infty} \sum_{n=0}^{\infty} \Pr\{O_n(H) = r\}z^n u^r$$

that are defined for $|z| \leq 1$ and $|u| \leq 1$. We adopt here the presentation of Régnier and Szpankowski [359], however, the reader is also advised to inspect Nicodéme, Salvy, and Flajolet [326].

We use the unlabeled combinatorial structures approach to find $T(z, u)$. In the case of strings, such structures are known as *languages*, and we use this term throughout this section. We extend, however, the definition of languages to include languages defined on a Markovian source. Therefore, we adopt the following definition.

Definition 7.7 *For any language $\mathcal{L}$ we define its generating function $L(z)$ as*

$$L(z) = \sum_{w \in \mathcal{L}} P(w)z^{|w|},$$

where $P(w)$ is the stationary probability of word w occurrence, $|w|$ is the length of w, and we assume that $P(\epsilon) = 1$, where ϵ is the empty word. In addition, we define H-conditional generating function of $\mathcal{L}$ as

$$L_H(z) = \sum_{w \in \mathcal{L}} P(w|w_{-m} = h_1 \cdots w_{-1} = h_m) z^{|w|}$$

$$= \sum_{w \in \mathcal{L}} P(w|w_{-m}^{-1} = H) z^{|w|},$$

where w_{-i} stands for a symbol preceding the first character of w at distance i.

Since we allow overlaps, the structure of the pattern has a profound impact on the number of occurrences. To capture this, we introduce the autocorrelation language and the autocorrelation polynomial.

Definition 7.8 *Given a string H, we define the autocorrelation set $\mathcal{A}$ as:*

$$\mathcal{A} = \{H_{k+1}^m : H_1^k = H_{m-k+1}^m\}, \tag{7.72}$$

*and by HH we denote the set of positions k satisfying $H_1^k = H_{m-k+1}^m$. The generating function of language $\mathcal{A}$ is denoted as $A(z)$ and we call it the **autocorrelation polynomial**. Its H-conditional generating function is denoted $A_H(z)$. In particular,*

$$A_H(z) = \sum_{k \in HH} P(H_{k+1}^m | H_k) z^{m-k} \tag{7.73}$$

for a Markov source.

Before we proceed, we present a simple example illustrating the definitions we introduced so far.

Example 7.26: *Illustration of Definition 7.8*   Let us assume that $H = 101$ over a binary alphabet $\mathcal{S} = \{0, 1\}$. Observe that $HH = \{1, 3\}$ and $\mathcal{A} = \{\epsilon, 01\}$, where ϵ is the empty word. Thus, for the unbiased memoryless source, we have $A(z) = 1 + \frac{z^2}{4}$, while for the Markovian model of order one, we obtain $A_{101}(z) = 1 + p_{10}p_{01}z^2$.   ∎

As mentioned above, we want to estimate the number of pattern occurrences in a text. Alternatively, we can seek the generating function of the language $\mathcal{T}$ that consists of words containing at least one occurrence of H. We will follow this approach, and introduce some other languages to describe $\mathcal{T}$.

Definition 7.9 *Given a pattern H:*

(i) *Let T_r be a language that consists of all words containing exactly $r \geq 1$ occurrences of H.*

(ii) *Let $\mathcal{R}$ be a set of words containing only one occurrence of H, located at the right end. We also define $\mathcal{U}$ as*

$$\mathcal{U} = \{u : H \cdot u \in T_1\}, \tag{7.74}$$

where the operation $\cdot$ means concatenation of words. In other words, a word $u \in \mathcal{U}$ if $H \cdot u$ has exactly one occurrence of H at the left end of $H \cdot u$.

(iii) *Let $\mathcal{M}$ be the language:*

$$\mathcal{M} = \{w : H \cdot w \in T_2 \text{ and } H \text{ occurs at the right end of } H \cdot w\},$$

that is, $\mathcal{M}$ is a language such that $H \cdot \mathcal{M}$ has exactly two occurrences of H at the left and right end of a word from $\mathcal{M}$.

We now can describe languages T and T_r in terms of $\mathcal{R}$, $\mathcal{M}$, and $\mathcal{U}$. This will further lead to a simple formula for the generating function of $O_n(H)$.

Theorem 7.10 *Language T satisfies the fundamental equation*

$$T = \mathcal{R} \cdot \mathcal{M}^* \cdot \mathcal{U}. \tag{7.75}$$

Notably, language T_r can be represented for any $r \geq 1$ as follows:

$$T_r = \mathcal{R} \cdot \mathcal{M}^{r-1} \cdot \mathcal{U}. \tag{7.76}$$

Here, by definition $\mathcal{M}^0 := \{\epsilon\}$ and $\mathcal{M}^* := \bigcup_{r=0}^{\infty} \mathcal{M}^r$.

Proof. We first prove (7.76) and obtain our decomposition of T_r as follows: The first occurrence of H in a word belonging to T_r determines a prefix p that is in $\mathcal{R}$. Then one concatenates a nonempty word w that creates the second occurrence of H. Hence, w is in $\mathcal{M}$. This process is repeated $r - 1$ times. Then one adds after the last H occurrence a suffix u that does not create a new occurrence of H. Equivalently, Hu is such that u is in $\mathcal{U}$, and w is a proper subword of Hu. Finally, a word belongs to T if for some $1 \leq r < \infty$ it belongs to T_r. The set union $\bigcup_{r=1}^{\infty} \mathcal{M}^{r-1}$ yields precisely $\mathcal{M}^*$. ∎

We now prove the following result that summarizes relationships between the languages $\mathcal{R}$, $\mathcal{M}$, and $\mathcal{U}$.

Theorem 7.11 *The languages $\mathcal{M}, \mathcal{U},$ and $\mathcal{R}$ satisfy*

$$\bigcup_{k\geq 1} \mathcal{M}^k = \mathcal{W} \cdot H + \mathcal{A} - \{\epsilon\}, \tag{7.77}$$

$$\mathcal{U} \cdot S = \mathcal{M} + \mathcal{U} - \{\epsilon\}, \tag{7.78}$$

$$H \cdot \mathcal{M} = S \cdot \mathcal{R} - (\mathcal{R} - H), \tag{7.79}$$

where $\mathcal{W}$ is the set of all words, S is the alphabet set and $+$ and $-$ are disjoint union and subtraction of languages. In particular, a combination of (7.78) and (7.79) gives

$$H \cdot \mathcal{U} \cdot S - H \cdot \mathcal{U} = (S - \epsilon)\mathcal{R}. \tag{7.80}$$

Additionally, we have:

$$T_0 \cdot H = \mathcal{R} \cdot \mathcal{A}. \tag{7.81}$$

Proof. All the relations above are proved in a similar fashion. We first deal with (7.77). Let k be the number of H occurrences in $\mathcal{W} \cdot H$. By definition, $k \geq 1$ and the last occurrence is on the right: This implies that $\mathcal{W} \cdot H \subseteq \bigcup_{k\geq 1} \mathcal{M}^k$. Furthermore, a word w in $\bigcup_{k\geq 1} \mathcal{M}^k$ is not in $\mathcal{W} \cdot H$ if and only if its size $|w|$ is smaller than $|H|$. Then the second H occurrence in Hw overlaps with H, which means that w is in $\mathcal{A} - \epsilon$.

Let us turn now to (7.78). When one adds a character s right after a word u from $\mathcal{U}$, two cases may occur. Either Hus still does not contain a second occurrence of H, which means that us is a nonempty word of $\mathcal{U}$. Or a new H appears, clearly at the right end. Then us is in $\mathcal{M}$. Furthermore, the whole set $\mathcal{M} + (\mathcal{U} - \epsilon)$ is attained (i.e., a strict prefix of $\mathcal{M}$ cannot contain a new H occurrence). Hence, it is in $\mathcal{U}$, and a strict prefix of a $\mathcal{U}$-word is in $\mathcal{U}$.

We now prove (7.79). Let $x = sw$ be a word in $H \cdot \mathcal{M}$ where s is a symbol from S. As x contains exactly two occurrences of H located at its left and right ends, w is in $\mathcal{R}$ and x is in $\mathcal{A} \cdot \mathcal{R} - \mathcal{R}$. Reciprocally, if a word swH from $S \cdot \mathcal{R}$ is not in $\mathcal{R}$, then swH contains a second H occurrence starting in sw. As wH is in $\mathcal{R}$, the only possible position is at the left end, and then x is in $H \cdot \mathcal{M}$. We now rewrite:

$$S \cdot \mathcal{R} - \mathcal{R} = S \cdot \mathcal{R} - (\mathcal{R} \cap S \cdot \mathcal{R}) = S \cdot \mathcal{R} - (\mathcal{R} - H),$$

which yields $H \cdot \mathcal{M} - H = (S - \epsilon) \cdot \mathcal{R}$.

Finally, we leave the derivation of (7.81) to the reader as an exercise. ∎

Now we translate the language relationships into the associated generating functions. We should point out, however, that one must be careful with products of two languages (i.e., concatenation of two languages). For the memoryless source, the situation is quite simple. Indeed, let $\mathcal{L}_1$ and $\mathcal{L}_2$ be two arbitrary languages with generating functions $L_1(z)$ and $L_2(z)$, respectively. Then for memoryless sources the language $\mathcal{L} = \mathcal{L}_1 \cdot \mathcal{L}_2$ is transferred into the generating function $L(z)$ such that

$$L(z) = L_1(z)L_2(z), \tag{7.82}$$

since $P(wv) = P(w)P(v)$ for $w \in \mathcal{L}_1$ and $v \in \mathcal{L}_2$. In particular, the generating function $L(z)$ of $\mathcal{L} = S \cdot \mathcal{L}_1$ is $L(z) = zL_1(z)$, where S is the alphabet set, since $S(z) = \sum_{s \in S} P(s)z = z$.

For Markovian sources $P(wv) \neq P(w)P(v)$; thus property (7.82) is no longer true. We have to replace it with a more sophisticated one. We have to condition $\mathcal{L}_2$ on symbols preceding a word from $\mathcal{L}_2$ (i.e., belonging to $\mathcal{L}_1$). In general, for a K order Markov chain, one must distinguish V^K ending states for $\mathcal{L}_1$ and V^K initial states for $\mathcal{L}_2$. For simplicity of presentation, we consider only first-order Markov chains (i.e., $K = 1$), and we write $\ell(w)$ for the last symbol of a word w. In particular, to rewrite property (7.82) we must introduce the following conditional generating function for a language $\mathcal{L}$:

$$L_i^j(z) = \sum_{w \in \mathcal{L}} P(w, \ell(w) = j | w_1 = i) z^{|w|}.$$

Let now $\mathcal{L} = \mathcal{W} \cdot \mathcal{V}$. Then

$$L_k^l(z) = \sum_{i,j \in S} p_{ji} W_k^j(z) V_i^l(z), \tag{7.83}$$

where $W_k^j(z)$ and $V_i^l(z)$ are conditional generating functions for $\mathcal{W}$ and $\mathcal{V}$ respectively. To prove this, let $w \in \mathcal{W}$ and $v \in \mathcal{V}$ and observe

$$P(wv) = \sum_{j \in S} P(wv, \ell(w) = j)$$

$$= \sum_{j \in S} P(w, \ell(w) = j)P(v|\ell(w) = j)$$

$$= \sum_{j \in S} \sum_{i \in S} P(w, \ell(w) = j)p_{ji} P(v|v_1 = i).$$

After conditioning on the first symbol of $\mathcal{W}$ and the last symbol of $\mathcal{V}$, we prove (7.83).

The lemma below translates (7.77)–(7.79) into generating functions.

Lemma 7.12 *The generating functions associated with languages $\mathcal{M}, \mathcal{U}$, and $\mathcal{R}$ satisfy*

$$\frac{1}{1 - M_H(z)} = A_H(z) + P(H)z^m \left(\frac{1}{1-z} + F(z) \right), \qquad (7.84)$$

$$U_H(z) = \frac{M_H(z) - 1}{z - 1}, \qquad (7.85)$$

$$R(z) = P(H)z^m \cdot U_H(z), \qquad (7.86)$$

provided the underlying Markov chain is stationary. The function $F(z)$ is defined below in the proof and explicitly in (7.93) of Theorem 7.13.

Proof. We first prove (7.85). Interestingly, it does not need the stationarity assumption. Let us consider the language relationship (7.78) from Theorem 7.11, which we rewrite as $\mathcal{U} \cdot \mathcal{S} - \mathcal{U} = \mathcal{M} - \epsilon$. Observe that $\sum_{j \in S} p_{i,j} z = z$. Hence, set $\mathcal{U} \cdot \mathcal{S}$ yields (conditioning on the left occurrence of H):

$$\sum_{w \in \mathcal{U}} \sum_{j \in S} P(wj|H)z^{|wj|} = \sum_{i \in S} \sum_{w \in \mathcal{U}, \ell(w)=i} P(w|H)z^{|w|} \sum_{j \in S} p_{i,j} z = U_H(z) \cdot z.$$

Of course, $\mathcal{M} - \epsilon$ and $\mathcal{U}$ translate into $M_H(z) - 1$ and $U_H(z)$, and (7.85) is proved.

We now turn our attention to (7.86), and we use relationship (7.79) of Theorem 7.11. Observe that $\mathcal{S} \cdot \mathcal{R}$ can be rewritten as

$$\sum_{j,i \in S^2} \sum_{iw \in \mathcal{R}} P(jiw)z^{|jiw|} = z^2 \sum_{j \in S} \sum_{i \in S} \pi_j p_{j,i} \sum_{iw \in \mathcal{R}} P(w|w_{-1} = i)z^{|w|}.$$

But due to the stationarity of the underlying Markov chain $\sum_j \pi_j p_{j,i} = \pi_i$. As $\pi_i P(w|w_{-1} = i) = P(iw)$, we get $zR(z)$. Furthermore, in (7.79) $H \cdot \mathcal{M} - H$ translates into $P(H)z^m \cdot (M_H(z) - 1)$. By (7.85), this becomes $P(H)z^m \cdot U_H(z)(z - 1)$, and after a simplification, we prove (7.86).

Finally, we deal with (7.84), and prove it using (7.77) from Theorem 7.11. The left-hand side of (7.77) involves language $\mathcal{M}$, hence we must condition on the left occurrence of H. In particular, $\bigcup_{r \geq 1} \mathcal{M}^r + \epsilon$ of (7.77) translates into $\frac{1}{1 - M_H(z)}$. Now we deal with $\mathcal{W} \cdot H$ of the right-hand side of (7.77). *Conditioning* on the left occurrence of H, the generating function $W(z)H(z)$ of $\mathcal{W} \cdot H$ becomes

$$W_H(z)H(z) = \sum_{n \geq 0} \sum_{|w|=n} z^{n+m} P(wH|w_{-1} = \ell(H))$$

$$= \sum_{n \geq 0} \sum_{|w|=n} z^n P(wh_1|w_{-1} = \ell(H)) P(v = h_2 \ldots h_m|v_{-1} = h_1)z^m.$$

We have $P(v = h_2 \ldots h_m | v_{-1} = h_1) z^m = \frac{1}{\pi_{h_1}} z^m P(H)$, and for $n \geq 0$:

$$\sum_{|w|=n} P(wh_1 | w_{-1} = \ell(H)) = [\mathsf{P}^{n+1}]_{\ell(H), h_1}$$

where, we recall, $\ell(H) = h_m$ is the last character of H. In summary: The language $\mathcal{W} \cdot H$ contributes $P(H) z^m \left[\frac{1}{\pi_{h_1}} \sum_{n \geq 0} \mathsf{P}^{n+1} z^n \right]_{\ell(H), h_1}$, while language $\mathcal{A} - \{\epsilon\}$ introduces $A_H(z) - 1$. We now observe that for any symbols i and j

$$\left[\frac{1}{\pi_j} \sum_{n \geq 0} \Pi z^n \right]_{i,j} = \sum_{n \geq 0} z^n = \frac{1}{1-z}.$$

Using the equality $\mathsf{P}^{n+1} - \Pi = (\mathsf{P} - \Pi)^{n+1}$ (which follows from a consecutive application of the identity $\Pi \mathsf{P} = \Pi$), we finally obtain the sum in (7.84), where $F(z) = \frac{1}{\pi_{h_1}} \sum_{n \geq 0} (P - \Pi)^{n+1} z^n$ for $|z| < (\|P - \Pi\|)^{-1}$. ∎

The lemma above together with Theorem 7.10 suffice to establish the following main result.

Theorem 7.13   (Régnier and Szpankowski, 1998)   *Let H be a given pattern of size m, and T be a random text of length n generated according to a stationary Markov chain over a V-ary alphabet S. The generating functions $T^{(r)}(z)$ and $T(z, u)$ satisfy*

$$T^{(r)}(z) = R(z) M^{r-1}(z) U_H(z), \quad r \geq 1, \tag{7.87}$$

$$T(z, u) = R(z) \frac{u}{1 - u M_H(z)} U_H(z), \tag{7.88}$$

where

$$M_H(z) = 1 + \frac{z-1}{D_H(z)}, \tag{7.89}$$

$$U_H(z) = \frac{1}{D_H(z)}, \tag{7.90}$$

$$R(z) = z^m P(H) \frac{1}{D_H(z)}. \tag{7.91}$$

with

$$D_H(z) = (1 - z) A_H(z) + z^m P(H)(1 + (1 - z) F(z)). \tag{7.92}$$

The function $F(z)$ is defined for $|z| \leq R$ where $R = \frac{1}{||P-\Pi||}$ as follows

$$F(z) = \frac{1}{\pi_{h_1}}[(P - \Pi)(I - (P - \Pi)z)^{-1}]_{h_m,h_1}, \qquad (7.93)$$

where h_1 and h_m are the first and the last symbols of H, respectively.
For memoryless sources, $F(z) = 0$, and hence

$$D(z) = (1 - z)A_H(z) + z^m P(H)$$

with the other formulas as above.

Theorem 7.13 is a starting point of the next findings concerning the average and the variance of O_n. The limiting distribution of O_n is established in Example 8.8.8 of Chapter 8.

Theorem 7.14   (Régnier and Szpankowski, 1998)   *Let the hypotheses of Theorem 7.13 be fulfilled and $nP(H) \to \infty$. The expectation $\mathbf{E}[O_n(H)]$ satisfies for $n \geq m$:*

$$\mathbf{E}[O_n(H)] = P(H)(n - m + 1), \qquad (7.94)$$

while the variance becomes for some $r > 1$

$$\mathbf{Var}[O_n(H)] = nc_1 + c_2 + O(r^{-n}), \qquad (7.95)$$

where

$$c_1 = P(H)(2A_H(1) - 1 - (2m - 1)P(H) + 2P(H)E_1)), \qquad (7.96)$$

$$c_2 = P(H)((m - 1)(3m - 1)P(H) - (m - 1)(2A_H(1) - 1) - 2A'_H(1))$$
$$- 2(2m - 1)P(H)^2 E_1 + 2E_2 P(H)^2, \qquad (7.97)$$

and the constants E_1, E_2 are

$$E_1 = \frac{1}{\pi_{h_1}}[(P - \Pi)Z]_{h_m,h_1}, \qquad (7.98)$$

$$E_2 = \frac{1}{\pi_{h_1}}[(P^2 - \Pi)Z^2]_{h_m,h_1}, \qquad (7.99)$$

where $Z = (I - (P - \Pi))^{-1}$ is the fundamental matrix of the underlying Markov chain.

For the memoryless source, $E_1 = E_2 = 0$ since $\mathsf{P} = \Pi$, and (7.95) reduces to an equality for $n \geq 2m - 1$. Thus

$$\mathbf{Var}[O_n(H)] = nc_1 + c_2 \tag{7.100}$$

with

$$c_1 = P(H)(2A(1) - 1 - (2m - 1)P(H)),$$

$$c_2 = P(H)((m - 1)(3m - 1)P(H) - (m - 1)(2A(1) - 1) - 2A'(1)).$$

Proof. We evaluate the first two moments of $T(z, u)$ at $u = 1$, that is, $\mathbf{E}[O_n] = [z^n]T'(z, 1)$ and $\mathbf{E}[O_n(O_n - 1)] = T''(z, 1)$. From Theorem 7.13 we conclude that

$$T_u(z, 1) = \frac{z^m P(H)}{(1 - z)^2},$$

$$T_{uu}(z, 1) = \frac{2z^m P(H)M_H(z)D_H(z)}{(1 - z)^3},$$

where $T_u(z, 1)$ and $T_{uu}(z, 1)$ are the first and the second derivatives of $T(z, u)$ at $u = 1$. Now we observe that both expressions admit as a numerator a function that is analytic beyond the unit circle. This allows for a very simple computation of the expectation and variance based on the following basic formula

$$[z^n](1 - z)^{-P} = \frac{\Gamma(n + p)}{\Gamma(p)\Gamma(n + 1)}. \tag{7.101}$$

To obtain $\mathbf{E}[O_n]$ we proceed as follows for $n \geq m$:

$$\mathbf{E}[O_n] = [z^n]T_u(z, 1) = P(H)[z^{n-m}](1 - z)^{-2} = (n - m + 1)P(H).$$

Let

$$\Phi(z) = 2z^m P(H)M_H(z)D_H(z),$$

which is a polynomial for the memoryless source. Observe that

$$\Phi(z) = \Phi(1) + (z - 1)\Phi'(1) + \frac{(z - 1)^2}{2}\Phi''(1) + (z - 1)^3 f(z),$$

where $f(z)$ is a polynomial of degree $2m - 2$. It follows that $[z^n](z - 1)f(z)$ is 0 for $n \geq 2m - 1$ and, using formula (7.101), we get

$$\mathbf{E}[O_n(O_n-1)] = [z^n]T_{uu}(z, 1) = \Phi(1)\frac{(n + 2)(n + 1)}{2} - \Phi'(1)(n+1) + \frac{1}{2}\Phi''(1).$$

Observing that $M_H(z)D_H(z) = D_H(z) + (1-z)$, we obtain (7.95). In the Markov case, we have to compute the additional term

$$[z^n]\frac{2(z^{2m}P(H)^2 F(z))}{(1-z)^2},$$

where $F(z)$ is analytic beyond the unit circle for $|z| \leq R$, with $R > 1$. The Taylor expansion of $F(z)$ is $E_1 + (1-z)E_2 + O((1-z)^2)$, and applying (7.101) again yields the result. ∎

In passing, we add that Nicodéme, Salvy, and Flajolet [326] extended these results to patterns that are regular expressions.

7.6.3 Delange's Formula for a Digital Sum

Let $v(k)$ be the number of 1-digits in the binary representation of the integer k. Define

$$S(n) = \sum_{k=1}^{n-1} v(k),$$

that is, the total number of 1-digits in the binary representations of the integers $k = 1, 2, \ldots, n-1$. We want to derive an exact formula for $S(n)$. We adopt here the presentation of Flajolet *et al.* [134].

Let $v_2(k)$ be the exponent of 2 in the prime decomposition of k. We first prove that for all $k \geq 1$

$$v(k) - v(k-1) = 1 - v_2(k). \tag{7.102}$$

Indeed, let $k = (d_m, d_{m-1}, \ldots, 1, 0, 0, \ldots, 0)_2$ be the binary representation of k with digits $d_i \in \{0, 1\}$. Observe that 1 is at position $v_2(k)$. Then $k - 1 = (d_m, d_{m-1}, \ldots, 0, 1, 1, \ldots, 1)_2$, where 0 is at position $v_2(k)$. Let M be the number of 1's among the digits $d_m, \ldots, d_{v_2(k)-1}$. Then $v(k) = M + 1$ and $v(k-1) = M + v_2(k)$, and (7.102) follows.

Now we proceed as follows. Summing up (7.102) from $k = 1$ to some $m - 1$, and then again summing from $m = 1$ to $n - 1$ we obtain

$$S(n) = \sum_{m=1}^{n-1}(S(m) - S(m-1)) = \frac{n(n-1)}{2} - \sum_{k=1}^{n-1}(n-k)v_2(k). \tag{7.103}$$

But the sum on the right-hand side can be handled by the Perron-Mellin formula with $m = 1$. Having this in mind, we shall prove the following result.

Theorem 7.15   (Delange, 1975)   *The sum-of-digits function $S(n)$ satisfies*

$$S(n) = \frac{1}{2}n \log_2 n + n \left(\frac{\log_2 \pi}{2} - \frac{1}{2 \ln 2} - \frac{3}{4} \right) + n F(\log_2 n), \qquad (7.104)$$

where $F(x)$ is representable by the Fourier series

$$F(x) = \sum_{k \in \mathbb{Z} \setminus \{0\}}^{\infty} f_k e^{2\pi i k x} = \sum_{k \neq 0} f_k e^{2\pi i k x}$$

with

$$f_k = -\frac{1}{\ln 2} \frac{\zeta(\chi_k)}{\chi_k(\chi_k + 1)}, \qquad \chi_k = \frac{2\pi i k}{\ln 2}, \qquad k \neq 0.$$

Proof. Expression (7.103) suggests to apply the Perron-Mellin formula with $a_k = v_2(k)$ and $m = 1$. In Example 7.18 we proved that the Dirichlet series $V(s)$ of $v_2(k)$ is

$$V(s) = \sum_{k=1}^{\infty} \frac{v_2(k)}{k^s} = \frac{\zeta(s)}{2^s - 1},$$

hence by the Perron-Mellin formula (7.52) we obtain

$$S(n) = \frac{n(n-1)}{2} - \frac{n}{2\pi i} \int_{2-i\infty}^{2+i\infty} \frac{n^s \zeta(s)}{2^s - 1} \frac{ds}{s(s+1)}. \qquad (7.105)$$

We now evaluate the integral in (7.105) by the closing-the-box method. We shift the line of the integration to the left such that the new contour $\mathcal{C}$ consists of vertical lines at $\Re(s) = 2$ and $\Re(s) = -\frac{1}{4}$ and closed on the North and South at high $\pm T$. This contour will enclose residues of the integrand at $s = 1$, $s = 0$ (a double pole) and simple poles at $s = \chi_k$ for integers $k \neq 0$. To compute the residue at the double pole $s = 0$, we use the following expansions

$$\zeta(s) = -\frac{1}{2} - s\frac{1}{2} \ln(2\pi) + O(s^2),$$

$$n^s = 1 + s \ln(n) + O(s^2),$$

$$\frac{1}{2^s - 1} = \frac{1}{s \ln 2} - \frac{1}{2} + O(s),$$

$$\frac{1}{s(s+1)} = \frac{1}{s} - 1 + O(s).$$

Multiplying the above and finding the coefficient at s^{-1} we compute that the residue is equal to

$$-\frac{1}{2}\log_2 n - \log_2 \sqrt{\pi} + \frac{1}{2\ln 2} - \frac{1}{4}.$$

The residue at χ_k for $k \neq 0$ gives $F(\log_2 n)$, that is,

$$F(\log_2 n) = \frac{1}{\ln 2} \sum_{k \in \mathbb{Z}\setminus\{0\}} \frac{\zeta(\chi_k)}{\chi_k(\chi_k + 1)} n^{\chi_k}.$$

Summing up, we obtain

$$\frac{1}{2\pi i} \int_C \frac{n^s \zeta(s)}{2^s - 1} \frac{ds}{s(s+1)}$$
$$= \frac{1}{2} n \log_2 n + n\left(\frac{\log_2 \pi}{2} - \frac{1}{2\ln 2} - \frac{3}{4}\right) nF(\log_2 n) - nR(n).$$

It remains to prove that the integral over the South/North line vanishes at $T \to \infty$, and that $R(n) \equiv 0$.

Let us first estimate the integral over the horizontal lines at positions $\pm T$. Due to the bound (2.47) for the $\zeta(s)$ function, we have

$$\left|\int_{-\frac{1}{4} \pm iT}^{2 \pm iT} \frac{n^s \zeta(s)}{2^s - 1} \frac{ds}{s(s+1)}\right| \leq c|\zeta(s)| \left(\frac{n}{2}\right)^2 \frac{1}{T^2} = O(T^{-5/4}).$$

Thus the integral vanishes at $T \to \infty$.

To prove $R(n) \equiv 0$ we use (7.54). We have

$$\begin{aligned}
R(n) &= \frac{1}{2\pi i} \int_{-1/4-i\infty}^{-1/4+i\infty} \frac{n^s \zeta(s)}{2^s - 1} \frac{ds}{s(s+1)} \\
&= \frac{1}{2\pi i} \int_{-1/4-i\infty}^{-1/4+i\infty} \frac{n^s \zeta(s)}{s(s+1)} ds \sum_{k \geq 0}\left(-2^{ks}\right) \\
&= \sum_{k \geq 0} \frac{1}{2\pi i} \int_{-1/4-i\infty}^{-1/4+i\infty} \zeta(s)(2^k n)^s \frac{ds}{s(s+1)} \\
&\overset{(7.54)}{=} 0,
\end{aligned}$$

where the second line is justifiable for $\Re(s) < 0$ since both the series and the integral converge absolutely. This proves the theorem.   ∎

7.7   EXTENSIONS AND EXERCISES

7.1   Prove (7.16).

7.2   Derive a closed-form formula for the Bell numbers.

7.3   Derive (7.35) from the Lagrange formula.

7.4   Let

$$S_n = \sum_{k=0}^{n} \binom{n}{k} a_k p^k q^{n-k},$$

where $p+q = 1$, and a_k is a given sequence. Let $S(z)$ and $A(z)$ be ordinary generating functions of S_n and a_n, respectively. Prove that

$$S(z) = \frac{1}{1 - qz} A\left(\frac{pz}{1 - qz}\right).$$

7.5   Consider the following sum

$$S_{n,k} = \sum_{i=0}^{n} \binom{n-k}{i} \left(\frac{i}{n}\right)^i \left(1 - \frac{i}{n}\right)^{n-i}.$$

Define

$$s_k(z) = \sum_{n \geq 0} S_{n,k} \frac{n^n}{n!} z^n.$$

Prove that

$$s_k(z) = z^k B(z) B^{(k)}(z),$$

where $B(z) = 1/(1 - T(z))$ is defined in (7.35), and $B^{(k)}(z)$ denotes the kth derivative of $B(z)$.

7.6   Let for $|z| < 1$

$$M(z) = \sum_{n \geq 1} z^n \log n.$$

Prove the following:

$$M_1(z) = \sum_{n \geq 1} z^n \log n! = \frac{M(z)}{1 - z},$$

$$M_2(z) = \sum_{n \geq 1} z^n \log(2n)! = \frac{1}{2}\left(\frac{M(\sqrt{z})}{1 - \sqrt{z}} + \frac{M(-\sqrt{z})}{1 + \sqrt{z}}\right).$$

7.7 ⚠ Let $0 < p < 1$. For fixed integer M and all $0 \le \ell < M$, define

$$\pi(\ell) = \sum_{k \ge 0} \binom{n}{\ell + kM} p^{\ell + kM} (1 - p)^{n - \ell - kM}.$$

Prove that for all $0 \le \ell \le M$ we have

$$\pi(\ell) = \frac{1}{M} + O(\rho^n),$$

where $\rho < 1$.

7.8 ⚠ Prove that the variance **Var**$[X_n]$ of the quicksort complexity (discussed in Example 7.17) is

$$\mathbf{Var}[X_n] = 7n^2 - 2(n + 1)H_n + 13n - 4(n + 1)^2 H_n^{(2)},$$

where $H_n^{(2)} = \sum_{k=1}^n \frac{1}{k^2}$.

7.9 ⚠ Let $A(s) = \sum_{n \ge 1} \frac{a_n}{n^s}$ converge absolutely for $\sigma = \Re(s) > \sigma_c$. Prove that for $\sigma > \sigma_c$

$$\lim_{T \to \infty} \frac{1}{2T} \int_{-T}^{T} A(\sigma + it) n^{\sigma + it} dt = a_n$$

for all $n \ge 1$.

7.10 Prove (7.61) and (7.65).

7.11 Show under what conditions on the function $a(s)$ the following is true:

$$\sum_{k=m}^{\infty} (-1)^k a(k) = \frac{1}{2\pi i} \int_{m-1/2-i\infty}^{m-1/2+i\infty} a(s) \frac{\pi ds}{\sin(\pi s)}.$$

7.12 ⚠ Let $\mathcal{H}$ be a set of patterns (e.g., set of all strings that are within given distance from a given pattern H). For such a set we define the correlation language $\mathcal{A}$ and the correlation matrix $A(z)$ as follows: Given two strings H and F, let $\mathcal{A}$ be the set of words:

$$\mathcal{A} = \{F_{k+1}^m : H_{m-k+1}^m = F_1^k\},$$

and the set of positions k satisfying $F_1^k = H_{m-k+1}^m$ is denoted as HF. Let $A(z) = \{A_{H_i H_j}(z)\}_{i,j=1,M}$ be the *correlation matrix*, where $A_{H_i H_j}(z)$ is the correlation polynomial for H_i and H_j. Let $O_n(\mathcal{H})$ be the number of patterns from $\mathcal{H}$ occurring in a random text of length n generated by a memoryless source. As in Section 7.6.2 we write

$$T^{(r)}(z) = \sum_{n \geq 0} \Pr\{O_n(\mathcal{H}) = r\}z^n,$$

$$T(z, u) = \sum_{r=1}^{\infty} T^{(r)}(z)u^r = \sum_{r=1}^{\infty} \sum_{n=0}^{\infty} \Pr\{O_n(\mathcal{H}) = r\}z^n u^r.$$

Prove the following theorem.

Theorem 7.16   (Régnier and Szpankowski, 1997)   *Let $\mathcal{H}$ be a given set of patterns of length m, and T be a random text of length n generated by a memoryless source. The generating functions $T^{(r)}(z)$ and $T(z, u)$ can be computed as follows:*

$$T^{(r)}(z) = \mathbf{R}^t(z)\mathsf{M}^{r-1}U(z),$$

$$T(z, u) = \mathbf{R}^t(z)u(\mathsf{I} - u\mathsf{M}(z))^{-1}U(z),$$

where

$$\mathsf{M}(z) = (\mathsf{D}(z) + (z - 1)\mathsf{I})[\mathsf{D}(z)]^{-1},$$

$$(\mathsf{I} - \mathsf{M}(z))^{-1} = \mathsf{A}(z) + \frac{z^m}{1-z}\mathbf{1} \cdot \mathbf{H}^t,$$

$$U(z) = \frac{1}{1-z}(\mathsf{I} - \mathsf{M}(z)) \cdot \mathbf{1},$$

$$\mathbf{R}^t(z) = \frac{z^m}{1-z}\mathbf{H}^t \cdot (\mathsf{I} - \mathsf{M}(z)),$$

and

$$\mathsf{D}(z) = (1 - z)\mathsf{A}(z) + z^m \mathbf{1} \cdot \mathbf{H}^t.$$

In the above, $\mathbf{H} = (P(H_1), \ldots, P(H_M))^t$, upper index t means transpose, and $\mathsf{A}(z) = \{A_{H_i, H_j}(z)\}_{i,j=1..M}$ is the matrix of the correlation polynomials of patterns from the set $\mathcal{H}$.

Derive from the above the average and the variance of $O_n(\mathcal{H})$.

8

Complex Asymptotic Methods

Asymptotic methods play an essential role in the analysis of algorithms, information theory, and other problems of engineering and science. In this chapter, we primarily focus on the methods of *precise* asymptotic estimates. Precise analysis of algorithms was launched in 1962 by D.E. Knuth, who analyzed in his "Notes on Open Addressing" hashing tables with linear probing. We discuss a variety of (mostly analytic) asymptotic techniques such as the Euler-Maclaurin summation formula, methods of applied mathematics (i.e., the WKB method and matched asymptotics), asymptotics of generating functions (polar singularities and algebraic singularities), saddle point method, methods of evaluating finite sums (e.g., Rice's method), and limiting distributions. In the application section, we enumerate words with approximate self-overlaps, discuss minimax redundancy for memoryless sources, and derive the limiting distribution of the depth in a digital search tree.

In the analysis of algorithms we often aim at predicting the rates of growth of time or space complexities for large inputs n. Such analyses rarely produce exact expressions that are easy to use or provide insight into the dependence of the algorithms performance on the parameters of the problem. In fact, quite often we end up with a complicated sum, an integral, a recurrence, or even a functional equation. In such situations, one may turn to an asymptotic analysis. It allows one to focus on those terms of the solution that contribute most to the answer. The extra insight from the asymptotic analysis may improve the performance of an algorithm since we simply focus on precisely those parts of the algorithm that affect the leading terms of the asymptotic solution.

Asymptotic analysis can be approached from many angles. We have already used "big-oh" notation. Basically, $f(n) = O(g(n))$ if $|f(n)/g(n)|$ is bounded from above for n sufficiently large. A more precise expression is Θ notation that assures $|f(n)/g(n)|$ is bounded from below and above for sufficiently large n. There are situations, however, when such asymptotics are still too crude and one must turn to *asymptotic expansions*.

Let us consider an example from data compression. Assume that we are to compress a binary string of length n. Knowing that the compression ratio (of the lengths of the compressed string and the original string) is $\Theta(1)$ is not too helpful! As a matter of fact, we would like to know how close this ratio is to the entropy of the source (see Chapter 6) in order to estimate how far away we are from the optimal compression. Thus we need at least the first-order asymptotic expansion. But Jacob Ziv in his *1997 Shannon Lecture* [459] presented compelling arguments for backing off to a certain degree from the first-order asymptotic analysis of information systems, in order to predict the behavior of real systems, where we always face *finite* (and often small) lengths (of sequences, files, programs, codes, etc.). One way of overcoming these difficulties is to increase the accuracy of asymptotic analysis and replace first-order analyses (e.g., a leading term of the average code length) by more complete asymptotic expansions, thereby extending their range of applicability to smaller values while providing more accurate analyses (like constructive error bounds, large deviations, local or central limit laws).

Another example provides a vast area of combinatorial optimization (see Section 1.5). Let us consider the problem that we have already discussed in Section 1.4, namely, the shortest common superstring problem. It is an NP-hard problem, but in Section 6.5.2 we proved that with high probability the ratio of the optimal solution to a greedy solution tends to 1 as the size of the problem $n \to \infty$. In this case, it is again important that the ratio is close to 1, not just bounded. How good is this approximation? Well, we then need more terms in the asymptotic expansions.

The analysis of algorithms that aims at precise information about algorithm performance was launched on July 27, 1963 by Donald E. Knuth in his "Notes on Open Addressing." Generally, an asymptotic expansion we have in mind is a series that represents a complicated expression in a certain limiting sense. More precisely, if x_n denotes a quantity of interest with input size n, we may look for a simple explicit function a_n (e.g., $a_n = \log n$ or $a_n = \sqrt{n}$) such that $x_n \sim a_n$ (i.e., $\lim_{n \to \infty} x_n/a_n = 1$) or we may be aiming at a very precise asymptotic expansion such as $x_n = a_n^1 + a_n^2 + \cdots + o(a_n^k)$, where for each $1 \leq i \leq k$ we require that $a_n^{i+1} = o(a_n^i)$. The "little oh" notation is another example of "less precise" asymptotic information and it means that $a_n^{i+1}/a_n^i \to 0$ as $n \to \infty$. Hereafter, we mostly discuss *analytic methods* of asymptotic analysis. As argued by Odlyzko [330]: "analytic methods are extremely powerful and when they apply, they often yield estimates of unparalleled precision."

This chapter begins with definitions of asymptotic notation. We used some of them already in this book, but here we extend them to complex functions and introduce more rigor. Following that, we present a few tools used in asymptotic analysis. We start with elementary methods such as the Euler-Maclaurin summation formula, methods of applied mathematics (e.g., the WKB and matched asymptotics), and sequences distributed modulo 1. We illustrate every method by an example taken from analysis of algorithms or information theory. Then comes our main topic, namely, how to extract asymptotic information from generating functions that are known either exactly or approximately. We first deal with the *small singularities* of analytic functions, that is, either polar singularities or algebraic singularities. Then we turn our attention to large singularities and discuss saddle point methods. We also discuss certain finite sums (e.g., alternating sums) that occur in analysis of algorithms and are not easy to evaluate by direct methods. We represent these sums by complex integrals and use the tools of this chapter to compute them (we have already seen this approach in Section 7.6.3). Finally, we study limiting distributions through generating functions and analytic tools described in this chapter. In the application section, we discuss how to enumerate words that approximately overlap, deal with interesting problems of code redundancy, and finally derive the limiting distribution for the depth in a digital search tree.

The standard reference in asymptotic analysis is De Bruijn [84]. The reader is also referred to an excellent recent survey by Odlyzko [330] on asymptotic methods. Detailed coverage of many topics considered here may be found in Knuth, Graham and Patashnik [169], Greene and Knuth [171], Hofri [197], Knuth [267, 268, 269], Sedgewick and Flajolet [383], Olver [331], and Wong [450]. The reader will find many interesting methods and examples of a combinatorial nature in the forthcoming book by Flajolet and Sedgewick [149]. The "little" book by Greene and Knuth [171] is a joy to read and highly recommended.

8.1 INTRODUCTION TO ASYMPTOTIC EXPANSIONS

In this section, we formally introduce the Poincaré notion of an asymptotic expansion. This concept is quite important since it enables us to manipulate a large class of divergent series — often arising in the analysis of algorithms—in much the same way as convergent power series. Even more important, in the spirit of Ziv's remark mentioned in the introduction, asymptotic expansions enable us to obtain numerical and quantitative results of increasing accuracy.

8.1.1 Definitions and Notations

Let S be a point set in the complex plane and z_0 a limit point of S. For two complex functions $f(z)$ and $g(z)$ defined on S we write

$$f(z) = O(g(z)), \quad z \to z_0 \tag{8.1}$$

to mean that there exists a constant K and a neighborhood $U(z_0)$ of z_0 such that $|f(z)| \leq K|g(z)|$ for all $z \in S \cap U(z_0)$. If $|f(z)| \geq K'|g(z)|$ for a constant $K' > 0$, then $f(z) = \Omega(g(z))$. Furthermore, if $f(z) = O(g(z))$ and $f(z) = \Omega(g(z))$, then $f(z) = \Theta(g(z))$. We shall also use the little-oh notation

$$f(z) = o(g(z)), \quad z \to z_0 \tag{8.2}$$

if for every $\varepsilon > 0$ there is a neighborhood $U_\varepsilon(z_0)$ of z_0 such that $|f(z)| \leq \varepsilon|g(z)|$ for all $z \in S \cap U_\varepsilon(z_0)$. Finally, if $f(z)/g(z) \to 1$ for $z \to z_0$, then

$$f(z) \sim g(z), \quad z \to z_0. \tag{8.3}$$

Observe that (8.3) can be also written as $f(z) = (1 + o(1))g(z)$.

We mostly have either $z_0 = 0$ or $z_0 = \infty$ and restrict S either to a sector

$$S_{\alpha,\beta} = \{z : 0 < |z| < \infty, \ \alpha < \arg(z) < \beta\},$$

in the complex plane, to the set of nonnegative integers $S = \mathbb{N}$, or to the set of all positive reals $S = \mathbb{R}^+$. We write $S_\theta = S_{-\theta,\theta}$ for a symmetric sector in the complex plane. Finally, we say that the relations O or o are to hold *uniformly* in a set of parameters λ, if the constant K in the big-Oh does not depend on λ in this set.

Here are some examples:

$$T(z) - 1 + \sqrt{2(1 - ez)} + \frac{2}{3}(1 - ez) = O((1 - ez)^{3/2}), z_0 = e^{-1}, \ S = S_{\pi/2},$$

$$\log(1 + a\sqrt{x} + bx + cx^{3/2}) - a\sqrt{x} - (b - \frac{1}{2}a^2)x = O(x^{3/2}), z_0 = 0, \ \mathcal{S} = \mathbb{R}^+,$$

$$n! \sim n^{n+1/2}e^{-n}\sqrt{2\pi}, z_0 = \infty, \mathcal{S} = \mathbb{N},$$

where $T(z)$ is the tree function defined in Chapter 7 (see Example 7.15).

In many applied problems one must deal with divergent series that nevertheless give an excellent numerical approximation. To handle it, Poincaré introduced in 1886 asymptotic power series. For $z \in \mathcal{S}_\theta$ when $z_0 = \infty$ Poincaré defined: *A power series $\sum_{n=0}^{\infty} a_n z^{-n}$ is called an asymptotic expansion of $f(z)$ defined on $\mathcal{S}_\theta$ if for every fixed integer $N \geq 0$*

$$f(z) = \sum_{n=0}^{N} a_n z^{-n} + O(z^{-(N+1)}), \quad z \to \infty. \tag{8.4}$$

If (8.4) holds, we write

$$f(z) \sim \sum_{n=0}^{\infty} a_n z^{-n}, \quad z \to \infty.$$

Poincaré's asymptotic expansions are too restrictive for our applications. For example, the average height in a PATRICIA trie built from n independent strings generated by an unbiased memoryless source is $\log_2 n + \sqrt{2\log_2 n} + \cdots$, while the average height in a digital search tree is $\log_2 n + \sqrt{2\log_2 n} - \log_2 \sqrt{2\log_2 n} + \cdots$ as $n \to \infty$ (cf. [259, 260]). These are legitimate asymptotic expressions, but they do not fall under the Poincaré definition. To encompass them, we define below *generalized* Poincaré-type asymptotic expansions.

A sequence of functions $\phi_k(z)$ is said to be an *asymptotic sequence* as $z \to z_0$ in $\mathcal{S}$ if

$$\phi_{k+1}(z) = o(\phi_k(z)), \quad z \to z_0, \tag{8.5}$$

for all k. The original Poincaré definition is recovered if one sets $\phi_k(z) = 1/z^k$ and $z_0 = \infty$. The function $f(z)$ is said to have an *asymptotic expansion* of order N with respect to the asymptotic sequence $\{\phi_k(z)\}$ if there exist constants a_k such that

$$f(z) = \sum_{k=0}^{N} a_k \phi_k(z) + o(\phi_N(z)), \quad z \to z_0. \tag{8.6}$$

If the above holds for all N, then we write

$$f(z) \sim \sum_{k=0}^{\infty} a_k \phi_k(z), \quad z \to z_0.$$

Observe that the above expansion is unique since one can determine the constants a_k by means of

$$a_k = \lim_{z \to z_0} \left(f(z) - \sum_{i=0}^{k-1} a_i \phi_i(z) \right) / \phi_k(z).$$

If the constants $a_k = 0$ for all k, then such an asymptotic expansion is called *asymptotically zero* with respect to a given asymptotic series $\{\phi_k(z)\}$. For example,

$$e^{-x} \sim 0 \cdot 1 + 0 \cdot x^{-1} + 0 \cdot x^{-2} + \cdots, \quad x \to \infty$$

is asymptotically (exponentially) zero as $x \to \infty$ in $\mathcal{S} = \mathbb{R}^+$. If $f(z)$ is asymptotically zero, we shall write $f(z) \sim 0$. It is usual to replace such functions by zero in any stage of a proof establishing the validity of Poincaré asymptotic expansion. We shall follow this approach in this book.

Here are some examples of asymptotic expansions as $x \to \infty$ in $\mathcal{S} = \mathbb{R}^+$

$$\frac{1}{1+x} \sim \sum_{k=1}^{\infty} \frac{(-1)^{k-1}}{x^k},$$

$$\int_0^{\infty} (t+x)^{-1} e^{-t} dt \sim \sum_{k=1}^{\infty} \frac{(-1)^{k-1}(k-1)!}{x^k},$$

$$x^{-\log x} \sim 0.$$

It is worth pointing out that the series on the right in the second example above is *not* convergent, but we get better approximations (for larger value of x) to the integral by taking more terms of the asymptotic series.

Asymptotic expansions can be manipulated as convergent series; however, some caution must be exercised with differentiation. In Exercises 1 and 2 we ask the reader to prove some of these facts.

We also should point out that asymptotic expansions of a complex function f in a sector S_θ may crucially depend on θ. For example, $\exp(-1/z) \sim 0$ for $z \to 0$ in any sector contained in $S_{\pi/2}$, but for purely imaginary $z = ix$ we have $|\exp(-1/z)| = 1$, so the asymptotics break down.

8.1.2 From Taylor Expansion to Asymptotic Series

We end this section with a useful result that allows us to compute asymptotic series of a complex function $f(z)$ as $z \to \infty$ from its Taylor series at $z \to 0$. We shall follow Ford [159].

Let $f(z)$ be defined by the power series

$$f(z) = \sum_{n=0}^{\infty} a_n z^n, \quad z \to 0 \tag{8.7}$$

with a positive radius of convergence. We aim at finding conditions under which there exist coefficients b_n such that

$$f(z) \sim \sum_{n=1}^{\infty} \frac{b_n}{z^n}, \quad z \to \infty \tag{8.8}$$

in a sector of the complex plane. Throughout, we assume that

(A) *there exists an analytic function $a(w)$ in the right half plane $\Re(w) > 0$ such that*

$$a(w)|_{w=n} = a_n \tag{8.9}$$

where $w = x + iy$.

As the first step, we represent the function $f(z)$ as a complex integral. We have already seen this trick in Section 7.6.3. This method is surprisingly successful while dealing with finite sums and infinite series. The thrust of this approach is to evaluate the integral through the Cauchy residue theorem (see Section 2.3). Under certain conditions such an integral can be evaluated inside and outside the contour of integration leading to two different representations. We shall see how it works in the problem at hand (see Section 8.5 for a more detailed discussion).

We start with a simple observation: If $P(w)$ is analytic around a given point a and $Q(w)$ has a simple pole at a, then by Cauchy's theorem

$$\frac{1}{2\pi i} \int_{C(a)} \frac{P(w)}{Q(w)} dw = \frac{P(a)}{Q'(a)}, \tag{8.10}$$

where the integration takes place around a sufficiently small closed contour $C(a)$ encircling the point $w = a$ in the positive direction. Let us now set

$$P(w) = \pi a(w)(-z)^w, \quad Q(w) = \sin \pi w,$$

where $(-z)^w$ is understood as a single-valued function in the sector $0 < \arg(z) < 2\pi$ (which excludes the positive half of the real axis), defined as follows

$$(-z)^w = \exp(w \log(-z))$$
$$= \exp(w(\log|z| + i(\arg(z) - \pi))), \quad 0 < \arg(z) < 2\pi. \quad (8.11)$$

We write $\overline{S}_\theta = \{z : 0 < |z| < \infty, \ \theta < \arg(z) < 2\pi - \theta\}$ for some $0 < \theta < \pi/2$; that is, $\overline{S}_\theta$ is a set in the complex plane that excludes the sector S_θ around the positive half of the real axis (e.g., for $\theta = 0$ we exclude only the real positive half axis).

Let $C_{n,l}$ be a contour, shown in Figure 8.1, that surrounds each of the following points: $w = -l, -l+1, \ldots, -1, 0, 1, \ldots, 2n-1, 2n$, where l is a positive integer. Since for any integer $k \in \mathbb{Z}$

$$\frac{\pi}{\sin \pi w} = \frac{(-1)^k}{(w-k)} + O(1),$$

the Cauchy residue theorem, (cf. (8.10)) leads to

$$\frac{1}{2\pi i} \int_{C_{n,l}} \frac{\pi a(w)(-z)^w}{\sin \pi w} dw = \sum_{m=1}^{l} \frac{a(-m)}{z^m} + \sum_{k=0}^{2n} a(k)z^k. \quad (8.12)$$

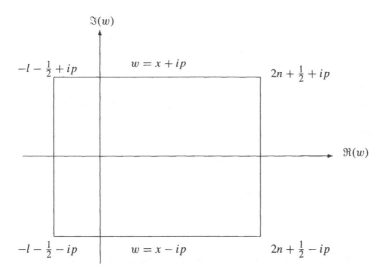

Figure 8.1. The integration contour $C_{n,l}$.

We now evaluate the integral in (8.12) along the four line segments of the curve $C_{n,l}$ shown in Figure 8.1, namely, the lines $w = -l - \frac{1}{2} + iy$, $w = 2n + \frac{1}{2} + iy$ and $w = x \pm ip$, where $-\infty < x, y < \infty$ and we allow p to be positive and large. We denote by D_p and D_{-p} the contributions of the integral along the upper and lower horizontal lines, respectively. For D_p we observe that for $w = x + ip$

$$\sin \pi w = \sinh \pi p (\sin \pi x \coth \pi p + i \cos \pi x),$$

which yields

$$D_p = \frac{(-z)^{ip}}{2i \sinh \pi p} \int_{2n+\frac{1}{2}}^{-l-\frac{1}{2}} \frac{a(x + ip)(-z)^x}{\sin \pi x \coth \pi p + i \cos \pi x} dx.$$

To assure that $D_p \to 0$ as $p \to \infty$, we adapt another assumption regarding the function $a(w)$:

(B) *for sufficiently large y and any positive $\varepsilon > 0$*

$$|a(x + iy)| < Ke^{\varepsilon|y|}, \tag{8.13}$$

where K is a constant that may depend only on ε.

In other words, $a(w)$ cannot grow faster than exponentially along any line parallel to the imaginary axis.

Suppose for the present that $z \in \overline{S}_\theta$ is *real and negative* so that $|-z^{ip}| = 1$. Then

$$|D_p| < \frac{Ke^{\varepsilon p}}{2 \sinh \pi p} \int_{-l-\frac{1}{2}}^{2n+\frac{1}{2}} (-z)^x dx,$$

since

$$|\sin \pi x \coth \pi p + i \cos \pi x|^2 = \sin^2 \pi x \cosh^2 \pi p + \cos^2 \pi x > 1.$$

In summary,

$$\lim_{p \to \infty} D_p = 0.$$

In a similar manner we prove that

$$\lim_{p \to \infty} D_{-p} = 0.$$

Next, let us consider the integral along the vertical right line $w = 2n + \frac{1}{2} + iy$, which we denote as D_n. Since for $w = 2n + \frac{1}{2} + iy$

$$\sin \pi w = \sin(\pi/2 + i\pi y) = \cos i\pi y = \cosh \pi y,$$

we arrive at (imposing $p \to \infty$)

$$D_n = \frac{(-z)^{2n+\frac{1}{2}}}{2} \int_{-\infty}^{\infty} \frac{a(2n + \frac{1}{2} + iy)(-z)^{iy}}{\cosh \pi y} dy.$$

Under hypothesis (B) we deduce that the integral above converges absolutely. *Restricting further z to the real segment $-1 < z < 0$ we certainly have*

$$\lim_{n \to \infty} D_n = 0$$

for $-1 < z < 0$. Soon we extend the validity of these estimates to $\overline{S}_\theta$.

Finally, we deal with the left vertical line $w = -l - \frac{1}{2} + iy$, which contribution we denote as D_l. We obtain

$$D_l = (-1)^l \frac{(-z)^{-l-\frac{1}{2}}}{2} \int_{-\infty}^{\infty} \frac{a(-l - \frac{1}{2} + iy)(-z)^{iy}}{\cosh \pi y} dy.$$

Under hypothesis (B) this integral converges absolutely for $-1 < z < 0$.

In view of the above we conclude that (8.12) becomes

$$f(z) = (-1)^l \frac{(-z)^{-l-\frac{1}{2}}}{2} \int_{-\infty}^{\infty} \frac{a(-l - \frac{1}{2} + iy)(-z)^{iy}}{\cosh \pi y} dy - \sum_{m=1}^{l} \frac{a(-m)}{z^m} \quad (8.14)$$

for $n \to \infty$ as long as $-1 < z < 0$.

We now analytically continue the solution (8.14) to the sector $\overline{S}_\theta$ (excluding real positive z). For this we define a subsector $\overline{S}_{\theta+\delta} \subset \overline{S}_\theta$ for any $\delta > 0$ in which we prove that the integral in (8.14) converges. It suffices to show that the integral converges for $y \in (-\infty, -d)$ and $y \in (d, \infty)$ for arbitrarily large d. Let $z = |z|(\cos \phi + i \sin \phi) \in \overline{S}_{\theta+\delta}$. Then under hypothesis (B) and (8.11), the above two integrals are bounded by

$$K \int_{-\infty}^{-d} \frac{e^{-(\phi - \pi + \varepsilon)y}}{\cosh \pi y} dy; \quad K \int_{d}^{\infty} \frac{e^{-(\phi - \pi - \varepsilon)y}}{\cosh \pi y} dy,$$

which converge uniformly for $z \in \overline{S}_{\theta+\delta}$. Thus the integral in (8.14) defines an analytic function and this representation is valid inside $\overline{S}_\theta$. But the factor $(-z)^{-l-\frac{1}{2}}$ will assure that the last omitted term in (8.14) is $o(z^{-l})$ leading to an asymptotic expansion. More precisely, it remains but to note that (8.14) is of the form

$$f(z) = -\sum_{m=1}^{l} \frac{a(-m)}{z^m} + \frac{\eta(z,l)}{z^l},$$

where $\eta(z,l) = O(1/\sqrt{z}) \to 0$ as $|z| \to \infty$.

In summary, we proved the following theorem.

Theorem 8.1 *Let $f(z)$ be analytic and have the following power series representation*

$$f(z) = \sum_{n=0}^{\infty} a_n z^n, \quad z \to 0$$

with a positive radius of convergence. Assume a_n possesses an analytic continuation $a(w)$ with $w = x + iy$ such that hypotheses (A) and (B) are satisfied, that is,

(A) *there exists an analytic function $a(w)$ in the right half plane $\Re(w) > 0$ such that $a(w)|_{w=n} = a_n$*

(B) *for sufficiently large y and any positive $\varepsilon > 0$*

$$|a(x+iy)| < Ke^{\varepsilon|y|},$$

where K is a constant that may depend only on ε.

Then the function $f(z)$ defined in any sector $\overline{S}_\theta$ (for arbitrary small $\theta > 0$) that excludes the positive half of the real axis can be represented by the following asymptotic series

$$f(z) \sim -\sum_{m=1}^{\infty} \frac{a(-m)}{z^m}, \quad z \to \infty \tag{8.15}$$

in $\overline{S}_\theta$.

In passing, we point out that if $a(w)$ is not analytic but has some singularities, then one must subtract this contribution from the left-hand side of (8.15).

Example 8.1: *Asymptotic Expansion of a Series* Let us consider

$$f(z) = \sum_{n=0}^{\infty} \frac{z^n}{n+\omega}, \quad z \to 0,$$

where ω is a constant not equal to zero or a negative integer. In this case, we have $a(w) = 1/(w + \omega)$, which clearly satisfies hypotheses (A) and (B) everywhere except $w = -\omega$, where there is a pole of $a(w)$. Proceeding as above, we end up with an equation like (8.12) with additional term on the right-hand side coming from the residue at $w = -\omega$. This residue depends on whether ω is a positive integer; if it is, we have a double pole at $w = \omega$, otherwise $w = \omega$ is a simple pole. Let us handle first the latter situation; we assume that $\omega \notin \mathbb{N}$. Then

$$\mathrm{Res}\left(\frac{\pi(-z)^w}{(w + \omega)\sin \pi w}, w = -\omega\right) = -\frac{\pi(-z)^{-\omega}}{\sin \pi \omega}.$$

For $\omega \in \mathbb{N}$, say $\omega = K$, we have at $w = -K$ a double pole of $1/\sin \pi K$ and $1/(w + \omega)$. To find the residue of

$$\frac{\pi(-z)^w}{(w + K)\sin \pi w}$$

at $w = -K$, we expand the functions involved in the above expression into the Taylor series around $w = -K$. Note that

$$(w + K)\sin \pi w = (-1)^K \pi (w + K)^2 \left(1 + \frac{(w + K)^2}{3!} + O((w + K)^4)\right),$$

$$\pi(-z)^w = \pi(-z)^{-K}\left(1 + (w + K)\log(-z) + O((w + K)^2)\right).$$

Upon finding the coefficient at $(w + K)^{-1}$ we arrive at

$$\mathrm{Res}\left(\frac{\pi(-z)^w}{(w + K)\sin \pi w}, w = -K\right) = z^{-K}\log(-z).$$

In summary, we obtain

$$f(z) \sim \frac{\pi(-z)^{-\omega}}{\sin \pi \omega} - \sum_{n=1}^{\infty} \frac{1}{(\omega - n)z^n}, \quad \omega \notin \mathbb{N}$$

and

$$f(z) \sim -\frac{\log(-z)}{z^K} - \sum_{n \neq K} \frac{1}{(K - n)z^n}, \qquad \omega = K \in \mathbb{N},$$

where the last sum is over all nonnegative integers not equal to K. ∎

It is to be noted, finally, that if instead of the original series $f(z) = \sum_{n \geq 0} a_n z^n$ we start with

$$f(z) = \sum_{n=0}^{\infty} a_n(-z)^n,$$

where $a(w)$ satisfies hypotheses (A) and (B), and if we replace $\pi a(w)(-z)^w$ by $\pi a(w)z^w$ defined in the sector $\mathcal{S}_\pi$, then the asymptotic expansion

$$f(z) \sim -\sum_{m=1}^{\infty} \frac{a(-m)}{(-z)^m}$$

is valid in any sector $\mathcal{S}_\theta$ that excludes the *negative* half of the real axis. The reader is asked in Exercise 3 to prove the following:

$$\log(1+x) = \sum_{k=1}^{\infty} \frac{(-1)^{k-1}}{k} x^k, \qquad\qquad x \to 0,$$

$$= \log x + \sum_{k=1}^{\infty} \frac{(-1)^{k-1}}{k} x^{-k}, \quad x \to \infty.$$

Observe that above we have $=$ (not $\sim$) for both limits.

8.2   BASIC METHODS

In this section we discuss asymptotic methods that are not based on extracting singularities of generating functions. We start with the Euler-Maclaurin summation formula, followed by methods of applied mathematics such as the WKB method and matched asymptotics. Finally, we deal with sequences distributed modulo 1 in order to estimate certain sums involving fractional parts.

8.2.1   Euler-Maclaurin Summation Formula

From a first course of calculus we know that there is a relationship between integrals and series. In fact, when the function f is monotone, then $\sum_{1 \le k \le n} f(k)$ is well approximated by $\int_1^n f(x)dx$. (The reason that one prefers integrals over the sums lies in the fact that such integrals are usually much easier to evaluate.) The Euler-Maclaurin summation formula makes this statement very precise for sufficiently smooth functions. It finds plenty of applications in discrete mathematics, combinatorics, and analysis of algorithms.

Let f be a differentiable function defined on the interval $[m, n]$, where $m < n$ are integers. Observe that for any $m \le j < n$

$$\frac{f(j) + f(j+1)}{2} = \int_j^{j+1} \frac{d}{dx} [((x - j - 1/2) f(x)] dx$$

$$= \int_j^{j+1} f(x) dx + \int_j^{j+1} (x - j - 1/2) f'(x) dx, \quad (8.16)$$

where $B_1'(x - j) = \frac{d}{dx}(x - j - 1/2) = 1$ and $B_1(1) = -B_1(0) = 1/2$; $B_1(x)$ is the first Bernoulli polynomial. Definition and properties of Bernoulli numbers and Bernoulli polynomials are reviewed in Table 8.1. If we write $B_1(x - \lfloor x \rfloor) = x - \lfloor x \rfloor - 1/2 = x - j - 1/2$ for $j \leq x < j + 1$, and sum (8.16) over $j = m, m + 1, \ldots, n - 1$, we immediately obtain

$$\sum_{k=m+1}^{n-1} f(k) + \frac{f(m) + f(n)}{2} = \int_m^n f(x) dx + \int_m^n B_1(x - \lfloor x \rfloor) f'(x) dx. \quad (8.17)$$

This is a particular case of the Euler-Maclaurin formula.

Example 8.2: *Moments of Nonnegative Random Variables* Let M_n be a sequence of nonnegative random variables having all moments satisfying $E[M_n^k] = o(E[M_n^{k+1}])$ with respect to n. For fixed k, we calculate the $(k + 1)$st moment of M_n as

$$E[M_n^{k+1}] = \sum_{m=0}^{\infty} m^{k+1} \Pr\{M_n = m\}$$

$$= \sum_{m=0}^{\infty} m^{k+1} (\Pr\{M_n > m - 1\} - \Pr\{M_n > m\})$$

$$= \sum_{m=0}^{\infty} ((m + 1)^{k+1} - m^{k+1}) \Pr\{M_n > m\}$$

$$= (k + 1) \sum_{m \geq 0} m^k \Pr\{M_n > m\} + O(E[M_n^k]),$$

where the last line follows from the identity: $a^k - b^k = (a - b)(a^{k-1} + a^{k-2}b + \cdots + b^{k-1})$. Using the Euler-Maclaurin formula we shall show that

$$\sum_{m \geq 0} m^k \Pr\{M_n > m\} = \int_0^{\infty} x^k \Pr\{\tilde{M}_n > x\} dx + O(E[M_n^k]),$$

TABLE 8.1. Bernoulli Numbers and Polynomials

1. *The Bernoulli numbers* B_m *for* $m = 0, 1, \ldots$ *are defined by the expression*

$$\frac{t}{e^t - 1} = \sum_{k=0}^{\infty} \frac{B_k}{k!} t^k, \quad |t| < 2\pi. \tag{8.18}$$

Since $\frac{t}{e^t - 1} = -t + \frac{-t}{e^{-t} - 1}$, hence all Bernoulli numbers with odd index $m \geq 3$ are zero.

2. *Bernoulli polynomials* $B_m(x)$ *of order* $m = 0, 1, 2, \ldots$ *are defined as*

$$\frac{t e^{xt}}{e^t - 1} = \sum_{k=0}^{\infty} \frac{B_k(x)}{k!} t^k, \quad |t| < 2\pi. \tag{8.19}$$

In particular, $B_0(x) = 1$, $B_1(x) = x - \frac{1}{2}$ and $B_2(x) = x^2 - x + \frac{1}{6}$.

3. *Properties* of Bernoulli polynomials directly following from the definition:

$$B_m(1 - x) = (-1)^m B_m(x), \tag{8.20}$$

$$B_m(0) = B_m, \tag{8.21}$$

$$B'_m(x) = m B_{m-1}(x), \quad m \geq 1, \tag{8.22}$$

$$|B_{2n}(x)| \leq |B_{2n}|, \quad 0 < x < 1, \tag{8.23}$$

The first property follows from $\frac{t e^{xt}}{e^t - 1} = \frac{(-t) e^{(1-x)(-t)}}{e^{-t} - 1}$ while the other two are obvious. The last is a consequence of the fact that $B_{2n}(x)$ for $x \in (0, 1)$ achieves maximum at $x = \frac{1}{2}$ and $B_{2n}(\frac{1}{2}) = (2^{1-2n} - 1) B_{2n}$.

4. *Asymptotics.* Consider now $n \geq 1$ and evaluate B_{2n} by Cauchy's formula

$$\frac{B_{2n}}{(2n)!} = \frac{1}{2\pi i} \oint \frac{dz}{z^{2n}(e^z - 1)}.$$

Computing the residue at $z = \pm 2\pi i k$, $k = 1, 2, \ldots$ we find

$$B_{2n} = 2(-1)^{n-1} \zeta(2n) \frac{(2n)!}{(2\pi)^{2n}} \sim 2(-1)^{n-1} \frac{(2n)!}{(2\pi)^{2n}} \tag{8.24}$$

where the asymptotics follow from $\zeta(n) = 1 + O(2^{-n})$ as $n \to \infty$.

5. *The Euler functions* $P_m(x)$ *are defined as periodic functions of period 1 such that*

$$P_m(x) = \frac{B_m(x)}{m!}, \quad 0 \leq x < 1, \tag{8.25}$$

$$P_m(0) = P_m(1) = \frac{B_m}{m!}, \quad m \neq 1. \tag{8.26}$$

By (8.22) we have $P'_{m+1}(x) = P_m(x)$ for $m \geq 1$.

TABLE 8.2. Abel's Partial Summation Formula

One often has to deal with sums like $\sum_i a_i b_i$ where $a_i, b_i, 1 \le i \le n$ are two real-valued sequences. A basic technique to handle such sums is *Abel's partial summation formula*. Let

$$A(k) = \sum_{j=1}^{k} a_j,$$

$$\widetilde{A}(k) = \sum_{j=n-k}^{n} a_j.$$

Then

$$\sum_{j=1}^{n} a_j b_j = \sum_{k=1}^{n-1} A(k)(b_k - b_{k-1}) + A(n)b_n \qquad (8.27)$$

$$= \sum_{i=0}^{n-1} \widetilde{A}(n - i - 1)(b_{i+1} - b_i), \qquad (8.28)$$

where $b_0 = 0$. The formulas work fine for $n = \infty$ provided the series involved converge. In this case the last term in (8.28) is not necessary. A derivation of Abel's formula is pretty straightforward and left as an exercise.

---

where, for illustrative purposes, we assume that $\widetilde{M}_n$ is a continuous approximation to M_n. Indeed, after applying (8.17) with $m = 0$, $n = \infty$ and $f(x) = x^k \Pr\{\widetilde{M}_n > x\}$ we arrive at

$$\sum_{m \ge 0} m^k \Pr\{M_n > m\} = \int_0^\infty x^k \Pr\{\widetilde{M}_n > x\} dx + O\left(\int_0^\infty |f'(x)| dx\right).$$

But

$$\int_0^\infty \left(x^k \frac{d}{dx} \Pr\{\widetilde{M}_n > x\} + k x^{k-1} \Pr\{\widetilde{M}_n > x\}\right) dx = O(\mathrm{E}[M_n^k]).$$

In summary,

$$\mathrm{E}[M_n^{k+1}] = (k + 1) \int_0^\infty x^k \Pr\{\widetilde{M}_n > x\} dx + O(\mathrm{E}[M_n^k])$$

for fixed k. ∎

We now refine (8.17) by having a closer look at the last term, that is,

$$\int_j^{j+1} B_1(x - \lfloor x \rfloor) f'(x) dx$$

for $j = m, \ldots, n - 1$. It turns out that instead of working with the Bernoulli polynomials $B_m(x - \lfloor x \rfloor)$ it is more convenient to use Euler's function $P_m(x) = B_m(x - \lfloor x \rfloor)/m!$ (cf. (8.25)-(8.26)) of Table 8.1) since they are periodic with period 1 and such that $P'_{m+1}(x) = P_m(x)$. Integrating by parts the above integral gives

$$\int_j^{j+1} P_1(x - \lfloor x \rfloor) f'(x) dx = P_2(0)(f'(j + 1) - f'(j)) - \int_j^{j+1} P_2(x) f''(x) dx$$

since $P_m(0) = P_m(1) = B_m/m!$ for $m > 1$ (cf. (8.26)). Summing up over $j = m, \ldots, n - 1$, and plugging into (8.17), we finally obtain

$$\sum_{k=m+1}^{n-1} f(k) + \frac{f(m) + f(n)}{2} = \int_m^n f(x) dx + \frac{B_2}{2!}(f'(n) - f'(m))$$

$$- \frac{1}{2} \int_m^n B_2(x - \lfloor x \rfloor) f''(x) dx.$$

Using mathematical induction and properties of the Bernoulli numbers and polynomials (in particular, (8.23) and (8.24)) we can generalize the above procedure leading to the following theorem.

Theorem 8.2   (Euler-Maclaurin)   *Suppose f has continuous derivatives up to order s. Then*

$$\sum_{k=m+1}^{n} f(k) = \int_m^n f(x) dx + \sum_{l=1}^{s} (-1)^l \frac{B_l}{l!} \left(f^{(l-1)}(n) - f^{(l-1)}(m) \right) \quad (8.29)$$

$$+ \frac{(-1)^{s-1}}{s!} \int_m^n B_s(x - \lfloor x \rfloor) f^{(s)}(x) dx \quad (8.30)$$

$$= \int_m^n f(x) dx + 1/2 f(n) - 1/2 f(m)$$

$$+ \sum_{\ell=1}^{S} \frac{B_{2\ell}}{(2\ell)!} \left(f^{(2\ell-1)}(n) - f^{(2\ell-1)}(m) \right)$$

$$+ O\left((2\pi)^{-2S}\right) \int_m^n |f^{(2S)}(x)| dx,$$

where $S = \lfloor \frac{s}{2} \rfloor$.

Proof. The first part (8.29) follows directly from our discussion above. To derive the error bound in (8.30) we observe that only even terms in the above sum contribute since $B_{2k+1} = 0$ for $k \geq 1$. We set $S = \lfloor \frac{s}{2} \rfloor$ and use (8.22) together with (8.24) to prove (8.30). ∎

Example 8.3: *Asymptotic Expansion of the Euler Gamma Function* Let us compute $\log \Gamma(z)$ for $z \in \mathbb{C} - (-\infty, 0)$. We start with the Euler representation for the gamma function,

$$\Gamma(z) = \lim_{n \to \infty} \prod_{k=1}^n \frac{k}{z+k-1} \left(\frac{k+1}{k}\right)^{z-1},$$

which implies

$$\log \Gamma(z) = \lim_{n \to \infty} \left((z-1)\log(n+1) - \sum_{k=1}^n \log \frac{z+k-1}{k}\right).$$

We apply the Euler-Maclaurin formula to the above sum; that is, we assume $m = 0$ and $f(x) = \log(x + z - 1) - \log x$ in (8.29). This leads to

$$\sum_{k=1}^n \log \frac{z+k-1}{k} = \log z + \int_1^n (\log(x + z - 1) - \log x) dx$$

$$+ \sum_{j=1}^m \frac{B_{2j}}{2j(2j-1)} \left(\frac{1}{(n+z-1)^{2j-1}}\right.$$

$$\left. - \frac{1}{n^{2j-1}} - \frac{1}{z^{2j-1}} + 1\right)$$

$$+ 1/2 \left(\log(n + z - 1) - \log n - \log z\right)$$

$$+ \frac{1}{2m} \int_0^n B_{2m}(x - \lfloor x \rfloor) \left(\frac{1}{(z+x-1)^{2m}} - \frac{1}{x^{2m}}\right) dx.$$

In the above we used $B_1 = 1/2$ and $B_{2k+1} = 0$ for $k \geq 1$. After computing the first integral and taking the limit $n \to \infty$ we finally arrive at

$$\log \Gamma(z) = (z - 1/2) \log z - z + 1 + \sum_{j=1}^{m} \frac{B_{2j}}{2j(2j-1)} \left(\frac{1}{z^{2j-1}} - 1 \right)$$

$$- \frac{1}{2m} \int_{0}^{n} B_{2m}(x - \lfloor x \rfloor) \left(\frac{1}{(z+x-1)^{2m}} - \frac{1}{x^{2m}} \right) dx.$$

Further simplification can be obtained for $z \to \infty$ knowing that

$$\lim_{z \to \infty} (\log \Gamma(z) - (z - 1/2) \log z + z) = 1/2 \log 2\pi.$$

Then one obtains Stirling's formula

$$\log \Gamma(z) = (z - 1/2) \log z - z + \sum_{j=1}^{m} \frac{B_{2j}}{2j(2j-1)} \frac{1}{z^{2j-1}}$$

$$+ 1/2 \log 2\pi - \frac{1}{2m} \int_{0}^{n} \frac{B_{2m}(x - \lfloor x \rfloor)}{(z+x)^{2m}} dx$$

for $z \to \infty$ in $\mathbb{C} - (-\infty, 0)$. This leads directly to the Stirling asymptotic formula for $n!$, namely,

$$n! \sim \sqrt{2\pi n} \left(\frac{n}{e} \right)^n \left(1 + \frac{1}{12n} + \frac{1}{288n^2} - \frac{139}{5140n^3} + \cdots \right) \qquad (8.31)$$

as $n \to \infty$.   ■

8.2.2 Matched Asymptotics and the WKB Method

Numerous problems in mathematics, physics, combinatorics, and analysis of algorithms reduce to either functional/differential equations or recurrence equations with a small (large) parameter $\varepsilon > 0$ ($\lambda > 0$). We discuss here two methods of applied mathematics that are routinely used to solve such problems. We have in mind the **matched asymptotics** and the **WKB approximation** named after physicists Wentzel, Kramers, and Brillouin (however, Olver in his book [331] rightfully calls this method the *Liouville-Green approximation* because they had already proposed it in 1837). We also briefly mention another method, **linearization**, that finds many applications in the analysis of algorithms. We must add, however, that the WKB method makes certain assumptions about the forms of the asymptotic expansions without a rigorous justification of the assumed forms. In particular, it does not address the issue of the existence of such asymptotic expansions.

   We will not discuss the methods in depth, referring the interested reader to many excellent books such as Bender and Orszag [39], Nayfeh [324], and Olver

[331]. We start with a general and brief overview of the two methods. Then we illustrate them by deriving asymptotic distributions for the heights in tries and b-tries (see Section 1.1).

Let $\varepsilon > 0$ be a small number and $\lambda > 0$ a large number. Consider the following two differential equations (arising in physics)

$$\varepsilon y''(x) + (1 + \varepsilon^2)y'(x) + (1 - \varepsilon^2)y(x) = 0, \quad y(0) = \alpha, \quad y(1) = \beta, \quad (8.32)$$

$$\frac{1}{\lambda^2}y''(x) - q(x)y(x) = 0, \quad (8.33)$$

where α and β are two given constants. The first problem above is a simple boundary-value problem, and can be solved exactly. However, we shall try to solve it approximately to illustrate the idea of the *matched asymptotic expansions*. We look for a solution in the form

$$y(x; \varepsilon) = \sum_{j=0}^{\infty} \varepsilon^j y_j(x), \quad (8.34)$$

where $y_j(x)$ are unknown functions. For the second problem above, we shall seek a (non-trivial) asymptotic solution in the following form

$$y(x; \lambda) = e^{\lambda \Phi(x)} \sum_{j=0}^{\infty} A_j(x)\frac{1}{\lambda^j}, \quad (8.35)$$

where $\Phi(x)$ and $A_0(x), A_1(x), \ldots$ are unknown functions. The form (8.35) is known as the *WKB approximation*. Here's what Fedoryuk [121] has to say about such approximations:

> ...It is necessary first of all to guess (and no other word will do) in what form to search for the asymptotic form. Of course, this stage—guessing the form of the asymptotic form—is not subject to any formalization. Analogy, experiments, numerical simulation, physical considerations, intuitions, random guesswork; these are the arsenal of means used by any research worker.

To illustrate the methods, we discuss a concrete example. Let us first illustrate the *matching principle*. We shall follow Nayfeh [324] and start with the boundary problem (8.32) and assume that (8.34) is its solution. Looking only at the leading term $y_0(x)$ we find, after some simple algebra, that $y_0(x) = c_0 e^{-x}$, where c_0 is a constant that we should determine from the boundary condition. We get either $c_0 = \alpha$ or $c_0 = \beta e$, and usually both cannot be satisfied. Therefore, we must choose one, say the value at $x = 1$, and ignore for now the other boundary value. We then obtain a full asymptotic expansion as

$$y(x) = \beta e^{1-x} + \varepsilon\beta(1-x)e^{1-x} + \cdots, \tag{8.36}$$

which works fine everywhere but near the origin, where $y(0) = \beta e(1 + \varepsilon + O(\varepsilon^2)) \neq \alpha$. In order to overcome these difficulties, we consider another scale, namely $\xi = x/\varepsilon$. Keeping ξ fixed and solving the equation we obtain

$$y(x) = \beta e(1-x) + (\alpha - \beta e)(1+x)e^{-x/\varepsilon} + \cdots \tag{8.37}$$

$$= \beta e + (\alpha - \beta e)e^{-\xi} + \varepsilon[(\alpha - \beta e)e^{-\xi} - \beta e\xi] + O(\varepsilon^2) \tag{8.38}$$

Observe that at the origin $y(0) = \alpha$, as desired.

In conclusion, we got two solutions (8.36) and (8.38). One is good in the interior and near the right end while the other is near the left end of the interval $[0, 1]$. In order to obtain a *uniform* expansion over $[0, 1]$, we may try to blend or match these two solutions. The basic idea of a *matched asymptotic expansion* is that an approximate solution to a given problem is sought not as a single expansion in terms of a single scale but as two or more separate expansions in terms of two or more scales each of which is valid in part of the domain. We chose the scales so that the overall expansion covers the whole domain and that the domains of validity of neighboring expansions *overlap*. Because the domains overlap, we can *match* or blend the neighboring expansions. If this is possible, the resulting solution is called the *matched asymptotic expansion*. The *uniform* expansion is obtained adding (8.36) and (8.38) and subtracting their "common" part, so it is not counted twice. This common part is either the expansion of (8.36) as $x \to 0$ or that of (8.38) as $\xi \to \infty$. The matching condition insures that the two are equal.

We illustrate this basic idea by an example dealing with the height distribution in a trie. The reader may want to go back to Section 1.1 to review some definitions related to tries.

Example 8.4: *Limiting Distribution of the Height in a Trie* Let H_n be the height of a trie built from n binary strings generated by independent memoryless unbiased sources, and let $h_n^k = \text{Pr}\{H_n \leq k\}$. Observe that h_n^k satisfies the following recurrence

$$h_n^{k+1} = 2^{-n} \sum_{i=0}^{n} \binom{n}{i} h_i^k h_{n-i}^k, \qquad k \geq 0 \tag{8.39}$$

$$h_0^0 = h_1^0 = 1, \quad \text{and} \quad h_n^0 = 0, \, n > 1. \tag{8.40}$$

This follows from $H_n = \max\{H_i^{LT}, H_{n-i}^{RT}\} + 1$, where H_i^{LT} and H_{n-i}^{RT} denote, respectively, the left subtree and the right subtree of sizes i and $n - i$, which happens with probability $2^{-n}\binom{n}{i}$ for memoryless unbiased sources. To solve this recurrence, we set

$$H^k(z) = \sum_{n \geq 0} h_n^k \frac{z^n}{n!}$$

to be the exponential generating function of h_n^k. Then (8.39) implies (see Table 7.3 in Chapter 7)

$$H^k(z) = \left(H^0(z2^{-k}) \right)^{2^k}, \tag{8.41}$$

where $H^0(z) = 1 + z$. Thus by Cauchy's formula

$$
h_n^k = \frac{n!}{2\pi i} \oint (1 + z2^{-k})^{2^k} z^{-n-1} dz
$$

$$
= \begin{cases}
0, & n > 2^k \\[2mm]
\dfrac{(2^k)!}{2^{nk}(2^k - n)!}, & 0 \leq n \leq 2^k
\end{cases}, \tag{8.42}
$$

where the second line follows directly from the binomial formula. Applying Stirling's formula to (8.42), we prove the following.

Theorem 8.3 **(Flajolet, 1983; Devroye, 1984; Pittel, 1986)** *The distribution of the height in a trie has the following asymptotic expansions:*

(i) *Right-Tail Region: $k \to \infty$, $n = O(1)$*

$$\Pr\{H_n \leq k\} = h_n^k = 1 - n(n-1)2^{-k-1} + O(2^{-2k}).$$

(ii) *Central Region: $k, n \to \infty$ with $\xi = n2^{-k}$, $0 < \xi < 1$*

$$h_n^k \sim A(\xi)e^{n\phi(\xi)},$$

where

$$\phi(\xi) = \left(1 - \frac{1}{\xi} \right) \log(1 - \xi) - 1,$$

$$A(\xi) = (1 - \xi)^{-1/2}.$$

(iii) *Left-Tail Region: $k, n \to \infty$ with $2^k - n = j = O(1)$*

$$h_n^k \sim n^j \frac{e^{-n-j}}{j!} \sqrt{2\pi n}.$$

This shows that there are three ranges of k and n, where the asymptotic form of h_n^k is different.

We now use matched asymptotics to show that these three regions cover the whole domain, where either k or n are large. If we expand (i) of Theorem 8.3 for n large, we obtain $1 - h_n^k \sim n^2 2^{-k-1}$. For $\xi \to 0$ we have $A(\xi) \sim 1$ and $\phi(\xi) \sim -\xi/2$ so that the result in (ii) becomes

$$A(\xi)e^{n\phi(\xi)} \sim e^{-n\xi/2} = \exp\left(-\frac{1}{2}n^2 2^{-k}\right) \sim 1 - \frac{1}{2}n^2 2^{-k},$$

where the last approximation assumes that $n, k \to \infty$ in such a way that $n^2 2^{-k} \to 0$. Since (ii) agrees precisely with the expansion of (i) as $n \to \infty$, we say that (i) and (ii) asymptotically match. To be precise, we say they match the leading order; higher-order matchings can be verified by computing higher-order terms in the asymptotic series in (i) and (ii). We can also easily show that the expansion of (ii) as $\xi \to 1^-$ agrees with the expansion of (iii) as $j \to \infty$, so that (ii) and (iii) also asymptotically match. Indeed, we observe that in this case $n(1/\xi - 1) = j$ and

$$n^j \frac{e^{-n-j}}{j!} \sqrt{2\pi n} \to_{j\to\infty} \exp\left(n\left(1 - \frac{1}{\xi}\right)\log(1-\xi) - 1\right) \frac{1}{\sqrt{1-\xi}}$$

$$= A(\xi)e^{n\Phi(\xi)}.$$

The matching verifications imply that, at least to leading order, there are no "gaps" in the asymptotics. In other words, one of the results in (i)–(iii) applies for any asymptotic limit which has k and/or n large.   ∎

We now briefly discuss the WKB approximation. We shall follow Nayfeh [324] and Fedoryuk [121]. In general, the WKB approximation assumes that an asymptotic solution is of the form

$$y(x; \lambda) \sim e^{\lambda\Phi(x)}\left(A(x) + \frac{1}{\lambda}A_1(x)\cdots\right)$$

or equivalently

$$y(x; \lambda) \sim \exp\left(\lambda(\Phi_0(x) + \lambda^{-1}\Phi_1(x) + \cdots)\right),$$

where $\Phi_i(x)$ are unknown functions that one tries to determine from the equation. For example, consider differential equation (8.33) and assume it has the above asymptotic expansion. Plugging it into (8.33), after some algebra, one arrives at the following system of equations

$$\Phi_0'^2(x) - q(x) = 0,$$
$$\Phi_0''(x) + 2\Phi_0'(x)\Phi_1'(x) = 0,$$

which for $q(x) > 0$ yields

$$\Phi_0(x) = \pm \int_{x_0}^x \sqrt{q(x)}dx,$$

$$\Phi_1(x) = -\log\sqrt{|\Phi_0'(x)|} = -\frac{1}{4}\log|q(x)|$$

for some constant x_0. Substituting this into the postulated asymptotic solution we finally show that

$$y(x; \lambda) \sim q^{-1/4}(x) \exp\left(\pm\lambda \int_{x_0}^x \sqrt{q(x)}dx\right).$$

We close this section with a more sophisticated application of the WKB method, matched asymptotics, and *linearization*. This is a continuation of Example 8.4.

Example 8.5: *Limiting Distribution of the Height in b-Tries* We consider an extension of tries discussed in the previous example, namely, b-tries. In these tries the external node is allowed to store up to b strings. The parameter b is assumed to be fixed. We study the height H_n (i.e., the longest path in such a tree), and let $h_n^k = \Pr\{H_n \le k\}$. As before, h_n^k satisfies the recurrence equation (8.39) with a new initial condition,

$$h_n^0 = 1, \ n = 0, 1, 2, \ldots, b; \qquad \text{and} \qquad h_n^0 = 0, \ n > b.$$

The exponential generating function $H^k(z)$ of h_n^k satisfies the same equation (8.41) except that now $H^0(z) = 1 + z + \cdots + z^b/b!$. In other words,

$$h_n^k = n![z^n]\left(1 + z2^{-k} + \frac{z^2}{2!}2^{-2k} + \cdots + \frac{z^b}{b!}2^{-bk}\right)^{2^k}. \qquad (8.43)$$

We can use the saddle point method (see Section 8.4 and Exercise 7) to extract asymptotic expansion of h_n^k from the above; however, in this example we apply indirect methods of WKB, matched asymptotics, and linearization to solve the problem. It will pay off in problems where one cannot obtain explicit expressions for the generating function (e.g., heights of PATRICIA tries and digital search trees, as analyzed in [259, 260]).

We consider first the case $k, n \to \infty$ with $\xi = n2^{-k}$ fixed. We set

$$h_n^k = G(\xi; n) = G(n2^{-k}; n)$$

and note that $h_i^{k-1} = G(2i\xi/n; i)$ and $h_{n-i}^{k-1} = G(2\xi(1 - i/n); n - i)$. From the original recurrence (8.39) we obtain

$$G(\xi; n) = \left(\frac{1}{2}\right)^n \sum_{i=0}^{n} \binom{n}{i} G\left(\frac{2i}{n}\xi; i\right) G\left(2\left(1 - \frac{i}{n}\right)\xi; n - i\right).$$

We shall seek an asymptotic solution of the WKB form

$$G(\xi; n) \sim e^{n\phi(\xi)} \left[A(\xi) + \frac{1}{n}A^{(1)}(\xi) + \frac{1}{n^2}A^{(2)}(\xi) + \cdots\right].$$

By symmetry, the major contribution to the sum will come from $i \approx n/2$. We also note that Stirling's formula (8.31) yields, for $i = xn$ ($0 < x < 1$),

$$2^{-n}\binom{n}{i} = \frac{e^{nf_0(x)}}{\sqrt{2\pi n}}\frac{1}{\sqrt{x(1-x)}}\left[1 + \frac{1}{12n}\left(1 - \frac{1}{x} - \frac{1}{1 - x}\right) + O(n^{-2})\right],$$

$$\tag{8.44}$$

where $f_0(x) = -\log 2 - x \log x - (1 - x)\log(1 - x)$. For $x = 1/2 + y/\sqrt{n}$ (i.e., $i = n/2 + O(\sqrt{n})$), (8.44) simplifies to the Gaussian form

$$2^{-n}\binom{n}{n/2 + y\sqrt{n}} = \sqrt{\frac{2}{\pi n}}e^{-2y^2}\left[1 + \frac{1}{n}\left(-\frac{1}{4} + 2y^2 - \frac{4}{3}y^4\right) + O(n^{-2})\right].$$

Thus we are led to

$$A(\xi)e^{n\phi(\xi)} \sim \sum_{i=0}^{n} \sqrt{\frac{2}{\pi n}}e^{-2y^2} A\left(\frac{2i}{n}\xi\right) A\left(\frac{2(n - i)}{n}\xi\right)$$

$$\times \exp\left[i\phi\left(\frac{2i}{n}\xi\right) + (n - i)\phi\left(2\left(1 - \frac{i}{n}\right)\xi\right)\right].$$

For $x = i/n$, we set $\psi(x) = x\phi(2x\xi) + (1 - x)\phi(2(1 - x)\xi)$ and expand this function about $x = 1/2$. We have $\psi(1/2) = \phi(\xi)$, $\psi'(1/2) = 0$ and $\psi''(1/2) =$

$8\xi\phi'(\xi) + 4\xi^2\phi''(\xi)$. Thus the exponent in the above expression becomes

$$\exp(n\psi(x)) = \exp\left(n\phi(\xi) + y^2(4\xi\phi'(\xi) + 2\xi^2\phi''(\xi)) + O(y^3)\right).$$

Using the Euler-Maclaurin formula, substituting $y = y'/\sqrt{n}$, and approximating the above sum by an integral we obtain as $n \to \infty$

$$e^{n\phi(\xi)}A(\xi) \sim \sqrt{\frac{2}{\pi}}e^{n\phi(\xi)}A^2(\xi)\int_{-\infty}^{\infty}\exp\left(-2y^2 + y^2(4\xi\phi'(\xi) + 2\xi^2\phi''\xi)\right)dy.$$

The exponential factors $e^{n\phi}$ cancel and we have

$$1 = \sqrt{\frac{2}{\pi}}A(\xi)\sqrt{\frac{\pi}{2 - 4\xi\phi'(\xi) - 2\xi^2\phi''(\xi)}}. \tag{8.45}$$

Thus the asymptotic series of the WKB approximation is known, up to the function $\phi(\xi)$. It does not seem to be possible to determine ϕ using only recurrence (8.39). This function is apparently very sensitive to the initial conditions. However, using matched asymptotics, we shall show how to determine $\phi(\xi)$ for $\xi \to 0$ (in a similar manner, one can find $\phi(\xi)$ as $\xi \to b$).

For this, we turn our attention to the case $n = O(1)$ and $k \to \infty$. We set $h_n^k = 1 - G_n^k$. Then G_n^k satisfies

$$G_n^{k+1} = 2\left(\frac{1}{2}\right)^n\sum_{i=0}^{n}\binom{n}{i}G_{n-i}^k - \left(\frac{1}{2}\right)^n\sum_{i=0}^{n}\binom{n}{i}G_i^kG_{n-i}^k \tag{8.46}$$

$$= 2\left(\frac{1}{2}\right)^n\sum_{i=0}^{n-b-1}\binom{n}{i}G_{n-i}^k - \left(\frac{1}{2}\right)^n\sum_{i=b+1}^{n-b-1}\binom{n}{i}G_i^kG_{n-i}^k.$$

Here we have used the fact that $G_n^k = 0$ for $0 \leq n \leq b$. For $b + 1 \leq n \leq 2b + 1$, the nonlinear term in (8.46) vanishes and we are left with a *linear* recurrence. We can solve (8.46) exactly by first solving the linear problem for $n \in [b+1, 2b+1]$, and using this solution to compute explicitly the nonlinear term in (8.46) for $n \in [2b+2, 3b+2]$, etc. However, the resulting expressions become complicated and we only need asymptotics of G_n^k. For ranges of k, n, where h_n^k is *asymptotically close to* 1, we can replace (8.46) by the asymptotic relation (thus *linearize* the recurrence), which yields

$$G_n^{k+1} \sim 2^{1-n}\sum_{i=0}^{n-b-1}\binom{n}{i}G_{n-i}^k.$$

This has the following asymptotic solution

$$G_n^k \sim 2^{-b}\binom{n}{b+1}, \quad n \geq b+1.$$

Thus for $k \to \infty$ and $n = O(1)$ we have

$$\Pr\{H_n \leq k\} = h_n^k \sim 1 - \binom{n}{b+1}2^{-kb}.$$

As in Example 8.4, we now match the two expansions just derived, that is, we require

$$1 - \frac{n^{b+1}}{(b+1)!}2^{kb}\bigg|_{n\to\infty} \sim A(\xi)e^{n\phi(\xi)]}\bigg|_{\xi\to 0}.$$

Since $A(\xi) \to 1$ as $\xi \to 0$, we obtain

$$\phi(\xi) \sim -\frac{\xi^b}{(b+1)!}, \quad \xi \to 0,$$

and hence

$$\Pr\{H_n^T \leq k\} \sim A(\xi)e^{n\phi(\xi)]} \sim \exp\left(-\frac{n\xi^b}{(b+1)!}\right) \quad \xi \to 0.$$

In passing we should point out that this range is the most interesting since the probability mass is concentrated around $k = (1 + 1/b)\log_2 n + x$, where $x = O(1)$. In fact,

$$\Pr\{H_n^T \leq (1 + 1/b)\log_2 n + x\} = \Pr\{H_n^T \leq \lfloor(1 + 1/b)\log_2 n + x\rfloor\}$$

$$\sim \exp\left(-\frac{1}{(1+b)!}2^{-bx+b\langle(1+b)/b\cdot\log_2 n+x\rangle}\right),$$

where $\langle x \rangle$ is the fractional part of x, that is, $\langle x \rangle = x - \lfloor x \rfloor$. Due to the erratic behavior of $\langle\log_2 n\rangle$ (see the next section), the limiting distribution of the height H_n does not exist! The reader can find detailed analyses on this and similar topics in a series of recent papers by Knessl and Szpankowski [258, 259, 260, 261]. ■

8.2.3 Uniform Distribution of Sequences

In the analysis of algorithms and information theory one often deals with sequences like $\langle\log n\rangle$ and $\langle\alpha n\rangle$, where $\langle x \rangle = x - \lfloor x \rfloor$ is the fractional part of x. We just saw it in Example 8.5. In fact, we are often interested in asymptotics of sums like

$$\sum_{k \geq 0} p_{n,k} f(\langle x_k \rangle)$$

as $n \to \infty$, where $p_{n,k}$ is a probability distribution (i.e., $\sum_k p_{n,k} = 1$) and f is a function. For example, the following sum

$$\sum_{k=0}^{n} \binom{n}{k} p^k (1-p)^{n-k} \left\langle \log \left(p^k (1-p)^{n-k} \right) \right\rangle$$

arises in coding theory. Our goal here is to present certain tools from the theory of *uniform distribution of sequences* to extract asymptotics of such sums. We shall follow Drmota and Tichy [109].

We start with some definitions.

Definition 8.4   (P-u.d. mod 1)   *A sequence $x_n \in \mathbb{R}$ is said to be P-uniformly distributed modulo 1 (P-u.d. mod 1) with respect to the probability distribution $P = \{p_{n,k}\}_{k \geq 0}$ if*

$$\lim_{n \to \infty} \sum_{k \geq 0} p_{n,k} I_A(\langle x_k \rangle) = \lambda(A) \qquad (8.47)$$

holds uniformly for every interval $A \subset [0, 1]$, where $I_A(x_n)$ is the characteristic function of A (i.e., it equals 1 if $x_n \in A$ and 0 otherwise) and $\lambda(A)$ is the Lebesgue measure of A.

Two special cases are of interest to us:

1. *Uniform distributed sequences* modulo 1 (u.d. mod 1) in which case $p_{n,k} = 1/n$, that is, (8.47) becomes

$$\lim_{n \to \infty} \frac{1}{n} \sum_{k=0}^{n} I_A(\langle x_k \rangle) = \lambda(A). \qquad (8.48)$$

2. *Bernoulli distributed sequences* modulo 1 (B-u.d. mod 1) that assumes the binomial distribution for $p_{n,k}$, that is, for $p > 0$ (8.47) becomes

$$\lim_{n \to \infty} \sum_{k=0}^{n} \binom{n}{k} p^k (1-p)^{n-k} I_A(\langle x_k \rangle) = \lambda(A). \qquad (8.49)$$

The following result summarizes the main property of P-u.d. modulo 1 sequences. It provides the leading term of asymptotics for sums like $\sum_k p_{n,k} f(\langle x_k + y \rangle)$, where x_k is P-u.d. mod 1 and y is a shift.

Theorem 8.5 *Suppose that the sequence x_n is P-uniformly distributed modulo 1. Then for every Riemann integrable function $f : [0, 1] \to \mathbb{R}$*

$$\lim_{n \to \infty} \sum_{k \geq 0} p_{n,k} f(\langle x_k + y \rangle) = \int_0^1 f(t)\, dt, \tag{8.50}$$

where the convergence is uniform for all shifts $y \in \mathbb{R}$.

Proof. The proof is standard but details are quite tedious. The main idea is to notice that by definition (8.50) holds for characteristic functions $I_A(x_k)$. Next, we approximate f by a step function (i.e., a linear combination of characteristic functions) and use the definition of the Riemann integral to bound the integral from below and above, proving (8.50).

Details are here. Let $\mathcal{L}$ be the linear space of all (Lebesgue) integrable functions $f : [0, 1]^k \to \mathbb{C}$ satisfying

$$\lim_{n \to \infty} \sup_{y \in \mathbb{R}} \left| \sum_{k \geq 0} p_{n,k} f(\langle x_k + y \rangle) - \int_0^1 f(t)\, dt \right| = 0. \tag{8.51}$$

Of course, $\mathcal{L}$ contains all constant functions (i.e. $\mathcal{L} \neq \emptyset$). Next, we prove the following closure property of $\mathcal{L}$.

Fact A. Suppose $f : [0, 1] \to \mathbb{R}$ is an integrable function such that for every $\varepsilon > 0$ there exist functions $g_1, g_2 \in \mathcal{L}$ with $g_1 \leq f \leq g_2$ and

$$\int_0^1 (g_2(t) - g_1(t))\, dt < \varepsilon.$$

Then $f \in \mathcal{L}$.

For this purpose, let $m_n(f, y)$ denote the positive linear functionals

$$m_n(f, y) := \sum_{k \geq 0} p_{n,k} f(\langle x_k + y \rangle)$$

and $m(f)$ the positive linear functional

$$m(f) := \int_0^1 f(t)\, dt.$$

By $m_n(g_1, y) \le m_n(f, y) \le m_n(g_2, y)$ and $m(g_1) \le m(f) \le m(g_2)$ we immediately get

$$
\begin{aligned}
m(g_1) &= \liminf_{n\to\infty} m_n(g_1, y) &\le& \liminf_{n\to\infty} m_n(f, y) \\
&\le \limsup_{n\to\infty} m_n(f, y) &\le& \limsup_{n\to\infty} m_n(g_2, y) \\
&= m(g_2),
\end{aligned}
$$

which implies

$$\left| m(f) - \liminf_{n\to\infty} m_n(f, y) \right| < \varepsilon$$

and

$$\left| m(f) - \limsup_{n\to\infty} m_n(f, y) \right| < \varepsilon$$

for every $\varepsilon > 0$. Thus

$$\lim_{n\to\infty} \sup_{y\in\mathbb{R}} |m_n(f, y) - m(f)| = 0,$$

and $f \in \mathcal{L}$.

Next, suppose that $f : [0, 1] \to \mathbb{R}$ is a continuous function. Then by the Weierstrass approximation theorem, for every $\varepsilon > 0$ there exist two trigonometric polynomials g_1, g_2 with $g_1 \le f \le g_2$ and

$$\int_0^1 (g_2(t) - g_1(t))\, dt < \varepsilon.$$

Thus by Fact A, all continuous functions $f : [0, 1] \to \mathbb{R}$ are contained in $\mathcal{L}$.

Now it is easy to derive that every characteristic function $f = I_A$ of any interval $A \subseteq [0, 1]$ is in $\mathcal{L}$, too. It is obvious that for every $\varepsilon > 0$ there exist two continuous functions g_1, g_2 with $g_1 \le I_A \le g_2$ and

$$\int_0^1 (g_2(t) - g_1(t))\, dt < \varepsilon.$$

Again by Fact A all step functions (i.e. finite linear combinations of characteristic functions) are in $\mathcal{L}$.

Finally, if $f : [0, 1] \to \mathbb{R}$ is a Riemann integrable function, then by definition for every $\varepsilon > 0$ there exist two step functions g_1, g_2 with $g_1 \le f \le g_2$ and

$$\int_0^1 (g_2(t) - g_1(t))\, dt < \varepsilon,$$

and by Fact A this completes the proof.   ∎

In passing we observe that Theorem 8.5 holds if f is continuous. Furthermore, due to $\langle x_k \rangle$, we may restrict f to complex-valued continuous function with period 1.

To apply Theorem 8.5, one needs easy criteria to verify whether a sequence x_k is P-u.d. mod 1. Fortunately, such a result exists and is due to H. Weyl.

Theorem 8.6   (Weyl, 1916)   *A sequence x_n is P-u.d. mod 1 if and only if*

$$\lim_{n \to \infty} \sum_{k \ge 0} p_{n,k} e^{2\pi i m x_k} = 0 \tag{8.52}$$

holds for all $m \in \mathbb{Z} \setminus \{0\}$.

Proof. The proof is standard and we only sketch it here (for details see [109, 282]). First observe that $f(x) = e^{2\pi i m x}$ satisfies (8.50). To prove sufficiency, we note that by Weierstrass's *approximation theorem* every Riemann integrable function f of period 1 can be uniformly approximated by a trigonometric polynomial (i.e., a finite combination of functions of the type $e^{2\pi i m x}$). That is, for any $\varepsilon > 0$ there exists a trigonometric polynomial $\Psi(x)$ such that

$$\sup_{0 \le x \le 1} |f(x) - \Psi(x)| \le \varepsilon. \tag{8.53}$$

Then

$$\left| \int_0^1 f(x)\, dx - \sum_{k \ge 0} p_{n,k} f(x_k) \right| \le \left| \int_0^1 (f(x) - \Psi(x))\, dx \right|$$

$$+ \left| \int_0^1 \Psi(x)\, dx - \sum_{k \ge 0} p_{n,k} \Psi(x_k) \right|$$

$$+ \left| \sum_{k \ge 0} p_{n,k} \left(\Psi(x_k) - f(x_k) \right) \right|$$

$$\le 3\varepsilon$$

since the first and the third term are bounded by ε due to (8.53) and the second by ε due to (8.52). This completes the proof. ∎

One of the most important and well-studied sequences is $\langle \alpha n \rangle_{n \geq 0}$ for α irrational. We discuss its properties in the example below.

Example 8.6: *Weyl's Sequence* $\langle \alpha n \rangle_{n \geq 0}$ Let us consider $\langle \alpha n \rangle_{n \geq 0}$ for α irrational. We prove that it is u.d mod 1 and B-u.d. mod 1. For the former, observe that by Weyl's criterion for every integer $m \neq 0$

$$\left| \frac{1}{n} \sum_{k=1}^{n} e^{2\pi i k m \alpha} \right| = \frac{|e^{2\pi i m n \alpha} - 1|}{n|e^{2\pi i m \alpha} - 1|} \leq \frac{1}{n|\sin \pi m \alpha|} \to 0, \qquad n \to \infty$$

as long as α is not rational. Thus $\langle \alpha n \rangle_{n \geq 0}$ is uniformly distributed sequence modulo 1. Consider now the B-u.d. mod 1 case. We have by Weyl's criterion (where, as usual, $q = 1 - p$)

$$\lim_{n \to \infty} \sum_{k=0}^{n} \binom{n}{k} p^k q^{n-k} e^{2\pi i m(k\alpha)} = \lim_{n \to \infty} \left(p e^{2\pi i m \alpha} + q \right)^n = 0$$

provided α is irrational. ∎

Let us stay with sequences $\langle \alpha n \rangle_{n \geq 0}$, but now we assume that α is rational, say $\alpha = N/M$ such that $\gcd(N, M) = 1$. Can we say something meaningful about the sum $\sum_k p_{n,k} f(\langle \alpha k \rangle)$ in this case? To simplify our presentation we restrict the analysis to the binomial distribution $p_{n,k} = \binom{n}{k} p^k q^{n-k}$ ($q = 1 - p$). We prove the following general result.

Theorem 8.7 *Let $0 < p < 1$ be a fixed real number and suppose that $\alpha = \frac{N}{M}$ is a rational number with $\gcd(N, M) = 1$. Then for every bounded function $f :$ $[0, 1] \to \mathbb{R}$, we have for some $\rho < 1$*

$$S_n(f) = \sum_{k=0}^{n} \binom{n}{k} p^k (1 - p)^{n-k} f(\langle k\alpha + y \rangle)$$

$$= \frac{1}{M} \sum_{l=0}^{M-1} f\left(\frac{l}{M} + \frac{\langle My \rangle}{M} \right) + O(\rho^n)$$

uniformly for all $y \in \mathbb{R}$.

Proof. Observe that (with $p_{n,k} = \binom{n}{k} p^k q^{n-k}$)

$$S_n(f) = \sum_{k=0}^{n} p_{n,k} f\left(\left\langle k\frac{N}{M} + y\right\rangle\right)$$

$$= \sum_{\ell=0}^{M-1} \sum_{m:\, k=\ell+mM\leq n} p_{n,k} f\left(\left\langle \ell\frac{N}{M} + N + y\right\rangle\right)$$

$$= \sum_{\ell=0}^{M-1} \sum_{m:\, k=\ell+mM\leq n} p_{n,k} f\left(\left\langle \frac{\ell}{M} + y\right\rangle\right)$$

$$= \sum_{\ell=0}^{M-1} f\left(\left\langle \frac{\ell}{M} + y\right\rangle\right) \sum_{m:\, k=\ell+mM\leq n} \binom{n}{k} p^k q^{n-k}.$$

To proceed, we need to evaluate the second sum. This sum takes every Mth term from the binomial distribution. In Section 7.6.1 we learned how to handle such sums through the Mth root of unity $\omega_k = e^{2\pi i k/M}$ for $k = 0, 1, \ldots, M - 1$. In particular, applying (7.13) we easily obtain

$$\sum_{m:\, k=\ell+mM\leq n} \binom{n}{k} p^k q^{n-k} = \frac{1 + (p\omega_1 + q)^{n-\ell} + \cdots + (p\omega_{M-1} + q)^{n-\ell}}{M}$$

$$= \frac{1}{M} + O(\rho^n), \tag{8.54}$$

where $\rho < 1$ since $|(p\omega_r + q)| = p^2 + q^2 + 2pq\cos(2\pi r/M) < 1$ for $r \neq 0$. Thus

$$S_n(f) = \frac{1}{M} \sum_{\ell=0}^{M-1} f\left(\left\langle \frac{\ell}{M} + y\right\rangle\right) + O(\rho^n).$$

To complete the proof, we observe that the function

$$F(y) = \frac{1}{M} \sum_{l=0}^{M-1} f\left(\left\langle \frac{l}{M} + y\right\rangle\right)$$

is periodic with period $1/M$. Thus for $0 \leq y < \frac{1}{M}$ we have the representation

$$F(y) = \frac{1}{M} \sum_{l=0}^{M-1} f\left(\frac{l}{M} + y\right).$$

For general $y \in \mathbb{R}$ we need only use the relation $F(y) = F(\langle My \rangle / M)$ to finish the proof. (In Exercises 8 and 9 we ask the reader to use another method, that of Fourier series, to establish the above result.) ∎

We complete this section with an interesting example arising in source coding.

Example 8.7: *Redundancy of the Shannon Code*   We consider here a very simple code, known as the Shannon code, that assigns code length $\lceil - \log_2 p^k (1 - p)^{n-k}) \rceil$ to a sequence x_1^n of length n occurring with probability $p(k) = p^k q^{n-k}$, where k is the number of "1" in x_1^n (cf. [75]). From Shannon Theorem 6.18 we know that the average code length cannot be smaller than the entropy of the source, which in our case is $- \sum_{k=0}^{n} \binom{n}{k} p^k q^{n-k} \log_2(p^k q^{n-k})$. From Chapter 6, we know that the difference between the average code length and the entropy is known as the average *redundancy* (see Section 8.7.2 for a more in-depth discussion of redundancy issues). We denote it as $\bar{R}_n$. For the Shannon block code of length n generated by a memoryless source we have

$$\bar{R}_n = \sum_{k=0}^{n} \binom{n}{k} p^k q^{n-k} \left(\lceil - \log_2 p(k) \rceil + \log_2 p(k) \right)$$

$$= 1 + \sum_{k=0}^{n} \binom{n}{k} p^k q^{n-k} \left(\log_2 p(k) + \lfloor - \log_2 p(k) \rfloor \right)$$

$$= 1 - \sum_{k=0}^{n} \binom{n}{k} p^k q^{n-k} \langle \alpha k + \beta n \rangle, \qquad (8.55)$$

where $\alpha = \log_2((1 - p)/p)$ and $\beta = \log_2(1/(1 - p))$. But the above sum is the one we analyzed in Theorems 8.5 and 8.7. If α is irrational by Theorem 8.5 (applied to the binomial distribution with $f(t) = t$) we obtain $\bar{R}_n = \frac{1}{2} + o(1)$. For $\alpha = N/M$, we apply Theorem 8.7 to arrive at

$$\bar{R}_n = 1 - \frac{1}{M} \sum_{l=0}^{M-1} \left(\frac{l}{M} + \frac{\langle M\beta n \rangle}{M} \right) + O(\rho^n)$$

$$= 1 - \frac{M-1}{2M} + \frac{\langle M\beta n \rangle}{M} + O(\rho^n)$$

$$= \frac{1}{2} - \frac{1}{M} \left(\langle M\beta n \rangle - \frac{1}{2} \right) + O(\rho^n).$$

In summary, we prove the following.

Theorem 8.8 **(Szpankowski, 2000)**   *Consider the Shannon block code of length n binomially(n,p) distributed over a binary alphabet. Then for $p < \frac{1}{2}$ as $n \to \infty$*

$$
\bar{R}_n =
\begin{cases}
\frac{1}{2} + o(1) & \alpha \quad \text{irrational} \\[2ex]
\frac{1}{2} - \frac{1}{M}\left(\langle Mn\beta \rangle - \frac{1}{2}\right) + O(\rho^n) & \alpha = \frac{N}{M}, \ \gcd(N, M) = 1 ,
\end{cases}
\tag{8.56}
$$

where $\rho < 1$.

8.3   SMALL SINGULARITIES OF ANALYTIC FUNCTIONS

It is time to deliver what the title of this section says, namely, complex asymptotics. From now on we shall deal with the generating function

$$
A(z) = \sum_{n=0}^{\infty} a_n z^n,
$$

and study various methods of extracting asymptotic expansions of a_n from an explicit knowledge of $A(z)$, and even more often, from partial knowledge of the generating function $A(z)$ around its singularities. Often the generating functions we work with (in combinatorics, analysis of algorithms, information theory, etc.) are *analytic functions*[1]. Therefore, by Cauchy's formula the coefficient a_n can be evaluated as

$$
a_n = \frac{1}{2\pi i} \int_C \frac{f(z)}{z^{n+1}} dz,
\tag{8.57}
$$

where C is a closed contour that encircles the origin in the positive direction (i.e., counterclockwise). One may ask why we use Cauchy's integral formula for a_n when simpler expressions exist (e.g., $a_n = \frac{1}{n!} A^{(n)}(z)|_{z=0}$, where $A^{(n)}(z)$ is the nth derivative of $A(z)$). The point to observe is that integration preserves many nice properties of analytic function; the contour C can be deformed in many ways to take advantage of analytic continuations of $A(z)$; and finally this integral is usually easily computable through the Cauchy residue theorem (see Theorem 2.5).

We shall argue that complex methods are successful in extracting asymptotics of a_n since the *asymptotic growth of a_n is determined by the location of the singularities closest to the origin*. To see this, we first recall that a singularity of an

---

[1] The reader may want to review elements of complex analysis discussed in Section 2.3. There are also excellent books on the subject; e.g., Henrici [195], Hille [196], Titchmarsh [424], and Whittaker and Watson [448].

analytic function is a point, where the function ceases to be analytic. Next, we recall from Hadamard's Theorem 7.1 that the radius of convergence of the series involved in $A(z)$ is

$$R^{-1} = \limsup_{n \to \infty} |a_n|^{1/n},$$

or informally $\frac{1}{n} \log |a_n| \sim -\log R$. That is, for every $\varepsilon > 0$ there exists N such that for $n > N$ we have

$$|a_n| \le (R^{-1} + \varepsilon)^n \ ;$$

and for infinitely many n we have

$$|a_n| \ge (R^{-1} - \varepsilon)^n.$$

Thus the exponential growth of a_n is determined by R^{-n}. A classical theorem (attributed to Pringsheim) says that an analytic function of finite radius of convergence must have a singularity on the boundary of its disk of convergence. Moreover, as we have already seen in Section 7.3.3, a bound on a_n is easily achievable through Cauchy's Integral Theorem 2.6 (cf. (2.22) in Section 2.3), namely,

$$|a_n| \le \frac{M(r)}{r^n}, \tag{8.58}$$

where $M(r)$ is the maximum value of $|A(z)|$ for any circle $r < R$. This is the *Cauchy bound*, and we shall use it often in this chapter. In summary, locations of singularities of $A(z)$ determine the asymptotic behavior of its coefficients. To obtain a more refined information about such asymptotics we must more carefully study various types of singularities, which we do next.

Finally, one of the most important aspects of complex asymptotics is to provide the *transfer theorems*, under which the following central implication is valid:

$$A(z) \sim f(z) \quad \Longrightarrow \quad [z^n]A(z) \sim [z^n]f(z), \tag{8.59}$$

where $A(z)$ is an implicitly known generating function and $f(z)$ is an asymptotic expansion of $A(z)$ near a singularity. (The above is not true in general; e.g., $A(z) = 1/(1-z) \sim 2/(1-z^2) = 1/(1-z) + 1/(1+z) =: B(z)$ at $z = 1$ but $[z^n]A(z) = 1$ while $[z^n]B(z) = 1 + (-1)^n$.) Following Odlyzko [330], we shall talk about *small singularity* if $A(z)$ has polynomial growth around such a singularity; otherwise it is called a *large singularity*. For example, $(1-z)^{3/2} \log(1-z)$ has a small singularity at $z_0 = 1$, while $\exp(1/(1-z))$ is said to have a large singularity at $z_0 = 1$. These are not precise concepts, but still clearly different methods are used to handle them. Therefore, we divide our discussion into two parts: In this section, we deal with methods for small singularities (i.e., poles and algebraic singularities) while in the next section we turn our attention to large singularities.

8.3.1 Polar Singularities

Here we deal with polar singularities, which we define precisely below. Let $A(z)$ be an analytic function in a ring $0 < |z - z_0| < R \leq \infty$. Then the Laurent expansion applies and one finds (cf. [196, 424])

$$A(z) = \sum_{n=-M}^{-1} a_n(z - z_0)^n + \sum_{n=0}^{\infty} a_n(z - z_0)^n, \tag{8.60}$$

where the coefficients a_n, $n = -M, -M + 1, \ldots, 0, 1, \ldots$ are computed by the Cauchy formula (8.57). The point z_0 is called:

- *Removable singularity* if $a_{-n} = 0$ for all $n = 1, \ldots, M$ and then $\lim_{z \to z_0} A(z) = a_0$;

- A *pole* of order M, if M is finite and $M > 0$;

- An *essential singularity* if if $M = \infty$.

We only deal here with poles since a function behaves very weirdly near essential singularities (e.g., according to the Casorati-Weierstrass theorem the function can take *any* value around such a singularity).

A function $A(z)$ is called **meromorphic** if its only singularities are poles. A special case of meromorphic functions are **rational functions**, which are the ratios of two polynomials.

We first discuss **rational functions**; that is, we assume that $A(z) = N(z)/D(z)$, where both $N(z)$ and $D(z)$ are polynomials. Hence, we can decompose $A(z)$ into a finite sum like

$$A(z) = \sum_{\rho,r} \frac{a_{\rho,r}}{(z - \rho)^r},$$

where the summation is over all poles (roots) of $A(z)$ (of $D(z)$) and r ranges over a set of integers determined by the largest multiplicity of a pole of A. To extract the coefficient a_n, it suffices to study asymptotics of $1/(z-\rho)^r$. But in fact we can do it *exactly*, not only asymptotically. Indeed, from Table 7.2, Entry 1, we find

$$[z^n] \frac{1}{(z - \rho)^j} = [z^n] \frac{(-1)^j}{\rho^j(1 - z/\rho)^j}$$

$$= (-1)^j \binom{n+j-1}{n} \rho^{-(n+j)} = (-1)^j \binom{n+j-1}{j-1} \rho^{-(n+j)}.$$

In summary, *if $A(z)$ is rational which is analytic at zero and has poles at points $\rho_1, \ldots, \rho_m$, then there exist polynomials $P_j(n)$ of degree equal to the order of pole ρ_j minus one such that exactly*

$$a_n := [z^n]A(z) = \sum_{j=1}^{m} P_j(n)\rho_j^{-n}. \tag{8.61}$$

We illustrate this result with a simple example which is a continuation of the pattern occurrence problem discussed in Section 7.6.2.

Example 8.8: *Frequency of a Given Pattern Occurrence* Let H be a *given* pattern of size m, and consider a random text of length n generated by a memoryless source. In Section 7.6.2 we studied the probability generating function $T_r(z) = \sum_{n=0}^{\infty} \Pr\{O_n = r\}z^n$ of the number O_n of pattern H occurrences in the text. In Theorem 7.13 we proved (for a memoryless source) that

$$T_r(z) = \frac{z^m P(H)(D(z) + z - 1)^{r-1}}{D^{r+1}(z)},$$

where $D(z) = P(H)z^m + (1 - z)A_H(z)$ and $A(z)$ is the *autocorrelation polynomial* (a polynomial of degree m). Let now $\rho < \rho_1 < \rho_2 < \cdots < \rho_{m-1}$ be m distinct roots of $D(z) = 0$. Then an easy application of the above analysis leads to

$$\Pr\{O_n(H) = r\} = \sum_{j=1}^{r+1} (-1)^j a_j \binom{n}{j-1} \rho^{-(n+j)}$$

$$+ \sum_{k=1}^{m-1} \sum_{i=1}^{r+1} (-1)^i b_{ki} \binom{n}{i-1} \rho_k^{-(n+i)]}$$

$$= \sum_{j=1}^{r+1} (-1)^j a_j \binom{n}{j-1} \rho^{-(n+j)} + O(\rho_1^{-n}), \tag{8.62}$$

where $a_{r+1} = \rho^m P(H) (\rho - 1)^{r-1} (D'(\rho))^{-r-1}$, while the other coefficients a_j and b_{ki} can be computed when needed. ∎

Let us now turn our attention to a more general case, namely, **meromorphic functions**. We derive an asymptotic expansion for the coefficients a_n similar to the one for rational functions except that this time we only obtain asymptotic approximation, not an exact formula. There are two approaches: One is called *subtracted singularities* and the other is known as the *contour integral* method.

We start with the subtracted singularities method. To make our discussion more concrete we study the following function

$$A(z) = \sum_{j=1}^{r} \frac{a_{-j}}{(z-\rho)^j} + \sum_{j=0}^{\infty} a_j (z-\rho)^j.$$

We assume that $A(z)$ has a Laurent expansion around the pole ρ of multiplicity r. We further assume that the pole ρ is the closest to the origin; that is, the radius of convergence R is $R = |\rho|$, and there are no other poles on the circle of convergence. In other words, the second term of $A(z)$, which we denote for simplicity as $A_1(z)$, is analytic in the circle $|z| \le |\rho|$, and its radius of convergence is $R_1 > |\rho|$. Thus coefficients of $A_1(n)$ are bounded by $O(R_1^{-n})$ due to Cauchy's bound (8.58). Using the exact expansion of the first part of $A(z)$, which is a rational function, we arrive at

$$[z^n]A(z) = \sum_{j=1}^{r} (-1)^j a_{-j} \binom{n}{j-1} \rho^{-(n+j)} + O(R_1^{-n})$$

for $R_1 > \rho$. Observe that in the last example formula (8.62) is presented in this form.

The main thrust of the method just described is as follows: Imagine we are interested in the asymptotics for coefficients a_n of a function $A(z)$ whose circle of convergence is R. Let us also assume that we can find a simpler function, say $\bar{A}(z)$, that has the same singularities as $A(z)$ (e.g., in the above example $\bar{A}(z) = \sum_{j=1}^{r} \frac{a_{-j}}{(z-\rho)^j}$). Then $A_1(z) = A(z) - \bar{A}(z)$ is analytic in a larger disk, of radius $R_1 > R$, say, and its coefficients are bounded by $O(R_1^{-n})$.

We now rederive the above formula using the *contour integral* approach. Define

$$I_n = \frac{1}{2\pi i} \int_{|z|=r} \frac{A(z)}{z^{n+1}} dz,$$

where r is such that the circle of radius r contains inside all singularities of $A(z)$. After applying Cauchy's formula we arrive at

$$I_n = \text{Res}[A(z); z=0] + \sum_{\rho} \text{Res}[A(z)z^{-n-1}; z=\rho],$$

where the summation is over all singularities ρ contained inside the circle $|z| = r$.

But by Cauchy's bound (8.58) we also know that $|I_n| = O(r^{-n})$. Thus

$$a_n = [z^n]A(z) = -\sum_\rho \text{Res}[A(z)z^{-n-1}; z = \rho] + O(r^{-n}).$$

This leads to the same estimate as above.

We can summarize our discussion in the following theorem.

Theorem 8.9 *Let $A(z)$ be a meromorphic function for $|z| \leq r$ with poles at $\rho_1, \rho_2, \ldots, \rho_m$ and analytic on $|z| = r$ and $z = 0$. Then there exist polynomials $P_i(n)$ such that*

$$a_n := [z^n]A(z) = \sum_{j=1}^m P_j(n)\rho_j^{-n} + O(r^{-n}),$$

where the degree of P_j is equal to the order of the pole at ρ_j minus one.

We finish this section with an example that will combine what we have learned in the last two sections.

Example 8.9: *Enumeration of (d, k) Sequences*   For the run-length coding the (d, k) *sequences* ($d \leq k$) are of prime importance. A binary string is a (d, k) sequence if it does not contain any runs of 0's shorter than d and longer than k. In other words, between any pair of 1's there must be at least d and at most k zeros. We want to enumerate all such (d, k) sequences. We first construct the ordinary generating function $W_{d,k}(z) = \sum_{w \in \mathcal{W}_{d,k}} z^{|w|}$ of all (d, k) words denoted as $\mathcal{W}_{d,k}$. We apply the symbolic methods of Chapter 7. Let $\mathcal{A}_{d,k}$ be the set of all words (strings) consisting only of 0's whose length is between d and k. The generating function $A(z)$ is clearly equal to

$$A(z) = z^d + z^{d+1} + \cdots + z^k = z^d \frac{1 - z^{k-d+1}}{1 - z}.$$

We now observe that $\mathcal{W}_{d,k}$ can be symbolically written as

$$\mathcal{W}_{d,k} = \mathcal{A}_{d,k}\left(\{1\} \times \epsilon + \bar{\mathcal{A}}_{d,k} + \bar{\mathcal{A}}_{d,k} \times \bar{\mathcal{A}}_{d,k} + \cdots + \bar{\mathcal{A}}_{d,k}^k + \cdots\right), \quad (8.63)$$

where $\bar{\mathcal{A}}_{d,k} = \{1\} \times \mathcal{A}_{d,k}$. Above basically says that the collection of (k, d) sequences, $\mathcal{W}_{d,k}$, is a concatenation of $\{1\} \times \mathcal{A}_{d,k}$. Thus (8.63) translates into the generating functions $W_{d,k}(z)$ as follows

$$W_{d,k}(z) = A(z)\frac{1}{1 - zA(z)} = \frac{z^d(1 - z^{k+1-d})}{1 - z - z^{d+1} + z^{k+2}}$$

$$= \frac{z^d + z^{d+1} + \cdots + z^k}{1 - z^{d+1} - z^{d+2} - \cdots - z^{k+1}}. \tag{8.64}$$

Of particular interest is the *capacity* $C_{d,k}$ defined as (cf. [26])

$$C_{d,k} = \lim_{n \to \infty} \frac{\log[z^n]W_{d,k}(z)}{n}.$$

If ρ is the smallest root in absolute value of $1 - z^{d+1} - z^{d+2} - \cdots - z^{k+1} = 0$, then clearly

$$C_{d,k} = -\log \rho.$$

But we can do much better. The function $W_{d,k}(z)$ is rational and we can compute $[z^n]W_{d,k}(z)$ exactly. Let us consider a particular case, namely, $d = 1$ and $k = 3$. Then the denumerator in (8.64) becomes $1 - z^2 - z^3 - z^4$, and its roots are

$$\rho_{-1} = -1, \quad \rho_0 = 0.682327\ldots,$$

$$\rho_1 = -0.341164\ldots + i1.161541\ldots, \quad \rho_2 = \bar{\rho}_1.$$

Using the contour integral approach and computing the residues we obtain

$$[z^n]W_{1,3}(z) = \frac{\rho_0 + \rho_0^2 + \rho_0^3}{(\rho_1 + 1)(\rho_0 - \rho_1)(\rho_0 - \bar{\rho}_1)}\rho_0^{-n-1}$$

$$+ (-1)^{n+1}\frac{1}{(\rho_0 + 1)(\rho_1 + 1)(\bar{\rho}_1 + 1)} + O(r^{-n}),$$

where $r \approx 0.68$. More specifically,

$$[z^n]W_{1,3}(z) = 0.594(1.465)^{n+1} + 0.189(-1)^{n+1} + O(0.68^n).$$

If we wish, we could compute the big-oh term more precisely.   ■

8.3.2 Algebraic Singularities: Singularity Analysis

In the last subsection, we considered one type of singularity, namely, poles. For such isolated singularities, a function was well defined in its whole punctured neighborhood of, say z_0. In particular, in $0 < |z - z_0| < R$ the function $A(z)$ has its Laurent series expansion. In this subsection, we study algebraic singularities for which Laurent's expansion does not apply since around these singularities the function is not defined in the whole ring $0 < |z - z_0| < R$ of z_0.

We first review some simple facts from complex analysis. Let us consider a particular example. Say, $w := f(z) = \sqrt{z}$. It is known that this function is multivalued. For $z = re^{i\theta}$ there are exactly two different values of w corresponding to z, namely, $w_1 = \sqrt{r}e^{i\theta/2}$ and $w_2 = \sqrt{r}e^{i\theta/2+\pi} = -w_1$. These two functions are known as *branches*. Observe the following phenomenon: When we start from a given point z and choose one of the branches, say w_1; then when traversing any closed contour we come back to w_1, *unless* we enclose the origin. If the contour encloses the origin, then w attains the value w_2 after one circle, and it comes back to w_1 after the second loop. We should conclude that the multivalued function is well defined *except* at $z = 0$, where the function interchanges branches. Such a point is called a *branch point*. The function ceases to be analytic at this point, and hence it has a singularity (in fact, there is another explanation why $\sqrt{z}$ must have a singularity at $z_0 = 0$: its derivative does not exist in this point). We shall call it an *algebraic singularity*. To make $\sqrt{z}$ analytic, we must restrict the domain of definition to $\mathbb{C} - (-\infty, 0)$, that is, complex plane *cut* along the negative real axis (in fact, it can be cut along any ray emanating from the origin). This way we prevent z encircling the origin and thus avoid ambiguity. As a consequence, we cannot expect to develop $\sqrt{z}$ in the Laurent series around $z = 0$; however, another expansion due to Puiseux will work, as we shall soon see. Naturally, $\sqrt{z}$ is not an exception. Functions like z^α for α real and not an integer, $\log z$, etc. belong to this category.

Let us now more precisely define a branch point and its order. The reader may review this material from any standard textbook on complex analysis (e.g., [196]). Consider an analytic function $A(z)$ that can be analytically continued along any path in the ring $0 < |z - z_0| < R$. If the function (or a branch of the function) returns to itself after its analytic continuation along any closed path encircling z_0, then A is a single-valued function and z_0 is a pole or an essential singularity. If, however, after a continuation along a closed path the function does not return to its original branch (as in the case of $\sqrt{z}$), then the function is multivalued and z_0 is a branch point. To further characterize it, let C be a circle $|z - z_0| = r$ contained in the ring $0 < |z - z_0| < R$, and pC be a closed contour that encircles p times z_0 along C. The point z_0 is called a branch point of order $p - 1$, if the function returns to its original branch along pC, but not along any closed path mC, where $m < p$. For example, $(z - 1)^{1/3}$ has a branch point at $z_0 = 1$ which is of order two; the function $\sqrt{z(z - 1)}$ has two branch points at 0 and 1, each of order 1.

Although the Laurent expansion does not apply for a function around its algebraic singularity, another expansion due to Puiseux works fine (cf. [196]).

Theorem 8.10 **(Puiseux, 1850)** *Let $A(z)$ be an analytic (multivalued) function and z_0 its branch point of order $p - 1$. Then in the neighborhood of z_0, $A(z)$ can be represented as the following series*

$$A(z) = \sum_{k=-\infty}^{\infty} a_k (z - z_0)^{\frac{k}{p}} \qquad (8.65)$$

that converges in a neighborhood of z_0.

The idea of the proof of Puiseux's expansion is quite simple. It suffices to substitute $z = z_0 + w^p$, and observe that $A(z)$ as a function of w is a singled-valued function defined in a ring $0 < |z - z_0| < R$; hence one can apply Laurent's expansion to it yielding

$$A(z_0 + w^p) = \sum_{k=-\infty}^{\infty} a_k w^k$$

which is the desired (8.65).

Puiseux's expansion suggests that for algebraic singularities, one must study asymptotics of functions like $(z - z_0)^{k/p}$. In general, in applications we often deal with functions like $(z - z_0)^{k/p} w(z)$, where $w(z)$ is either analytic in a neighborhood of z_0 or not (e.g., consider $A(z) = \frac{e^{-z-z^2/2}}{\sqrt{1-z}}$ and $\sqrt{1-z} \log \frac{1}{1-z}$). We next discuss in some depth how to obtain precise asymptotic expansions for such functions around its algebraic singularities. There are many techniques to deal with such singularities. However, in analysis of algorithms the *singularity analysis* of Flajolet and Odlyzko is the most useful, and we devote some time to it in the sequel.

Before we proceed with general discussions, we observe that because

$$[z^n]A(z) = \rho^n [z^n]A(z/\rho),$$

one only needs to study singularities at, say, $z = 1$. That's what we will do next. We shall follow Flajolet and Odlyzko [140]. The reader will also find an excellent account of it in Odlyzko [330] and Flajolet and Sedgewick [149]. Actually, because of these excellent references, we focus here on explaining things rather than on detailed proofs.

We start with obtaining asymptotic series for $(1 - z)^{-\alpha}$, where α is real and $\alpha \notin \{0, -1, -2, \ldots\}$. First, we observe that for $\alpha = k \in \{1, 2, \ldots\}$ we have from Entry 1 of Table 7.2

$$(1 - z)^{-k} = \sum_{n=0}^{\infty} \binom{n + k - 1}{n} z^n.$$

It is then natural to expect that

$$[z^n](1-z)^{-\alpha} = \binom{n+\alpha-1}{n} = \frac{\Gamma(n+\alpha)}{\Gamma(\alpha)\Gamma(n+1)}$$

$$= \frac{n^{\alpha-1}}{\Gamma(\alpha)}\left(1 + \frac{\alpha(\alpha-1)}{2n} + O\left(\frac{1}{n^2}\right)\right)$$

provided $\alpha \notin \{0, -1, -2, \ldots\}$. We will prove it using Hankel's contour, which we review in Table 8.3.

TABLE 8.3. Hankel's Contour Integral for Gamma Function

Let $\mathcal{H}$ be a contour starting at $+\infty$ in the upper half-plane, winding counterclockwise around the origin, and proceeding back toward $+\infty$ in the lower half-plane. (Figure 8.2). We denote the integral along such a loop as $\int_{\mathcal{H}}$ or $\int_{+\infty}^{(0)}$. The following theorem is a particular result for the Euler gamma function, but it plays an important role in the analysis of algebraic singularities.

Theorem 8.11   (Hankel's Contour Integral)   *For all $s \in \mathbb{C}$*

$$\frac{1}{\Gamma(s)} = -\frac{1}{2\pi i} \int_{+\infty}^{(0)} (-w)^{-s} e^{-s}\, dw \qquad (8.66)$$

where the function $(-w)^{-s}$ has its principal value, that is,

$$(-w)^{-s} = \exp(-s \ln r - si(\theta - \pi))$$

with $w = re^{i\theta}$.

To see why (8.66) is true, let us specify Hankel's contour as follows (Figure 8.2).

$$\mathcal{H} = \begin{cases} \mathcal{H}^+ & w = t + \delta i & t \geq 0 \\ \mathcal{H}^- & w = t - \delta i & t \geq 0 \\ \mathcal{H}^o & w = \delta e^{i\phi} & \phi \in [-\pi/2, \pi/2]. \end{cases}$$

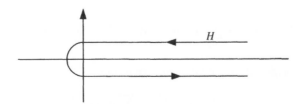

Figure 8.2. Hankel's contour.

TABLE 8.3. Continued

where $\delta \to 0$. Easy calculations show that for $\delta \to 0$

$$\int_{\mathcal{H}+} (-w)^{(-s)} e^{-w} \, dw = e^{is\pi} \int_{\infty}^{\delta} e^{-t} t^{-s} \, dt,$$

$$\int_{\mathcal{H}-} (-w)^{(-s)} e^{-w} \, dw = e^{-is\pi} \int_{\delta}^{\infty} e^{-t} t^{-s} \, dt,$$

$$\int_{\mathcal{H}^o} (-w)^{(-s)} e^{-w} \, dw = i\delta^{1-s} \int_{-\pi/2}^{\pi/2} \exp\left(-\delta^{i\phi} - si(\phi - \pi)\right) d\phi.$$

The last integral converges to zero for $\delta \to 0$ as long as $\Re(1 - s) > 0$. Then, the first two integrals sum up to

$$-\frac{e^{-i\pi s} - e^{i\pi s}}{2\pi i} \int_0^{\infty} e^{-t} t^{-s} \, dt = \frac{\sin \pi s}{\pi} \Gamma(1 - s) = \frac{1}{\Gamma(s)}$$

where the last implication follows from the fundamental relationship (2.33) for the gamma function (see Section 2.4.1). Thus, we prove (8.66) for $\Re(1 - s) > 0$. By analytic continuation (since both sides of (8.66) are analytic) we can extend it to all $s \in \mathbb{C}$.

Observe that if we traverse the Hankel loop in the reverse order (i.e., from $+\infty$ in the lower plane, winding the origin clockwise and moving toward $+\infty$ in the upper plane) we must reverse the sign in (8.66).

Theorem 8.12   (Flajolet and Odlyzko, 1990)   *Let $\alpha \notin \{0, -1, -2, \ldots\}$. Then*

$$[z^n](1 - z)^{-\alpha} \sim \frac{n^{\alpha-1}}{\Gamma(\alpha)} \left(1 + \sum_{k=1}^{\infty} \frac{e_k(\alpha)}{n^k}\right) \tag{8.67}$$

$$= \frac{n^{\alpha-1}}{\Gamma(\alpha)} \left(1 + \frac{\alpha(\alpha - 1)}{2n} + \frac{\alpha(\alpha - 1)(\alpha - 2)(3\alpha - 1)}{24n^2} + \cdots\right),$$

where

$$e_k(\alpha) = \sum_{j=k}^{2k} c_{k,j}(\alpha - 1)(\alpha - 2) \cdots (\alpha - j),$$

and $c_{k,j}$ are coefficients in the following expansion

$$e^x (1 + yt)^{-1-1/y} = \sum_{k,j} c_{k,j} x^j y^k. \tag{8.68}$$

Furthermore, for any β

$$[z^n](1-z)^{-\alpha} \left(\frac{1}{z} \log \frac{1}{1-z} \right)^{\beta} \sim \frac{n^{\alpha-1}}{\Gamma(\alpha)} (\log n)^{\beta} \left(1 + \sum_{k=1}^{\infty} \frac{C_k(\alpha, \beta)}{(\log n)^k} \right), \quad (8.69)$$

where

$$C_k(\alpha, \beta) = \binom{\beta}{k} \Gamma(\alpha) \frac{d^k}{ds^k} \frac{1}{\Gamma(s)} \big|_{s=\alpha}.$$

Proof. We give a detailed proof of (8.67) and only a sketch of (8.69). By Cauchy's formula

$$[z^n](1-z)^{-\alpha} = \frac{1}{2\pi i} \int_C (1-z)^{-\alpha} \frac{dz}{z^{n+1}},$$

where C is a small contour encircling the origin (e.g., $C = \{z : |z| = \frac{1}{2}\}$). We deform it in such a way that the new contour, say C_1, consists of a large circle of radius R with a notch that comes back as close to $z = 1$ as possible (see Figure 8.3). By Cauchy's bound the integral on the large circle can be bounded by $O(R^{-n-\alpha})$, and we are left with estimating the integral along the notch that resembles the Hankel contour except that it is traversed in the opposite direction (see Table 8.3). We denote it by $\bar{\mathcal{H}}$ and specify as:

$$\bar{\mathcal{H}} = \begin{cases} \mathcal{H}^- & w = t - \frac{i}{n} & t \geq 0 \\ \mathcal{H}^+ & w = t + \frac{i}{n} & t \geq 0 \\ \mathcal{H}^o & w = 1 - \frac{e^{i\phi}}{n} & \phi \in [-\pi/2, \pi/2] \end{cases}$$

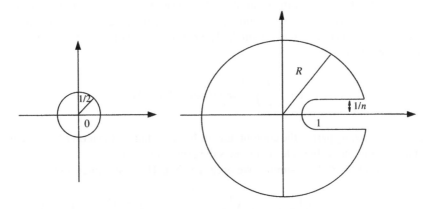

Figure 8.3. Contours Used in Theorem 8.12.

Then after a change of variable $z = 1 + t/n$, we arrive at

$$[z^n](1-z)^{-\alpha} = \frac{n^{\alpha-1}}{2\pi i} \int_{\tilde{\mathcal{H}}} (-t)^{-\alpha} \left(1 + \frac{t}{n}\right)^{-n-1} dt + O(R^{-n-\alpha}).$$

We now proceed formally, omitting $O(R^{-n-\alpha})$, and later justify it. Using (8.68) we have

$$e^t \left(1 + \frac{t}{n}\right)^{-1-n} = \sum_{k,j} c_{k,j} t^j n^{-k}, \tag{8.70}$$

which yields

$$
\begin{aligned}
[z^n](1-z)^{-\alpha} &= \frac{n^{\alpha-1}}{2\pi i} \int_{\tilde{\mathcal{H}}} (-t)^{-\alpha} e^{-t} \sum_{k,j} c_{k,j} t^j n^{-k} dt \\
&= n^{\alpha-1} \sum_{k,j} c_{k,j} n^{-k} \frac{1}{2\pi i} \int_{\tilde{\mathcal{H}}} (-t)^{-\alpha+j} e^{-t} dt \\
&\overset{(8.66)}{=} n^{\alpha-1} \sum_{k,j} c_{k,j} n^{-k} \frac{1}{\Gamma(\alpha-j)} \\
&= \frac{n^{\alpha-1}}{\Gamma(\alpha)} \sum_{k,j} c_{k,j} \frac{(\alpha-1)(\alpha-2)\cdots(\alpha-j)}{n^k},
\end{aligned}
$$

which formally establishes our result. To justify it rigorously, we must prove convergence of the above series in order to interchange the integral and the sum.

Actually, this is not difficult at all. All we need to do is to split the integral according to, say $|t| \leq \log^2 n$ and $|t| \geq \log^2 n$. This is necessary in order to justify (8.70), that is, to make t small. The part corresponding to $|t| \geq \log^2 n$ is negligible since

$$\left(1 + \frac{t}{n}\right)^{-n} = O(\exp(-\log^2 n)).$$

On the remaining part of the contour, that is, for $|t| \leq \log^2 n$, the term t/n is small and (8.70) is applicable. This completes the proof of (8.67).

The proof of (8.69) is similar. We only sketch it. The basic expansion is

$$z^{-n-1}(1-z)^{-\alpha}\left(\frac{1}{z}\log\frac{1}{1-z}\right)^{\beta}\bigg|_{z=1+t/n} \sim e^{-t}\left(\frac{n}{-t}\right)^{\alpha}\log^{\beta}\left(\frac{-n}{t}\right)$$

$$= \frac{e^{-t}(-t)^{\alpha}}{n^{\alpha}} \log^{\beta} n \left(1 - \frac{\log(-t)}{\log n} \right)^{\beta}$$

$$= \frac{e^{-t}(-t)^{\alpha}}{n^{\alpha}} \log^{\beta} n \left(1 - \beta \frac{\log(-t)}{\log n} + \frac{\beta(\beta - 1)}{2} \left(\frac{\log(-t)}{\log n} \right)^2 + \cdots \right).$$

By considering $|t| \le \log^2 n$ and $|t| \ge \log^2 n$, we can justify the expansion. The last observation is that we end up with the Hankel integrals of the form

$$\frac{1}{2\pi i} \int_{\tilde{\mathcal{H}}} (-t)^{-(\alpha-j)} e^{-t} \log^k(-t) dt,$$

which reduces to the derivatives of $1/\Gamma(\alpha - j)$. This completes the sketch of the proof for (8.69). For details the reader is referred to [140]). ∎

The most important aspect of the singularity theory comes next: In the analysis of algorithms, analytic combinatorics, information theory, and many engineering problems, we rarely have an explicit expression for the generating function $A(z)$; but often we get an expansion of $A(z)$ around its singularities. For example, we are often left with $A(z) \sim (1 - z)^{-\alpha} = (1 - z)^{-\alpha} + o((1 - z)^{-\alpha})$. In order to evaluate a_n as $n \to \infty$ we need a "transfer theorem" that will allow us to pass to coefficients of $o((1 - z)^{-\alpha})$ under the little-oh notation. These transfer theorems are jewels of Flajolet and Odlyzko singularity analysis [140], and we discuss them next. However, in order not to clutter our expositions too much, we consider a particular example adopted from Odlyzko [330].

Let $A(z) \sim \sqrt{1 - z} = \sqrt{1 - z} + f(z)$, where $f(z) = o(\sqrt{1 - z})$. From Theorem 8.12 we conclude that

$$[z^n](1 - z)^{\frac{1}{2}} = -\frac{1}{\sqrt{\pi} n^{3/2}} \left(\frac{1}{2} + \frac{3}{16n} + O(n^{-2}) \right).$$

Hence, we would like to prove that

$$[z^n]f(z) = [z^n]o(\sqrt{1 - z}) = o(n^{-3/2}) \qquad (8.71)$$

as $n \to \infty$. The analyticity of $f(z)$ in a region around $z_0 = 1$ is crucial for obtaining any further information. Let us consider a few different scenarios:

- Assume $f(z)$ is analytic in a larger disk, say, of radius $1 + \delta$ for some $\delta > 0$. Then by Cauchy's bound we immediately conclude that $[z^n]f(z) = O((1 + \delta')^{-n})$ for $0 < \delta' < \delta$, a conclusion that is much stronger than the desired (8.71). Furthermore, this assumption will not allow us to handle expressions like $\sqrt{1 - z} \log \frac{1}{1-z}$, which often arises in analysis of algorithms.

- If we only know that $f(z) = o(\sqrt{1-z})$ in $|z| \leq 1$, then we may only conclude that $[z^n]f(z) = O(1)$. This is true since $|f(z)| \leq C$ in $|z| < 1$, where C is a constant, and for all $r < 1$ and $n \geq 0$ we know that by Cauchy's bound $[z^n]f(z) \leq Cr^{-n}$. But this only suggests that $[z^n]f(z) = O(1)$, a result that is far from what is required. However, if we know some smoothness condition for the function f on the circle $|z| = 1$, then some further progress can be achieved. This is due to Darboux and we briefly discuss it at the end of this section.

We now consider the third scenario analyzed by Flajolet and Odlyzko [140]. Let us define a domain around $z_0 = 1$ as follows

$$\Delta(R, \phi) = \{z : |z| < R, \ z \neq 1, \ |\arg(z - 1)| > \phi\} \qquad (8.72)$$

for some $R > 1$ and $0 < \phi < \pi/2$. The domain Δ is an extended disk around $z = 1$ with a circular part rooted at $z = 1$ deleted, as shown in Figure 8.4. In our case, we assume that $f(z)$ can be analytically continued to $\Delta(R, \pi/4)$. We apply Cauchy's formula to extract the coefficients of $f(z)$, and as the contour $\mathcal{C}$ we choose (see Figure 8.4)

$$\mathcal{C}_1 = \{z : |z - 1| = 1/n, \ |\arg(z - 1) \geq \pi/4\},$$
$$\mathcal{C}_2 = \{z : z = 1 + re^{i\pi/4}, \ 1/n \leq r \leq \delta\},$$
$$\mathcal{C}_3 = \{z : |z - 1| = \delta, \ |\arg(z - 1) \geq \pi/4\},$$
$$\mathcal{C}_4 = \{z : z = 1 + re^{-i\pi/4}, \ 1/n \leq r \leq \delta\},$$

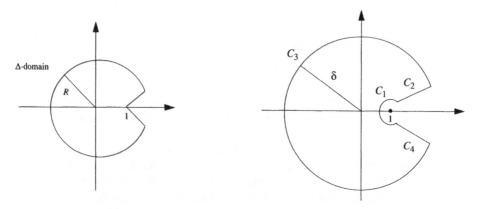

Figure 8.4. The Δ domain and the contour used in the singularity analysis.

where $0 < \delta < 1/2$. We evaluate the following four integrals for $j = 1, 2, 3, 4$

$$f_j = \frac{1}{2\pi i} \int_{C_j} f(z) \frac{dz}{z^{n+1}}.$$

The easiest to handle is f_3, since $f(z)$ is bounded on C_3 (outer large circle). By Cauchy's bound

$$f_3 = O((1+\delta)^{-n}).$$

On C_1 (inner small circle), we have $|f(z)| = o(n^{-1/2})$ and the length of the circle is $2\pi/n$, so that

$$f_1 = o(n^{-3/2})$$

as $n \to \infty$. Finally, we are left with C_2 and C_4, which can be handled in a similar manner. Let us consider C_2. Observe that for $z = 1 + re^{i\pi/4}$

$$|z^{-n}| = |1 + r\cos \pi/4 + ir\sin \pi/4|^{-n}$$
$$\le (1+r)^{-n/2} \le \exp(-nr/10)$$

for $0 \le r < 1$. Since $f(z) = o(\sqrt{1-z})$, for any $\varepsilon > 0$ we have

$$|f(1 + re^{i\pi/4})| \le \varepsilon\sqrt{r}.$$

This yields, after the change $nr = z$,

$$|f_2| \le \varepsilon \int_0^\delta \sqrt{r}\exp(-nr/10)dr \le \varepsilon n^{-3/2}\int_0^\infty \sqrt{z}\exp(-z/10)dz = o(n^{-3/2})$$

as desired. In a similar manner we show that $f_4 = o(n^{-3/2})$. Thus we prove that if $f(z)$ is expendable to the Δ-domain, then $f_n := [z^n]f(z) = o(n^{-3/2})$. We should observe that the critical derivation was the one on the rectilinear curves C_2 and C_4.

The method just described extends to big-oh notation as long as one can analytically continue a function to the $\Delta(R, \phi)$ domain. For example, let $f(z) = O((1-z)^{-\alpha})$. Considering the contour C as above, and after bounding the integral on C_1 and C_3 through Cauchy's bound, we are left with estimating on C_2

$$|f_2| \le \frac{1}{2\pi n}\int_1^\infty K\left(\frac{t}{n}\right)^{-\alpha}|1 + e^{i\phi}t/n|^{-n-1}dt,$$

where K is the constant from the big-oh notation, that is, such that $|f(z)| \le K(1-z)^{-\alpha}$. Since $|1 + e^{i\phi}t/n| \ge 1 + \Re(e^{i\phi}t/n) = 1 + n^{-1}t\cos\phi$ we finally obtain

$$|f_2| \leq \frac{K}{2\pi} J_n n^{\alpha-1},$$

where $J_n = \int_1^\infty t^{-\alpha}(1 + t\frac{1}{n}\cos\phi)^{-n} dt$ is bounded for $0 < \phi < \pi/2$. We just established one of the transfer theorems of Flajolet and Odlyzko.

In summary, the singularity analysis of Flajolet and Odlyzko is based on the following transfer theorems. The proof follows in the footsteps of the above derivations, and details can be found in [140].

Theorem 8.13 (Flajolet and Odlyzko, 1990)   *Let $A(z)$ be Δ-analytical, where $\Delta(R, \phi)$ is defined (8.72). If $A(z)$ satisfies in $\Delta(R, \pi)$ either*

$$A(z) = O\left((1-z)^{-\alpha}\log^\beta(1-z)^{-1}\right)$$

or

$$A(z) = o\left((1-z)^{-\alpha}\log^\beta(1-z)^{-1}\right),$$

then either

$$[z^n]A(z) = O\left(n^{\alpha-1}\log^\beta n\right)$$

or

$$[z^n]A(z) = o\left(n^{\alpha-1}\log^\beta n\right),$$

respectively.

A classical example of singularity analysis is the Flajolet and Odlyzko [139] analysis of the height of binary trees. However, we finish this subsection with a simpler application that illustrates the theory quite well.

Example 8.10:   *Certain Sums from Coding Theory*   In coding theory the following sum is of interest (cf. [255]):

$$S_n = \sum_{i=0}^n \binom{n}{i}\left(\frac{i}{n}\right)^i\left(1-\frac{i}{n}\right)^{n-i}.$$

Let $s_n = n^n S_n$ and $S(z)$ be the exponential generating function of s_n. In addition, let

$$B(z) = \frac{1}{1 - T(z)},$$

where $T(z)$ is the "tree function" defined in Example 7.15 (Chapter 7) (cf. (7.31) and (7.32)). Since $[z^n]T(z) = n^{n-1}/n!$, hence $[z^n]B(z) = n^n/n!$. Thus by (7.17) of Table 7.3 we find that

$$S(z) = (B(z))^2.$$

From Example 7.15 we also know that $T(z)$ is singular at $z = e^{-1}$. Using MAPLE we find (see also [73] for a more detailed analysis)

$$T(z) - 1 = \sqrt{2(1 - ez)} + \frac{2}{3}(1 - ez) + \frac{11\sqrt{2}}{36}(1 - ez)^{3/2}$$

$$+ \frac{43}{135}(1 - ez)^2 + O((1 - ez)^{5/2})$$

in a Δ domain around $z = e^{-1}$. Indeed, expanding $z = ye^{-y}$ in a power series around $z = e^{-1}$ we have

$$z = e^{-1} - \frac{e^{-1}}{2}(y - 1)^2 + \frac{e^{-1}}{3}(y - 1)^3 + O((y - 1)^4),$$

and after solving it for y we obtain the above expansion for $T(z)$. Now let $h(z) = \sqrt{1 - ez}$. Then

$$S(z) = \frac{1}{2h(z)\left(1 + \frac{\sqrt{2}}{3}\sqrt{h(z)} + \frac{11}{36}h(z) + O(h^{3/2}(z))\right)^2}$$

$$= \frac{1}{2(1 - ez)} + \frac{\sqrt{2}}{3\sqrt{(1 - ez)}} + \frac{1}{36} + \frac{\sqrt{2}}{540}\sqrt{1 - ez} + O(1 - ez).$$

An application of the singularity analysis (i.e., Theorems 8.12 and 8.13) yields

$$S_n = \sqrt{\frac{n\pi}{2}} + \frac{2}{3} + \frac{\sqrt{2\pi}}{24}\frac{1}{\sqrt{n}} - \frac{4}{135}\frac{1}{n} + O(1/n^{3/2}),$$

as shown in [413], from where this example is taken. We shall discuss another example in the applications Section 8.7.2. ∎

What if $A(z)$ cannot be analytically continued to a Δ domain? We still have some options. One is real analysis, another is the Tauberian theorem. However, the most successful seems to be the Darboux method that we briefly discuss next.

We start with a quick review of the *Riemann-Lebesgue lemma*, which is often useful, as an intermediate result, in establishing asymptotics.

Lemma 8.14   *Let $f(t)$ be continuous in $(0, \infty)$. Then*

$$\lim_{x \to \infty} \int_0^\infty f(t) e^{ixt} dt = 0 \tag{8.73}$$

provided the integral converges uniformly at 0 and ∞ for all sufficiently large x.

The proof of the Riemann-Lebesgue lemma can be found in [331, 450]. We just observe that it follows from a few simple facts. First of all, by the hypothesis, there are positive a and b such that

$$\left| \int_0^a f(t) e^{ixt} dt \right| \le \frac{\varepsilon}{3}, \quad \left| \int_b^\infty f(t) e^{ixt} dt \right| \le \frac{\varepsilon}{3}$$

for any $\varepsilon > 0$. In addition, since $f(t)$ is continuous, it is bounded in $[a, b]$. Approximating f by its value at a partition $a = t_0 < t_1 < \ldots < t_n = b$ (e.g., $|f(t) - f(t_j)| \le \varepsilon/(6(b-a))$) we arrive at

$$\int_a^b f(t) e^{ixt} dt = \sum_{j=0}^{n-1} \int_{t_j}^{t_{j+1}} |f(t) - f(t_j)| e^{ixt} dt + \sum_{j=0}^{n-1} f(t_j) \int_{t_j}^{t_{j+1}} e^{itx} dt.$$

But the first sum is bounded by, say $\varepsilon/6$, while the second is bounded by $2n \max\{f\}/x < \varepsilon/6$, hence can be made as small as desired for large x, which completes the sketch of the proof.

Now we present the Darboux method. In this case, we assume only that $A(z)$ is smooth enough on the disk of convergence.

Theorem 8.15   (Darboux, 1878)   *Assume that $A(z)$ is continuous in the closed disk $|z| \le 1$, and in addition k times continuously differentiable on $|z| = 1$. Then*

$$[z^n] A(z) = o(n^{-k}). \tag{8.74}$$

Proof. By the continuity assumption on $|z| = 1$ we have

$$a_n = [z^n] A(z) = \frac{1}{2\pi i} \int_{|z|=1} A(z) \frac{dz}{z^{n+1}}$$

$$\overset{z = e^{i\theta}}{=} \frac{1}{2\pi} \int_0^{2\pi} A(e^{i\theta}) e^{-ni\theta} d\theta.$$

By the Riemann-Lebesgue lemma we must conclude that $a_n = o(1)$ as $n \to \infty$. This proves the result for $k = 0$. To get a stronger result we integrate the above by parts k times to see that

$$[z^n]A(z) = \frac{1}{2\pi (in)^k} \int_0^{2\pi} A^{(k)}(e^{i\theta})e^{-ni\theta}d\theta,$$

where $A^{(k)}$ is the kth derivative of A. Again, the proof follows from the Riemann-Lebesgue lemma. ∎

As an example, consider $A(z) = v(z)/\sqrt{1-z}$, where $v(z)$ is analytic at $z = 1$. Then

$$A(z) = \frac{v(1)}{\sqrt{1-z}} - v'(1)\sqrt{1-z} + O((1-z)^{3/2}).$$

Since $(1-z)^{3/2}$ is once continuously differentiable on $|z| \leq 1$, we can apply Darboux's Theorem with $k = 1$ to show that $[z^n]O((1-z)^{3/2}) = o(1/n)$.

Finally, we should mention that so far we have dealt only with a single singularity. Multiple singularities can be handled in a similar fashion, however, some complications arise. The reader is referred to Flajolet and Sedgewick [149] for details.

8.4 LARGE SINGULARITIES: SADDLE POINT METHOD

This section presents asymptotic methods for generating functions whose dominant singularities are large. We recall that, roughly speaking, a singularity is large if the growth of the function near such a singularity is rapid, say, larger than any polynomial. We shall see that methods of the previous section that work fine for small singularities are useless for large singularities. To focus, we concentrate on asymptotics of the Cauchy integral and hence coefficients of some generating functions. In short, we shall investigate integrals like

$$I(\lambda) = \int_C f(z)e^{-\lambda h(z)}dz, \qquad (8.75)$$

where $f(z)$ and $h(z)$ are analytic functions, and λ is a large positive parameter (often we assume that $\lambda = n$ with n being a natural number and the coefficients of f and h are nonnegative).

We start with the Laplace method which assumes that the path C in (8.75) is an interval $[a, b]$, and $f(z)$ and $h(z)$ are real-valued functions. Then we discuss the *saddle point method* (also known as the *steepest descent method*) which by far is the most useful and popular technique to extract asymptotic information of rapidly growing functions. It is one of the most complicated methods, therefore, given the purpose and limitations, we do not present a full discussion of it. The reader is referred to De Bruijn [84], Olver [331], and Wong [450] for a complete and insightful presentations. However, we describe here a very useful (and largely forgotten) method of van der Waerden [443], and a well-known method of Hayman [192]. We complete this section with an application of the saddle point method to large deviations. The reader is also referred to an excellent survey by Odlyzko [330] and the book by Flajolet and Sedgewick [149].

8.4.1 Laplace Method

Let us consider a special case of (8.75), namely

$$I(x) = \int_a^b f(t)e^{-xh(t)}dt, \tag{8.76}$$

where $-\infty \leq a < b \leq \infty$ and x is a large positive parameter. Furthermore, let the minimum of $h(x)$ be attained at the point $a < t_0 < b$. Assume that $f(t_0) \neq 0$ and $h''(t_0) > 0$. Roughly speaking, as $x \to \infty$

$$f(t) \exp\left((-x(h(t) - h(t_0)))\right) \to f(t_0)\delta(t - t_0), \tag{8.77}$$

where $\delta(t)$ is the Dirac-delta function taking the value ∞ at $t = 0$ and zero otherwise. More precisely, we approximate $I(x)$ by

$$I(x) \sim \int_a^b f(t_0) \exp\left(-x(h(t_0) + 1/2(t - t_0)^2 h''(t_0))\right) dt$$

$$\sim f(t_0)e^{-xh(t_0)} \int_{-\infty}^{\infty} \exp\left(-1/2x(t - t_0)^2 h''(t_0)\right) dt$$

$$= f(t_0)e^{-xh(t_0)} \sqrt{\frac{2\pi}{xh''(t_0)}}.$$

The second line is a consequence of (8.77) while the last line follows from the Gauss integral that we recall in the next lemma.

Lemma 8.16   (Gauss Integral)   *The following identities hold:*

$$\frac{1}{\sqrt{2\pi}} \int_{-\infty}^{\infty} x^k e^{-tx^2} dx = \begin{cases} 0 & k = 0, 1, 3, 5, \ldots \\ \frac{t^{-1/2-k/2}k!}{(k/2)!2^{k+1/2}} & k = 2, 4, 6, \ldots \end{cases} \tag{8.78}$$

and for fixed k and α > 0

$$\int_{\theta}^{\infty} x^k e^{-\alpha x^2} dx = O\left(e^{-\alpha \theta^2}\right) \tag{8.79}$$

as $\theta \to \infty$.

The first identity above is nothing but moments of the Gaussian distribution. For the latter observe that

$$\int_{\theta}^{\infty} x^k e^{-\alpha x^2} dx = e^{-\alpha \theta^2} \int_{\theta}^{\infty} x^k e^{-\alpha (x^2 - \theta^2)} dx = O\left(e^{-\alpha \theta^2}\right)$$

since the integral converges for $\alpha > 0$.

On the other hand, if the minimum of $h(x)$ is achieved at the end of the interval, say at $t = a$, then for $f(a) \neq 0$ we obtain

$$I(x) \sim \int_a^b f(a) \exp\left(-x(h(a) + (t-a)h'(a))\right) dt$$

$$\sim f(a) e^{-xh(a)} \int_a^b \exp\left(-x(t-a)h'(a)\right) dt$$

$$= \frac{f(a) e^{-xh(a)}}{xh'(a)}. \tag{8.80}$$

Finally, before we provide a rigorous proof of Laplace's method, let us say a word about the **principle of stationary phase** that applies to integrals of the form

$$F(x) = \int_a^b f(t) e^{ixh(t)} dt,$$

where x is assumed to be large and positive. The underlying principle, due to Lord Kelvin, is the assertion that the major asymptotic contribution comes from points where the *phase of* $h(t)$ *is stationary*, that is, where $h'(t)$ vanishes. We should add that oscillations near such points slow down. Heuristically, if $t_0 \in (a, b)$ is such that $h'(t_0) = 0$, then

$$F(x) \sim \int_{t_0-\varepsilon}^{t_0+\varepsilon} f(t) e^{ixh(t)} dt, \quad \varepsilon > 0,$$

$$\sim f(t_0) e^{ixh(t_0)} \int_{-\infty}^{\infty} e^{ixh''(t_0)(t-t_0)^2/2} dt$$

$$= f(t_0) \sqrt{\frac{2\pi}{x|h''(t_0)|}} e^{i(xh(t_0) + \frac{\pi}{4} \operatorname{sgn} h''(t_0))},$$

where $\text{sgn}(x) = 1$ for $x > 0$ and $\text{sgn}(x) = -1$ for $x < 0$. The last equation above follows from the Fresnel integral (cf. [2])

$$\int_{-\infty}^{\infty} e^{1/2it^2} dt = \sqrt{2\pi} e^{i\pi/4}.$$

As mentioned before it is not difficult to make the above analyses completely rigorous. We now focus on the Laplace method and prove the following theorem.

Theorem 8.17 (Laplace's Method)   *Let*

$$I(x) = \int_a^b f(t) e^{-xh(t)} dt,$$

where $-\infty \leq a < b \leq \infty$ and x is a large positive parameter. Assume that

    (i) *$h(t)$ has only one minimum inside (a, b) at point $t = t_0$, that is, $h'(t_0) = 0$ and $h''(t_0) > 0$.*

    (ii) *$h(t)$ and $f(t)$ are continuously differentiable in a neighborhood of t_0 with $f(t_0) \neq 0$ and*

$$h(t) = h(t_0) + 1/2h''(t_0)(t - t_0)^2 + O((t - t_0)^3);$$

    (iii) *The integral $I(x)$ exists for sufficiently large x, say for $x \geq \xi$.*

Then

$$\int_a^b f(t) e^{-xh(t)} dt = f(t_0) \sqrt{\frac{2\pi}{xh''(t_0)}} e^{-xh(t_0)} \left(1 + O\left(\frac{1}{\sqrt{x}}\right)\right) \qquad (8.81)$$

for $x \to \infty$.

Proof. We adopt here the proof from Lauwerier [283]. Without loss of generating we take $t_0 = 0$ (hence $a < 0$ and $b > 0$) and $h(t_0) = 0$ (otherwise substitute $t' = t - t_0$). We first assume $f(t) \equiv 1$ to succinctly spell out the main idea. From (ii) we conclude that there exist constants c and δ such that

$$|h(t) - 1/2h''(0)t^2| \leq c|t|^3, \qquad |t| < \delta. \qquad (8.82)$$

We split the integral into two parts, namely,

$$I(x) = \int_{-\delta}^{\delta} e^{-xh(t)} dt + \int_{t \notin [-\delta, \delta]} e^{-xh(t)} dt. \qquad (8.83)$$

The latter integral can be evaluated as follows: By (i), $\mu(\delta) = \inf_{|t|>\delta} h(t) > 0$ and hence

$$\left| \int_{t \notin [-\delta,\delta]} e^{-xh(t)} dt \right| \leq \int_{t \notin [-\delta,\delta]} e^{-\xi h(t) - (x-\xi)h(t)} dt$$

$$\leq e^{-(x-\xi)\mu(\delta)} \int_{t \notin [-\delta,\delta]} e^{-\xi h(t)} dt = O(e^{-x\beta})$$

for some $\beta > 0$, where the last assertion follows from (iii).

We now deal with the first integral of (8.83). Let $\omega = h''(t_0)$. Then

$$I_1(x) = \int_{-\delta}^{\delta} e^{-xh(t)} dt$$

$$= \int_{-\delta}^{\delta} e^{-1/2x\omega t^2} dt + \int_{-\delta}^{\delta} e^{-1/2x\omega t^2} \left[\exp\left(-x(h(t) - 1/2\omega t^2) \right) - 1 \right] dt.$$

From Lemma 8.16 we estimate the first integral as

$$\int_{-\delta}^{\delta} e^{-1/2x\omega t^2} dt = \sqrt{\frac{2\pi}{h''(0)x}} + O(e^{-x\delta^2}).$$

With the help of

$$|e^t - 1| < |t| e^{|t|}$$

and using (8.82) several times, we evaluate the second integral in $I_1(x)$ as

$$\left| \int_{-\delta}^{\delta} e^{-1/2x\omega t^2} \left[\exp\left(-x(h(t) - 1/2\omega t^2) \right) - 1 \right] dt \right|$$

$$\leq 2cx \int_{0}^{\delta} t^3 \exp\left(-\left(1/2\omega x t^2 - cxt^3 \right) \right) dt$$

$$\leq 2cx \int_{0}^{\delta} t^3 \exp\left(-xt^2 (1/2\omega - \delta c) \right) dt$$

$$\leq 2cx \int_{0}^{\delta} t^3 e^{-xbt^2} dt \leq \frac{2c}{b^2 x} \int_{0}^{\infty} u^3 e^{-u^2} du = O(x^{-1}),$$

where $b = 1/2\omega - \delta c > 0$ for sufficiently small δ. Combining everything we obtain (8.81) for $f(t) \equiv 1$. To extend to general f, we just follow the steps of the above proof using $f(t) = f(t_0) + O(t - t_0)$. This completes the proof. ∎

Laplace's method has many applications in the analysis of algorithms. We illustrate it in the evaluation of a certain integral arising from the analysis of b-tries with large b.

Example 8.11: *Height in b-Tries for Large b* In Example 8.5 we studied the height of a b-tries. In particular, the height distribution $h_n^k = \Pr\{H_n \le k\}$ is given by (8.43), where the following truncated exponential is involved

$$1 + z2^{-k} + \cdots + \frac{z^b 2^{-kb}}{b!} = e^{z2^{-k}} \int_{z2^{-k}}^{\infty} e^{-w} \frac{w^b}{b!} dw$$

$$= e^{z2^{-k}} \left[1 - \int_0^{z2^{-k}} e^{-w} \frac{w^b}{b!} dw \right].$$

We evaluate the asymptotic behavior of the above integral, and this is summarized in the lemma below.

Lemma 8.18 *We let*

$$I = I(A, b) = \frac{1}{b!} \int_0^A e^{b \log w - w} dw = \frac{e^{-b} b^{b+1}}{b!} \int_0^{A/b} e^{b(\log u - u + 1)} du.$$

Then the asymptotic expansions of I are as follows:

(i) $b, A \to \infty$, $\alpha = b/A > 1$

$$I = e^{-A} \frac{A^b}{b!} \left[\frac{1}{b/A - 1} - \frac{b}{A^2} \frac{1}{(b/A - 1)^3} + O(A^{-2}) \right].$$

(ii) $b, A \to \infty$, $b/A < 1$

$$I = 1 - e^{-A} \frac{A^b}{b!} \left[\frac{1}{1 - b/A} - \frac{b}{A^2} \frac{1}{(1 - b/A)^3} + O(A^{-2}) \right].$$

(iii) $b, A \to \infty$, $A - b = \sqrt{b}B$, $B = O(1)$

$$I = \frac{1}{\sqrt{2\pi}} \left(\int_{-\infty}^B e^{-x^2/2} dx - \frac{1}{3\sqrt{b}} (B^2 + 2) e^{-B^2/2} \right.$$

$$\left. + \frac{1}{b} \left(-\frac{B^5}{18} - \frac{B^3}{36} - \frac{B}{12} \right) e^{-B^2/2} + O(b^{-3/2}) \right).$$

Proof. To establish this result we note that I is a Laplace-type integral (8.81). Setting $h(u) = \log u - u + 1$ we see that $h(u)$ is maximal at $u = 1$. For $A/b < 1$ we have $h'(u) > 0$ for $0 < u \leq A/b$ and thus the major contribution to the integral comes from the upper endpoint (more precisely, from $u = A/b - O(b^{-1})$). Then (8.80) yields part (i) of Lemma 8.18. If $A/b > 1$ we write $\int_0^{A/b}(\cdots) = \int_0^\infty(\cdots) - \int_{A/b}^\infty(\cdots)$, evaluate the first integral exactly, and use Laplace's method on the second integral. Now $h'(u) < 0$ for $u \geq A/b$ and the major contribution to the second integral is from the lower endpoint. Obtaining the leading two terms leads to (ii) in Lemma 8.18.

To derive part (iii), we scale $A - b = \sqrt{b}B$ to see that the main contribution will come from $u - 1 = O(b^{-1/2})$. We thus set $u = 1 + x/\sqrt{b}$ and obtain

$$
I = \frac{e^{-b}b^{b+1}}{b!} \int_{-\sqrt{b}}^{B} \exp\left(b\left[\log\left(1 + \frac{x}{\sqrt{b}} \right) - \frac{x}{\sqrt{b}} \right] \right) \frac{dx}{\sqrt{b}} \tag{8.84}
$$

$$
= \frac{b^b \sqrt{b} e^{-b}}{b!} \int_{-\infty}^{B} e^{-x^2/2} \left[1 + \frac{x^3}{3\sqrt{b}} + \frac{1}{b}\left(-\frac{x^4}{4} + \frac{x^6}{18} \right) + O(b^{-3/2}) \right] dx.
$$

Evaluating explicitly the above integrals and using Stirling's formula in the form $b! = \sqrt{2\pi b}\, b^b e^{-b}(1 + (12b)^{-1} + O(b^{-2}))$, we obtain part (iii) of Lemma 8.18. This example is adopted from Knessl and Szpankowski [261]. ∎

8.4.2 Method of Steepest Descent

Here again we shall analyze (8.75), that is,

$$
I(n) = \int_C f(z)e^{-nh(z)}dz \tag{8.85}
$$

for n large, where $f(z)$ and $h(z)$ are *analytic* functions and C is a path in a complex plane. We deform C into another contour C' such that the region between C and C' does not contain singularities of f or h. The idea is to find such a path C' that transforms the complex integral $I(n)$ into the real case considered in the previous section, that is, to reduce $I(n)$ to the one manageable by the Laplace method. In general, this is not an easy task and it requires quite a bit of experience. Given the length of this chapter, we restrict ourselves to a short account of the general method, explaining the underlying principle, and later focus only on Cauchy's coefficient formula, which is easier to handle than the general case.

To explain the method in somewhat more detail, we start with the general case. We assume that $f(z) \equiv 1$, and $h(z) = u(x, y) + iv(x, y)$ with $z = x + iy$. Our goal is to find such a path C' that

(i) the imaginary part of $h(z)$ on C' is constant.

This condition allows us to reduce the complex integral to the real case that we know how to handle. Indeed, assuming $z(\tau)$ is a parametric description of the path C', and z_0 is a fixed point on the path; hence, $v(x, y) = \text{constant} = \Im h(z_0)$, the integral $I(n)$ becomes

$$I(n) = e^{-nh(z_0)} \int_a^b \exp\left[-n(h(z(\tau)) - h(z_0))\right] \frac{dz(\tau)}{d\tau} d\tau$$

$$= e^{-nh(z_0)} \int_a^b \exp\left[-n\Re(h(z(\tau) - h(z_0)))\right] \frac{dz(\tau)}{d\tau} d\tau$$

for some $a \leq \tau \leq b$. Clearly, the latter integral can be treated by the Laplace method. In particular, we expect the main contribution to come from a small region around a point z_0 that minimizes the exponent, that is, $u_x(x_0, y_0) = u_y(x_0, y_0) = 0$. But by the Cauchy-Riemann equations (i.e., $u_x = v_y$ and $u_y = -v_x$; cf. [195]) at point z_0 also $h'(z_0) = u_x(x_0, y_0) - iu_y(x_0, y_0) = 0$. The point z_0 such that $h'(z_0) = 0$ is called the **saddle point** of $h(z)$. In view of this, we add one more condition to (i), namely,

(ii) the path C' passes through the saddle point(s) $h'(z_0) = 0$,

which reduces (8.85) to the Laplace integral. This constitutes the backbone of the saddle point method.

We now work out more details concerning the curve C', which is always the hardest part of the method. Let us try to understand condition (i) and its relationship to condition (ii). The Laplace method will work only if on the curve C' near the saddle point the value of h drops significantly. But this is the case, and here is why. Let us consider $u(x, y)$ as a function of (x, y). Notice that $|e^{-nh(z)}| = e^{-nu(x,y)}$ so the absolute value of the integral $I(n)$ depends on the behavior of $u(x, y)$ along C'. Let now $x(t)$, $y(t)$ $u(t)$ be a parametric description of a curve on the surface $u(x, y)$. It is known that (details can be found in Wong [450])

$$\frac{du}{dt} = u_x \cos\theta + u_y \sin\theta$$

for some θ. To measure the steepness on the surface $u(x, y)$ along the path $u(t)$, one computes the cosine of an angle between the path and the u-axis. It can be shown that the derivative of this cosine is zero (i.e., maximized or minimized) when

$$-u_x \sin\theta + u_y \cos\theta = 0.$$

But then, again by the Cauchy-Riemann equation, this implies

$$\frac{dv}{dt} = v_x \cos\theta + v_y \sin\theta = 0$$

on the path. Thus the steepest curve (where the derivative of $\cos\alpha$ is zero) is also where the imaginary part of $h(z)$ is constant (i.e., $\Im(h(z)) = v(s)$ =constant). In other words, the path C' postulated in (i) is the steepest curve on $u(x, y)$. Simultaneously, the curve C' on $u(x, y)$ reaches the maximum/minimum value at the saddle point z_0, where $u_x = u_y = 0$. Thus "luckily" the second condition (ii) is also satisfied, and one expects most of the contribution to the integral to be coming from a small region around the saddle point z_0 if C' is the steepest descent curve.

Example 8.12: *Steepest Descent Method*   Let

$$I(n) = \int_{-\infty}^{\infty} e^{-n(z^2 - 2zi)]} f(z)dz.$$

The saddle point is at $z_0 = i$ and $\Im(z^2 - 2zi) = 2x(y-1)$ =constant represents the steepest descent curves, where $z = x + iy$. In particular, $y = 1$ is the steepest descent curve that passes through the saddle point. Accordingly, we deform the path $(-\infty, \infty)$ to the line $\Im z = 1$, reducing the integral to

$$I(n) = e^{-n} \int_{-\infty}^{\infty} e^{-nt^2} f(t+i)dt$$

provided $f(z)$ is well-behaved (e.g., no singularities in the strip $0 \le \Im z \le 1$). Then by Theorem 8.17 we directly find that

$$I(n) \sim e^{-n} f(i)\sqrt{\frac{\pi}{n}}.$$

If, however, $f(z)$ has a singularity in $0 \le \Im z \le 1$, we must add this contribution. For example, assuming that

$$f(z) = \frac{1}{z - p} + \cdots,$$

where $p \ne i$, we must add $2\pi i \exp\left(-n(p^2 - 2pi)\right)$. The situation is even more interesting when $p = i$ (i.e., the pole coincides with the saddle point). We shall discuss this case later in this chapter (in particular, in Example 8.22).   ∎

Hereafter, we concentrate on the Cauchy integral and study asymptotics of

$$a_n := [z^n]A(z) = \frac{1}{2\pi i} \oint \frac{A(z)}{z^{n+1}} dz = \frac{1}{2\pi i} \oint e^{h(z)} dz, \qquad (8.86)$$

where

$$h(z) = \log A(z) - (n+1)\log z,$$

and $A(z)$ is a generating function of rapid growth with positive coefficients. Since in this case, $|A(z)| \leq A(|z|)$ we have

$$\max_{|z|=r} |A(z)| = A(r) \tag{8.87}$$

for some $r < R$, where it is assumed that A is analytic for $|z| < R \leq \infty$ (and often $R = \infty$). The saddle point z_0 of (8.86) is located at $h'(z) = 0$, that is

$$\frac{A'(z_0)}{A(z_0)} = \frac{n+1}{z_0}. \tag{8.88}$$

In view of (8.87), we may also compute the saddle point by observing that it must minimize $r^{-n}A(r)$.

We summarize what we learned so far in the next example, where we also work out all details of the method that we later generalize to establish the Hayman result.

Example 8.13: *Asymptotic of* $\exp(z^\alpha)$ Let us assume that $A(z) = \exp(z^\alpha)$ with $\alpha > 0$. In particular, for $\alpha = 1$ we have $[z^n]A(z) = 1/n!$, so we can verify our analysis. Since $h(z) = z^\alpha - (n+1)\log z$, the saddle point is at

$$z_0 = \left(\frac{n+1}{\alpha}\right)^{1/\alpha} \sim \left(\frac{n}{\alpha}\right)^{1/\alpha}.$$

Thus writing $n_1 = n/\alpha$ we have

$$[z^n]e^{z^\alpha} = \frac{1}{2\pi i} \int_{|z|=n_1} \frac{e^{z^\alpha}}{z^{n+1}} dz$$

$$= \frac{n_1^{-n_1}}{2\pi} \int_{-\pi}^{\pi} \exp\left(n_1 e^{i\alpha\theta} - ni\theta\right) d\theta,$$

where we substituted $z = z_0 e^{i\theta}$. We now choose such $\theta_0 \in (0, \pi)$ that for $-\theta_0 \leq \theta \leq \theta_0$ the following expansion is valid

$$e^{i\alpha\theta} = 1 + i\alpha\theta - 1/2\alpha^2\theta^2 + O(\theta^3). \tag{8.89}$$

For the above to hold, we need ensure that $\theta_0 = o(1)$. For such θ_0, we split the integral into three parts as follows:

$$I_1 = \frac{n_1^{-n_1}}{2\pi} \int_{-\theta_0}^{\theta_0} \exp\left(n_1 - 1/2n\alpha\theta^2 + O(n\theta^3)\right) d\theta$$

$$I_2 = \frac{n_1^{-n_1}}{2\pi} \int_{\theta_0}^{\pi} \exp\left(n_1 e^{i\alpha\theta} - ni\theta\right) d\theta$$

$$I_3 = \frac{n_1^{-n_1}}{2\pi} \int_{-\pi}^{-\theta_0} \exp\left(n_1 e^{i\alpha\theta} - ni\theta\right) d\theta,$$

where certainly I_2 and I_3 are of the same order, so we only estimate I_2.

We first evaluate I_1. Our goal is to reduce it to the Gauss integral, so we shall require that

$$n\theta_0^2 \to \infty, \tag{8.90}$$

$$n\theta_0^3 \to 0, \tag{8.91}$$

and, of course, $\theta_0 \to 0$ to assure validity of (8.89). A good choice is $\theta_0 = n^{-2/5}$. Then $\exp\left(n_1 - n\theta^2/2 + O(n\theta^3)\right) = (1 + O(n^{-1/5})) \exp\left(n_1 - n\theta^2/2\right)$ and this yields

$$I_1 = \left(1 + O(n^{-1/5})\right) \left(\frac{n}{\alpha}\right)^{-n/\alpha} e^{n/\alpha} \left(\int_{-\infty}^{\infty} e^{-1/2\alpha n\theta^2} d\theta - 2\int_{\theta_0}^{\infty} e^{-1/2\alpha n\theta^2} d\theta\right)$$

$$= \left(1 + O(n^{1/5})\right) \left(\frac{n}{\alpha}\right)^{-n/\alpha} e^{n/\alpha} \left(\sqrt{\frac{2\pi}{\alpha n}} - O(e^{-\alpha n^{1/5}/2})\right)$$

$$= \left(1 + O(n^{-1/5})\right) \left(\frac{n}{\alpha}\right)^{-n/\alpha} e^{n/\alpha} \sqrt{\frac{2\pi}{\alpha n}}.$$

Now we estimate I_2. For this, we observe $|\exp(z^\alpha)| = \exp(\Re(z^\alpha))$, hence

$$|I_2| \leq \frac{n_1^{-n_1}}{2\pi} \int_{\theta_0}^{\pi} e^{n_1 \cos\alpha\theta} d\theta.$$

We want to show that I_2 contributes negligibly. Since $\theta_0 \leq |\theta| \leq \pi$ we have

$$\cos\alpha\theta \leq \cos\alpha\theta_0 = 1 - \alpha^2\theta_0^2/2 + O(\theta_0^4)$$

provided $\alpha\theta \leq 2\pi - \theta_0$ since otherwise the above inequality would not hold. To assure this condition will hold, we now restrict $\alpha \leq 1$ (later we shall discuss the case $\alpha > 1$). Then we can proceed with our calculation to obtain

$$|I_2| \le \frac{1}{2\pi} \le \left(\frac{n}{\alpha}\right)^{-n/\alpha} e^{n/\alpha} e^{-\alpha n^{1/5}/3} = O(I_1 e^{-\alpha n^{1/5}/3}), \quad \alpha \le 1,$$

since $n\alpha^{-1} \cos \alpha\theta \le n\alpha^{-1} - \alpha n^{1/5}/2 + O(n^{-3/5})$. This finally proves that

$$[z^n]e^{z^\alpha} = \left(1 + O(n^{-1/5})\right) \left(\frac{n}{\alpha}\right)^{-n/\alpha} e^{n/\alpha} \sqrt{\frac{2\pi}{\alpha n}}$$

for $\alpha \le 1$. When $\alpha = 1$ we obtain Stirling's approximation, as expected.   ∎

The above example sheds some light on some complications that may arise when the saddle point method is used but we do not use the steepest descent contour through the saddle. The reader should notice that in the above we were unable to show that I_2 is small for $\alpha > 1$. A question arises whether we can overcome this difficulty by applying a more sophisticated saddle point approximation. It turns out that we can, and let us illustrate it for $\alpha = 2$. Then

$$e^{z^2} = \sum_{n=0}^{\infty} \frac{z^{2n}}{n!}$$

so that

$$[z^n]e^{z^2} = \left\{ \begin{array}{ll} \frac{1}{k!} & n = 2k \\ 0 & \text{otherwise} \end{array} \right.$$

or simply

$$[z^n]e^{z^2} = \frac{1 + (-1)^n}{2} \frac{1}{(n/2)!}.$$

In other words, the coefficients $a_n = [z^n]e^{z^2}$ oscillate. Why then could we not handle it by the saddle point method? The culprit lies in the *multiplicity* of the saddle points. For $\alpha = 2$ we have two saddle points, at $z_0 = (n/\alpha)^{1/\alpha}$ and at $-z_0$. Multiple saddle points almost always lead to some periodicity in the asymptotics coefficients, and one must be more careful with the selection of the steepest descent contour. We shall not discuss it in this book; the reader is referred to De Bruijn [84] or Wong [450] for a detailed discussion.

Let us now generalize Example 8.13. Consider

$$a_n = \frac{1}{2\pi i} \int_{|z|=r} \frac{A(z)}{z^{n+1}} dz = \frac{1}{2\pi r^n} \int_{-\pi}^{\pi} A(re^{i\theta}) e^{-in\theta} d\theta$$

$$= \frac{1}{2\pi r^n} \int_{-\pi}^{\pi} \exp\left(\log A(re^{i\theta}) - in\theta\right) d\theta. \tag{8.92}$$

Formally,

$$\log A(re^{i\theta}) = \log A(r) + \sum_{k=1}^{\infty} \alpha_k(r) \frac{(i\theta)^k}{k!}, \tag{8.93}$$

where

$$\alpha_1(r) := a(r) = r\frac{d}{dr} \log A(r) = r\frac{A'(r)}{A(r)}, \tag{8.94}$$

$$\alpha_2(r) := b(r) = \left(r\frac{d}{dr}\right)^2 \log A(r) = ra'(r), \tag{8.95}$$

$$\alpha_k(r) := k\alpha_{k-1}(r). \tag{8.96}$$

Let us continue our formal derivation and assume that the whole contribution to a_n concentrates near one saddle point, say r_0, determined by

$$\frac{d}{d\theta}\left(\log A(re^{i\theta}) - in\theta\right) = 0,$$

that is,

$$a(r_0) = n. \tag{8.97}$$

Then, after an application of the Gauss integral (8.78), we formally obtain the **saddle point approximation**

$$a_n \sim \frac{A(r_0)}{2\pi r_0^n} \int_{-\theta_0}^{\theta_0} e^{-b(r_0)\theta^2/2} d\theta \sim \frac{A(r_0)}{r_0^n \sqrt{2\pi b(r_0)}} \tag{8.98}$$

provided θ_0 is appropriately chosen and $b(r_0) \neq 0$. The main ingredients of the saddle point approximation are summarized in Table 8.4. We shall justify them below.

However, one must be very careful with a blind application of this heuristic. We have already seen when the saddle point method breaks down due to the impossibility of establishing smallness of the integral for $\theta \in (-\pi, -\theta_0) \cup (\theta_0, \pi)$. Now we discuss an example when the main contribution from $\theta \in (-\theta_0, \theta_0)$ does not give the right asymptotic approximation and the above heuristic again breaks down.

Example 8.14: *Invalid Application of the Saddle Point Method* Let us consider $[z^n](1 - z)^{-\alpha}$ for α not being a negative integer. We "blindly" apply the saddle point approximation. We find that $r_0 = n/(n+\alpha)$ and $b(r_0) = n(n+\alpha)/\alpha$.

TABLE 8.4. Summary of the Saddle Point Approximation

Input: A function $g(z)$ analytic in $|z| < R$ $(0 < R < +\infty)$ with nonnegative Taylor coefficients and "fast growth" as $z \to R^-$. Let $h(z) := \log g(z) - (n+1) \log z$.

Output: The asymptotic formula (8.107) for $g_n := [z^n] g(z)$ derived from the Cauchy coefficient integral

$$g_n = \frac{1}{2i\pi} \int_\gamma g(z) \frac{dz}{z^{n+1}} = \frac{1}{2i\pi} \int_\gamma e^{h(z)} dz \qquad (8.99)$$

where γ is a loop around $z = 0$.

(SP1). *Saddle point contour.* Assume that $g'(z)/g(z) \to +\infty$ as $z \to R^-$. Let $r = r(n)$ be the unique positive root of the saddle point equation

$$h'(r) = 0 \qquad \text{or} \qquad \frac{rg'(r)}{g(r)} = n+1, \qquad (8.100)$$

so that $r \to R$ as $n \to \infty$. The integral (8.99) is evaluated on $\gamma = \{z : |z| = r\}$.

(SP2). *Basic split.* Assume that $h'''(r)^{1/3} h''(r)^{-1/2} \to 0$. Define $\delta = \delta(n)$ called the "range" of the saddle point by

$$\delta = \left| h'''(r)^{-1/6} h''(r)^{-1/4} \right|, \qquad (8.101)$$

so that

$$\delta \to 0, \qquad (8.102)$$

$$h''(r)\delta^2 \to \infty, \qquad (8.103)$$

$$h'''(r)\delta^3 \to 0. \qquad (8.104)$$

Split $\gamma = \gamma_0 \cup \gamma_1$, where

$$\gamma_0 = \{z \in \gamma : \arg(z)| \le \delta\}, \quad \gamma_1 = \{z \in \gamma : \arg(z)| \ge \delta\}.$$

(SP3) *Elimination of tails.* Assume that $|g(re^{i\theta})| \le |g(re^{i\delta})|$ on γ_1. Then, the tail integral satisfies the trivial bound,

$$\left| \int_{\gamma_1} e^{h(z)} dz \right| = O\left(|e^{-h(re^{i\delta})}| \right). \qquad (8.105)$$

(SP4) *Local approximation.* Assume that $h(re^{i\theta}) - h(r) - \frac{1}{2} r^2 \theta^2 h''(r) = O(|h'''(r)\delta^3|)$ on γ_0. Then, the central integral is asymptotic to a complete Gaussian integral, and

$$\frac{1}{2i\pi} \int_{\gamma_0} e^{h(z)} dz = \frac{g(r)r^{-n}}{\sqrt{2\pi h''(r)}} \left(1 + O(|h'''(r)\delta^3|) \right). \qquad (8.106)$$

(SP5) *Collection.* Assumptions $(SP1)$, $(SP2)$, $(SP3)$, $(SP4)$, imply the estimate:

$$[z^n]g(z) = \frac{g(r)r^{-n}}{\sqrt{2\pi h''(r)}} \left(1 + O(|h'''(r)\delta^3|) \right) \sim \frac{g(r)r^{-n}}{\sqrt{2\pi h''(r)}}. \qquad (8.107)$$

Using (8.98) we should obtain

$$[z^n](1-z)^{-\alpha} \sim n^{\alpha-1} \left(\frac{e}{\alpha}\right)^\alpha \sqrt{\frac{\alpha}{2\pi}}$$

which is not equal to $n^{\alpha-1}/\Gamma(\alpha)$ as proved in Theorem 8.12. For example, for $\alpha = 1$ we have $e/\sqrt{2\pi} = 1.0844\ldots \ne 1 = [z^n](1-z)^{-1}$. ∎

Now we are ready to rigorously establish the saddle point approximation (8.98) and generalize Example 8.13. After Hayman [192] (cf. also [191]), we define *admissible functions* $A(r)$ that satisfy the following three conditions:

(H1) there exists a finite number R_0 such that $A(r) > 0$ for $R_0 < r < \infty$;
(H2) $\lim_{r \to \infty} b(r) = +\infty$, where $b(r)$ is defined in (8.95);
(H3) for some $0 < \delta(r) < \pi$ we have with $a(r)$ defined in (8.94)

$$A(re^{i\theta}) \sim A(r) \exp\left(i\theta a(r) - 1/2\theta^2 b(r)\right), \quad r \to \infty \qquad (8.108)$$

uniformly in $|\theta| \le \delta(r)$, and

$$A(re^{i\theta}) = o\left(\frac{A(r)}{\sqrt{b(r)}}\right), \quad r \to \infty \qquad (8.109)$$

uniformly in $\delta(r) \le |\theta| \le \pi$.

We observe that by (8.93) condition (H3) implies

$$\sum_{k=3}^{\infty} \alpha_k(r) \frac{(i\theta)^k}{k!} \to 0$$

as $r \to \infty$ for $\delta(r) < \pi$. By (H2) we also have

$$\delta^2(r)b(r) \to \infty$$

as $r \to \infty$. To evaluate $[z^n]A(z)$ for A satisfying (H1)–(H3) we use Cauchy's integral (8.92) split at $\delta = \delta(r)$, that is, $a_n r^n = I_1 + I_2$, where

$$I_1 = \frac{1}{2\pi} \int_{-\delta}^{\delta} A(re^{i\theta})e^{-in\theta} d\theta$$

$$I_2 = \frac{1}{2\pi} \int_{-\delta}^{2\pi-\delta} A(re^{i\theta})e^{-in\theta} d\theta.$$

By (H3) we have $I_2 = o(A(r)/\sqrt{b(r)})$ as $r \to \infty$. After another application of (H3) the first integral becomes

$$
\begin{aligned}
I_1 &= \frac{A(r)}{2\pi} \int_{-\delta}^{\delta} \exp\left(i(a(r)-n)\theta - b(r)\theta^2/2\right) d\theta + o(f(r)/\sqrt{b(r)}) \\
&\overset{y=\theta\sqrt{b(r)/2}}{=} \frac{A(r)}{\pi\sqrt{2b(r)}} \exp\left(-\frac{(a(r)-n)^2}{2b(r)}\right) \int_{-\delta}^{\delta} \exp\left[-(y-icy)^2\right] d\theta \\
&\sim \frac{A(r)}{\sqrt{2\pi b(r)}} \exp\left(-\frac{(a(r)-n)^2}{2b(r)}\right),
\end{aligned}
$$

where $c = (a(r)-n)\sqrt{2/b(r)}$. We formulate the final result as the following theorem.

Theorem 8.19   (Hayman, 1956)   *Let* $A(z) = \sum_{n\geq 0} a_n z^n$.

(i) *For admissible functions satisfying conditions (H1)-(H3) we have uniformly as* $n \to \infty$

$$
a_n = \frac{A(r)}{r^n \sqrt{2\pi b(r)}} \left(\exp\left(-\frac{(a(r)-n)^2}{2b(r)}\right) + o(1)\right) \tag{8.110}
$$

as $r \to \infty$.

(ii) *If* $A(z)$ *and* $B(z)$ *are admissible, then* $e^{A(z)}$ *and* $A(z)B(z)$ *are admissible. If* $P(z)$ *is a polynomial and* $A(z) + P(z)$ *is admissible, then* $A(z)P(z)$ *and* $P(A(z))$ *are admissible, too. Finally,* $e^{P(z)}$ *is admissible, if* $[z^n]e^{P(z)} \in \mathbb{R}^+$.

In passing, we observe that to get the saddle point approximation from Hayman's theorem one defines r_n as a unique solution of $a(r_n) = n$. Such a unique solution exists since by (H2) $b(r)ra'(r) \to +\infty$, hence $a'(r) > 0$ and $a(r)$ must be an increasing function in some interval. This suffices to prove the assertion.

Example 8.15:   *Asymptotics of the Bell Numbers*   In Example 7.10 we defined Bell numbers b_n as

$$
\exp(e^z - 1) = \sum_{n=0}^{\infty} b_n \frac{z^n}{n!}.
$$

By Theorem 8.19(ii) $\exp(e^z - 1)$ is admissible with $a(r) = re^r$ and $b(r) = (r^2 + r)e^r$. By Theorem 8.19(i) we have

$$b_n \sim \frac{n! \exp(e^{r_n} - 1)}{r_n^n \sqrt{2\pi(r_n + 1)r_n \exp(r_n)}},$$

where r_n is defined by $r_n \exp(r_n) = n$.   ∎

Finally, we discuss one systematic approach to the saddle point method due to **van der Waerden** [443] that is particularly useful when the saddle point coincides with a pole. We present this method in a rather formal way, but it is not difficult to make it rigorous, as with Hayman's method, by stating necessary restrictive conditions on the admissible functions. When discussing this method we again return to the general integral representation (8.85). We shall follow here the presentation of Lauwerier [283].

Consider again the integral

$$I(n) = \int_C f(z)e^{-nh(z)}dz,$$

where C is a contour that starts and ends at infinity. We define the van der Waerden transformation as

$$w = h(z), \tag{8.111}$$

where $w = u + iv$. The importance of this transformation lies in changing the steepest descent lines into the horizontal lines in the (u, v) plane. Furthermore, this transformation is singular at the saddle points of $h(z)$. We assume now that:

(W1) $f(z)$ and $h(z)$ are algebraic functions of z;
(W2) The function

$$P(w) = f(z(w))\frac{dz}{dw}$$

does not grow faster than an exponential function, that is, $|P(w)| \leq Be^{c|w|}$ for some positive constants B and c.

Under this assumption, any saddle point $h'(z) = 0$ becomes a branch point in the complex plane. Then also the integral becomes

$$I(n) = \int_{C'} e^{-nw} f(z(w))\frac{dz}{dw}dw, \tag{8.112}$$

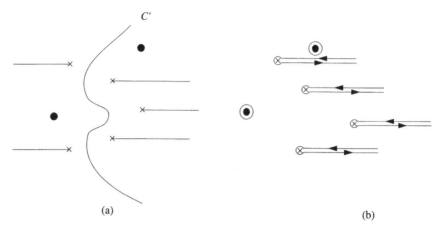

Figure 8.5. A contour in the van der Waerden method.

where C' is a contour that meanders between branch points and poles of the w plane, as illustrated in Figure 8.5(a). We should observe that apart from branch points and poles due to the factor dz/dw there might be some singularities coming from $f(z(w))$.

To obtain the saddle point approximation we deform the contour C' into another one C'' that will minimize $|e^{-nw}| = e^{-nu}$. This means to shift C' to the right as much as possible. It should be clear that the best contour satisfying this condition is one that consists of loops surrounding the branch points and poles, coming from infinity in straight lines from the right and going back to infinity to the right, as illustrated in Figure 8.5(b).

Let us discuss a concrete situation and assume that $w = 0$ is a single branch point of order one. A local substitution $w = s^2$ will transform the integrand function into

$$F(s) := f(z(w))\frac{dz}{dw} = f(z(s^2))\frac{1}{2s}\frac{dz}{ds},$$

so that it has a pole of the first order at $s = 0$. This implies that $F(s)$ admits an expansion

$$F(s) = a_{-1}s^{-1} + a_0 + a_1 s + \cdots,$$

which yields

$$I(n) = \int_{-\infty}^{\infty} e^{-ns^2}\left(a_{-1} + a_0 s + a_1 s^2 + \cdots\right) ds = \sqrt{\frac{\pi}{n}}\left(2a_{-1} + \frac{a_1}{n} + \cdots\right).$$

More interesting, if the saddle point (i.e., branch point at $w = s^2$) coincides with the pole at $s = 0$, then $F(s)$ becomes

$$F(s) = a_{-2}s^{-2} + a_{-1}s^{-1} + a_0 + a_1 s + \cdots,$$

which is easily computable through the following formula (cf. [2])

$$\frac{1}{2\pi i} \int_{-\infty}^{\infty} \frac{e^{ax^2}}{x - ib} dx = e^{ab^2}(1 - \Phi(b\sqrt{2a})) \tag{8.113}$$

provided $a > 0$, where

$$\Phi(x) = \frac{1}{\sqrt{2\pi}} \int_{-\infty}^{x} e^{-t^2/2} dt$$

is the standard normal distribution.

Example 8.16:   *Illustration of the van der Waerden's Method*   Consider the integral

$$I(n) = \int_{-\infty}^{\infty} \frac{e^{-n(z^2 - 2zi)}}{(z^2 + 1)} dz.$$

The saddle point is at $z = i$, which coincides with the pole of $1/(z^2 + 1)$. According to van der Waerden's approach we consider the transformation

$$z^2 - 2zi = 1 + s^2,$$

which simply becomes $z = i + s$. Then

$$\frac{1}{z^2 + 1} = \frac{1}{2is} + \frac{1}{4} + \frac{is}{8} + O(s^2)$$

and

$$I(n) = e^{-n} \int_{-\infty}^{\infty} e^{-ns^2} \left(\frac{1}{2is} + \frac{1}{4} + \frac{is}{8} + O(s^2) \right) ds$$

$$= e^{-n} \left(\frac{\pi}{2} + \frac{1}{4}\sqrt{\frac{\pi}{n}} + O(n^{-3/2}) \right),$$

where the last line follows from (8.78) and (8.113).   ∎

In the next section (see Example 8.22 below) we present another example that even better illustrates the power of the van der Waerden method.

In passing we should observe that other complications with the saddle point method may arise. For example, let the saddle point z_0 be of order m; that is,

$$h'(z_0) = h''(z_0) = \cdots = h^{(m-1)}(z_0) = 0, \quad \text{and} \quad h^{(m)}(z_0) \neq 0.$$

In this case one must apply the generalized Laplace method discussed in Exercise 15. More difficult to handle are *coalescing saddle points*: Imagine that the integral $I(n, \alpha)$ defined in (8.85) depends on the parameter α such that for $\alpha \neq \alpha_0$ there are two distinct saddle points z_+ and z_- of multiplicity one. For $\alpha = \alpha_0$ these two points coincide to a single saddle point z_0 of multiplicity two. Based of what we have learned so far, we conclude that (under appropriate assumptions) for $\alpha \neq \alpha_0$

$$I(n, \alpha) \sim f(z_+)e^{-nh(z_+)} \left[\frac{2\pi}{nh''(z_+)}\right]^{1/2} + f(z_-)e^{-nh(z_-)} \left[\frac{2\pi}{nh''(z_-)}\right]^{1/2}.$$

For $\alpha = \alpha_0$ the asymptotic behavior of $I(n, \alpha_0)$ differs radically since $h''(z_0) = 0$. Following the approach from Exercise 15, one arrives at

$$I(n, \alpha_0) \sim Af(z_0)e^{-nh(z_0)}\Gamma\left(\frac{4}{3}\right)\left[\frac{3!}{nh'''(z_+)}\right]^{1/3},$$

where A is a constant that depends on the contour $\mathcal{C}$. Thus the subexponential term changes discontinuously from $n^{1/2}$ to $n^{1/3}$. The interested reader is refereed to Wong [450], and Bleistein and Handelsman [51] for detailed discussions (cf. also Banderier et al. [32] for an application of coalescing saddle points to planar maps).

8.4.3 Local Central Limit Theorem and Large Deviations Revisited

The saddle point method finds many applications in probability theory. In particular, it can be used to derive a local central limit theorem and a local large deviations result. The reader may want to review and compare Chapter 5 before reading this section. Following Greene and Knuth [171] we give here a short presentation (see also Section 8.6). Further readings on this topic are Bender [36], Bender and Richmond [40], and Daniels [81].

Let us consider $S_n = X_1 + \cdots + X_n$, where X_i are i.i.d. discrete random variables with the probability generating function $G(z) = \mathbf{E}[z^{X_1}] = \sum_{k \geq 0} \Pr\{X_1 = k\}z^k$. Let $\mu = \mathbf{E}[X_1]$ and $\sigma^2 = \mathbf{Var}[X_1]$. Clearly, the probability generating function of S_n is $\mathbf{E}[z^{S_n}] = G^n(z)$. Then

$$\Pr\{S_n = k\} := [z^k]\mathbb{E}[z^{S_n}] = \frac{1}{2\pi i} \oint \frac{G^n(z)}{z^{k+1}} dz.$$

Of particular interest are two cases: (i) $k = n\mu + r$ with $r = o(\sqrt{n})$ (central limit regime), and (ii) $k = n(\mu + \Delta) = na$, where $\Delta > 0$ or $a > \mu$ (large deviations regime).

We adopt the following three assumptions:

(K1) The generating function $G(z)$ is defined in $|z| < 1 + \delta$, so that

$$G(e^t) = \exp\left(\mu t + \frac{\sigma^2 t^2}{2} + \frac{\kappa_3 t^3}{3!} + \frac{\kappa_4 t^4}{4!} + \cdots\right), \tag{8.114}$$

where κ_j is the jth cumulant of X_1 (cf. Section 5.4).

(K2) $G(0) \neq 0$.

(K3) X_1 is not periodic, that is, the greatest common divisor of all k with $\Pr\{X_1 = k\} \neq 0$ is one.

Under these assumptions, we observe that the Cauchy integral above can be evaluated by the saddle point method with the saddle point at $z = 1 - O(1/n) \sim 1$; hence

$$\Pr\{S_n = \mu n + r\} = \frac{1}{2\pi} \int_{-\pi}^{\pi} \frac{G^n(e^{it})}{e^{it(\mu n + r)}} dt.$$

This integral can be evaluated by the Laplace method after noting that under (K3) for $\delta < |t| < \pi$ there exists $\alpha < 1$ such that $|G(e^{it})| < \alpha$, and

$$\frac{G^n(e^{it})}{e^{it(\mu n + r)}} = \exp\left(irt - \frac{\sigma^2 n t^2}{2} + \frac{\kappa_3 n t^3}{3!} + \frac{\kappa_4 n t^4}{4!} + \cdots\right).$$

If we choose $\delta^2 n \to \infty$ (say $\delta = n^{1/2+\varepsilon}$), then the Laplace method yields, after a careful evaluation,

$$\Pr\{S_n = \mu n + r\} = \frac{1}{\sigma\sqrt{2\pi n}} \exp\left(\frac{-r^2}{2\sigma^2 n}\right)\left(1 - \frac{\kappa_3}{2\sigma^4}\left(\frac{r}{n}\right) + \frac{\kappa_3}{6\sigma^6}\left(\frac{r^3}{n^2}\right)\right)$$

$$+ O(n^{-3/2}). \tag{8.115}$$

We mentioned this result in Section 5.4.

This is the *local central limit* law, from which we can obtain the central limit law after summing over r. Unfortunately, like all central limit results, the formula suffers from weakness since the range of its applicability is limited to $r = o(\sqrt{n})$ due to the fact that the error term is only a polynomial of $n^{-1/2}$. But there is a remedy: To expand the range of its validity one should shift the distribution to a new location. This is called the method of *shifted mean*. In Section 5.4 we used a variant of it called the *exponential change of measure* to obtain large deviations results. Below we shall show how to recover a stronger large deviations result for discrete distributions by the shifted mean method.

The idea is to shift the mean of S_n to a new value such that (8.115) is valid again. Let us define a new random variable $\widetilde{X}$ whose generating function is

$$\widetilde{G}(z) = \frac{G(z\alpha)}{G(\alpha)},$$

where α is a constant that is to be determined. Observe that

$$\mathrm{E}[\widetilde{X}] = \alpha \frac{G'(\alpha)}{G(\alpha)}, \tag{8.116}$$

$$\mathrm{Var}[\widetilde{X}] = \alpha^2 \frac{G''(\alpha)}{G(\alpha)} + \alpha \frac{G'(\alpha)}{G(\alpha)} - \alpha^2 \left(\frac{G'(\alpha)}{G(\alpha)}\right)^2. \tag{8.117}$$

If one needs a large deviations result around $m = na = n(\mu + \Delta)$, where $\Delta \neq 0$, then clearly (8.115) cannot be applied directly, but a proper choice of α can help. Let us select α such that the new $\widetilde{S}_n = \widetilde{X}_1 + \cdots + \widetilde{X}_n$ has mean $m = n(\mu + \Delta)$. This results in setting α to be a solution of

$$\frac{\alpha G'(\alpha)}{G(\alpha)} = \frac{m}{n} = a = \mu + \Delta. \tag{8.118}$$

Then a change of variable yields

$$[z^m]G^n(z) = \frac{G^n(\alpha)}{\alpha^m}[z^m]\left(\frac{G(\alpha z)}{G(\alpha)}\right)^n. \tag{8.119}$$

Now we can apply (8.115) (with $r = 0$) to $[z^m]\left(\frac{G(\alpha z)}{G(\alpha)}\right)^n$ since $m = an$ and $\mathrm{E}[\widetilde{X}_1] = a = \mu + \Delta$ so that $\mathrm{E}[\widetilde{S}_n] = an$. We rewrite the above as follows

$$\Pr\{S_n = na\} = \frac{1}{\widetilde{\sigma}\sqrt{2\pi n}}e^{-nI(a)}\left(1 + O(1/n)\right), \tag{8.120}$$

where $\widetilde{\sigma}^2 = \mathrm{Var}[\widetilde{X}_1]$ given by (8.117) and $I(a) = a\log\alpha - \log G(\alpha)$. This is a stronger version of Theorem 5.13 if one sets $\lambda_a = \log\alpha$.

Example 8.17: *Large Deviations for the Binomial Distribution*   This is taken
from Greene and Knuth [171] (a more sophisticated example is presented in Section 8.6). Let S_n be binomially distributed with parameter $1/2$; that is, $G^n(z) = ((1 + z)/2)^n$. We want to estimate the probability $\Pr\{S_n = n/3\} = [z^{n/3}](z/2 + 1/2)^n$ which is far away from its mean $\mathbf{E}[S_n] = n/2$. Shift of mean yields α such
that

$$\frac{\alpha G'(\alpha)}{G(\alpha)} = \frac{\alpha}{1 + \alpha} = \frac{1}{3}$$

(i.e., $\alpha = 1/2$). Using (8.115) we obtain

$$[z^{n/3}]\left(\frac{2}{3} + \frac{1}{3}z\right)^n = \frac{3}{2\sqrt{\pi n}}\left(1 - \frac{7}{24n}\right) + O(n^{-5/2})$$

and an application (8.120) gives

$$[z^{n/3}](z/2 + 1/2)^n = \left(\frac{3 \cdot 2^{1/3}}{4}\right)^n \frac{3}{2\sqrt{\pi n}}\left(1 - \frac{7}{24n} + O(n^{-2})\right)$$

for $n \to \infty$.   ■

8.5   FINITE SUMS AS COMPLEX INTEGRALS

In Section 8.1.2 we saw how a series may be represented by a complex integral.
Such a representation is often quite useful, and often can be easily evaluated by
the residue theorem. Here is another representative example of this kind

$$\sum_{k=m}^{n}(-1)^k f_k = \frac{1}{2\pi i} \int_{m-1/2-i\infty}^{m-1/2+i\infty} f(s)\frac{\pi}{\sin \pi s}ds,$$

which is true under suitable conditions on $f(s)$ (e.g., polynomial growth), where
$f(s)$ is an analytic continuation of f_k, that is, $f(s)_{s=k} = f_k$. The above is a direct
consequence of the residue theorem and the fact that $\mathrm{Res}[\pi/\sin(\pi s), \ s = k] = (-1)^k$, provided $f(s)$ does not have any other singularity right to the line of the
integration.

   In this section, we discuss an important finite sum that often arises in the analysis of algorithms (see Section 7.6.1). Namely, we deal with the following alternating sum

$$S_n[f] = \sum_{k=m}^{n}(-1)^k \binom{n}{k} f_k,$$

where f_k is a known, but otherwise, general sequence. Due to the alternating sign this sum is not easy to evaluate by traditional methods. In fact, we shall show that there are some fluctuations involved in its asymptotics that can be easily treated by complex asymptotic methods. We will not dwell too much on this topic since there is an excellent survey on such sums by Flajolet and Sedgewick [148]. The reader is referred to it (and other references such as Knuth [269] and Szpankowski [407]) for more examples.

Theorem 8.20 (Rice's Formula) *Let $f(s)$ be an analytic continuation of $f(k) = f_k$ that contains the half line $[m, \infty)$. Then*

$$S_n[f] := \sum_{k=m}^{n} (-1)^k \binom{n}{k} f_k = \frac{(-1)^n}{2\pi i} \int_C f(s) \frac{n!}{s(s-1)\cdots(s-n)} ds, \quad (8.121)$$

where C is a positively oriented curve that encircles $[m, n]$ and does not include any of the integers $0, 1, \ldots, m-1$.

Proof. It is a direct application of the residue calculus after noting that the poles of the integrand are at $k = m, m+1, \ldots, n$ and

$$\mathrm{Res}\left[\frac{n!}{s(s-1)\cdots(s-n)} f(s); \ s = k\right] = (-1)^{n-k} \binom{n}{k} f(k).$$

This completes the proof.   ∎

There is another representation of $S_n[f]$ that is sometimes easier to handle computationally. (It is called by Knuth [269] the gamma method.) In this case, however, we must restrict f to have a *polynomial growth*, that is, $f(s) = O(|s|^k)$ for some k as $s \to \infty$. In fact, when evaluating the integral in the Rice method we usually must impose some growth condition on f over large circles.

Theorem 8.21 (Knuth, 1973; Szpankowski, 1988) *Let $f(z)$ be of a polynomial growth at infinity and analytic left to the vertical line $(\frac{1}{2} - m - i\infty, \frac{1}{2} - m + i\infty)$. Then*

$$S_n[f] = \frac{1}{2\pi i} \int_{\frac{1}{2}-m-i\infty}^{\frac{1}{2}-m+i\infty} f(-z) B(n+1, z) dz \qquad (8.122)$$

$$
= \frac{1}{2\pi i} \int_{\frac{1}{2}-m-i\infty}^{\frac{1}{2}-m+i\infty} f(-z) n^{-z} \Gamma(z) \left(1 - \frac{z(z+1)}{2n} \right.
\tag{8.123}
$$

$$
\left. + \frac{z(1+z)}{24n^2} (3(1+z)^2 + z - 1) + O(n^{-3}) \right) dz, \quad \Re(z) > 0,
$$

where $B(x, y) = \Gamma(x)\Gamma(y)/\Gamma(x+y)$ is the beta function.

Proof. We again use the residue theorem, the fact that $\mathrm{Res}[\Gamma(z), z = k] = (-1)^k/k!$ and $\mathrm{Res}[B(n+1, z), z = k] = (-1)^k \binom{n}{k}$. However, some care is needed. Let us consider a large rectangle $R_{\alpha,M}$ left to the line of the integration, with corners at four points $(1/2 - M \pm i\alpha, c \pm i), \alpha > 0, M$ is a positive integer, and $c = 1/2 - m$. By Cauchy's theorem the integral in (8.122) is equal to the sum of residues in $R_{\alpha,M}$ minus the values of the integral on the bottom, top and left lines of the rectangle. We must prove now that the integrals along left, top, and bottom lines are small. This is not difficult due to the behavior of the gamma function for imaginary values (cf. (2.34)) and the polynomial growth of f. We consider the integral along the top line, which we denote as I_T. Using the estimate (2.35) for the gamma function we obtain

$$
|I_T| = O\left(e^{-\pi\alpha/2} \int_{-\infty}^{c} (\alpha/n)^x |f(-x - i\alpha)| dx \right) \to 0.
$$

Let now I_L be the integral along the left line. We need a different estimate here. Using successively $\Gamma(z+1) = z\Gamma(z)$ we first observe that

$$
\Gamma(1/2 - M + iy) = \Gamma(1/2 + iy)/O((M-1)!).
$$

Then

$$
|I_L| = O\left(\frac{n^M}{(M-1)!} \int_{-\infty}^{\infty} |\Gamma(1/2 + iy)| |f(M - 1/2 - iy)| dy \right),
$$

which also tends to zero as $M \to \infty$ since the integral converges due to the exponential smallness of $\Gamma(z)$ along the imaginary line and the polynomial growth of $f(z)$. This completes the proof. ∎

Example 8.18: *Asymptotics of an Alternating Sum* In Section 7.6.1 we encountered alternating sums of the following type

$$S_{n+r} = \sum_{k=2}^{n+r} (-1)^k \binom{n+r}{k} \binom{k}{r} \frac{1}{1 - p^{-k} - q^{-k}}, \qquad (8.124)$$

where $p + q = 1$. We now use Theorem 8.21 to obtain asymptotics of S_{n+r} as n becomes large and r is fixed. We first slightly generalize Rice's method by observing that for any sequence f_k

$$S_{n+r}[f] = \sum_{k=2}^{n+r} (-1)^k \binom{n+r}{k} \binom{k}{r} f_k$$

$$= (-1)^r \binom{n+r}{r} \sum_{k=[2-r]^+}^{n} (-1)^k \binom{n}{k} f(k+r),$$

where $x^+ = \max\{0, x\}$. Thus by (8.123) the sum (8.124) becomes

$$S_{n+r} = \left(1 + O(n^{-1})\right) \frac{(-1)^r}{r!} \frac{1}{2\pi i} \int_{\frac{1}{2}-[2-r]^+ - i\infty}^{\frac{1}{2}-[2-r]^+ + i\infty} n^{r-z} \Gamma(z) \frac{1}{1 - p^{r-z} - q^{r-z}} dz + e_n,$$

where e_n is an error term that we discuss later. Naturally, we evaluate the integral by the residue theorem. However, first we observe that the function under the integral has infinitely many poles that are roots of

$$1 = p^{r-z} + q^{r-z}.$$

This equation appears many times in this book, and we need to study the locations of its roots in order to find asymptotic expansions of S_{n+r}. The reader is asked in Exercise 20 to provide a full proof of the following lemma (see Lemma 10.25 for a generalization).

Lemma 8.22   (Jacquet 1989, Schachinger 1993)   *Let z_k for $k = \mathbb{Z} = 0, \pm 1, \pm 2, \ldots$ be solutions of*

$$p^{-z+r} + q^{-z+r} = 1, \qquad (8.125)$$

where $p + q = 1$ and z is complex.

   (i) *For all $k \in \mathbb{Z} = \{0, \pm 1, \pm 2, \ldots\}$*

$$-1 + r \leq \Re(z_k) \leq \sigma_0 + r, \qquad (8.126)$$

where σ_0 is a positive solution of $1 + q^{-s} = p^{-s}$. Furthermore,

$$\frac{(2k-1)\pi}{\log p} \leq \Im(z_k) \leq \frac{(2k+1)\pi}{\log p}.$$

(ii) *If $\Re(z_k) = -1 + r$ and $\Im(z_k) \neq 0$, then $\log p / \log q$ must be rational. More precisely, if $\frac{\log p}{\log q} = \frac{s}{t}$, where $\gcd(s, t) = 1$ for $s, t \in \mathbb{Z}$, then*

$$z_k = -1 + r + \frac{2ks\pi i}{\log p} \tag{8.127}$$

for all $k \in \mathbb{Z}$.

Proof. We give only a sketch of the proof for $r = 1$. Let us assume that $z = -1 + it$ is a solution of $p^{-z} + q^{-z} = 1$. Then

$$|p^{-z} + q^{-z}| = 1 = p + q = |p^{1-it}| + |q^{1-it}|$$
$$= |p^{-z}| + |q^{-z}|.$$

Now we know that the first and the last expression are equal and both equal to 1. But for complex numbers u and v the equality $|u+v| = |u|+|v|$ holds if and only if $u = \lambda v$, with some real $\lambda \geq 0$. Therefore, p^{-z} and q^{-z} have the same argument in the complex plane. But $p^{-z} + q^{-z} = 1$, hence the argument in question must be zero. This means that p^{-z} as well as q^{-z} are nonnegative real numbers as soon as z is a solution. Their imaginary parts are zero, which yields

$$\sin(t \log p) = \sin(t \log q) = 0.$$

Therefore, $t \log p$ as well as $t \log q$ are an integer multiple of π, from which it follows immediately, that the ratio of $\log p$ and $\log q$ has to be rational, if t is different from zero. Also, if this ratio is not in $\mathbb{Q}$, then there does not exist a solution with real part -1 and imaginary part different from zero. ∎

Based on the above lemma, we conclude that either $\Re(z_k) > r - 1$ for $\log p / \log q \notin \mathbb{Q}$ or $z_k = r - 1 + 2\pi iks / \log p$ when $\log p / \log q = s/t \in \mathbb{Q}$. In any case, the line $\Re(z_k)$ lies to the *right* of the line of the integration ($\frac{1}{2} - [2 - r]^+ - i\infty$, $\frac{1}{2} - [2 - r]^+ + i\infty$). We now apply the closing-the-box method discussed in Section 7.6.3. We consider a big rectangle with the left side being the line of integration, the right side positioned at $\Re(z) = M$ (where M is a large positive number), and the bottom and top sides positioned at $\Im(z) = \pm A$, say. We further observe that the right side contributes only $O(n^{r-M})$ due to the factor n^{r-M} in the integral. Both the bottom and top sides contribute negligibly, too, since the gamma function decays exponentially fast with the increase of the imaginary part when $A \to \infty$. This fact was already used in the proof of Theorem 8.21. In summary, the integral is equal to a circular integral (around the rectangle) plus a negligible part $O(n^{r-M})$. But then by Cauchy's residue theorem the latter integral is equal to minus the sum of all residues at z_k, that is,

$$S_{n+r} = -\frac{(-1)^r}{r!} \sum_{k=-\infty}^{\infty} \mathrm{Res}\left(\frac{n^{r-z}\Gamma(z)}{1 - p^{r-z} - q^{r-z}}, \; z = z_k\right) + O(n^{r-M}). \quad (8.128)$$

The minus sign above indicates that the contour is encircled in the clockwise direction. When $\log p/\log q \notin \mathbb{Q}$, then $\Re(z_k) > r - 1$, hence $S_{n+r} = o(n)$.

Now we consider $\log p/\log = s/t \in \mathbb{Q}$. In this case, the main contribution to the asymptotics comes from $z_0 = r - 1$, which is a double pole. We shall use the following expansions for $w = z - z_0$

$$n^{r-z} = n(1 - w \ln n + O(w^2)),$$

$$(1 - p^{r-z} - q^{r-z})^{-1} = -w^{-1}h^{-1} + \frac{1}{2}h_2 h^{-2} + O(w),$$

$$\Gamma(z) = (-1)^{r+1}(w^{-1} - \gamma + \delta_{r,0}) + O(w) \quad r = 0, 1,$$

where $h = -p \ln p - q \ln q$, $h_2 = p \ln^2 p + q \ln^2 q$, and $\gamma = 0.577215\ldots$ is the Euler constant. Considering, in addition, the residues coming from z_k for $k \neq 0$ we finally arrive at

$$S_{n+r} = \begin{cases} \frac{1}{h}n(\ln n + \gamma - \delta_{r,0} + \frac{1}{2}h_2) + \frac{(-1)^r}{r!}n P_r(n) + e_n & r = 0, 1 \\[2mm] n\frac{(-1)^r}{r(r-1)h} + (-1)^r n P_r(n) + e_n & r \geq 2, \end{cases}$$

provided $\log p/\log = s/t \in \mathbb{Q}$. In the above $P_r(n)$ is a contribution coming from z_k for $k \neq 0$, and one computes

$$P_r(n) = \frac{1}{h} \sum_{k \in \mathbb{Z}\setminus\{0\}} \Gamma(r - 1 + 2\pi k s/\log p) \exp(2\pi i k s \log_p n)$$

$$= \frac{1}{h} \sum_{k \neq 0} \Gamma(r - 1 + 2\pi k s/\log p) \exp(2\pi i k s \log_p n),$$

where throughout we write $\sum_{k \neq 0} := \sum_{k \in \mathbb{Z}\setminus\{0\}}$. We observe that $P_r(n)$ is a periodic function of $\log_p n$ with period 1, mean 0, and small amplitude for small r. For example, when $p = q = 1/2$ we have

$$P_r(n) = \frac{1}{\ln 2} \sum_{k \neq 0} \Gamma(r - 1 + 2\pi i k / \log 2) \exp(-2\pi i k \log_2 n).$$

The absolute value of $P_r(n)$ is shown in Table 8.5 for some r. Observe that the amplitude of $P_r(n)$ is very small for small r but rapidly increases even for moderate r.

Concerning the error term e_n; it can be represented as

$$e_n = O\left(\frac{1}{2n} \frac{1}{2\pi i} \frac{(-1)^n}{r!} \int_{\frac{1}{2} - [2-r]^+ - i\infty}^{\frac{1}{2} - [2-r]^+ + i\infty} n^{r-z}(z(z+1)\Gamma(z) \frac{1}{1 - p^{r-z} - q^{r-z}} dz\right);$$

hence the same procedure as above may be used to evaluate the integral. Clearly, $e_n = O(1)$. ∎

Alternating sums arise in many computations, and we often must use various asymptotic methods to evaluate the integral involved. The reader is referred to Flajolet and Sedgewick [148] and exercises at the end of this chapter for more examples. In particular, if $F(z)$ denotes the ordinary generating function of f_k and $S(z)$ the ordinary generating function of $S_n[f]$, then one obtains, as in Example 7.11 of Chapter 7,

$$S(z) = \frac{1}{1 - z} F\left(-\frac{z}{1 - z}\right).$$

Thus, if $F(z)$ has algebraic singularities, so $S(z)$ does, and one can use the singularity analysis to derive asymptotics of S_n.

TABLE 8.5. Upper Bound on $|P_r(n)|$ as a Function of r

r	$|P_r(n)|$
-2	$.1725 10^{-6}$
-1	$.1573 10^{-5}$
0	$.1426 10^{-4}$
1	$.1300 10^{-3}$
2	$.1207 10^{-2}$
3	$.1153 10^{-1}$
4	$.1142 10^{-0}$
5	1.1823
6	12.8529
7	147.2071
8	1779.8207

Furthermore, this approach works fine also for the **binomial sums** defined as

$$\bar{S}_n[f] = \sum_{k=0}^{n} \binom{n}{k} p^k q^{n-k} f_k, \quad q = 1 - p.$$

Again, the ordinary generating function $\bar{S}(z)$ of $\bar{S}_n[f]$ can be expressed in terms of the generating function $F(z)$ of f_k as

$$\bar{S}(z) = \frac{1}{1 - qz} F\left(\frac{pz}{1 - qz}\right).$$

Binomial sums were studied by Flajolet [129] using singularity analysis. We shall return to them in the final chapter of this book, where they are solved by the depoissonization tool (see Example 10.7).

We complete this section with the analysis of an interesting alternating sum that requires an application of the saddle point method in the Rice formula. Moreover, in this case the saddle point coincides with a pole.

Example 8.19:	*Digital Search Tree*

In the analysis of digital search trees Louchard [290] considered the following sum

$$F_n(k) = \sum_{i=0}^{k} \binom{k}{i} p^i q^{k-i} (1 - p^i q^{k-i})^n, \tag{8.129}$$

where $k = O(\log n)$ and $p + q = 1$. This is related to the depth in a digital search tree, and therefore of particular interest is the case when $k = \frac{1}{h} \log n + x \sqrt{\log n}$ and $x = O(1)$. As before, let $h = -p \log p - q \log q$ be the entropy. First, observe that

$$F_n(k) = \sum_{i \geq 0} \binom{k}{i} p^i q^{k-i} \sum_{\ell \geq 0} \binom{n}{\ell} (-1)^{n-\ell} p^{\ell i} q^{\ell(k-i)}$$

$$= \sum_{\ell \geq 0} \binom{n}{\ell} (-1)^\ell \sum_{i \geq 0} \binom{k}{i} p^{i(\ell+1)} q^{(k-i)(\ell+1)}$$

$$= \sum_{\ell \geq 0} \binom{n}{\ell} (-1)^\ell (p^{\ell+1} + q^{\ell+1})^k.$$

The last sum is an alternating sum and can be treated by the Rice method. Using Theorem 8.21 we arrive at the following;

$$F_n(k) = \frac{1}{2\pi i}(1 + O(1/n))\int_{1/2-i\infty}^{1/2+i\infty} n^{-s}\Gamma(s)(p^{1-s}+q^{1-s})^k ds \qquad (8.130)$$

that we evaluate by the saddle point method.

Theorem 8.23 *Let $h_2 = p\log^2 p + q\log^2 q$ and $\sigma^2 = (h_2-h^2)/h^3$ with $p \neq q$. For $k = \frac{1}{h}\log n + y\sigma\sqrt{\log n}$ we have as $n \to \infty$*

$$F_n(k) \sim \Phi(y) = \frac{1}{\sqrt{2\pi}}\int_{-\infty}^{y} e^{-t^2/2} dt \qquad (8.131)$$

provided $y = O(1)$.

Proof. We apply the saddle point method, and estimate the integral

$$I(n) = \frac{1}{2\pi i}\int_{1/2-i\infty}^{1/2+i\infty} n^{-s}\Gamma(s)(p^{1-s}+q^{1-s})^k ds$$

for $k = c_1 \log n + x\sqrt{\log n}$, where $c_1 = 1/h$ and $x = y\sigma$. Clearly,

$$I(n) = \frac{1}{2\pi i}\int_{1/2-i\infty}^{1/2+i\infty} \Gamma(s)\exp\left(-(s\log n - k\log(p^{1-s}+q^{1-s}))\right) ds.$$

Observe that

$$s\log n - k\log(p^{1-s}+q^{1-s}) = -\frac{(s-s_0)^2}{2}k(h_2-h^2)$$

$$+ \frac{(\log n - kh)^2}{2k(h_2-h^2)} + O(ks^3),$$

where

$$s_0 = \frac{\log n - kh}{k(h_2-h^2)} = \frac{-y}{h\sigma\sqrt{\log n}}.$$

Let now $s = s_0 + it$. Since s_0 is also close to a pole of $\Gamma(s)$ we must use

$$\Gamma(s) \sim \frac{1}{s_0+it} = \frac{s_0}{s_0^2+t^2} - \frac{it}{s_0^2+t^2}.$$

Substituting $u = t^2$ and $x = y\sigma$, and putting everything together, we finally obtain

$$I(n) \sim \frac{1}{\pi i} e^{-y^2/2} \int_0^\infty e^{t^2 h^2 \sigma^2 \log n/2} \frac{s_0}{s_0^2 + t^2} (1 + O(x/\sqrt{\log n})) dt$$

$$= (1 + O(x/\sqrt{\log n})) \frac{s_0}{\pi i} e^{-y^2/2} \int_0^\infty e^{-u\beta} \frac{1}{\sqrt{u}(s_0^2 + u)} du,$$

where $\beta = h^2 \sigma^2 \log n/2$. But it is known that (cf. [2])

$$\frac{1}{\pi i} \int_0^\infty e^{-u\beta} \frac{1}{\sqrt{u}(s_0^2 + u)} du = \frac{1}{2s_0} e^{\beta s_0^2} \text{Erfc}(s_0\sqrt{\beta}),$$

where

$$\text{Erfc}(x) = \frac{2}{\sqrt{\pi}} \int_x^\infty e^{-t^2} dt.$$

Then

$$I(n) \sim \frac{1}{2} \text{Erfc}(-y/\sqrt{2}) = 1 - \Phi(-y) = \Phi(y),$$

where Φ is the standard normal distribution. The reader is asked in Exercise 19 to use the van der Waerden method to prove this result.   ■

8.6 LIMITING DISTRIBUTIONS

In this section we shall apply the asymptotic methods learned so far to derive limiting distributions of a sequence of random variables from its sequence of generating functions. We shall mostly deal with normal approximation since the Poisson approximation (used to estimate rare or exceptional occurrences) is a subject of a recent book by Barbour, Holst and Janson [33].

8.6.1 Discrete Limit Laws Through PGF

From now on, we shall deal with a sequence of nonnegative discrete random variables $X_n \geq 0$ and the corresponding sequence of generating functions $G_n(z) = E[z^{X_n}]$. A natural question to ask is whether one can infer from the limit of $G_n(z)$ ($n \to \infty$), if it exists, whether the coefficients $p_{n,k} = [z^k]G_n(z)$ (for any fixed k) tend to a limit, say $p_k = \lim_{n\to\infty} \Pr\{X_n = k\}$ such that $\sum_{k\geq 0} p_k = 1$. If the latter holds, then X_n is said to satisfy a *discrete limit law*. A *continuity theorem* for the probability generating functions (PGF) answers the above question in the positive. Interestingly enough, convergence to a discrete law, if exists, is always *uniform*, as we shall see below.

Let $G_n(z) = \mathbf{E}[z^{X_n}]$ be a sequence of generating functions. Observe that $G_n(z)$ are uniformly bounded since $|G_n(z)| \leq 1$ for $|z| < 1$. If, in addition, $G_n(z) \to G(z)$, then Vitali's theorem (cf. [424] pp. 168) applies. (Vitali's theorem asserts that: *Let $f_n(z)$ be a sequence of functions, each analytic in a region D; let $|f_n(z)| \leq M$ for every n and z in D; and let $f_n(z)$ tend to a limit as $n \to \infty$, at a set of points having a limit-point inside D. Then $f_n(z)$ tends uniformly to a limit in any region bounded by a countour interior to D, the limit being, therefore, an analytic function of z.*)

Theorem 8.24 (Continuity of PGF) *Let $G(z) = \sum_{k \geq 0} p_k z^k$ together with the sequence $G_n(z)$ of generating functions be defined for $|z| < 1$. If*

$$\lim_{n \to \infty} G_n(z) = G(z) \tag{8.132}$$

pointwise for $z \in \Omega \subset \{z : |z| < 1\}$, then for every k we have

$$\lim_{n \to \infty} [z^k] G_n(z) = p_k. \tag{8.133}$$

Proof. Following Flajolet and Sedgewick [149], we apply Cauchy's theorem enforced by Vitali's theorem to obtain

$$
\begin{aligned}
p_k &= \frac{1}{2\pi i} \oint \frac{G(z)}{z^{k+1}} dz \\
&= \lim_{n \to \infty} \frac{1}{2\pi i} \oint \frac{G_n(z)}{z^{k+1}} dz \\
&= \lim_{n \to \infty} [z^k] G_n(z),
\end{aligned}
$$

where the second, crucial line is justified by the Vitali theorem. ∎

Example 8.20: *Poisson Theorem* Let X_n have the *Binomial*(n, p) distribution, that is, $\Pr\{X_n = k\} = \binom{n}{k} p^k q^{n-k}$, where $q = 1 - p$. We shall assume that p depends on n such that $\lambda = \lim_{n \to \infty} np$ is a constant. Observe that

$$G_n(z) = (1 - p + pz)^n,$$

hence

$$\lim_{n \to \infty} G_n(z) = \lim_{n \to \infty} \left(1 + \frac{\lambda(z-1)}{n}\right)^n = e^{\lambda(z-1)}.$$

Since $e^{\lambda(z-1)}$ is PGF for the Poisson distribution, by Continuity Theorem 8.24 we proved Poisson's theorem, which asserts that the binomial distribution converges to the Poisson when $np \to \lambda$.   ∎

In passing, we should point out that one must be careful with limiting distributions of discrete random variables since often such a distribution might not exist even if an *asymptotic formula* for the "limiting" distribution can be derived. This problem is notorious for the *extreme distributions* that we already touched on in Section 3.3.2. To see this, let us consider a sequence of i.i.d. discrete random variables, say, $X_1, \ldots, X_n$. Define

$$M_n = \max\{X_1, \ldots, X_n\}.$$

The question we ask is whether one can find sequences a_n and b_n such that

$$\frac{M_n - a_n}{b_n}$$

converges to a nondegenerate random variable. Anderson [12] proved that this is not the case as long as the probability mass $p_k = \Pr\{X_i = k\}$ is such that

$$\frac{p_k}{\sum_{j=k}^{\infty} p_j} \tag{8.134}$$

does not converge to 0 or ∞ as $k \to \infty$. We illustrate this point with an example.

Example 8.21: *Extreme Distribution for i.i.d. Geometric Distributions*   Let $\Pr\{X_i = k\} = p^k(1 - p)$ for each $1 \le i \le n$ have *Geometric(p)* distribution. Then (8.134) converges to $1 - p > 0$; hence a limiting distribution cannot exist. But an asymptotic formula for $\Pr\{M_n < \log_{1/p} n + x\}$ can be found for fixed x and large n as follows

$$\Pr\{M_n < \log_{1/p} n + x\} \sim \exp\left(-p^{x - \langle \log_{1/p} n + x \rangle}\right),$$

where $\langle t \rangle = t - \lfloor t \rfloor$. Indeed, let $\Pr\{X_i \ge k\} = 1 - F(k) = p^k$. Then

$$\Pr\{M_n < k\} = (F(k))^n = (1 - (1 - F(k)))^n$$
$$= \exp(n \log(1 - (1 - F(k))))$$
$$\sim \exp(-n(1 - F(k)) \qquad 1 - F(k) = p^k \to 0 \quad \text{as} \quad k \to \infty.$$

Thus for x real and $k = \lfloor \log_{1/p} n + x \rfloor$ we have

$$\Pr\{M_n < \lfloor \log_{1/p} n + x \rfloor\} = \Pr\{M_n < \log_{1/p} n + x\}$$
$$\sim \exp\left(-np^{\lfloor \log_{1/p} n + x \rfloor}\right)$$
$$= \exp\left(-p^{x - \langle \log_{1/p} n + x \rangle}\right).$$

Observe that the limiting distribution of M_n does not exist. ∎

8.6.2 Continuous Limit Laws by Integral Transforms

In this book, we treat mostly discrete structures, and hence deal mostly with discrete random variables. However, when considering limit laws, a discrete random variable after a normalization may converge to a continuous random variable. The most famous example is the central limit law that asserts that the sequence $(X_n - E[X_n])/\sqrt{\text{Var}[X_n]}$ convergences in distribution to the standard normal distribution. We have already seen it in Section 8.4.3.

Convergence *in distribution* was already discussed in Section 2.2.2. To recall, we say that X_n converges in distribution to X, and we write $X_n \xrightarrow{d} X$, if $F_n(x) \to F(x)$ for each point of continuity of $F(x)$, where $F_n(x)$ and $F(x)$ are distribution functions of X_n and X, respectively. If X is a continuous random variable, then instead of PGF we should deal with the *moment generating function* (MGF), also known as the Laplace transform but strictly speaking it is the double-sided standard Laplace transform with s replaced by $-s$,

$$M(s) := E[e^{sX}] = \int_{-\infty}^{\infty} e^{sx} dF(x);$$

or the *characteristic function* (CF), which is really the Fourier transform,

$$\phi(t) := E[e^{itX}] = \int_{-\infty}^{\infty} e^{itx} dF(x).$$

Clearly, the characteristic function is always defined for any distribution, while the moment generating function might not exist for heavily tailed distributions. In fact, we proved in Section 5.4 that the existence of MGF near the origin implies an exponential tail.

We finish these introductory comments with a few remarks. First, the standard normal distribution has the following MGF and CF;

$$M(s) = e^{s^2/2}, \qquad \phi(t) = e^{-t^2/2}.$$

Second, we recall that the rth moment of X, if it exists, can be computed as follows

$$\mathbf{E}[X^r] = \frac{d^r}{ds^r} M(s)\Big|_{s=0} = i^{-r} \frac{d^r}{dt^r} \phi(t)\Big|_{t=0}.$$

Third, for $\mu = \mathbf{E}[X]$ and $\sigma^2 = \mathbf{Var}[X] > 0$, if we denote by

$$X^* = \frac{X - \mu}{\sigma},$$

then the normalized random variable has

$$\phi_{X^*}(t) = e^{-i\mu t/\sigma} \phi_X(t/\sigma),$$
$$M_{X^*}(s) = e^{-s\mu/\sigma} M_X(s/\sigma).$$

With this in mind, we can move to more interesting considerations. As in the case of discrete random variables, we would like to infer convergence in distribution of X_n from convergence of MGF or CF. Fortunately, this is possible and known as Lévy's continuity theorem. We state it below, and the reader is referred to Feller [123] or Durrett [117] for a proof.

Theorem 8.25 (Lévy's Continuity Theorem) *Let X_n and X be random variables with characteristic functions $\phi_n(t)$ and $\phi(t)$, and moment generating functions $M_n(s)$, $M(s)$ (if they exist), respectively.*

(i) CHARACTERISTIC FUNCTION. *A necessary and sufficient condition for $X_n \xrightarrow{d} X$ is*

$$\lim_{n\to\infty} \phi_n(t) = \phi(t)$$

pointwise for all real t.

(ii) LAPLACE TRANSFORM. *If $M_n(t)$ and $M(t)$ exist in a common interval $[-t_0, t_0]$, and the following holds pointwise for all real $t \in [-t_0, t_0]$*

$$\lim_{n\to\infty} M_n(t) = M(t),$$

then $X_n \xrightarrow{d} X$.

In combinatorics and the analysis of algorithms, the following three corollaries are very useful. They take advantage of the fact that often generating functions of discrete structures are analytic for all complex z in a neighborhood of zero. We know that analytic functions have all derivatives; hence the underlying random variables should possess all moments.

Corollary 8.26 **(Goncharov)** *Let X_n be a sequence of discrete random variables with generating function $G_n(z) = \mathbb{E}[z^{X_n}]$, mean $\mu_n = \mathbb{E}[X_n]$ and variance $\text{Var}[X_n] = \sigma_n^2 > 0$. If*

$$\lim_{n \to \infty} e^{-\tau \mu_n / \sigma_n} G_n(e^{\tau / \sigma_n}) = e^{\tau^2 / 2} \qquad (8.135)$$

for all $\tau = it$, and $-\infty < t < \infty$, then $X_n \xrightarrow{d} N(0, 1)$, where $N(0, 1)$ stands for the random variable having the standard normal distribution.

Proof. It follows directly from the Lévy continuity theorem. ∎

Corollary 8.27 *If (8.135) holds for $\tau \in [-\theta, \theta]$, $\theta > 0$, then there exists $\alpha > 0$ such that*

$$\Pr\left\{ \frac{X_n - \mu_n}{\sigma_n} > k \right\} \le e^{-\alpha k} \qquad (8.136)$$

uniformly for all n and k.

Proof. It follows from Markov's inequality. Let $X_n^* = (X_n - \mu_n)/\sigma_n$. Since $\mathbb{E}[e^{tX_n^*}] < C$ for $t \in [-\theta, \theta]$, where C is a constant, we obtain for any $k > 0$

$$\Pr\{X_n^* > k\} \le e^{-tk} \mathbb{E}[e^{tX_n^*}] \le C e^{-k\theta},$$

as needed. ∎

Corollary 8.28 *Let X_n and X be respectively a sequence of real random variables and a real random variable such that $M_n(t) = \mathbb{E}[e^{tX_n}]$ and $M(t) = \mathbb{E}[e^{tX}]$ are their moment generating functions defined in a real neighborhood of $t = 0$. Suppose that $\lim_{n \to \infty} M_n(t) = M(t)$ for t belonging to such a real neighborhood of 0. Then X_n converges to X both in distribution and in moments (i.e., $\mathbb{E}[X_n^r] \to \mathbb{E}[X^r]$ for any $r > 0$).*

Proof. The convergence in distribution follows directly from Lévy's theorem. We concentrate on proving the second part. Let us first extend the definition of $M_n(t)$ to a complex neighborhood of $t = 0$. Observe that in such a neighborhood

$$|e^{tX_n}| = e^{\Re(t)X_n}.$$

Therefore, $M_n(t)$ exists and is bounded in a complex neighborhood of 0. Clearly, $M_n(t)$ and $M(t)$ are analytic functions. We know that $|M_n(t)|$ is uniformly bounded in this neighborhood by, say, a number $A > 0$. We redefine this neighborhood by removing all points with distance smaller than $\varepsilon > 0$ of the boundary of the former neighborhood, where ε is arbitrarily small. Due to Cauchy's estimate the derivatives $|M_n'(t)|$ are also uniformly bounded by $A\varepsilon^{-1}$, and therefore $M_n(t)$ are bounded and uniformly continuous. By *Ascoli's theorem*, from every sequence of $M_n(t)$ we can extract a convergent subsequence. This limit function can only be the unique analytic continuation of $M(t)$; thus $M_n(t)$ converges to $M(t)$ in this complex neighborhood of 0. The convergence in moments follows since convergence of analytic functions implies the convergence of their derivatives.   ∎

We should add one more comment. When deriving generating functions of discrete structures, we can often obtain an error term. To translate such an analytic error term into a distribution error term, we often use the Cauchy estimate. But in some situations the following estimate of Berry and Essén is very useful. The proof can be found in Durrett [117] or Feller [123]. We write $\| f \|_\infty := \sup_x |f(x)|$.

Lemma 8.29   (Berry-Essén Inequality)   *Let F and G be distribution functions with characteristic functions $\phi_F(t)$ and $\phi_G(t)$. Assume that G has a bounded derivative. Then*

$$\| F - G \|_\infty \leq \frac{1}{\pi} \int_{-T}^{T} \left| \frac{\phi_F(t) - \phi_G(t)}{t} \right| dt + \frac{24}{\pi} \frac{\| G' \|_\infty}{T} \tag{8.137}$$

for any $T > 0$.

These analytic tools allow us to derive many convergence results in probability theory, analytic combinatorics, and analysis of algorithms. For example, consider the *central limit theorem*. It concerns the sum $S_n = X_1 + \cdots + X_n$ of i.i.d. random variables with mean μ and variance σ^2. Consider the normalized random variable

$$S_n^* = \frac{1}{\sigma\sqrt{n}} \sum_{i=1}^{n} \bar{X}_i,$$

where $\bar{X}_i = X_i - \mu_i$. The CF of S_n^* is

$$\phi_{S_n^*}(t) = \phi_{\bar{X}}^n \left(\frac{t}{\sigma\sqrt{n}} \right).$$

But Taylor's expansion of $\phi_{\bar{X}}(\tau)$ around $\tau = 0$ gives

$$\phi_{\bar{X}}^n\left(\frac{t}{\sigma\sqrt{n}}\right) = \left(1 - \frac{t^2}{2n} + o(t^2/2n)\right)^n \rightarrow e^{-t^2/2},$$

which proves the central limit theorem.

In the analysis of algorithms, we often deal with weakly dependent random variables $X_1, \ldots, X_n, Y$ such that

$$Z_n = Y + X_1 + \cdots + X_n$$

satisfies the following asymptotic formula

$$G_{Z_n}(z) = G_Y(z)[G_X(z)]^{\beta_n}\left(1 + O(\kappa_n^{-1})\right),$$

where $\beta_n, \kappa_n \rightarrow \infty$ as $n \rightarrow \infty$ and $G_X(z)$ is the probability generating function of X_1. Can we infer the central limit law from this asymptotic expression? The answer is *yes* as demonstrated by Bender [36], Bender and Richmond [40], Hwang [201], and Flajolet and Soria [151].

Theorem 8.30 (Hwang, 1994) *Assume that the moment generating functions* $M_n(s) = \mathbf{E}[e^{sX_n}]$ *of a sequence of random variables X_n are analytic in the disc* $|s| < \rho$ *for some $\rho > 0$, and satisfy there the expansion*

$$M_n(s) = e^{\beta_n U(s) + V(s)}\left(1 + O\left(\frac{1}{\kappa_n}\right)\right) \tag{8.138}$$

for $\beta_n, \kappa_n \rightarrow \infty$ as $n \rightarrow \infty$, and $U(s), V(s)$ are analytic in $|s| < \rho$. Assume also that $U''(0) \neq 0$. Then

$$\mathbf{E}[X_n] = \beta_n U'(0) + V'(0) + O(\kappa_n^{-1}), \tag{8.139}$$

$$\mathbf{Var}[X_n] = \beta_n U''(0) + V''(0) + O(\kappa_n^{-1}), \tag{8.140}$$

and for any fixed x

$$\Pr\left\{\frac{X_n - \beta_n U'(0)}{\sqrt{\beta_n U''(0)}} \leq x\right\} = \Phi(x) + O\left(\frac{1}{\kappa_n} + \frac{1}{\sqrt{\beta_n}}\right), \tag{8.141}$$

where $\Phi(x)$ is the distribution function of the standard normal distribution.

Proof. We shall follow Hwang [201]. The convergence in distribution and in moments follow from Theorem 8.25 and Corollary 8.28, so we need only derive the rate of convergence. We shall use the Berry-Essén estimate with $T_n = \sqrt{\beta_n}$ to prove that

$$\| F_n - \Phi \|_\infty \leq \frac{1}{\pi} \int_{T_n}^{T_n} \left| \frac{M_n^*(it) - e^{-t^2/2}}{t} \right| dt + \frac{24}{\pi} \frac{c}{T_n} = O\left(\kappa_n^{-1} + \beta_n^{-1/2} \right),$$

(8.142)

where c is a constant, and $M_n^*(s)$ stands for the MGF of $X_n^* = (X_n - \beta_n U'(0))/\sqrt{\beta_n U''(0)}$. Indeed, setting $\sigma_n = \sqrt{\beta_n U''(0)}$, and using (8.138), we obtain

$$M_n^*(it) = \exp\left(-\frac{\beta_n U'(0)}{\sigma_n} it + U\left(\frac{it}{\sigma_n} \right) \beta_n + V\left(\frac{it}{\sigma_n} \right) \right) \left(1 + O(\kappa_n^{-1}) \right)$$

$$= \exp\left(-\frac{t^2}{2} + O\left(\frac{|t| + |t|^3}{\sigma_n} \right) \right) \left(1 + O(\kappa_n^{-1}) \right) \quad |t| \leq \sqrt{\beta_n}.$$

In the above, we used the following Taylor expansions valid for $|t| \leq \sqrt{\beta_n}$

$$U\left(\frac{it}{\sigma_n} \right) = \frac{it}{\sigma_n} U'(0) - \frac{t^2}{2\sigma_n} U''(0) + O\left(\frac{|t|^3}{\sigma_n} \right),$$

$$V\left(\frac{it}{\sigma_n} \right) = O\left(\frac{|t|}{\sigma_n} \right).$$

Since $|e^w - 1| \leq |w| e^{|w|}$, we find

$$|M_n^*(it) - e^{-t^2/2}| = O\left(\left(\frac{|t| + |t|^3}{\sigma_n} + \frac{1}{\kappa_n} \right) \exp\left(-\frac{t}{2} + O\left(\frac{|t| + |t|^3}{\sigma_n} + \frac{1}{\kappa_n} \right) \right) \right)$$

$$= O\left(\left(\frac{|t| + |t|^3}{\sigma_n} + \frac{1}{\kappa_n} \right) e^{-t^2/4} \right) \quad |t| \leq \beta_n.$$

After simple algebra, this leads to (8.142) and the proof is complete.   ∎

Example 8.22: *Pattern Occurrences Revisited*   We shall continue the discussion of the problem from Section 7.6.2, where we analyzed the number of occurrences O_n of a given pattern H in a random text. In particular, we extend Theorem 7.14 and prove the following: *Let $r = E[O_n] + x\sqrt{Var[O_n]}$ for $x = O(1)$, then*

$$\Pr\{O_n(H) = r\} = \frac{1}{\sqrt{2\pi c_1 n}} e^{-\frac{1}{2}x^2} \left(1 + O\left(\frac{1}{\sqrt{n}} \right) \right),$$

(8.143)

where c_1 is defined in (7.96) of Theorem 7.14. Let $\mu_n = \mathbf{E}[O_n(H)] = (n - m + 1)P(H)$ and $\sigma_n^2 = \mathbf{Var}[O_n(H)] \sim c_1 n$. By Cauchy's theorem

$$T_n(u) = \frac{1}{2\pi i} \oint \frac{T(z, u)}{z^{n+1}} dz = \frac{1}{2\pi i} \oint \frac{u P(H)}{D_H^2(z)(1 - u M_H(z))z^{n+1-m}} dz,$$

where the integration is along a circle around the origin. To evaluate this integral we enlarge the circle of integration to a bigger one, say $R > 1$, such that the bigger circle contains the dominating pole of the integrand function. By Cauchy's estimate we conclude that the integral over the bigger circle is $O(R^{-n})$. We now need to evaluate the residues inside the circle. Let us now substitute $u = e^t$ and $z = e^\rho$ (so that $T_n(e^t)$ becomes the moment generating function of O_n). Then the poles of the integrand are the roots of the equation

$$1 - e^t M_H(e^\rho) = 0.$$

This equation implicitly defines, in some neighborhood of $t = 0$, a unique differentiable function $\rho(t)$, satisfying $\rho(0) = 0$. Notably, all other roots ρ satisfy $\inf |\rho| = \rho' > 0$. Then the residue theorem leads to

$$T_n(e^t) = C(t)e^{-(n+1-m)\rho(t)} + O(R^{-n}), \tag{8.144}$$

where

$$C(t) = \frac{P(H)}{D_H^2(\rho(t))M_H'(\rho(t))}.$$

But (8.144) satisfies the assumption of Theorem 8.30, which implies the desired result.

Finally, we observe that (8.144) can be used to obtain the large deviations result. Indeed, we conclude that

$$\lim_{n \to \infty} \frac{\log T_n(e^t)}{n} = -\rho(t).$$

Thus directly from the Gärtner-Ellis Theorem 5.15 we obtain for $a > P(H)$

$$\lim_{n \to \infty} \frac{\log \Pr\{O_n > na\}}{n} = -I(a),$$

where

$$I(a) = a\omega_a + \rho(\omega_a)$$

with ω_a being the solution of

$$-\rho'(\omega_a) := a.$$

In passing we observe that this can be also derived directly by the shift of mean method. In Exercise 23 we propose a stronger version of the large deviations result. ∎

8.7   APPLICATIONS

It is again time to apply what we have learned so far. Since this chapter is very long, we will be rather sketchy. We start with a combinatorial problem on words, where we find a set of words that approximately self-overlap. This problem combines methodologies of the previous two chapters. Then we plunge into singularity analysis to provide an asymptotic expansion for the minimax redundancy for memoryless sources. Finally, in our most sophisticated application we derive the limiting distribution of the depth in a digital search tree.

8.7.1   Variations on Approximate Self-Overlapping Words

We are interested here in the structure of a word $w_k := w_1^k \in \mathcal{A}^k$ of length k such that when shifted by, say, s, the shifted word is within a given Hamming distance from the original (unshifted word). More precisely, let the set of all words of length k be denoted as $\mathcal{W}_k := \mathcal{A}^k$. A prefix of length $q \le k$ of w_k is written as $\overline{w}_k(q)$ or simply $\overline{w}_k$ if there is no confusion. The distance between words is understood as the relative Hamming distance, that is, $d_n(x_1^n, \widetilde{x}_1^n) = n^{-1} \sum_{i=1}^n d_1(x_i, \widetilde{x}_i)$, where $d_1(x, \widetilde{x}) = 0$ for $x = \widetilde{x}$ and 1 otherwise $(x, \widetilde{x} \in \mathcal{A})$. We also write $M(x_1^n, \widetilde{x}_1^n) = nd_n(x_1^n, \widetilde{x}_1^n)$ for the number of mismatches between x_1^n and $\widetilde{x}_1^n$. We now fix $D \ge 0$. Consider a word $w_{k+s} = w_1^{k+s}$ of length $k + s$, and shift it by $s \le k$. The shifted word of length k is w_s^{k+s}. We are interested in the set $\mathcal{W}_{k,s}(D)$ of all words w_{k+s} such that

$$d(w_1^k, w_s^{k+s}) \le D. \tag{8.145}$$

This problem is well understood for the "faithful" (lossless) overlapping case, that is, when $D = 0$. Then for $m = \lfloor k/s \rfloor$ (cf. Lemma 4.9 and [289, 411, 412])

$$\mathcal{W}_{k,s}(0) = \{w_s \in \mathcal{W}_s : w_{k+s} = w_s^{(m+1)}\overline{w}_s$$

$$= \bigcup_{w_s \in \mathcal{W}_s} \{w_s^{(m+1)}\overline{w}_s\},$$

where $\overline{w}_s$ is a prefix of length $q = k - m \cdot s$, and $w_s^{(m)}$ is a concatenation of m words w_s.

We now construct all words w_{k+s} that belong to $\mathcal{W}_{k,s}(D)$. First, let us define an integer ℓ such that $\ell/k \le D < (\ell+1)/k$. Also, we write $k = s \cdot m + q$, where $0 \le q < s$. Take now $0 \le l \le \ell$, and partition the integer l into $m + 1$ integer terms as follows:

$$l = a_1 + a_2 + \cdots + a_m + \tilde{a}_{m+1} \qquad 0 \le a_i \le s \quad \text{for} \quad 1 \le i \le m \qquad (8.146)$$

and $0 \le \tilde{a}_{m+1} \le q$. Let the set of all such partitions be denoted as $\mathcal{P}_{k,s}(l)$. We now define recursively m sets $\mathcal{W}_s(a_i)$ for $i \le m$. We set $\mathcal{W}_s(a_0) := \mathcal{W}_s$, where $a_0 = 0$, and then

$$\mathcal{W}_s(a_k) = \{v_s \in \mathcal{W}_s : M(w_s, v_s) = a_k \quad \text{for} \quad w_s \in \mathcal{W}_s(a_{k-1})\},$$

together with

$$\overline{\mathcal{W}}_q(\tilde{a}_{m+1}) = \{v_q \in \mathcal{W}_q : M(\overline{w}_s(q), v_q) = \tilde{a}_{m+1} \quad \text{for} \quad w_s \in \mathcal{W}_s(a_m)\}.$$

Thus

$$\mathcal{W}_{k,s}(D) = \bigcup_{l=0}^{l} \{\mathcal{W}_{k,s}(l)\}$$

such that

$$\mathcal{W}_{k,s}(l) = \bigcup_{w_s(0) \in \mathcal{W}_s} \bigcup_{\mathcal{P}_{s,k}(l)} \bigcup_{w_s(1) \in \mathcal{W}_s(a_1)}$$
$$\cdots \bigcup_{w_s(m) \in \mathcal{W}_s(a_m)} \bigcup_{\overline{w}_s(m+1) \in \overline{\mathcal{W}}_s(\tilde{a}_{m+1})} w_s(0) w_s(1) \cdots w_s(m) \overline{w}_s(m+1),$$
$$(8.147)$$

where $w_s(0) w_s(1) \cdots w_s(m) \overline{w}_s(m+1)$ is the concatenation of words $w_s(0)$ and $\ldots \overline{w}_s(m+1)$. Summing over all ℓ we obtain $\mathcal{W}_{k,s}(D)$.

In some applications the cardinality of $\mathcal{W}_{k,s}(D)$ and $\mathcal{P}_{s,k}(l)$ are of interest. Let $G(z)$ be the generating function of the cardinality $|\mathcal{P}_{s,k}(l)|$ of $\mathcal{P}_{s,k}(l)$. By (8.146), we immediately obtain

$$G(z) = (1 + x + x^2 + \cdots + x^s)^m (1 + x + x^2 + \cdots + x^q) \qquad (8.148)$$
$$= \frac{(1 - x^{s+1})^m (1 - x^{q+1})}{(1 - x)^{m+1}}, \qquad (8.149)$$

where $m = \lfloor k/s \rfloor$ and $q = k - ms$. Let now $e_l = |\mathcal{P}_{s,k}(l)| = [z^l]G(z)$. As in Comtet [71] we introduce the *polynomial coefficients* $\binom{n,q}{k}$ defined as

$$(1 + x + \cdots + x^{q-1})^n = \sum_{k=0}^{\infty} \binom{n,q}{k} x^k \qquad (8.150)$$

for integer q. Note that $\binom{n,2}{k} = \binom{n}{k}$. Observe that

$$e_l = |\mathcal{P}_{s,k}(l)| = \sum_{j=0}^{q} \binom{m, s+1}{l-j}$$

$$= \sum_{j=0}^{q} \sum_{(s+1)i+t=l-j} (-1)^i \binom{m}{i} \binom{m+t}{m}, $$

where $m = \lfloor k/s \rfloor$ and $q = k - ms$.

The enumeration of $\mathcal{W}_{k,s}$ is even easier since

$$|\mathcal{W}_{k,s}(l)| = 2^s \sum_{a_1+a_2+\cdots+a_m+a_{m+1}=l} \binom{s}{a_1} \cdots \binom{s}{a_m} \binom{q}{a_{m+1}} = 2^s \binom{k}{l}.$$

The above directly follows from $(1+x)^s(1+x)^s \cdots (1+x)^s(1+x)^q = (1+x)^{ms+q} = (1+x)^k$.

The next interesting question is how to get asymptotics for e_l and $\binom{n,q}{k}$. We prove the following result. Let $g(z) = (\frac{1}{q} + \frac{z}{q} + \cdots + \frac{z^{q-1}}{q})$ be the probability generating function so that

$$q^n[z^k]g^n(z) = \binom{n,q}{k}.$$

From the Cauchy formula we have

$$\binom{n,q}{k} = \frac{q^n}{2\pi i} \oint \frac{g(z)^n}{z^{k+1}} dz,$$

where the path of integration encloses the origin. Judging from the binomial coefficients (i.e., $q = 2$) we should expect different asymptotics for various values of k (e.g., bounded k, k around the mean $n\mu = n(q-1)/2$, and $k = \alpha n$, where $\alpha \neq (q-1)/2$). This is summarized in Lemma 8.31 below.

Lemma 8.31 *For any q and large n the following holds.*

(i) *If $k = n(q-1)/2 + r$, where $r = o(\sqrt{n})$, then*

$$\binom{n,q}{k} \sim \frac{q^n}{\sigma\sqrt{2\pi n}} \exp\left(-\frac{r^2}{2n\sigma^2}\right), \qquad (8.151)$$

where $\sigma^2 = (q^2 - 1)/12$. In particular (cf. Comtet [71])

$$\sup_k \binom{n,q}{k} = \binom{n,q}{n(q-1)/2} \sim q^n \sqrt{\frac{6}{(q^2-1)\pi n}}. \qquad (8.152)$$

(ii) *If $k = \alpha n$, where $\alpha \neq (q-1)/2$, then*

$$\binom{n,q}{k} \sim \frac{g(\beta)^n}{\beta^{\alpha n}} \frac{1}{\sigma_\alpha \sqrt{2\pi n}}, \qquad (8.153)$$

where β is a solution of $\beta g'(\beta) = \alpha g(\beta)$ and $\sigma_\alpha^2 = \beta^2 g''(\beta)/g(\beta) + \alpha - \alpha^2$.

(iii) *If $k = O(1)$, then*

$$\binom{n,q}{k} \sim \frac{n^k}{k!} \qquad (8.154)$$

Proof. Part (i) directly follows from the saddle point method in exactly the same manner as we did in Section 8.4.3 so we omit details. Setting $r = 0$ in (8.151) we obtain (8.152). Comtet [71] suggested another derivation of it: Note that after the substitution $z = e^{ix}$, the Cauchy formula yields

$$\binom{n,q}{k} = \frac{1}{\pi} \int_{-\pi/2}^{\pi/2} \left(\frac{\sin(qx)}{\sin(x)}\right)^n \cos(x(n(q-1) - 2k))dx.$$

Observe that for $k = n(q-1)/2$ the cosine function achieves its maximum; hence by a simple application of Laplace's method we again obtain (8.152).

Part (ii) follows from (i) and the method of shifted mean as discussed in Section 8.4.3. More precisely, we use

$$[z^{\alpha n}](g(z))^n = \frac{g(\beta)^n}{\beta^{\alpha n}} \left(\frac{g(\beta z)}{g(\beta)}\right)^n$$

and find β to be $\beta g_1'(\beta) = \alpha g_1(\beta)$.

Part (iii) can be proved as follows. From the Cauchy integral we have, after substituting $z = w/n$,

$$
\binom{n, q}{k} = \frac{1}{2\pi i} \oint \frac{G(z)^n}{z^{k+1}} dz
$$

$$
= \frac{1}{2\pi i} \oint \frac{(1 + w/n + \cdots + (w/n)^{q-1})^n}{w^{k+1}} n^k dw
$$

$$
\rightarrow \frac{n^k}{2\pi i} \oint \frac{e^w}{w^{k+1}} = \frac{n^k}{k!}.
$$

This completes the proof.   ∎

8.7.2   Redundancy Rate for Memoryless Sources

We briefly discussed redundancy of source coding in Section 6.4, Example 7.16 of Chapter 7, and Example 8.7 of this chapter. Here, we show how the analytic tools presented in this chapter can be used to evaluate the minimax redundancy for memoryless sources.

We start with a brief introduction, but the reader may also want to review Chapter 6. A code $\mathscr{C}_n : \mathcal{A}^n \rightarrow \{0, 1\}^*$ is defined as a mapping from the set $\mathcal{A}^n$ of all sequences of length n over the finite alphabet $\mathcal{A}$ to the set $\{0, 1\}^*$ of all binary sequences. Given a probabilistic source model, we let $P(x_1^n)$ be the probability of the message x_1^n while for a given code $\mathscr{C}_n$ we denote by $L(\mathscr{C}_n, x_1^n)$ the code length for x_1^n.

Shannon's Theorem 6.18 asserts that the entropy

$$
H_n(P) = -\sum_{x_1^n} P(x_1^n) \log_2 P(x_1^n)
$$

is the lower bound on the expected code length. Hence, $-\log_2 P(x_1^n)$ can be viewed as the "ideal" code length. The next natural question is to ask by how much the length $L(\mathscr{C}_n, x_1^n)$ of a code differs from the ideal code length, either for individual sequences or on average. The *pointwise redundancy* $R_n(\mathscr{C}_n, P; x_1^n)$ and the *average redundancy* $\bar{R}_n(\mathscr{C}_n, P)$ are defined as

$$
R_n(\mathscr{C}_n, P; x_1^n) = L(\mathscr{C}_n, x_1^n) + \log_2 P(x_1^n)
$$

$$
\bar{R}_n(\mathscr{C}_n, P) = \mathbf{E}_P[R_n(\mathscr{C}_n, P; X_1^n)] = \mathbf{E}[L(\mathscr{C}_n), X_1^n] - H_n(P).
$$

Another natural measure of code performance is the *maximal* redundancy defined as

$$
R_n^*(\mathscr{C}_n, P) = \max_{x_1^n} \{R_n(\mathscr{C}_n, P; x_1^n)\}.
$$

Observe that while the pointwise redundancy can be negative, maximal and average redundancies cannot, by Shannon's Theorem 6.18 and Kraft's inequality (see Theorem 6.17).

In practice, the source probabilities are unknown; hence there is the desire to design codes for a whole class of source models $\mathcal{S}$. When the source is known, the redundancy can be as low as 1 bit, as demonstrated by Shannon codes (see Example 8.7). Therefore, for unknown probabilities, the redundancy rate also can be viewed as the penalty paid for estimating the underlying probability measure. More precisely, *universal codes* are those for which the redundancy is $o(n)$ for all $P \in \mathcal{S}$. The (asymptotic) *redundancy-rate problem* consists in determining for a class $\mathcal{S}$ of source models the rate of growth of the minimax quantities

$$R_n^*(\mathcal{S}) = \min_{\mathcal{C}_n} \sup_{P \in \mathcal{S}} \{R_n^*(\mathcal{C}_n, P)\}$$

as $n \to \infty$. This minimax redundancy can be viewed as the additional "price" on top of entropy incurred (at least) by any code in order to be able to cope with *all* sources.

In passing we would like to add that the redundancy rate problem is typical of a situation, where second-order asymptotics play a crucial role since the leading term of $\mathbf{E}[L(\mathcal{C}_n), X_1^n]$ is known to be nh, where h is the entropy rate. This problem is an ideal candidate for analytic information theory, which applies analytic tools to information theory.

Hereafter, we aim at establishing a precise asymptotic expansion of $R^*(\mathcal{M})$ for memoryless sources $\mathcal{M}$. First we review Shtarkov's result [390] concerning the minimax redundancy

$$R_n^*(\mathcal{S}) = \min_{\mathcal{C}_n} \sup_{P \in \mathcal{S}} \max_{x_1^n} \{L(\mathcal{C}_n, x_1^n) + \log_2 P(x_1^n)\}$$

for general class of sources $\mathcal{S}$, where the supremum is taken over all distributions $P \in \mathcal{S}$. Shtarkov proposed the following bound

$$\log_2 \left(\sum_{x_1^n} \sup_P P(x_1^n) \right) \le R_n^*(\mathcal{S}) \le \log_2 \left(\sum_{x_1^n} \sup_P P(x_1^n) \right) + 1. \qquad (8.155)$$

To prove it, Shtarkov first introduced a new probability distribution, namely,

$$q(x_1^n) := \frac{\sup_P P(x_1^n)}{\sum_{x_1^n} \sup_P P(x_1^n)}.$$

By Kraft's inequality (see Theorem 6.17) there exists $\tilde{x}_1^n$ such that

$$-L(\mathscr{C}_n, \tilde{x}_1^n) \le \log_2 q(\tilde{x}_1^n).$$

The above must be true since otherwise the Kraft inequality would be violated. The lower bound follows. For the upper bound, Shtarkov proposed the Shannon code $\widetilde{\mathscr{C}}_n$ for the distribution $q(x_1^n)$ defined above whose length is

$$L(\widetilde{\mathscr{C}}_n, x_1^n) = \left\lceil \log_2 \left(\sum_{x_1^n} \sup_P P(x_1^n) \right) - \log_2 \left(\sup_P P(x_1^n) \right) \right\rceil.$$

This gives the desired upper bound.

Now, we concentrate on studying

$$R_n := \log_2 \left(\sum_{x_1^n} \sup_P P(x_1^n) \right)$$

for a class of memoryless sources over an m-ary alphabet $\mathcal{A}(m)$. Observe that R_n is within distance one from the minimax redundancy R_n^*. Following Shtarkov we shall prove that

$$R_n = \log D_n(m),$$

where $D_n(m)$ satisfies

$$D_n(m) = \sum_{i=1}^{m} \binom{m}{i} D_n^*(i) \tag{8.156}$$

with $D_0^*(1) = 0$, $D_n^*(1) = 1$ for $n \ge 1$, and for $i > 1$ we have

$$D_n^*(i) = \sum_{k=1}^{n} \binom{n}{k} \left(\frac{k}{n}\right)^k \left(1 - \frac{k}{n}\right)^{n-k} D_{n-k}^*(i-1). \tag{8.157}$$

Indeed, we first observe that we can partition all sequences into m sets such that the ith set is built from sequences composed of symbols from the alphabet $\mathcal{A}(i)$ consisting of any i symbols from $\mathcal{A}(m)$. This implies (8.156) since

$$D_n(m) = \sum_{i=1}^{m} \binom{m}{i} \sum_{x^n \in \mathcal{A}(i)} \sup P(x^n) = \sum_{i=1}^{m} \binom{m}{i} D_n^*(i),$$

where $D_n^*(i)$ is defined above. To derive the recurrence of $D_n^*(i)$ we argue as follows: Consider an alphabet $\mathcal{A}(i-1)$ and assume that these $i-1$ symbols of $\mathcal{A}(i-1)$ occur on $n-k$ positions of x_1^n, which contributes $D_{n-k}^*(i-1)$ to the sum. On the remaining k positions we place the ith symbol with the maximum probability sup $P(x_1^n)$. This probability for memoryless sources is

$$\sup P(x_1^n) = \left(\frac{k}{n}\right)^k \left(1 - \frac{k}{n}\right)^{n-k}.$$

since, in the situation described above, one can view x_1^n as being built over a binary alphabet (i.e., $\{\mathcal{A}(i-1), i\}$) and clearly $\sup_x x^k(1-x)^{n-k} = (k/n)^k(1-k/n)^{n-k}$. This establishes the desired recurrence (8.157).

We now prove an asymptotic expansion of $D_n^*(m)$ and $D_n(m)$.

Theorem 8.32 (Szpankowski, 1998) *For fixed $m \geq 1$ the quantity $D_n^*(m)$ attains the following asymptotics*

$$D_n^*(m) = \frac{\sqrt{\pi}}{\Gamma(\frac{m}{2})} \left(\frac{n}{2}\right)^{\frac{m}{2}-\frac{1}{2}} - \frac{\sqrt{\pi}}{\Gamma(\frac{m}{2}-1/2)} \left(\frac{2m}{3}\right) \left(\frac{n}{2}\right)^{\frac{m}{2}-1} \quad (8.158)$$

$$+ \frac{\sqrt{\pi}}{\Gamma(\frac{m}{2})} \left(\frac{n}{2}\right)^{\frac{m}{2}-\frac{3}{2}} \left(\frac{3 + m(m-2)(8m-5)}{72}\right)$$

$$+ O(n^{\frac{m}{2}-2}).$$

Furthermore, for $m \geq 2$

$$R_n = \log D_n(m) = \frac{m-1}{2} \log \left(\frac{n}{2}\right) + \log \left(\frac{\sqrt{\pi}}{\Gamma(\frac{m}{2})}\right) \quad (8.159)$$

$$+ \frac{\Gamma(\frac{m}{2})m}{3\Gamma(\frac{m}{2}-1/2)} \cdot \frac{\sqrt{2}}{\sqrt{n}}$$

$$+ \left(\frac{3 + m(m-2)(2m+1)}{36} - \frac{\Gamma^2(\frac{m}{2})m^2}{9\Gamma^2(\frac{m}{2}-1/2)}\right) \cdot \frac{1}{n}$$

$$+ O\left(\frac{1}{n^{3/2}}\right)$$

for large n.

Proof. Observe that the recurrence (8.157) is the same as the one analyzed in Example 7.16 of Chapter 7. In particular, we prove there that the tree-like generating function

$$D_m^*(z) = \sum_{k=0}^{\infty} \frac{k^k}{k!} z^k D_k^*(m)$$

satisfies (cf. (7.38))

$$D_m^*(z) = (B(z) - 1)^m,$$

where $B(z)$ is related to the tree-function $T(z)$ and defined in (7.35). Observe that $B(z) = 1/(1 - T(z))$ and $[z^n]B(z) = n^n/n!$. Furthermore, by (7.39)

$$D_m(z) = B^m(z) - 1.$$

We now use the singularity analysis to obtain asymptotics of $D_n^*(m)$ and $\log_2 D_n(m)$. In Example 8.10 we derive asymptotic expansion of $T(z) - 1$ around $z = e^{-1}$ from which we directly obtain

$$B(z) = \frac{1}{\sqrt{2(1 - ez)}} + \frac{1}{3} - \frac{\sqrt{2}}{24}\sqrt{(1 - ez)}\frac{4}{135}(1 - ez)$$

$$+ -\frac{23\sqrt{2}}{1728}(1 - ez)^{3/2} + O((1 - ez)^2). \tag{8.160}$$

Then applications of Theorem 8.12 and Theorem 8.13 yield

$$e^{-n}[z^n](B(z) - 1)^m = \frac{n^{\frac{m}{2}-1}}{2^{\frac{m}{2}}\Gamma(\frac{m}{2})} - \frac{n^{\frac{m}{2}-\frac{3}{2}}}{2^{\frac{m}{2}-\frac{1}{2}}}\left(\frac{2m}{3\Gamma(\frac{m}{2} - 1/2)}\right)$$

$$+ \frac{n^{\frac{m}{2}-2}}{2^{\frac{m}{2}}}\left(\frac{m(m - 2)(8m - 5)}{36\Gamma(\frac{m}{2})}\right) + O(n^{\frac{m}{2}-\frac{5}{2}}).$$

This establishes the first part of the theorem. The second part immediately follows from

$$e^{-n}[z^n]B^m(z) = \frac{n^{\frac{m}{2}-1}}{2^{\frac{m}{2}}\Gamma(\frac{m}{2})} + \frac{n^{\frac{m}{2}-\frac{3}{2}}}{2^{\frac{m}{2}-\frac{1}{2}}}\left(\frac{m}{3\Gamma(\frac{m}{2} - 1/2)}\right)$$

$$+ \frac{n^{\frac{m}{2}-2}}{2^{\frac{m}{2}}}\left(\frac{m(m - 2)(2m + 1)}{36\Gamma(\frac{m}{2})}\right) + O(n^{\frac{m}{2}-\frac{5}{2}})$$

and

$$\log(1 + a\sqrt{x} + bx + cx^{3/2}) = a\sqrt{x} + (b - 1/2a^2)x + O(x^{3/2})$$

as $x \to 0$. This completes the proof. ∎

8.7.3 Limiting Distribution of the Depth in Digital Search Trees

We shall derive here the limiting distribution of the typical depth D_m in a digital search tree (see Section 1.1). We assume that m independent strings are generated by a binary memoryless source with p being the probability of emitting "0" and $q = 1 - p$. Recall that the typical depth is defined as follows. Let $D_m(i)$ be the depth of the ith node in a digital tree. Actually, observe that $D_m(i) = D_i(i)$ for $m \geq i$. Clearly, for various $i \leq m$ distributions of $D_m(i)$ are different, and therefore it makes sense to define the *typical depth* D_m as

$$\Pr\{D_m < x\} = \frac{1}{m} \sum_{i=1}^{m} \Pr\{D_m(i) < x\}.$$

Let B_m^k denote the number of internal nodes at level k in a digital tree. To derive a recurrence on D_m we introduce the *average profile* $\bar{B}_m^k := \mathbf{E}[B_m^k]$ to be the average number of internal nodes at level k in a digital tree. The following relationship between the depth D_m and the average profile $\bar{B}_m^k$ holds:

$$\Pr\{D_m = k\} = \frac{\bar{B}_m^k}{m}. \tag{8.161}$$

This follows from the definition of D_m and the definition of $\bar{B}_m^k$.

We shall work initially with the average profile, and we define the generating function $B_m(u) = \sum_{k=0}^{\infty} \bar{B}_m^k u^k$, which satisfies the following recurrence

$$B_{m+1}(u) = 1 + u \sum_{j=0}^{m} \binom{m}{j} p^j q^{m-j} (B_j(u) + B_{m-j}(u)) \tag{8.162}$$

with $B_0(u) = 0$. This recurrence arises naturally in our setting by considering the left and the right subtrees of the root, and noting that j strings will go to the left subtree with the probability $\binom{m}{j} p^j q^{m-j}$. A general recurrence of this type was discussed in Section 7.6.1 from which we conclude that

$$B_m(u) = m - (1 - u) \sum_{k=2}^{m} (-1)^k \binom{m}{k} Q_{k-2}(u), \tag{8.163}$$

where

$$Q_k(u) = \prod_{j=2}^{k+1} (1 - up^j - uq^j) \qquad Q_0(u) = 1. \tag{8.164}$$

Actually, the derivation of (8.163) is not too complicated, so we provide a sketch of the proof. Let us start with multiplying both sides of (8.163) by $z^m/m!$ to get $B_z'(z, u) = e^z + uB(pz, u)e^{qz} + uB(qz, u)e^{pz}$, where $B(z, u) = \sum_{m=0}^{\infty} B_m(u)\frac{z^m}{m!}$, and $B_z'(z, u)$ is the derivative of $B(z, u)$ with respect to z. We now multiply this functional equation by e^{-z} and introduce $\widetilde{B}(z, u) = B(z, u)e^{-z}$. This leads to a new functional equation, namely $\widetilde{B}'(z, u) + \widetilde{B}(z, u) = 1 + u(\widetilde{B}(zp, u) + \widetilde{B}(zq, u))$. Comparing now the coefficients at z^m one immediately obtains $\widetilde{B}_{m+1}(u) = \delta_{m,0} - \widetilde{B}_m(u)(1 - up^m - uq^m)$, where $\delta_{0,m}$ is the Kronecker symbol. To prove (8.163) it only suffices to note that $B_m(u) = \sum_{k=0}^{m} \binom{m}{k} \widetilde{B}_k(u)$.

Using again the idea from Section 7.6.1, Example 8.19 and the Rice method we shall sketch a proof of the following result. In Exercise 24 the reader is asked to provide details.

Theorem 8.33   (Kirschenhofer and Prodinger, 1988; Szpankowski, 1991)
Consider a digital search tree built over independently generated sequences.

(i) *The average* $\mathbf{E}[D_m]$ *of the depth attains the following asymptotics as* $m \to \infty$

$$\mathbf{E}[D_m] = \frac{1}{h}\left(\log m + \gamma - 1 + \frac{h_2}{2h} + \theta + \delta(m)\right) + O\left(\frac{\log m}{m}\right),$$
(8.165)

where h *is the entropy*, $h_2 = p\log^2 p + q\log^2 q$, $\gamma = 0.577\ldots$ *is the Euler constant, and*

$$\theta = -\sum_{k=1}^{\infty} \frac{p^{k+1}\log p + q^{k+1}\log q}{1 - p^{k+1} - q^{k+1}}.$$

The function $\delta(x)$ *is a fluctuating function with a small amplitude when* $\log p/\log q$ *is rational, and* $\delta(x) \to 0$ *as* $n \to \infty$ *for* $\log p/\log q$ *irrational. More precisely, for* $\log p/\log q = r/t$*, where* r, t *are integers,*

$$\delta_1(x) = \sum_{\substack{\ell=-\infty \\ \ell \neq 0}}^{\infty} \frac{\Gamma(s_0^\ell)Q(-2)}{Q(s_0^\ell - 1)} \exp\left(-\frac{2\pi i\ell r}{\log p}\log x\right),$$
(8.166)

where $s_0^\ell = -1 + 2\pi i\ell r/\log p$.

(ii) *The variance of* D_m *for large* m *satisfies*

$$\mathbf{Var}[D_m] = \frac{h_2 - h^2}{h^3}\log m + A + \Delta(m) + O(\log^2 m/m),$$
(8.167)

where A *is a constant (see Exercise 24) and* $\Delta(x)$ *is a fluctuating function with a small amplitude. In the symmetric case* $(p = q)$*, the coefficient at* $\log m$ *becomes zero, and then*

$$\text{Var}[D_m] = \frac{1}{12} + \frac{1}{\log^2 2} \cdot \frac{\pi^2}{6} - \alpha - \beta + \Delta(\log_2 m) - [\delta_1^2]_0 + O(\log^2 m/m),$$

(8.168)

where

$$\alpha = \sum_{j=1}^{\infty} \frac{1}{2^j - 1} \qquad \beta = \sum_{j=1}^{\infty} \frac{1}{(2^j - 1)^2},$$

the function $\Delta(x)$ is continuous with period 1 and mean zero, and $|[\delta_1^2]_0| < 10^{-10}$ is the mean of $\delta_1^2(x)$.

Sketch of Proof. It suffices to derive asymptotics of the average path length $l_m := m\mathbf{E}[D_m]$ and the second factorial moment of the path length $l_m^2 := m\mathbf{E}[D_m(D_m - 1)]$. But $l_m = B_m'(1)$ and $l_m^2 = B_m''(1)$. After some algebra, we arrive at the following two recurrences for all $m \geq 0$

$$l_{m+1} = m + \sum_{k=0}^{m} \binom{m}{k} p^k q^{m-k}(l_k + l_{m-k}),$$

(8.169)

$$l_{m+1}^2 = 2(l_{m+1} - m) + \binom{m}{k} p^k q^{m-k}(l_k^2 + l_{m-k}^2)$$

(8.170)

with $l_0 = l_0^2 = 0$. The recurrence (8.169) falls directly under the general recurrence (7.57) of Section 7.6.1. Its solution is

$$l_m = \sum_{k=2}^{m} (-1)^k \binom{m}{k} Q_{k-2},$$

(8.171)

where $Q_k = Q_k(1) = \prod_{j=2}^{k+1}(1 - p^j - q^j)$. This is an alternating sum with the kernel Q_k of a polynomial growth, so Rice's method applies. We only need analytic continuation of Q_k but this is easy. Indeed, define

$$Q(z) = \frac{P(0)}{P(z)},$$

where $P(z) = \prod_{j=2}^{\infty}(1 - p^{z+j} - q^{z+j})$. Then $Q_k = Q(z)|_{z=k}$ and by Rice's method or Theorem 8.21

$$l_m = \frac{1}{2\pi i} \int_{-\frac{3}{2}-i\infty}^{-\frac{3}{2}+i\infty} \Gamma(z) n^{-z} Q(-z - 2) dz + O(\log m).$$

The integral can be evaluated in the same manner as in Example 8.19, so we only sketch it. Considering singularities right to the line of the integration, we must take into account singularities of the gamma function at $z_0 = 0$ and $z_{-1} = -1$ together with all roots of

$$p^{j-2-z} + q^{j-2-z} = 1$$

that we denote as $z_{k,j}$. The main contribution comes from $z_{-1} = z_{0,2} = -1$. Using the expansions

$$\Gamma(z) = \frac{1}{z+1} + \gamma - 1 + O(z+1),$$

$$n^{-z} = n - n \log n(z+1) + O((z+1)^2),$$

$$\frac{1}{1 - p^{-z} - q^{-z}} = -\frac{1}{h}\frac{1}{z+1} + \frac{h_2}{2h^2} + O(z+1),$$

$$Q(-z-2) = -\frac{1}{h}\frac{1}{z+1} + \frac{\theta}{h} + \frac{h_2}{2h^2} + (z+1)\frac{\theta h_2}{2h^2} + O((z+1)^2),$$

and extracting the coefficient at $(z+1)^{-1}$ we establish the leading term of the asymptotics without $\delta_1(m)$. To obtain $\delta_1(m)$ we must consider the roots at $z_{k,j}$ other than $z_{-1} = -1$. Using Lemma 8.22 we finally prove (8.165).

The asymptotics of l_m^2 can be obtained in a similar manner but with more involved algebra. To use (7.70) of Section 7.6.1 we need to find the binomial inverse to l_m. But solution (8.171) implies that $\hat{l}_m = Q_{m-2}$. With this in mind we directly obtain

$$l_m^2 = -2 \sum_{k=2}^{m} \binom{m}{k}(-1)^k Q_{k-2} T_{k-2},$$

where

$$T_n = \sum_{i=2}^{n+1} \frac{p^i + q^i}{1 - p^i - q^i}.$$

To apply Rice's method we need analytic continuation of T_n. A "mechanical derivation" works fine here after observing that

$$T_{n+1} = T_n + \frac{p^{n+2} + q^{n+2}}{1 - p^{n+2} - q^{n+2}}.$$

Replacing n by z and solving the recurrence with respect to z we find

$$T(z) = \theta - \sum_{i=2}^{\infty} \frac{p^{z+i} + q^{z+i}}{1 - p^{z+i} - q^{z+i}},$$

where θ is defined in the statement of the theorem. The above derivation can be easily made rigorous by observing that all series involved do converge for $\Re(z) > -2$.   ∎

We are now ready to derive the limiting distribution of D_m for $p \neq q$. We prove the following result.

Theorem 8.34 **(Louchard and Szpankowski, 1995)** *For biased memoryless source $(p \neq q)$ the limiting distribution of D_m is normal, that is,*

$$\frac{D_m - E[D_m]}{\sqrt{\text{Var}[D_m]}} \to N(0, 1), \tag{8.172}$$

where $E[D_m]$ and $\text{Var}[D_m]$ are given by (8.165) and (8.167), respectively, and the moments of D_m converge to the appropriate moments of the normal distribution. More generally, for any complex ϑ

$$e^{-\vartheta c_1 \log m} E(e^{\vartheta D_m / \sqrt{c_2 \log m}}) = e^{\frac{\vartheta^2}{2}} \left(1 + O \left(\frac{1}{\sqrt{\log m}} \right) \right), \tag{8.173}$$

where $c_1 = 1/h$ and $c_2 = (h_2 - h^2)/h^3$.

Proof. We shall work with the probability generating function $D_m(u)$ for the depth which is equal to $B_m(u)/m$, that is,

$$D_m(u) = 1 - \frac{1 - u}{m} \sum_{k=2}^{m} (-1)^k \binom{m}{k} Q_{k-2}(u). \tag{8.174}$$

Let $\mu_m = E[D_m]$ and $\sigma_m^2 = \text{Var}[D_m]$. Theorem 8.33 implies $\mu_m \sim c_1 \log m$ and $\sigma_m^2 \sim c_2 \log m$, where $c_1 = 1/h$ and $c_2 = (h_2 - h^2)/h^3$. We use Goncharov's Theorem 8.26 to establish the normal distribution of D_m by showing that

$$\lim_{m \to \infty} e^{-\vartheta \mu_m / \sigma_m} D_m(e^{\vartheta / \sigma_m}) = e^{\vartheta^2 / 2}, \tag{8.175}$$

where ϑ is a complex number. By Corollary 8.28, this will establish both the limiting distribution and the convergence in moments for D_m.

We now derive asymptotics for the probability generating function $D_m(u)$ around $u = 1$. We assume $u = e^v$, and due to $\sigma_m = O(\sqrt{\log m})$, we define

$v = \vartheta/\sigma_m \to 0$. Note that $1 - D_m(u)$ given in (8.174) has the form of an alternating sum. We now apply the Rice method, more precisely, Theorem 8.21. To do so, however, we need an analytical continuation of $Q_k(u)$ defined in (8.164). Denote it as $Q(u, s)$, and observe that

$$Q(u, s) = \frac{P(u, 0)}{P(u, s)} = \frac{Q_\infty(u)}{P(u, s)}, \tag{8.176}$$

where $P(u, s) = \prod_{j=2}^{\infty}(1 - up^{s+j} - uq^{s+j})$. Using Theorem 8.21 we arrive at

$$1 - D_m(u) = \frac{1 - u}{m2\pi i} \int_{-3/2-i\infty}^{-3/2+i\infty} \Gamma(s)m^{-s}Q(u, -s - 2)ds + e_m, \tag{8.177}$$

where $e_m = O(1/m^2) \int_{-3/2-i\infty}^{-3/2+i\infty} \Gamma(s)m^{-s}sQ(u, -s - 2)ds = O(1/m)$, as we shall soon see. As usual, we evaluate the integral in (8.177) by the residue theorem. We compute residues *right* to the line of integration in (8.177). The gamma function has its singularities at $s_{-1} = -1$ and $s_0 = 0$, and in addition we have infinite number of zeros $s_{k,j}(v)$ ($j = 2, 3, \ldots, k = 0\pm1, \pm2, \ldots$) of $P(e^v, -s-2)$ of the denominator of $Q(e^v, -s - 2)$, where we substituted $u = e^v$. More precisely, $s_{k,j}(v)$ are zeros of

$$p^{-s-2+j} + q^{-s-2+j} = e^{-v}. \tag{8.178}$$

The dominating contribution to the asymptotics comes from $s_{0,j}(v)$ and s_1. Indeed, the contributions of the first two singularities at s_{-1} and s_0 are, respectively, $(1 - u)Q(u, -1) = 1$ and $(1 - u)Q(u, -2)/m = O(1/m)$. We now concentrate on the contribution coming from $s_{0,j}(v)$. In this case, one can solve equation (8.178) to derive

$$s_{0,j}(v) = j - 3 - \frac{v}{h} - \frac{1}{2}\left(\frac{1}{h} - \frac{h_2}{h^3}\right)v^2 + O(v^3) \tag{8.179}$$

for integer $j \geq 2$ and $v \to 0$. We also note that $\Im(s_{k,j}(v)) \neq 0$ for $k \neq 0$.

Let now $R_k^j(v)$ denote the residue of $(1 - e^v p^{-s_{k,j}-2+j} + e^v q^{-s_{k,j}-2+j})^{-1}$ at $s_{k,j}(v)$, and let $g(s) = \Gamma(s)Q(u, -s - 1)$. In the sequel, we use the following expansion

$$Q(u, -s - 2) = \frac{1}{1 - u(p^{-s} + q^{-s})} \cdot \frac{Q_\infty(u)}{P(u, -s - 1)}$$

$$= -\frac{w^{-1}}{h} - \frac{\theta}{h} + \frac{h_2}{2h^2} + w\frac{\theta h_2}{2h^2} + O(w^2),$$

where $w = s - s_{0,j}(v)$. Then

$$-D_m(e^v) = R_0^2(v)g(s_{0,2}(v))(1 - e^v)m^{-1}m^{-s_{0,2}(v)}$$

$$+ \sum_{j=3}^{\infty} R_0^j(v)g(s_{0,j}(v))(1 - e^v)m^{-1}m^{-s_{0,j}(v)}]$$

$$+ \sum_{\substack{k=-\infty \\ k \neq 0}}^{\infty} \sum_{j=2}^{\infty} R_k^j(v)g(s_{k,j}(v))(1 - e^v)m^{-1}m^{-s_{k,j}(v)} + O(1). \quad (8.180)$$

We consider the above three terms separately:

(a) $j = 2$ and $k = 0$

Set $v = \vartheta/\sigma_m = \vartheta/\sqrt{c_2 \log m}$. Then by (8.179)

$$m^{-s_{0,2}(v)} = m \exp\left(\frac{\vartheta}{h}\sqrt{\frac{\log m}{c_2}} + \frac{\vartheta^2}{2}\right).$$

In addition, the following holds: $R_0^2(v) = -1/h + O(v)$, and $g(s_{0,2}(v)) = -h/v + O(1)$, and finally $1 - e^{-v} = v + O(1)$. Therefore, we obtain

$$e^{-\vartheta\mu_m/\sigma_m}R_0^2(v)g(s_{0,2}(v))(1 - e^{-v})m^{-s_{0,2}(v)-1} \rightarrow -e^{\vartheta^2/2} \quad (8.181)$$

(b) $j \geq 3$ and $k = 0$

In this case we can repeat the analysis from case (a) to get

$$e^{-\vartheta\mu_m/\sigma_m}R_0^2(v)g(s_{0,2}(v))(1 - e^{-v})m^{-s_{0,2}(v)-1} \rightarrow O(m^{2-j}e^{\vartheta^2/2}), \quad (8.182)$$

so this term is of order magnitude smaller than the first term in (8.180).

(c) $k \neq 0$

Fix $j = 2$. Observe that

$$\sum_{\substack{k=-\infty \\ k \neq 0}}^{\infty} R_k^2(v)g(s_{k,2}(v))(1 - e^v)m^{-1}m^{-s_{k,2}(v)} = O(vm^{-1-\Re(s_{k,2}(v))}).$$

By the same argument as in Lemma 8.22 we conclude that $\Re(s_{k,j}) \geq j - 3 + O(v)$. If $\Re(s_{k,j}) > j - 3 + O(v)$, then the term $O(vm^{-1-\Re(s_{k,2}(v))})$ is negligible compared to $O(m^{2-j})$. Otherwise, that is, for $s_{k,j}$ such that $\Re(s_{k,j}) = j - 3 + O(v)$, we observe that the error term is $O(vm^{j-2})$. Actually, we can do better using Jacquet and Régnier [212] observation (the reader is asked to prove it in Exercise 25; for a generalization see Jacquet and Szpankowski [215]):

$$\Re(s_{k,2}(v)) \geq s_{0,2}(\Re(v)),$$

so the above sum becomes

$$\sum_{\substack{k=-\infty \\ k \neq 0}}^{\infty} R_k^2(v) g(s_{k,2}(v))(1 - e^v)m^{-1}m^{-s_{k,2}(v)} = m^{-1-\Re(s_{0,2}(v))} O(vm^{\Re(s_{0,2}(v))-s_{0,2}(\Re(v))})$$

$$= m^{-1-\Re(s_{0,2}(v))} O(vm^{-\beta v^2})$$

for some β. Finally, consider general $j \geq 3$. As in case (b), we note that $m^{-s_{k,j}(v)}$ contributes $O(m^{2-j})$, so this term is negligible.

Putting everything together, we note that as $v = O(1/\sqrt{m}) \to 0$ for $m \to \infty$

$$e^{-\vartheta \mu_m/\sigma_m} D_m(e^{\vartheta/\sigma_m}) = e^{\vartheta^2/2}(1 + O(vm^{-\beta v^2}) + O(1/m)) \to e^{\vartheta^2/2}, \quad (8.183)$$

which proves the theorem.   ∎

8.8   EXTENSIONS AND EXERCISES

8.1   Prove that if $f \sim \sum a_k \phi_k$ and $g \sim \sum b_k \phi_k$, then $\alpha f + \beta g \sim \sum(\alpha a_k + \beta b_k)\phi_k$.

8.2   Prove the following theorems:

Theorem 8.35   Let $f(x) \sim \sum_{k=0}^{\infty} a_k x^k$ as $x \to 0$.
(i) If $g(z) \sim \sum_{k=0}^{\infty} b_k x^k$ as $x \to 0$, then $f(x)g(x) \sim \sum_{n=0}^{\infty} c_n x^n$, where $c_n = \sum_{k=0}^{n} a_k b_{n-k}$.
(ii) We have

$$\int_0^x f(t)dt \sim \sum_{k=0}^{\infty} \frac{a_k}{k+1} x^{k+1}.$$

(iii) If $f(x)$ has a continuous derivative and if $f'(x)$ possesses an asymptotic expansion, then

$$f'(x) \sim \sum_{k=1}^{\infty} k a_k x^{k-1}$$

as $x \to 0$.

(iv) *For any power series $\sum_{k=0}^{\infty} a_k x^k$ there exists a function $f(x)$ with this as its asymptotic expansion.*

8.3 Using similar arguments to those presented in Section 8.1.2 show that

$$\log(1 + x) = \sum_{k=1}^{\infty} \frac{(-1)^{k-1}}{k} x^k, \qquad x \to 0,$$

$$= \log x + \sum_{k=1}^{\infty} \frac{(-1)^{k-1}}{k} x^{-k}, \qquad x \to \infty.$$

8.4 Prove Abel's summation formulas (8.27) and (8.28).

8.5 Consider again recurrence (8.39) for the height in a b-trie. In this exercise we will analyze it for

$$k = \frac{b+1}{b} \log_2 n + \beta$$

on the β-scale, where $\beta = O(1)$. We set

$$h_n^k = F(\beta; n) = F(k - (1 + 1/b) \log_2 n; n)$$

so that (8.39) becomes

$$F(\beta + 1; n) = \left(\frac{1}{2}\right)^n \sum_{i=0}^{n} \binom{n}{i} \cdot F\left(\beta - \left(1 + \frac{1}{b}\right) \log_2 \left(\frac{i}{n}\right); i\right)$$

$$\cdot F\left(\beta - \left(1 + \frac{1}{b}\right) \log_2 \left(1 - \frac{i}{n}\right); n - i\right).$$

Postulating that $F(\beta; n)$ assumes an expansion of the form

$$F(\beta; n) = F_0(\beta) + \frac{1}{n} F_1(\beta) + O(n^{-2}),$$

prove that F_0 satisfies the following nonlinear functional equation

$$F_0(\beta + 1) = [F_0(\beta + 1 + 1/b)]^2.$$

Next prove that a general solution of the above is

$$F_0(\beta) = \exp(-c2^{-b\beta}) = \exp(-ce^{-b\beta\log 2}).$$

Finally, using matched asymptotics show that

$$c = \frac{1}{(b+1)!}.$$

8.6 Knessl and Szpankowski analyzed in [259] the height of a PATRICIA trie. Among others, the following recurrence arises

$$(2 - 2^n)L_2(n) + 2nL_2(n-1) = 0, \quad n \ge 3,$$

with $L_2(2) = 1$. Show that it has the following solution

$$L_2(n) = n!2^{-n^2/2}2^{n/2}\prod_{m=3}^{n}\left(\frac{1}{1 - 2^{1-m}}\right), \quad n \ge 2.$$

Similarly, show that the recurrence

$$(2 - 2^n)L_3(n) + 2nL_3(n-1) + 2\binom{n}{2}L_2(n-2) = 0, \quad n \ge 5$$

with $L_3(3) = 1$ and $L_3(4) = 6$, leads to

$$L_3(n) = n!2^{-n^2/2}2^{3n/2}\left(\frac{1}{4} - \frac{n2^{-n}}{2} - \frac{1}{4}2^{-n}\right)\prod_{m=3}^{n}\left(\frac{1}{1 - 2^{1-m}}\right), \quad n \ge 4.$$

In general, the following recurrence for $n \ge 2j - 3$

$$(2 - 2^n)L_{j-1}(n) + 2\sum_{i=1}^{j-2}\binom{n}{i}L_{j-i}(n-i) = 0, \quad n \ge 2j - 3$$

has the following asymptotic solution

$$L_j(n) \sim \rho_0 K_j n!2^{-n^2/2}2^{(j-3/2)n}, \quad j \text{ fixed},$$

$$\rho_0 = \prod_{\ell=2}^{\infty}(1 - 2^{-\ell})^{-1},$$

where K_j is a constant.

8.7 ⚠ Using the saddle point method and (8.43) derived in Example 8.5, prove the following theorem.

Theorem 8.36 **(Knessl and Szpankowski, 2000)** *The distribution of the height of b-tries has the following asymptotic expansions for fixed b:*

(i) RIGHT-TAIL REGION: $k \to \infty$, $n = O(1)$:

$$\Pr\{H_n^T \leq k\} = h_n^k \sim 1 - \frac{n!}{(b+1)!(n-b-1)!} 2^{-kb}.$$

(ii) CENTRAL REGIME: $k, n \to \infty$ with $\xi = n2^{-k}$, $0 < \xi < b$:

$$h_n^k \sim A(\xi; b)e^{n\phi(\xi; b)},$$

where

$$\phi(\xi; b) = -1 - \log \omega_0$$
$$+ \frac{1}{\xi} \left(b \log(\omega_0 \xi) - \log b! - \log \left(1 - \frac{1}{\omega_0} \right) \right),$$

$$A(\xi; b) = \frac{1}{\sqrt{1 + (\omega_0 - 1)(\xi - b)}}.$$

In the above, $\omega_0 = \omega_0(\xi; b)$ is the solution to

$$1 - \frac{1}{\omega_0} = \frac{(\omega_0 \xi)^b}{b! \left(1 + \omega_0 \xi + \frac{\omega_0^2 \xi^2}{2!} + \cdots + \frac{\omega_0^b \xi^b}{b!} \right)}.$$

(iii) LEFT-TAIL REGION: $k, n \to \infty$ with $j = b2^k - n$

$$h_n^k \sim \sqrt{2\pi n} \frac{n^j}{j!} b^n \exp \left(-(n+j) \left(1 + b^{-1} \log b! \right) \right),$$

where $j = O(1)$.

8.8 Using the following Fourier series

$$\langle x \rangle = \frac{1}{2} - \sum_{m \in \mathbb{Z} \setminus \{0\}} c_m e^{2\pi i m x}, \qquad c_m = -\frac{i}{2\pi m},$$

to prove that

$$\sum_{k=0}^{n} \binom{n}{k} p^k (1-p)^{n-k} \left\langle k\frac{N}{M} + y \right\rangle = \frac{1}{2} - \frac{1}{M} \left(\frac{1}{2} - \langle yM \rangle \right) + O(\rho^n),$$

where $\rho < 1$.

8.9   Using the Fourier series of $2^{-\langle x \rangle}$ prove the following

$$\frac{1}{M} \sum_{l=0}^{M-1} 2^{-\langle l/M+y \rangle} = \frac{1}{2M(1-2^{-1/M})} 2^{-\langle yM \rangle /M}$$

for any real y.

8.10   Show that for any irrational α

$$\lim_{n \to \infty} \sum_{k=0}^{n} \binom{n}{k} p^k (1-p)^{n-k} 2^{-\langle \alpha k + \beta n \rangle} = \frac{1}{2 \log 2}$$

for any real β.

8.11   Analyze asymptotically the following sum

$$S_2(n, p) := \sum_{0 \le k \le n} \binom{n}{k} p^k (1-p)^{n-k} \langle -\log P_e(k, n-k) \rangle,$$

where

$$P_e(a, b) := \frac{\Gamma(a+\frac{1}{2})\Gamma(b+\frac{1}{2})}{\pi \Gamma(a, b)}.$$

8.12   Prove the following expansions

$$(1-z)^{3/2} = \frac{1}{\sqrt{\pi n^5}} \left(\frac{3}{4} + \frac{45}{32n} + O\left(\frac{1}{n^2}\right) \right),$$

$$\sqrt{1-z} \log(1-z)^{-1} = -\frac{1}{\sqrt{\pi n^3}} \left(\frac{1}{2} \log n + \frac{\gamma + 2\log 2 - 2}{2} \right.$$

$$\left. + O\left(\frac{\log n}{n}\right) \right), \log^2(1-z)^{-1}$$

$$= \frac{1}{n} \left(2\log n + 2\gamma + +O\left(\frac{1}{n}\right) \right),$$

$$(1-z)^{-1/2} = \frac{1}{\sqrt{\pi n}} \left(+\frac{8}{n} + O\left(\frac{1}{n^2}\right) \right),$$

$$(1-z)^{-3/2} = \frac{n}{\sqrt{\pi}} \left(2 + \frac{3}{4n} + O\left(\frac{1}{n^2}\right) \right),$$

$$(1-z)^{-3/2}\log(1-z)^{-1} = \frac{n}{\sqrt{\pi}}\left(2\log n + 2\gamma + 4\log 2 - 2\right.$$

$$\left. +O\left(\frac{\log n}{n}\right)\right).$$

8.13 Analyze asymptotically

$$A(z) = \frac{e^z}{\sqrt{1-z^2}}.$$

Notice that there are two algebraic singularities at $z = 1$ and $z = -1$. Prove that

$$[z^n]A(z) \sim \frac{1}{\sqrt{2\pi n}}\left(e + (-1)^n e^{-1}\right).$$

Find a full asymptotic expansion.

8.14 ⚠ Let for some $k \geq 0$

$$S_{n,k} = \sum_{i=1}^{n}\binom{n-k}{i}\left(\frac{i}{n}\right)^i\left(1 - \frac{i}{n}\right)^{n-i}.$$

(i) (Szpankowski 1995) Prove that for fixed k

$$S_{n,k} = \frac{1}{2^{2k}}\binom{2k}{k}\sqrt{\frac{n\pi}{2}} - \frac{2}{3} + O\left(\frac{1}{\sqrt{n}}\right)$$

for large n.

(ii) Find a *uniform* (with respect to k) asymptotic expansion of $S_{n,k}$ (cf. Hwang [208]).

8.15 ⚠ Consider the following extension of the Lapalce method: Let, as in Theorem 8.17,

$$I(x) = \int_a^b f(t)e^{-xh(t)}\,dt,$$

and we analyze this integral for $x \to \infty$. We adopt the same assumptions as in Theorem 8.17, except that the point t_0, where $h(t)$ is minimized, is of order m; that is,

$$h'(t_0) = h''(t_0) = \cdots = h^{(m-1)}(t_0) = 0, \quad h^{(m)}(t_0) > 0.$$

Prove that

$$I(x) \sim f(t_0)e^{-xh(t_0)}\Gamma\left(\frac{m+1}{m}\right)\left[\frac{m!}{xh^{(m)}(t_0)}\right]^{1/m}$$

as $x \to \infty$.

8.16 ⚠ Consider the following multidimensional extension of the Laplace method. Let $s = (s_1, \ldots, s_n)$ and for large positive x

$$I(x) = \int_S f(s)e^{-xh(s)}ds.$$

Let s_0 be a unique point, where $h(s)$ attains minimum. We further assume that second partial derivatives of h exist and are continuous and the Hessian matrix $H = \left(\frac{\partial^2 h(s_0)}{\partial s_i \partial s_j}\right)$ is positive definite. Prove the following (cf. [195]).

Theorem 8.37 *Under the above hypothesis (see also hypothesis of Theorem 8.17)*

$$I(x) \sim \frac{f(s_0)}{\sqrt{\det H}}e^{-xh(s_0)}\left(\frac{2\pi}{x}\right)^{n/2}$$

as $x \to \infty$.

8.17 ⚠ The *involution number* I_n has the following exponential generating functions

$$A(z) = e^{z+z^2/2}.$$

Using Hayman's method prove that

$$I_n = n!\frac{e^{-1/4}}{\sqrt{2\pi n}}n^{-n/2}e^{n/2+\sqrt{n}}$$

(originally due to Moser and Wyman).

8.18 Using the van der Waerden method prove that

$$I(n) = \int_{1/2-i\infty}^{1/2+i\infty} \exp(-n(z^2 + 2/z))dz$$

(the line of integration is $\Im z = 1/2$) yields

$$I(n) \sim \sqrt{\frac{\pi}{3n}} \exp\left(n\left(\frac{3}{2} + i\frac{3\sqrt{3}}{2}\right)\right).$$

8.19 Provide a rigorous proof of the result from Example 8.22 through an application of the van der Waerden method.

8.20 Prove all the claims of the Jacquet-Schachinger Lemma 8.22.

8.21 ⚠ (Flajolet and Sedgewick, 1995) Let

$$S_n(\lambda) = \sum_{k=1}^{n} (-1)^n \binom{n}{k} k^{-\lambda},$$

where $\lambda \notin \mathbb{Z}$. Prove the Flajolet and Sedgewick [148] result, namely

$$S_n(\lambda) = -(\log n)^\lambda \sum_{j=0}^{\infty} (-1)^j \frac{\Gamma^{(j)}(1)}{j!\Gamma(1+\lambda-j)} \frac{1}{\log^j n},$$

where $\Gamma^{(j)}(1)$ is the jth derivative of the gamma function at $z = 1$.

8.22 ⚠ (Flajolet and Sedgewick, 1995) Let

$$Y_n = \sum_{k=1}^{n} (-1)^n \binom{n}{k} \log k.$$

Prove that

$$Y_n = \log\log n + \gamma + \frac{\gamma}{\log n} - \frac{\pi^2 + 6\gamma^2}{12\log^2 n} + O(\log^{-3} n).$$

8.23 Consider the exact pattern matching problem of Section 7.6.2. Prove the following stronger version of the large deviations. Let $r = (1 + \delta)\mathbf{E}[O_n]$ and $a = (1 + \delta)P(H)$ with $\delta \neq 0$. For complex t, define $\rho(t)$ to be the root of

$$1 - e^t M_H(e^\rho) = 0,$$

while ω_a and σ_a are defined as

$$-\rho'(\omega_a) = a,$$
$$-\rho''(\omega_a) = \sigma_a^2.$$

Then

$$\Pr\{O_n(H) = (1 + \delta)\mathbf{E}[O_n]\}$$

$$= \frac{1}{\sigma_a \sqrt{2\pi(n - m + 1)}} e^{-(n-m+1)I(a)} \left(1 + O\left(\frac{1}{n}\right)\right),$$

where $I(a) = a\omega_a + \rho(\omega_a)$.

8.24 ⚠ (Szpankowski, 1991) Using Rice's method or Theorem 8.21 prove Theorem 8.33. In particular, establish a formula on the constant A in (8.167).

8.25 ⚠ (Jacquet and Régnier, 1986) Let $s_{j,k}$ ($j = 1, 2, \ldots$ and $k = 0, \pm1, \ldots$) be a solution of (8.178). Prove that

$$\Re(s_{k,2}(v)) \geq s_{0,2}(\Re(v))$$

8.26 ⚠ (Flajolet, 1999) Using singularity analysis and the method of Section 8.5 prove that

$$\sum_{k=0}^{n} \binom{n}{k} p^k q^{n-k} \frac{1}{4^k}\binom{2k}{k} = \frac{1}{\sqrt{pn}}\left(1 - \frac{3p - 2}{8pn} + O\left(\frac{1}{n^3}\right)\right),$$

$$\sum_{k=0}^{n} \binom{n}{k} p^k q^{n-k} H_k = p\log(pn) + \gamma + \frac{1}{2n} + O\left(\frac{1}{n^2}\right),$$

where H_k is the harmonic number.

8.27 Prove the following generalization of Theorem 8.24: *If for some $r > 0$, the PGFs $G_n(z)$ and $G(z)$ of discrete distributions are analytic in $|z| < r$ and continuous of $|z| = r$, then*

$$|p_{n,k} - p_k| \leq r^{-k} \sup_{|z|=r} |G_n(z) - G(z)|,$$

where $p_{n,k} = [z^k]G_n(z)$ and $p_k = [z^k]G(z)$.

8.28 🖔 Find the limiting distribution of the profile $B_m(k)$ (defined in Section 8.7.3) for $k = \frac{1}{h}\log m + O(\sqrt{\log m})$, where $B_m(k)$ is the number of nodes on level k in a digital search tree built from m strings generated by a biased memoryless source.

8.29 ⚠ Using techniques of this chapter, prove **Watson's Lemma**: *If $f(t)$ is a real or complex function of positive real variable t, and*

$$f(t) \sim t^{\lambda-1} \sum_{n=0}^{\infty} a_n t^n \quad \Re(\lambda) > 0,$$

and the integral (i.e., Laplace transform)

$$F(z) = \int_0^{\infty} f(t) e^{-zt} dt$$

is convergent for sufficiently large $\Re(z)$, then

$$F(z) \sim \sum_{n=0}^{\infty} \Gamma(n+\lambda) \frac{a_n}{z^{n+\lambda}}, \quad z \to \infty$$

in the sector $|\arg(z)| < \frac{1}{2}\pi$ (cf. [331, 422, 448]).

8.30 ⚠ (Jacquet and Régnier, 1986) Using Rice's formula, prove that the limiting distribution of the typical depth in a trie built from n independently generated binary sequences is normal. More precisely: *For biased memoryless source ($p \neq q$) the limiting distribution of the depth D_m in a trie is*

$$\frac{D_m - c_1 \log n}{\sqrt{c_2 \log n}} \to N(0, 1),$$

where $c_1 = 1/h$ and $c_2 = (h_2 - h^2)/h^3$ (as before, h is the entropy of the source and $h_2 = p \log^2 p + q \log^2 q$). Extend it to Markovian sources (cf. [215]).

9

Mellin Transform and Its Applications

The Mellin transform (Hjalman Mellin 1854–1933, Finish mathematician) is the most popular transform in the analysis of algorithms. It is closely related to the two-sided Laplace and Fourier transforms except that it has a polynomial kernel. D. E. Knuth, together with De Bruijn, introduced it in the orbit of discrete mathematics in the mid-1960s, however, Flajolet's school systematized and applied the Mellin transform to myriad problems of analytic combinatorics and analysis of algorithms. Recently, the Mellin transform found its way into information theory. The popularity of this transform stems from two important properties. It allows the reduction of certain functional equations to algebraic ones, and it provides a direct mapping between asymptotic expansions of a function near zero or infinity and the set of singularities of the transform in the complex plane. The latter asymptotic property, enriched in the singularity analysis or depoissonization (discussed in the next chapter), is crucial for applications. We discuss here some properties of the Mellin transform, and illustrate its applications with many examples.

The Mellin transform $f^*(s)$ of a complex-valued function $f(x)$ defined over positive reals is

$$\mathcal{M}[f(x); s] := f^*(s) = \int_0^\infty f(x)x^{s-1}dx$$

with s being a complex number. The Mellin transform can be viewed as the Laplace transform or the Fourier transform. However, it proves convenient to work with the Mellin transform rather than the Laplace-Fourier version. This is particularly true in the analysis of algorithms and analytic combinatorics, where one often deals with functional equations like

$$f(x) = a(x) + \alpha f(xp) + \beta f(xq),$$

where α, β are constants, and $a(x)$ is a known function. We have already encountered such functional equations in this book (e.g., Sections 7.6.1 and 8.7.3). We will see more of these in this chapter. The point is that the Mellin transform maps the above functional equation into an algebraic one that is easier to solve and hence allows us to recover $f(x)$, at least asymptotically as $x \to 0$ or $x \to \infty$.

The usefulness of the Mellin transform stems from its asymptotic properties. There is a direct mapping between asymptotic expansions of a function near zero or infinity and the set of singularities of the transform in the complex plane. This plays a crucial role in applications. For example, sums like

$$G(x) = \sum_{k=0}^\infty \left(1 - e^{-x/2^k}\right) \quad \text{and} \quad H(x) = \sum_{k=1}^\infty (-1)^k e^{-k^2 x} \log k$$

that are not easily computable, either numerically of asymptotically, can be treated by Mellin transforms for $x \to 0$ or $x \to \infty$. We study some representative examples in this chapter.

In the analysis of algorithms, $f(x)$ is often a generating function of a sequence f_n. Using the asymptotic properties of the Mellin transform, we are able to find an asymptotic expansion of $f(x)$, but we still need a tool to recover f_n (this is called a *two-step approach*). We may either use the singularity analysis discussed in the previous chapter or turn to another approach called *analytic depoissonization* that we shall introduce in the next chapter.

Over the last 30 years, myriad analyses of algorithms have been successfully accomplished through Mellin transforms. The list is quite long, and we mention here only: sorting and searching methods [269], digital trees such as tries [212, 213, 215, 243, 269, 250, 407, 406], PATRICIA trie [251, 269, 350, 408], digital search trees [147, 217, 220, 253, 269, 409] and suffix trees [216], string matching [28, 330], data compression [297, 217, 220], multidimensional searching [142], communication protocols [170, 311, 404], randomized data structures

[247], and probabilistic counting [254]. In some cases, notably digital search trees and sorting, alternative methods (e.g., Rice method) can be used. However, when the problem becomes more complicated, it is fair to say that the Mellin transform must be involved. We shall see it in this chapter, and even more in the next one.

We start with a short discussion of basic properties of the Mellin transform followed by a longer excursion into asymptotic properties. We also present a brief extension of the Mellin transform to the complex plane that is crucial for applications. As always, we finish with an applications section, where we discuss the average and the variance of the depth in an extension of digital search trees, and then we evaluate the minimax redundancy of a renewal process. The latter example is one of the most involved in this book.

We intend to make this chapter short since there are excellent and in-depth surveys on the Mellin transform. The recent extended survey by Flajolet, Gourdon, and Dumas [132] contains more than we plan to discuss here. We shall borrow from it freely, and refer the reader to it for further details. An even more detailed account on the method can be found in the forthcoming book of Flajolet and Sedgewick [149]. Brief discussions are also in Hofri [197] and Mahmoud [305].

9.1   BASIC PROPERTIES

Let $f(x)$ be a complex-valued function that exists over $(0, \infty)$ and is locally integrable. To avoid further complication, we assume throughout that f is continuous in $(0, \infty)$. The Mellin transform is defined as

$$\mathcal{M}[f(x); s] := f^*(s) = \int_0^\infty f(x)x^{s-1}dx. \tag{9.1}$$

In the sequel, we study the existence of this transform, compute the transform of some common functions, and investigate several functional properties of the transform.

(M1) Fundamental Strip   Let $f(x)$ be a continuous function on the interval $(0, \infty)$ such that

$$f(x) = \begin{cases} O(x^\alpha) & x \to 0 \\ O(x^\beta) & x \to \infty . \end{cases}$$

Then the Mellin transform $f^*(s)$ exists for any complex number s in the *fundamental strip* $-\alpha < \Re(s) < -\beta$, which we also denote as $\langle -\alpha, -\beta \rangle$. This is true since

$$\left| \int_0^\infty f(x) x^{s-1} dx \right| \leq \int_0^1 |f(x)| x^{\Re(s)-1} dx + \int_1^\infty |f(x)| x^{\Re(s)-1} dx$$

$$\leq c_1 \int_0^1 x^{\Re(s)+\alpha-1} dx + c_2 \int_1^\infty x^{\Re(s)+\beta-1} dx,$$

where c_1 and c_2 are constants. The first integral above exists for $\Re(s) > -\alpha$ and the second for $\Re(s) < -\beta$. This proves the claim. For example, $f_1(x) = e^{-x}$, $f_2(x) = e^{-x} - 1$, $f_3(x) = 1/(1+x)$ have the fundamental strips $\langle 0, \infty \rangle$, $\langle -1, 0 \rangle$, and $\langle 0, 1 \rangle$, respectively (see also Table 9.1). In passing, we observe that the Mellin transform is *analytic* inside its fundamental strip. We also notice that the Mellin transform of a polynomial does not exist. However, as we shall see, polynomials play a crucial role; they shift the fundamental strip (see Entries 1, 2, and 3 in Table 9.1).

Table 9.1 presents some of the most commonly used Mellin transforms with their corresponding fundamental strips. These formulas will be established in the course of the next few pages. For now we observe that Entry 1 of the table is the classical Euler's gamma function. Entries 5 and 7 follow from the classical beta

TABLE 9.1. Mellin Transform of Some Functions

Item	Function	Mellin Transform	Fund. Strip
1.	e^{-x}	$\Gamma(s)$	$\langle 0, \infty \rangle$
2.	$e^{-x} - 1$	$\Gamma(s)$	$\langle -1, 0 \rangle$
3.	$e^{-x} - 1 + x$	$\Gamma(s)$	$\langle -2, -1 \rangle$
4.	e^{-x^2}	$\frac{1}{2}\Gamma\left(\frac{s}{2}\right)$	$\langle 0, \infty \rangle$
5.	$\frac{1}{1+x}$	$\frac{\pi}{\sin \pi s}$	$\langle 0, 1 \rangle$
6.	$\log(1 + x)$	$\frac{\pi}{s \sin \pi s}$	$\langle -1, 0 \rangle$
7.	$\frac{1}{(1+x)^\nu}$	$B(s, \nu - s)$	$\langle 0, \nu \rangle, \quad \nu > 0$
8.	$H(x) = I(0 \leq x \leq 1)$	$\frac{1}{s}$	$\langle 0, \infty \rangle$
9.	$1 - H(x)$	$-\frac{1}{s}$	$\langle -\infty\, 0 \rangle$
10.	$x^a \log^k(x) H(x)$	$\frac{(-1)^k k!}{(s+a)^{k+1}}$	$\langle -a, \infty \rangle, \quad k \in \mathbb{N}$
11.	$x^a \log^k(x)(1 - H(x))$	$-\frac{(-1)^k k!}{(s+a)^{k+1}}$	$\langle -a, \infty \rangle, \quad k \in \mathbb{N}$
12.	$\frac{e^{-x}}{1-e^{-x}}$	$\Gamma(s)\zeta(s)$	$\langle 1, \infty \rangle$
13.	$\log\left(\frac{1-e^{-x}}{x}\right)$	$-\frac{\Gamma(s+1)\zeta(s+1)}{s}$	$\langle -1, 0 \rangle$
14.	$\log(1 - e^{-x})$	$-\Gamma(s)\zeta(s + 1)$	$\langle 0, \infty \rangle$

function. Indeed,

$$f^*(s) = \int_0^\infty x^{s-1}(1+x)^{-1}dx = \int_0^1 t^s(1-t)^{1-s}dt = \Gamma(s)\Gamma(1-s) = \frac{\pi}{\sin \pi s},$$

where the last equation is a consequence of (2.33). Entries 8–11 are trivial to prove, but they play the most important role in the asymptotic analysis of the Mellin transform. As we shall see in Section 9.2, these formulas remain valid even when $=$ is replaced by $\sim$ for $x \to 0$ and $x \to \infty$, respectively.

(M2) Smallness of Mellin transforms   Let $s = \sigma + it$. We first only assume that $f(x)$ is continuous. Observe that

$$f^*(\sigma + it) = \int_0^\infty f(x)e^{-\sigma x}e^{it \log x}dx,$$

thus by the Riemann-Lebesgue Lemma 8.13

$$f^*(\sigma + it) = o(1) \quad \text{as } t \to \pm\infty. \tag{9.2}$$

If $f(x)$ is r times differentiable, then by integration by parts we have

$$\mathcal{M}[f^{(r)}(x); s] = (-1)^r(s-r)^r f^*(s-r)$$

(cf. also (9.8) below). This, together with (9.2), yields

$$f^*(\sigma + it) = o(|t|^{-r}) \quad \text{as } t \to \pm\infty \tag{9.3}$$

for σ fixed.

(M3) Functional Properties   On the following page we list main properties of the Mellin transform. They are easy to establish in the corresponding fundamental strips. Formulas (9.4)–(9.11) are direct and simple consequences of the Mellin transform definition. For example, (9.9) follows from integration by parts

$$\mathcal{M}[x\frac{d}{dx}f(x); s] = f(x)x^s\big|_0^\infty - s\mathcal{M}[f(x); s] = -s\mathcal{M}[f(x); s]$$

with the validity in a strip, that is, a subset of the fundamental strip, dictated by the growth of f.

   Using the above properties, we may establish some more entries from Table 9.1. In particular, Entry 4 follows directly from Entry 1 and (9.6). To prove Entry 6 we apply property (9.10) to Entry 5. To establish Entries 12–14, we need to generalize property (9.11) to infinite $\mathcal{K}$, which we discuss next.

Function	$\Leftrightarrow$	Mellin Transform		
$x^{\nu} f(x)$	$\Leftrightarrow$	$f^*(s + \nu)$,		(9.4)
$f(\mu x)$	$\Leftrightarrow$	$\mu^{-s} f^*(s)$	$\mu > 0$,	(9.5)
$f(x^{\rho})$	$\Leftrightarrow$	$\dfrac{1}{\rho} f^*(s/\rho)$	$\rho > 0$,	(9.6)
$f(x) \log x$	$\Leftrightarrow$	$\dfrac{d}{ds} f^*(s)$,		(9.7)
$\dfrac{d}{dx} f(x)$	$\Leftrightarrow$	$-(s - 1) f^*(s - 1)$		(9.8)
$x \dfrac{d}{dx} f(x)$	$\Leftrightarrow$	$-s f^*(s)$		(9.9)
$\displaystyle\int_0^x f(t) dt$	$\Leftrightarrow$	$-\dfrac{1}{s} f^*(s + 1)$		(9.10)
$f(x) = \displaystyle\sum_{k \in \mathcal{K}} \lambda_k g(\mu_k x)$	$\Leftrightarrow$	$f^*(s) = g^*(s) \displaystyle\sum_{k \in \mathcal{K}} \lambda_k \mu_k^{-s}$,	$\mathcal{K}$ finite	(9.11)

The sum

$$F(x) = \sum_k \lambda_k f(\mu_k x) \tag{9.12}$$

is called the *harmonic sum*. After formally applying the scaling property (9.5), and assuming the corresponding series exist, we obtain

$$F^*(s) = f^*(s) \sum_k \frac{\lambda_k}{\mu_k^s}. \tag{9.13}$$

This formula is responsible, to a large extent, for usefulness of the Mellin transform and finds plenty of applications in discrete mathematics, analysis of algorithms, and analytic combinatorics. Theorem 9.1 below provides conditions under which it is true.

Theorem 9.1 (Harmonic Sum Formula) *The Mellin transform of the harmonic sum (9.12) is defined in the intersection of the fundamental strip of $f(x)$*

and the domain of absolute convergence of the Dirichlet series $\sum_k \lambda_k \mu_k^{-s}$ (which is of the form $\Re(s) > \sigma_a$ for some real σ_a), and it is given by (9.13).

Proof. Since $f^*(s)$ and the Dirichlet series $\sum_k \lambda_k \mu_k^{-s}$ are both analytic in the intersection, if not empty, of the corresponding convergence regions, the interchange of summation and integration is legitimate by Fubini's theorem. ∎

Using the above result, we can now prove Entries 12–14 of Table 9.1. In particular, to establish Entry 12 of Table 9.1 we proceed as follows. For any $x > 0$ we have

$$\frac{e^{-x}}{1 - e^{-x}} = \sum_{k=1}^{\infty} e^{-kx}.$$

By (9.13) its Mellin is $\mathcal{M}[e^{-x}; s] \sum_{k=1}^{\infty} k^{-s} = \Gamma(s)\zeta(s)$ for $\Re(s) > 1$. For Entry 14 we need to wrestle a little more, but

$$\mathcal{M}[\log(1 - e^{-x}); s] = \mathcal{M}\left[\int_0^x \frac{e^{-t}}{1 - e^{-t}} dt; s \right]$$

$$= -\frac{1}{s}\Gamma(s + 1)\zeta(s + 1) = -\Gamma(s)\zeta(s + 1).$$

As a consequence of this, we immediately prove that

$$\mathcal{M}\left[\sum_{k \geq 1} \log(1 - e^{-kx}); s \right] = -\Gamma(s)\zeta(s)\zeta(s + 1) \tag{9.14}$$

for $\Re(s) > 1$. The reader is asked to prove Entry 13 in Exercise 1.

(M4) Inverse Mellin Transform   Since the Mellin transform is a special case of the Fourier transform, we should expect a similar inverse formula. Indeed this is the case. Let $s = \sigma + it$ and set $x = e^{-y}$. Then

$$f^*(s) = -\int_{-\infty}^{\infty} f(e^{-y})e^{-\sigma y}e^{-ity} dy.$$

This links Mellin and Fourier transforms, and from the inverse Fourier transform we obtain (cf. Henrici [195]): *If $f(x)$ is continuous on $(0, \infty)$, then*

$$f(x) = \frac{1}{2\pi i} \int_{c-i\infty}^{c+i\infty} f^*(s)x^{-s} ds, \tag{9.15}$$

where $a < c < b$ and $\langle a, b \rangle$ is the fundamental strip of $f^*(s)$. We should remind the reader that we have already encountered an integral like the one in (9.15) in Section 7.6.3 and Section 8.5 (e.g., Theorem 8.20 and Rice's method).

The inverse formula is of prime importance to asymptotics. We illustrate it by proving Entry 2 of Table 9.1. We apply (9.15) with $\Gamma(s)$ defined in the strip $\langle -1, 0 \rangle$. Let $c = -\frac{1}{2}$ and consider a large rectangle *left* to the line of integration in (9.15) with the left line located at, say $-M - \frac{1}{2}$ for $M > 0$. As in the proof of Theorem 8.20 we show that the top and the bottom line contribute negligible while the left line contributes $O(x^{M+1/2})$. Thus

$$f(x) = \frac{1}{2\pi i} \int_{-\frac{1}{2}-i\infty}^{-\frac{1}{2}+i\infty} \Gamma(s) x^{-s} ds = \sum_{k=1}^{M} \text{Res}[\Gamma(s); s = -k] + O(x^{M+1/2})$$

$$= \sum_{k=1}^{M} \frac{(-1)^k}{k!} x^k + O(x^{M+1/2})$$

$$\to e^{-x} - 1 \qquad \text{as} \quad M \to \infty,$$

which proves Entry 2. In a similar manner we can prove Entry 3 (cf. also Exercise 2). This completes the derivations of all entries in Table 9.1.

In the introduction we mentioned that Mellin transforms are useful to solve functional equations. We shall illustrate this in Example 9.2 below, while in Example 9.1 we show that the Mellin transform can also be used to establish functional equations.

Example 9.1: *A Functional Equation Through the Mellin Transform* In the analysis of the variance of the partial match in a multidimensional search Kirschenhofer, Prodinger, and Szpankowski [252] studied the following function

$$F(x) = \sum_{k \geq 1} \frac{e^{-kx}}{1 + e^{-2kx}}.$$

As a matter of fact, it was already analyzed by Ramanujan (cf. [45]). Our goal is to prove the following functional equation on $F(x)$

$$F(x) = \frac{\pi}{4x} - \frac{1}{4} + \frac{\pi}{x} F\left(\frac{\pi^2}{x}\right). \tag{9.16}$$

Let

$$\beta(s) = \sum_{j \geq 0} (-1)^j \frac{1}{(2j+1)^s},$$

which resembles zeta function. Observe that

$$F(x) = \sum_{k \geq 1} \frac{e^{-kx}}{1 + e^{-2kx}} = \sum_{j \geq 0} (-1)^j \sum_{k \geq 1} e^{-k(2j+1)x}.$$

Then by the harmonic sum formula, the Mellin transform $F^*(s)$ of $F(x)$ is

$$F^*(s) = \Gamma(s)\zeta(s)\beta(s).$$

By the Mellin inversion formula this yields

$$F(x) = \frac{1}{2\pi i} \int_{3/2-i\infty}^{3/2+i\infty} \Gamma(s)\zeta(s)\beta(s)x^{-s}ds .$$

Now we take the two residues at $s = 1$ and $s = 0$ out from the above integral (notice that $\beta(0) = 1/2$ and $\beta(1) = \pi/4$; cf. [2]) and apply the duplication formula for $\Gamma(s)$ (see Section 2.4.1) to obtain

$$F(x) = \frac{\pi}{4x} - \frac{1}{4} + \frac{1}{2\pi i} \int_{-\frac{1}{2}-i\infty}^{-\frac{1}{2}+i\infty} \frac{1}{\sqrt{\pi}} 2^{s-1} \Gamma\left(\frac{s}{2}\right) \Gamma\left(\frac{s+1}{2}\right) x^{-s} \zeta(s)\beta(s)\, ds.$$

As before, the exponential smallness of the Γ-function along vertical lines justifies the shifting of the line integral. We now use the functional equations for $\zeta(s)$ and $\beta(s)$ (see Section 2.4), namely,

$$\Gamma\left(\frac{s}{2}\right) \zeta(s) = \pi^{s-\frac{1}{2}} \Gamma\left(\frac{1-s}{2}\right) \zeta(1-s), \tag{9.17}$$

and

$$\beta(1-s)\Gamma\left(1-\frac{s}{2}\right) = 2^{2s-1}\pi^{-s+\frac{1}{2}} \Gamma\left(\frac{s+1}{2}\right) \beta(s) . \tag{9.18}$$

Identity (9.17) is the Riemann functional equation for $\zeta(s)$, and (9.18) is an immediate consequence of the functional equation for Hurwitz's ζ-function $\zeta(s, a)$ (cf. [15]), and the fact that

$$\beta(s) = 4^{-s} \left[\zeta(s, \frac{1}{4}) - \zeta(s, \frac{3}{4}) \right].$$

Using (9.17) and (9.18), and substituting $1 - s = u$, we arrive at

$$F(x) = \frac{\pi}{4x} - \frac{1}{4} + \frac{1}{2\pi i} \int\limits_{\frac{3}{2} - i\infty}^{\frac{3}{2} + i\infty} \pi^{1-2u} \Gamma(u) x^{u-1} \zeta(u) \beta(u) du.$$

This proves (9.16).

In passing we mention that the functional equation can be used for a numerical evaluation of $F(x)$, as already observed by Kirschenhofer, Prodinger, and Schoissengeier [249] and further explored in Kirschenhofer and Prodinger [243].

For example, in applications the following series is of interest

$$F(\log 2) = \sum_{k \geq 1} \frac{2^k}{2^{2k} + 1}.$$

Since $\pi^2 / \log 2 \approx 14.2$ we expect $F(\pi^2 / \log 2)$ to be small so that the first two terms in (9.16) will determine the numerical value of $F(\log 2)$. As a matter of fact, pulling out one term from $F(\pi^2 / \log 2)$ and computing other terms we obtain $F(\log 2) = 0.8830930036(\pm 10^{-11})$.

In Exercise 4 we ask the reader to prove similar functional equations. ∎

9.2 ASYMPTOTIC PROPERTIES OF THE MELLIN TRANSFORM

We have said it many times, and we repeat it again, the usefulness of the Mellin transform stems from its asymptotic properties, which we shall examine in this section. We start with an informal discussion that will lead to the direct and reverse mapping theorems, the pillars of the Mellin transform asymptotics.

Let $f(x)$ be r times differentiable function with $r \geq 2$. The fundamental strip of $f^*(s)$ is assumed to be $\alpha < \Re(s) < \beta$. We restrict analysis to meromorphic functions $f^*(s)$ that can be analytically continued to $\beta \leq \Re(s) \leq M$ for any M. We also postulate that $f^*(s)$ has finitely many poles λ_k such that $\Re(\lambda_k) < M$. Then we claim that for $x \to \infty$

$$f(x) = - \sum_{\lambda_k \in \mathcal{K}} \text{Res}[f^*(s) x^{-s}, s = \lambda_k] + O(x^{-M}) \quad x \to \infty, \qquad (9.19)$$

where $\mathcal{K}$ is the set of singularities and M is as large as we want. (In a similar fashion one can continue the function $f^*(s)$ to the left to get an asymptotic formula

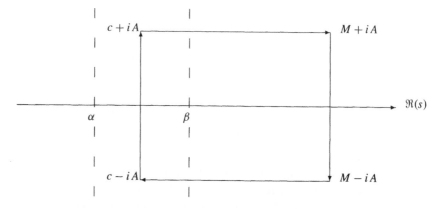

Figure 9.1. The fundamental strip of $f^*(s)$ and the integration contour.

for $x \to 0$.) This is the announced asymptotic property. Its proof is quite simple and we have already seen the main ingredients of it in Section 7.5.3. Due to its importance, we repeat below the main steps of the analysis.

Consider the rectangle $\mathcal{R}$ shown in Figure 9.1 with the corners as illustrated. Choose A so that the sides of $\mathcal{R}$ do not pass through any singularities of $f^*(s)x^{-s}$. When evaluating

$$\lim_{A \to \infty} \int_{\mathcal{R}} = \lim_{A \to \infty} \left(\int_{c-iA}^{c+iA} + \int_{c+iA}^{M+iA} + \int_{M+iA}^{M-iA} + \int_{M-iA}^{c-iA} \right),$$

the second and fourth integrals contribute $O(A^{-r})$ due to the smallness property (M2). The contribution of the fourth integral is computed as follows:

$$\left| \int_{M+i\infty}^{M-i\infty} f^*(s)x^{-s}ds \right| = \left| \int_{\infty}^{-\infty} f^*(M+it)x^{-M-it}dt \right|$$

$$\leq |x^{-M}| \int_{\infty}^{-\infty} |f^*(M+it)|dt = O(x^{-M}),$$

since f is at least twice differentiable so that the integral above exists. Using now the Cauchy residue theorem and taking into account the negative direction of $\mathcal{R}$ we finally obtain

$$-\sum_{\lambda_k \in \mathcal{H}} \text{Res}[f^*(s)x^{-s}, s = \lambda_k] = \frac{1}{2\pi i} \int_{c-i\infty}^{c+i\infty} f^*(s)x^{-s}ds + O(x^{-M}),$$

where $\alpha < c < \beta$. This proves (9.19).

In particular, if $f^*(s)$ has the following form

$$f^*(s) = \sum_{k=0}^{K} \frac{d_k}{(s-b)^{k+1}}, \tag{9.20}$$

and $f^*(s)$ can be analytically continued to $\beta \leq \Re(s) < M$, then Entry 10 of Table 9.1 implies that

$$f(x) = -\sum_{k=0}^{K} (-1)^k \frac{d_k}{k!} x^{-b} \log^k x + O(x^{-M}) \quad x \to \infty. \tag{9.21}$$

In a similar fashion, if for $-M < \Re(s) \leq \alpha$ the smallness condition of $f^*(s)$ holds and

$$f^*(s) = \sum_{k=0}^{K} \frac{d_k}{(s-b)^{k+1}}, \tag{9.22}$$

then

$$f(x) = \sum_{k=0}^{K} (-1)^k \frac{d_k}{k!} x^{-b} \log^k x + O(x^M) \quad x \to 0. \tag{9.23}$$

We return to these relationships below when we establish the direct and reverse mapping theorems.

Example 9.2: *Splitting Process Arising in Probabilistic Counting* Splitting processes arise in many applications that include probabilistic counting [254], selecting the loser [124, 347], estimating the number of questions necessary to identify distinct objects [339], searching algorithms based on digital tries [212, 213, 269, 305, 406], approximate counting [244], conflict resolution algorithms for multiaccess communication [214, 311, 405] and so forth. Let us focus on the probabilistic counting. Using a digital tree representation one can describe it in terms of this splitting process as follows: Imagine n persons flipping a fair coin, and those who get 1 discontinue throwing (move to the right in the digital tree representation) while the rest continue throwing (i.e., move to the left in the digital tree). The process continues until all remaining persons flip a 0. Consider now the following *more general version* of the splitting process, where the coin flipping ends as soon as at most d persons have flipped a 1 in the same round, where d is a given parameter. We denote the number of rounds by $1 + R_{n,d}$. Observe that $d = 0$ corresponds to the original situation.

It is easy to see (cf. also [254]) that the Poisson transform (see also the next chapter)

$$\tilde{L}(z) = e^{-z} \sum_{n \geq 0} \mathbf{E}[R_{n,d}] \frac{z^n}{n!}$$

satisfies the following functional equation

$$\tilde{L}(z) = f_d(z/2)\tilde{L}(z/2) + f_d(z/2),$$

where $f_d(z) = 1 - e_d(z)e^{-z}$ and

$$e_d(z) = 1 + \frac{z^1}{1!} + \cdots + \frac{z^d}{d!}.$$

After iterating this equation we arrive at (see Section 7.6.1)

$$\tilde{L}(z)\varphi(z) = \sum_{n=0}^{\infty} \varphi(z2^{-n-1}), \tag{9.24}$$

where

$$\varphi(z) = \prod_{j=0}^{\infty} f_d(z2^j) = \prod_{j=0}^{\infty} \left(1 - e_d(z2^j)e^{-z2^j}\right). \tag{9.25}$$

By d'Alembert's criterion, the final expression converges absolutely for all complex numbers z. We derive an asymptotic expansion of $\tilde{L}(z)$ for $z \to \infty$ (and for now assume that z is real).

Observe that (9.24) is a harmonic sum. Let $Q_1(z) = \tilde{L}(z)\varphi(z)$ and $Q_1^*(s)$ be its Mellin transform that exists in the strip $\Re(s) \in (-\infty, 0)$. By the harmonic sum formula we have $Q_1^*(s) = 2^s/(1 - 2^s)\varphi^*(s)$, where the Mellin transform $\varphi^*(s)$ of $\varphi(x)$ exists in $\Re(s) \in (-\infty, 0)$. However, there is a problem since the Mellin transform of $\varphi(s)$ does not exist at $s = 0$, and we need a more precise estimate of $\varphi^*(s)$ at $s = 0$. Thus, we proceed as follows. Define

$$\Phi(x) = \varphi(2x) - \varphi(x) = e_d(x)e^{-x}\varphi(2x).$$

Observe now that the Mellin transform of $\Phi(x)$ exists in $(-\infty, \infty)$, and

$$Q_1^*(s) = \frac{2^s}{1 - 2^s}\varphi^*(s) = \left(\frac{2^s}{1 - 2^s}\right)^2 \Phi^*(s), \tag{9.26}$$

where $\Phi^*(s)$ is an entire function, as we noticed above, defined as

$$\Phi^*(s) = \int_0^{\infty} e_d(x)e^{-x}\varphi(2x)x^{s-1}dx.$$

To evaluate $\tilde{L}(z)$ asymptotically we apply (9.19). Note that $\chi_k = 2\pi i k / \log 2$ ($k = 0, \pm 1, \pm 2, \ldots$) are the solutions of $1 - 2^{-s} = 0$, and the main contribution to the asymptotics comes from $\chi_0 = 0$. Since $\varphi(z) \sim 1 + O(z^{-M})$ for any $M > 0$ as $z \to \infty$, we obtain

$$\tilde{L}(z) = \log_2 z \cdot \frac{\Phi^*(0)}{\log 2} - \frac{\Phi^*(0)}{\log 2} - \frac{\Phi^{*\prime}(0)}{\log^2 2} + P_1(\log_2 z) + O(z^{-M}),$$

where

$$\Phi^*(0) = \int_0^\infty e^{-x} e_d(x)\varphi(2x)\frac{dx}{x},$$

$$\Phi^{*\prime}(0) = \int_0^\infty e^{-x} e_d(x)\varphi(2x)\frac{\log x}{x}dx,$$

and

$$P_1(x) = \sum_{k \neq 0} \left(\frac{\Phi^*(-\chi_k)(x-1)}{\log 2} - \frac{\Phi^{*\prime}(-\chi_k)}{\log^2 2} \right) e^{2k\pi i x}.$$

To complete the proof, we need to evaluate $\Phi^*(0), \Phi^*(\chi_k), \Phi^{*\prime}(0)$, and $\Phi^{*\prime}(\chi_k)$. We show here how to estimate $\Phi^*(0)$; the other evaluations can be found in [254]. Define the function $\mathbf{1}(x)$ as follows

$$\mathbf{1}(x) = \begin{cases} 1 & \text{if } x \geq 1 \\ 0 & \text{if } x < 1. \end{cases}$$

Then we can write

$$\Phi^*(0) = \int_0^\infty (\varphi(2x) - \varphi(x))\frac{dx}{x}$$

$$= \int_0^\infty (\varphi(2x) - \mathbf{1}(2x))\frac{dx}{x} + \int_0^\infty (\mathbf{1}(2x) - \mathbf{1}(x))\frac{dx}{x}$$

$$+ \int_0^\infty (\mathbf{1}(x) - \varphi(x))\frac{dx}{x}$$

$$= \int_0^\infty (\mathbf{1}(2x) - \mathbf{1}(x))\frac{dx}{x} = \int_{1/2}^1 \frac{dx}{x} = \log 2,$$

since after the substitution $u = 2x$ in the second line of the above display, the first and second integrals cancel. A similar derivation shows that $\Phi^*(\chi_k) = 0$. ∎

We are now in a position to precisely state results regarding asymptotic properties of the Mellin transform. We start with the *reverse mapping*, which maps the asymptotics of the Mellin transform into asymptotics of the original function for $x \to 0$ and $x \to \infty$. Following Flajolet, Gourdon, and Dumas [132] we introduce the *singular expansion*. It is basically a sum of the Laurent expansions around *all* poles truncated to the $O(1)$ term. For example, since

$$\frac{1}{s(s-1)} = \frac{-1}{s} - 1 + O(s), \quad s \to 0,$$

$$\frac{1}{s(s-1)} = \frac{1}{s-1} - 1 + O(s-1), \quad s \to 1,$$

we write

$$\frac{1}{s(s-1)} \asymp \left[\frac{-1}{s} - 1 \right]_{s=0} + \left[\frac{1}{s-1} - 1 \right]_{s=1}$$

as a singular expansion. In general, for a meromorphic function $f(s)$ with poles in $\mathcal{K}$ the singular expansion is

$$f(s) \asymp \sum_{w \in \mathcal{K}} \Delta_w(s),$$

where $\Delta_w(s)$ is the Laurent expansion of f around $s = w$ up to at most $O(1)$ term.

Based on our discussion above we formulate the following result.

Theorem 9.2   (Reverse Mapping)   *Let f be continuous with Mellin transform $f^*(s)$ defined in a non-empty strip $\langle \alpha, \beta \rangle$.*

(i) [ASYMPTOTICS FOR $x \to \infty$] *Assume that $f^*(s)$ admits a meromorphic continuation to $\Re(s) \in (\alpha, M)$ for some $M > \beta$ such that*

$$f^*(s) = O(|s|^{-r}), \quad s \to \infty \qquad (9.27)$$

for some $r > 1$ and

$$f^*(s) \asymp \sum_{k,b} \frac{c_{k,b}}{(s-b)^{k+1}}, \qquad (9.28)$$

where the sum is over a finite set of k and b. Then

$$f(x) = -\sum_{k,b}(-1)^k \frac{C_{k,b}}{k!} x^{-b} \log^k x + O(x^{-M}) \qquad (9.29)$$

as $x \to \infty$.

(ii) [ASYMPTOTICS FOR $x \to 0$] *Assume that $f^*(s)$ admits a meromorphic continuation to $\Re(s) \in (-M, \beta)$ for some $-M < \alpha$ such that (9.27) and (9.28) hold in $\Re(s) \in (-M, \beta)$. Then*

$$f(x) = \sum_{k,b}(-1)^k \frac{C_{k,b}}{k!} x^{-b} \log^k x + O(x^M) \qquad (9.30)$$

as $x \to 0$.

Proof. Part (i) we have already discussed above. It suffices to follow these steps tailored to (9.28). For part (ii) it is enough to observe that by (9.6) $M[f(1/x); s] = -M[f(x); s]$. An alternative proof is possible by following the above steps and expanding the Mellin integral to the left instead of to the right as we did for $x \to \infty$. ∎

Example 9.3: *A Recurrence Arising in the Analysis of the PATRICIA Height*
In [259] Knessl and Szpankowski analyzed the height of PATRICIA tries. Among others, the following equation arose

$$C_n = \frac{2^{2-n}}{n+1} \sum_{k=2}^{n} \binom{n+1}{k} C_k, \quad n \geq 2 \qquad (9.31)$$

with $C_2 = 1$. Certainly, the exponential generating function $C(z) = \sum_{n=0}^{\infty} \frac{z^n}{n!} C_n$ of C_n satisfies

$$C(z) = 8C\left(\frac{z}{2}\right) \frac{e^{z/2} - 1}{z}.$$

Setting $\tilde{C}(z) = e^{-z} C(z)$ (so that $\tilde{C}(z)$ is the Poisson transform of C_n) we arrive at

$$\tilde{C}(z) = \frac{8}{z}(1 - e^{-z/2}) \tilde{C}\left(\frac{z}{2}\right).$$

Observe that $\tilde{C}(z) \sim z^2/2$ as $z \to 0$ since $C_2 = 1$ and we define $C_0 = C_1 = 0$. Next we set $\tilde{C}(z) = z^2 G(z)/2$ and replace z by $2z$ in the above, which gives

$$G(2z) = G(z)\left(\frac{1 - e^{-z}}{z}\right)$$

with $G(0) = 1$, and after another change $F(z) = \log G(z)$ we obtain

$$F(2z) - F(z) = \log\left(\frac{1 - e^{-z}}{z}\right),$$

with $F(0) = 0$. The above functional equation is manageable by Mellin transforms. The Mellin transform $F^*(s)$ of $F(z)$ is

$$F^*(s) = \frac{2^s}{1 - 2^s}\mathcal{M}\left[\log\left(\frac{1 - e^{-z}}{z}\right)\right]$$

for $\Re(s) \in (-1, 0)$. In Exercise 1 we asked the reader to prove Entry 13 of Table 9.1, that is,

$$\mathcal{M}\log\left(\frac{1 - e^{-z}}{z}\right) = -\frac{\Gamma(s + 1)\zeta(s + 1)}{s}, \qquad -1 < \Re(s) < 0$$

leading to

$$F^*(s) = -\frac{\Gamma(s + 1)\zeta(s + 1)}{s(2^{-s} - 1)}$$

for $\Re(s) \in (-1, 0)$. Taking into account the singularities of $F^*(s)$, that is, a triple pole at $s = 0$ and simple poles along the imaginary axis at $\chi_k = -2\pi i k / \log 2$, and using the reverse mapping theorem, we arrive at

$$\exp[F(z)] \sim \sqrt{z}2^{-1/12}\exp\left(\frac{\gamma(1) + \gamma^2/2 - \pi^2/12}{\log 2}\right)\exp\left[-\frac{1}{2}\frac{\log^2(z)}{\log 2} + \Psi(\log_2 z)\right],$$

$$(9.32)$$

where

$$\Psi(\log_2 z) = \sum_{\substack{\ell=-\infty \\ \ell\neq 0}}^{\infty}\frac{1}{2\pi i\ell}\Gamma\left(1 - \frac{2\pi i\ell}{\log 2}\right)\zeta\left(1 - \frac{2\pi i\ell}{\log 2}\right)e^{2\pi i\ell\log_2 z}.$$

The above is true for $z \to \infty$ along the real axis, but we can analytically continue it to a cone around the real positive axis. In Section 9.3 we extend the Mellin transform to the complex plane, thus automatically obtaining the same conclusion. However, to obtain asymptotics of C_n one needs extra step. The singularity analysis discussed in Section 8.3.2 does not work well in this case. In the next chapter we develop a new tool called the analytic depoissonization that will allow us to recover C_n . In fact, we return to this problem in Example 10.10.8 of Chapter 10.   ∎

Not surprisingly, there exists a direct mapping from asymptotics of $f(x)$ near zero and infinity to a singular expansion of $f^*(s)$. We formulate it next.

Theorem 9.3 (**Direct Mapping**) *Let f be continuous with Mellin transform $f^*(s)$ defined in a nonempty strip $\langle \alpha, \beta \rangle$.*

(i) *Assume that $f(x)$ admits near $x \to 0$ the following asymptotic expansion for $-M < -b \leq \alpha$*

$$f(x) = \sum_{k,b} c_{k,b} x^b \log^k x + O(x^M), \tag{9.33}$$

where the sum is over a finite set of k and b. Then $f^*(s)$ is continuable to the strip $\langle -M, \beta \rangle$ and admits the singular expansion

$$f^*(s) \asymp \sum_{k,b} c_{k,b} \frac{(-1)^k k!}{(s+b)^{k+1}} \tag{9.34}$$

for $-M < \Re(s) < \beta$.

(ii) *Let $f(x)$ have the following asymptotic expansion for $x \to \infty$*

$$f(x) = \sum_{k,b} c_{k,b} x^{-b} \log^k x + O(x^{-M}), \tag{9.35}$$

where $\beta \leq b < M$. The Mellin transform $f^*(s)$ can be continued to the strip $\langle \alpha, M \rangle$ and

$$f^*(s) \asymp -\sum_{k,b} c_{k,w} \frac{(-1)^k k!}{(s-b)^{k+1}} \tag{9.36}$$

for $\alpha < \Re(s) < M$.

Proof. As before we need only concentrate on one case. For a change, we prove the case $x \to 0$. We shall follow Flajolet, Gourdon, and Dumas [132]. By assumption the function

$$g(x) = f(x) - \sum_{k,b} c_{k,b} x^b \log^k x$$

is $O(x^M)$. In the fundamental strip we also have

$$f^*(s) = \int_0^1 g(x) x^{s-1} + \int_0^1 \sum_{k,b} c_{k,b} x^{s+b-1} \log^k x + \int_1^\infty f(x) x^{s-1} dx.$$

The first integral above is defined in $\langle -M, \infty \rangle$ since $g(x) = O(x^M)$ for $x \to 0$. The third integral is analytic for $\Re(s) < \beta$, while the second integral is easily computable (Table 9.1). In summary, $f^*(s)$ exists in $\langle -M, \beta \rangle$ and has (9.34) as its singular expansion. ∎

The direct and reverse mapping theorems are at the heart of the Mellin transform method. They have myriad applications. The survey by Flajolet, Gourdon, and Dumas [132] is the best source of information. Chapter 7 of the forthcoming book by Flajolet and Sedgewick also has many interesting examples. Here, we just illustrate it with an example and a few exercises. We shall return to it in the next chapter.

Example 9.4: *Generalized Euler-Maclaurin Summation Formula* This example is adopted from Flajolet, Gourdon, and Dumas [132]. In Section 8.2.1 we discussed the Euler-Maclaurin summation formula that allows us to approximate discrete sums by integrals. Let us now consider

$$F(x) = \sum_{n=1}^{\infty} f(nx),$$

where

$$f(x) = \sum_{k=0}^{\infty} c_k x^k, \qquad x \to 0,$$

and for simplicity of presentation we postulate $f(x) = O(x^{-3/2})$ as $x \to \infty$. The Mellin transform $F^*(s)$ of $F(x)$ is

$$F^*(s) = f^*(s)\zeta(s)$$

for $\Re(s) > 1$. Taking into account the pole of $\zeta(s)$ at $s = 1$, we can write $F^*(s)$ as the following singular expansion

$$F^*(s) \asymp \frac{f^*(1)}{s-1} + \sum_{k=0}^{\infty} \frac{c_k \zeta(-k)}{s+k},$$

due to our assumptions regarding $f(x)$ as $x \to 0$. By the reverse mapping theorem we find

$$\sum_{n=1}^{\infty} f(nx) \sim \frac{1}{x} \int_0^{\infty} f(t)dt + \sum_{k=1}^{\infty} c_k \zeta(-k)x^k,$$

$$= \frac{1}{x} \int_0^{\infty} f(t)dt - \frac{1}{2}f(0) - \sum_{k=1}^{\infty} \frac{B_{2k}}{2k} c_{2k-1} x^{2k-1},$$

where B_k are Bernoulli numbers (see Table 8.1). The last line above is a consequence of $\zeta(-2k+1) = -B_{2k}/2k$.

The above is basically the Euler-Maclaurin formula. Interestingly, it can be generalized to fractional exponents of x. For example, let us consider

$$F(x) = \sum_{n=1}^{\infty} f(nx) \log n,$$

where

$$f(x) = \sum_{k=0}^{\infty} c_k x^{\beta_k}, \qquad x \to 0$$

with $-1 < \beta_0 < \beta_1 < \cdots$. This translates to

$$F^*(s) = -f^*(s)\zeta'(s).$$

The singular expansion, due to the double pole of $\zeta'(s)$ at $s = 1$, is

$$F^*(s) \asymp \frac{f^*(1)}{(s-1)^2} + \frac{f^{*'}(1)}{s-1} - \sum_{k=0}^{\infty} \frac{c_k \zeta'(-\beta_k)}{s + \beta_k}.$$

By the reverse mapping theorem we finally arrive at

$$F(x) \sim \frac{1}{x} \log \frac{1}{x} \int_0^{\infty} f(t)dt + \frac{1}{x} \int_0^{\infty} f(t) \log t\, dt - \sum_{k=0}^{\infty} c_k \zeta'(-\beta_k)x^{\beta_k}. \quad (9.37)$$

Using this, we can easily derive the Ramanujan formula, namely,

$$\sum_{k=1}^{\infty} e^{-kx} \log k = \frac{1}{x} \left(\log \frac{1}{x} - \gamma \right) + \log \sqrt{2\pi} + O(x).$$

In Exercises 7–10 we propose some more examples and generalizations. ∎

9.3 EXTENSION TO THE COMPLEX PLANE

In various applications of the Mellin transform one needs an extension of the transform to complex analytic functions. To motivate this, let us consider a generic example. Let f_n be a sequence such that it has the generating function $f(z)$ satisfying, say, the following functional equation

$$f(z) = a(z) + f(pz) + f(qz) \qquad (9.38)$$

for some $a(z)$ and $p + q = 1$. If z is real, the Mellin transform $f^*(s)$ becomes

$$f^*(s) = \frac{a^*(s)}{1 - p^{-s} - q^{-s}},$$ (9.39)

where $a^*(s)$ is the Mellin transform of $a(z)$. This can be used to obtain asymptotics of $f(x)$ for $x \to 0$ or $x \to \infty$. Assume that we obtained $f(x) = 1/\sqrt{1 - x} + O(1)$, and would like to infer asymptotics of f_n through singularity analysis. For this, however, we need *an extension* of $f(x)$ to complex z, which further would require an extension of the Mellin transform to complex analytic functions.

We shall discuss such an extension in this section; that is, we define the Mellin transform for complex functions $F(z)$ such that $f(x) = F(z) \mid_{z=x>0}$ for z complex. Its application will be illustrated in Section 9.4.2 and the next chapter.

Theorem 9.4 *Let (θ_1, θ_2) be an interval containing 0. Let $F(z)$ be an analytic function defined for z in a cone $S_{\theta_1,\theta_2} = \{z : \theta_1 < \arg(z) < \theta_2\}$ such that $F(z) = O(z^{-c})$ for $z \to 0$, and $F(z) = O(z^{-d})$ for $z \to \infty$ with z confined to the cone (or more generally: there is an interval (c, d) such that for all $b \in (c, d)$: $F(z) = O(z^{-b})$ as $z \to 0$ and $z \to \infty$ in the cone).*

(i) *Let $\mathcal{L}$ be a curve that starts at $z = 0$ and goes to ∞ inside the cone S_{θ_1,θ_2} (cf. Figure 9.2). For $\Re(s) \in (c, d)$ the integral*

$$\mathcal{M}(F, s) = F^*(s) := \int_{\mathcal{L}} F(z) z^{s-1} dz$$

*exists and does not depend on the curve $\mathcal{L}$. We call it the **complex Mellin transform**.*

(ii) *The complex Mellin transform $F^*(s)$ is identical to the real Mellin transform $f^*(s)$, that is,*

$$\int_0^{\infty} F(x) x^{s-1} dx = f^*(s) = F^*(s) = \int_{\mathcal{L}} F(z) z^{s-1} dz.$$

Furthermore, $\mathcal{M}(F(za), s) = a^{-s} f^*(s)$ defined on $S_{\theta_1 - \arg(a), \theta_2 - \arg(a)}$ with $a \in S_{\theta_1, \theta_2}$.

(iii) *For $b \in (c, d)$ and $\theta_1 < \theta < \theta_2$, $f^*(b + it) e^{\theta t}$ is absolutely integrable with respect to the variable t from $-\infty$ to $+\infty$. The inverse Mellin exists and becomes*

$$F(z) = \frac{1}{2i\pi} \int_{\Re(s)=b} z^{-s} f^*(s) ds$$

for $b \in (c, d)$.

Proof. For part (i), observe that the function $F(z)z^{s-1}$ is analytic, thus by Cauchy's formula

$$\oint F(z)z^{s-1}dz = 0,$$

where the integration is over any arbitrary closed curve inside the cone $\mathcal{S}_{\theta_1,\theta_2}$. Let z_1 and z_2 be two points on the curve $\mathcal{L}_0^\infty$ that goes from 0 to infinity inside the cone $\mathcal{S}_{\theta_1,\theta_2}$. Let also $r = |z_1|$, $R = |z_2|$ and let $\mathcal{L}_r^R$ be the part of $\mathcal{L}_0^\infty$ between z_1 and z_2. Furthermore, we denote by A_r and A_R the circular arcs connecting z_1 with r and z_2 with R. We integrate along the contour $\mathcal{C}$ formed by $\mathcal{L}_r^R$, the subinterval (r, R) of the real line and the two arcs A_r and A_R (see Figure 9.2). We have

$$\int_{\mathcal{L}_r^R} F(z)z^{s-1}dz = \int_r^R F(x)x^{s-1}dx + \int_{A_R} F(z)z^{s-1}dz + \int_{A_r} F(z)z^{s-1}dz$$

$$(9.40)$$

For $R \to \infty$ and $r \to 0$ we observe that $\int_{A_R} F(z)z^{s-1}dz = O(R^{\Re(s)-b})$ for any $b \in (\Re(s), d)$, and $\int_{A_r} F(z)z^{s-1}dz = O(r^{\Re(s)-a})$ for $a \in (c, \Re(s))$, respectively. Thus passing to the limits as $R \to \infty$ and $r \to 0$, we conclude that the contributions of the arcs vanish. This proves part (i) and the first part of (ii).

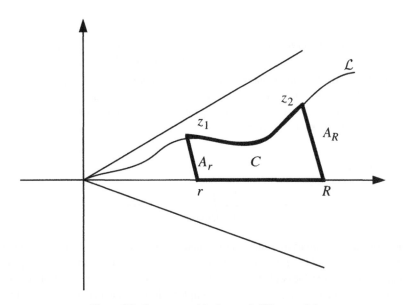

Figure 9.2. Contours used in the proof of Theorem 9.4

For part (ii), we immediately infer from (9.40), that is,

$$F^*(s) = \lim_{R \to \infty} \int_{\mathcal{L}_0^R} F(z) z^{s-1} dz = \lim_{R \to \infty} \int_0^R F(z) z^{s-1} dz = f^*(s) \,.$$

To prove the identity $\mathcal{M}(F(za), s) = a^{-s} f^*(s)$ one needs only make the change of variable $za = z'$ in the integration. Notice that the function $F(za)$ is now defined in the cone $S_{\theta_1 - \arg(a), \theta_2 - \arg(a)}$.

Point (iii) is a natural consequence of point (ii) by selecting $a = e^{i\theta}$. In this case, the (real) Mellin transform of the function $F(xe^{i\theta})$ becomes $e^{-si\theta} f^*(s)$, which is absolutely integrable. Therefore, for all $z \in S_{\theta_1, \theta_2}$ the function $z^{-s} f^*(s)$ is absolutely integrable with respect to $\Im(s)$. Finally, we conclude that the Mellin inverse identity holds in the complex case. ∎

The above extension enjoys all the "nice" properties of Mellin transforms of real x that we discussed so far. In particular, it can be used to obtain an asymptotic expansion of $F(z)$ for $z \to 0$ and $z \to \infty$.

Theorem 9.5   *Let $f(x)$ be a function of a real variable. Let $f^*(s)$ be its Mellin transform defined in the strip $\Re(s) \in (c, d)$. Let (θ_1, θ_2) be an interval containing 0. If for all $b \in (c, d)$ function $f^*(b + it)e^{\theta_1 t}$ and $f^*(b + it)e^{\theta_2 t}$ are absolutely integrable, then*

(i) *The analytic continuation $F(z)$ of $f(x)$ exists in cone S_{θ_1, θ_2} and $F(z) = O(z^{-b})$ for $|z| \to 0$ and $|z| \to \infty$ for all $b \in (c, d)$.*

(ii) *If the Mellin transform $f^*(s)$ has singularities that are regular isolated poles in the strip $\Re(s) \in (c', d')$, where $c' < c$ and $d' > d$, then the function $F(z)$ has the same asymptotic expansions as $f(x)$ up to $z^{-c'}$ at 0 and up to order $z^{-d'}$ at ∞.*

Proof. Defining

$$F(z) = \frac{1}{2\pi i} \int_{\Re(s)=b} z^{-s} f^*(s) ds$$

one introduces an analytic function that matches $f(x)$ on the real line. By an elementary estimate of the integrand we have $F(z) = O(z^{-b})$ due to the assumption concerning the smallness of $f^*(b + it)$ as $t \to \pm\infty$. Part (ii) is a simple consequence of (i), the fact that $f^*(s)$ is the complex Mellin transform of $F(z)$, and the inverse mapping theorem. ∎

To illustrate the above let us return to the functional equation (9.38) that has solution (9.38). One can obtain asymptotics of $f(z)$ by the residue theorem. In particular if $a^*(s) = \alpha(s)\Gamma(s)$ with $\alpha(s)$ of at most polynomial growth for $\pm i\infty$, then the asymptotic expansion remains valid in all cones strictly included in $\mathcal{S}_{-\pi/2,\pi/2}$ because $|\Gamma(it)|^2 = \pi(t\sinh(\pi t))^{-1} = O(e^{-\pi|t|})$ when real t tends to $\pm\infty$.

9.4 APPLICATIONS

We are again in the applications section. We shall derive here the variance of the depth in a generalized digital search tree, and the asymptotic expansion of the minimax redundancy for renewal processes. This last application is probably one of the most complicated examples discussed in this book; to solve this problem we will need all the knowledge we have gained so far.

9.4.1 Variance of the Depth in a Generalized Digital Search Tree

As in Section 8.7.3 we study here the depth in a digital search tree. However, this time we consider a generalized digital search tree in which every node can store up to b strings, known also as b-DST. The depth is denoted as D_m and we compute here the variance of D_m for the unbiased binary memoryless source in which m independent strings are generated, each being a sequence of zeros and ones occurring with the same probability. As before, let $\bar{B}_m(k)$ be the *expected* number of strings on level k of a b-DST tree (i.e., the average profile). We know that $\Pr\{D_m = k\} = \bar{B}_m(k)/m$ (cf. (8.141). Let $B_m(u) = \sum_{k\geq 0} \bar{B}_m(k)u^k$ be the generating function of $\bar{B}_m(k)$. As in Section 8.7.3, we observe that the Poisson generating function $\widetilde{B}(u, z) = e^{-z}\sum_{m=0}^{\infty} B_m(u)\frac{z^m}{m!}$ satisfies

$$\left(1 + \frac{\partial}{\partial z}\right)^b \widetilde{B}(u, z) = b + 2u\widetilde{B}(u, z/2), \tag{9.41}$$

where $\left(1 + \frac{\partial}{\partial z}\right)^b := \sum_{i=0}^{b} \binom{b}{i}\frac{\partial^i}{\partial z^i}$. The coefficients of $\widetilde{B}(u, z)$ can be computed by solving a linear recurrence discussed in Sections 7.6.1 and 8.7.3. Unfortunately, there is no easy way to solve such a recurrence unless $b = 1$ (see Section 7.6.1). To circumvent this difficulty, Flajolet and Richmond [145] reduced it to a certain functional equation on an *ordinary* generating function that is easier to solve. We proceed along this path.

We define a sequence $g_k(u) = k![z^k]\widetilde{B}(u, z)$, that is, $\widetilde{B}(u, z) = \sum_{k=0}^{\infty} g_k(u)\frac{z^k}{k!}$. Observe that $B_m(u) = \sum_{k=0}^{m} \binom{m}{k}g_k(u)$. Let $G(u, z) = \sum_{k=0}^{\infty} g_k(u)z^k$. We also need the ordinary generating function of $B_k(u)$ denoted as

$$F(u, z) = \sum_{k=0}^{\infty} B_k(u) z^k.$$

As in Example 7.7.11 of Chapter 7 we have

$$F(u, z) = \frac{1}{1-z} G\left(u, \frac{z}{1-z}\right),$$

which implies

$$F_u^{(k)}(u, z) = \frac{1}{1-z} G_u^{(k)}\left(u, \frac{z}{1-z}\right), \tag{9.42}$$

where $F_u^{(k)}(z, u)$ denotes the kth derivative of $F(z, u)$ with respect to u. Then from (9.41) we obtain

$$G(u, z)(1+z)^b = z(1+z)^b - z^{b+1} + 2uz^b G(u, \frac{z}{2}), \tag{9.43}$$

$$G_u'(u, z)(1+z)^b = 2z^b G(u, \frac{z}{2}) + 2uz^b G_u'(u, \frac{z}{2}), \tag{9.44}$$

$$G_u''(u, z)(1+z)^b = 4z^b G_u'(u, \frac{z}{2}) + 2uz^b G_u''(u, \frac{z}{2}). \tag{9.45}$$

In order to compute the variance, we need $L^1(z) := G_u'(u, z)|_{u=1}$ and $L^2(z) := G_u''(u, z)|_{u=1}$. From (9.44) and (9.45) we immediately obtain

$$L^1(z)(1+z)^b = z^{b+1} + 2z^b L^1(\frac{z}{2}),$$

$$L^2(z)(1+z)^b = 4z^b L^1(\frac{z}{2}) + 2z^b L^2(\frac{z}{2}).$$

Iterating these equations we find (cf. [145] and Section 7.6.1)

$$L^1(z) = \sum_{k=0}^{\infty} \frac{(2z^b)(2(\frac{z}{2})^b) \cdots (2(\frac{z}{2^k})^b)}{\left((1+z)(1+\frac{z}{2}) \cdots (1+\frac{z}{2^k})\right)^b} \frac{z}{2^{k+1}}, \tag{9.46}$$

$$L^2(z) = \sum_{k=0}^{\infty} \frac{(2z^b)(2(\frac{z}{2})^b) \cdots (2(\frac{z}{2^k})^b)}{\left((1+z)(1+\frac{z}{2}) \cdots (1+\frac{z}{2^k})\right)^b} 2L^1(\frac{z}{2^{k+1}}). \tag{9.47}$$

The next step is to transform the above sums (9.46) and (9.47) into harmonic sums. For this, we set $z = 1/t$ and define $Q(t) = \prod_{k=0}^{\infty}(1 + \frac{t}{2^k})$. Then (9.46)–

(9.47) become

$$\frac{tL^{\underline{1}}(\frac{1}{t})}{\left(Q(\frac{t}{2})\right)^b} = \sum_{k=0}^{\infty} \frac{1}{\left(Q(2^k t)\right)^b}, \tag{9.48}$$

$$\frac{tL^{\underline{2}}(\frac{1}{t})}{\left(Q(\frac{t}{2})\right)^b} = 2\sum_{k=0}^{\infty} \frac{2^{k+1} t L^{\underline{1}}(\frac{1}{2^{k+1}t})}{\left(Q(\frac{2^{k+1}t}{2})\right)^b}, \tag{9.49}$$

which are indeed harmonic sums. The Mellin transforms are

$$\mathcal{M}\left[\frac{tL^{\underline{1}}(\frac{1}{t})}{Q^b(\frac{t}{2})}; s\right] = \frac{1}{1-2^{-s}} I(s),$$

$$\mathcal{M}\left[\frac{tL^{\underline{2}}(\frac{1}{t})}{Q^b(\frac{t}{2})}; s\right] = \frac{2^{1-s}}{(1-2^{-s})^2} I(s),$$

where

$$I(s) = \int_0^{\infty} \frac{t^{s-1}}{Q^b(t)} dt = \frac{\pi}{\sin \pi s} J(s), \tag{9.50}$$

$$J(s) = \frac{1}{2\pi i} \int_{\mathcal{H}} \frac{(-t)^{s-1}}{Q^b(t)} dt \tag{9.51}$$

with $\mathcal{H}$ being the Hankel contour (see Table 8.3).

Applying the reverse mapping Theorem 9.2 we find that

$$L^{\underline{1}}(\frac{1}{t}) = \frac{1}{t} k(t) + bk(t) + O(t \log t^{-1}),$$

$$L^{\underline{2}}(\frac{1}{t}) = \frac{1}{t} K(t) + bK(t) + O(t \log^2 t^{-1}),$$

where

$$k(t) = \log_2 \frac{1}{t} + \frac{1}{2} + \frac{J'(0)}{\log 2} - \frac{1}{\log 2} \sum_{k=0}^{\infty} \frac{I(\chi_k)}{\chi_k} t^{-\chi_k},$$

$$K(t) = \log_2^2 \frac{1}{t} + \frac{2J'(0)}{\log^2 2} \log \frac{1}{t} - \left(\frac{1}{6} + \frac{J''(0)}{\log^2 2} - \frac{\pi^2}{3 \log^2 2}\right) + 8bt$$

$$- \frac{2}{\log^2 2} \sum_{k=0}^{\infty} \frac{I(\chi_k)}{\chi_k} t^{-\chi_k} \log \frac{1}{t} + \frac{2}{\log^2 2} \sum_{k=0}^{\infty} \left(\frac{I(\chi_k)}{\chi_k^2} - \frac{I'(\chi_k)}{\chi_k}\right) t^{-\chi_k},$$

with $\chi_k = 2\pi i k / \log 2$ for $k = 0, \pm 1, \ldots$ being the roots of $1 - 2^{-s} = 0$. The next step is to recover asymptotic expansions of $F'_u(1, z)$ and $F''_u(1, z)$ from which computing the mean $\mathbf{E}[D_m]$ and the variance $\mathbf{Var}[D_m]$ is routine. Using (9.42) and changing $t = 1/z$, the above translates to

$$
F'_u(1, z) = \frac{z}{(1-z)^2} \log_2 \frac{z}{1-z} + \frac{1}{\log 2}\left(\frac{\log 2}{2} + J'(0)\right)\frac{z}{(1-z)^2}
$$

$$
- \frac{1}{\log 2}\sum_{k\neq 0}\frac{I(\chi_k)}{\chi_k}\frac{z^{1+\chi_k}}{(1-z)^{2+\chi_k}} + \frac{b}{\log 2}\frac{1}{1-z}\left(\log\frac{z}{1-z}\right)
$$

$$
+ \frac{\log 2}{2} + J'(0) - \sum_{k\neq 0}\frac{I(\chi_k)}{\chi_k}\frac{z^{\chi_k}}{(1-z)^{\chi_k}}\Big) + O\left(\log\frac{1}{(1-z)}\right),
$$

$$
F''_{uu}(1, z) = \frac{z}{(1-z)^2}\log_2^2\frac{z}{1-z} + \frac{2J'(0)}{\log^2 2}\frac{z}{(1-z)^2}\log\frac{z}{1-z} - \left(\frac{1}{6} + \frac{J''(0)}{\log^2 2}\right.
$$

$$
\left. - \frac{\pi^2}{3\log^2 2}\right)\frac{z}{(1-z)^2} - \frac{2}{\log^2 2}\sum_{k\neq 0}\frac{I(\chi_k)}{\chi_k}\frac{z^{\chi_k}}{(1-z)^{\chi_k}}\log\frac{z}{1-z}
$$

$$
+ \frac{2}{\log^2 2}\sum_{k\neq 0}\left(\frac{I(\chi_k)}{\chi_k^2} - \frac{I'(\chi_k)}{\chi_k}\right)\frac{z^{\chi_k}}{(1-z)^{\chi_k}} + O\left(\frac{\log^2(1-z)}{(1-z)}\right).
$$

Having this, we apply the singularity analysis to recover the mean and the variance (through the second factorial moment $\mathbf{E}[D_m(D_m - 1)]$). However, strictly speaking we derived the above *only* for z real (since the Mellin transform is defined for z real) while we need an extension to the complex plane in order to apply Theorem 8.12 of the singularity analysis. Luckily enough, the extension of Mellin transforms to the complex plane was already presented in Section 9.3. With this in mind, we are allowed to apply the singularity analysis, in particular Theorem 8.11, and after some algebra, we find that

$$
\mathbf{E}[D_m] = \log_2 m + \frac{1}{\log 2}\left(\gamma - 1 + \frac{\log 2}{2} + J'(0) - \sum_{k\neq 0}\frac{I(\chi_k)}{\chi_k}\frac{m^{\chi_k}}{\Gamma(2+\chi_k)}\right)
$$

$$
+ O\left(\frac{\log m}{m}\right),
$$

$$
\mathbf{Var}[D_m] = \frac{1}{12} + \frac{1}{\log^2 2}\left(1 + \frac{\pi^2}{6}\right) + \frac{1}{\log^2 2}\left(J''(0) - (J'(0))^2\right)
$$

$$
+ \frac{1}{\log 2}\delta_2(\log_2 m) - [\delta_1^2]_0 + O\left(\frac{\log^2 m}{m}\right),
$$

where

$$J'(0) = \int_0^1 \left(\frac{1}{Q(t)^b} - 1 \right) \frac{dt}{t} + \int_0^\infty \frac{1}{Q(t)^b} \frac{dt}{t}, \tag{9.52}$$

$$J''(0) = -\frac{\pi^2}{3} + 2\int_0^1 \left(\frac{1}{Q(t)^b} - 1 \right) \frac{\log t}{t} dt + 2\int_0^\infty \frac{1}{Q(t)^b} \frac{\log t}{t} dt, \tag{9.53}$$

$\delta_2(\cdot)$ is a periodic function with mean zero and period 1, and $[\delta_1^2]_0$ is a very small constant (e.g., $[\delta_1^2]_0 \leq 10^{-10}$ for $b = 1$). More precisely, (cf. Hubalek [209])

$$[\delta_1^2]_0 = \frac{1}{L^2} \sum_{k \neq 0} \frac{I(\chi_k)I(-\chi_k)}{\Gamma(2 + \chi_k)\Gamma(2 - \chi_k)},$$

and

$$I(\chi_k) = \frac{1}{\chi_k} + \int_0^1 \left(Q^{-b}(t) - 1 \right) t^{\chi_k - 1} dt + \int_1^\infty Q^{-b}(t) t^{\chi_k} dt.$$

Numerical evaluation of $J'(0)$ and $J''(0)$ can be found in Hubalek [209].

9.4.2 Redundancy of Renewal Processes

We evaluate here the minimax redundancy of a renewal process. We shall follow Flajolet and Szpankowski [152]. The reader is referred to Section 8.7.2 for a brief review of the redundancy problem. We start with a precise definition of the class $\mathcal{R}_0$ of renewal processes and its associated sources. Let $T_1, T_2, \ldots$ be a sequence of i.i.d. (i.e., independently, identically distributed) nonnegative integer-valued random variables with distribution $Q(j) = \Pr\{T_1 = j\}$ and $\mathbf{E}[T_1] < \infty$. The quantities $\{T_i\}_{i=1}^\infty$ are the interarrival times, while T_0 is the initial waiting time. The process $T_0, T_0 + T_1, T_0 + T_1 + T_2, \ldots$ is then called a renewal process. With such a renewal process there is associated a *binary renewal sequence* that is a 0, 1-sequence in which the 1's occur exactly at the renewal epochs $T_0, T_0 + T_1, T_0 + T_1 + T_2$, etc. We consider renewal sequences that start with a "1". The minimax redundancy, R_n^*, of the renewal process was studied by Csiszár and Shields [79], who proved that $R_n^* = \Theta(\sqrt{n})$.

By Shtarkov's method discussed in Section 8.7.2 we conclude that the minimax redundancy $R_n^*(\mathcal{R}_0)$ is within one from

$$R_n' := \log_2 \left(\sum_{x_1^n} \sup_Q P(x_1^n) \right)$$

(i.e., $R_n' \leq R_n^*(\mathcal{R}_0) \leq R_n' + 1$). Instead of working with R_n' we estimate another quantity, R_n, that is within $O(\log n)$ from R_n', that is, $R_n = R_n' + O(\log n)$. In

[152] it is argued that R_n is the minimax redundancy of (non-stationary) renewal sequences that starts with "1". Let now

$$r_n = 2^{R_n}.$$

Solving a simple optimization problem, we show (cf. also [152]) that r_n can be represented by the following combinatorial sum.

Lemma 9.6 *The quantity $r_n = 2^{R_n}$ admits the combinatorial form*

$$\begin{cases} r_n & = \displaystyle\sum_{k=0}^{n} r_{n,k} \\ r_{n,k} & = \displaystyle\sum_{\mathcal{P}(n,k)} \binom{k}{k_0 \cdots k_{n-1}} \left(\frac{k_0}{k}\right)^{k_0} \cdots \left(\frac{k_{n-1}}{k}\right)^{k_{n-1}}, \end{cases}$$

where $\mathcal{P}(n, k)$ denotes the set of partitions of n into k summands, that is, the collection of tuples of nonnegative integers $(k_0, k_1, \ldots, k_{n-1})$ satisfying

$$n = k_0 + 2k_1 + \cdots + nk_{n-1}, \tag{9.54}$$

$$k = k_0 + k_1 + \cdots + k_{n-1} \tag{9.55}$$

for all $n \geq 0$ and $k \leq n$.

It can also be observed that the quantity r_n has an intrinsic meaning. Let $\mathcal{W}_n$ denote the set of all n^n sequences of length n over the alphabet $\{0, \ldots, n-1\}$. For a sequence w, take k_j to be the number of letters j in w. Then each sequence w carries a "maximum likelihood probability", $\pi_{ML}(w)$: this is the probability that w gets assigned in the memoryless model that makes it most likely. The quantity r_n is then

$$r_n = \sum_{w \in \mathcal{W}_n} \pi_{ML}(w).$$

Our goal is to asymptotically estimate r_n through asymptotics of $r_{n,k}$. A difficulty of finding such asymptotics stems from the factor $k!/k^k$ present in the definition of $r_{n,k}$. We circumvent this problem by analyzing a related pair of sequences, s_n and $s_{n,k}$, that are defined as

$$\begin{cases} s_n & = \displaystyle\sum_{k=0}^{n} s_{n,k} \\ s_{n,k} & = e^{-k} \displaystyle\sum_{\mathcal{P}(n,k)} \frac{k_0^{k_0}}{k_0!} \cdots \frac{k_{n-1}^{k_{n-1}}}{k_{n-1}!}. \end{cases} \tag{9.56}$$

The translation from s_n to r_n is most conveniently expressed in probabilistic terms. Introduce the random variable K_n whose probability distribution is $s_{n,k}/s_n$, that is,

$$\varpi_n : \qquad \Pr\{K_n = k\} = \frac{s_{n,k}}{s_n}, \tag{9.57}$$

where ϖ_n denotes the distribution. Then Stirling's formula yields

$$\frac{r_n}{s_n} = \sum_{k=0}^{n} \frac{r_{n,k}}{s_{n,k}} \frac{s_{n,k}}{s_n} = \mathbf{E}[(K_n)! K_n^{K_n} e^{-K_n}]$$

$$= \mathbf{E}[\sqrt{2\pi K_n}] + O(\mathbf{E}[K_n^{-\frac{1}{2}}]). \tag{9.58}$$

Thus the problem of finding r_n reduces to asymptotic evaluations of s_n, $\mathbf{E}[\sqrt{K_n}]$ and $\mathbf{E}[K_n^{-\frac{1}{2}}]$. The heart of the matter is the following lemma, which provides the necessary estimates. The somewhat delicate proof of Lemma 9.7 constitutes the core of the section.

Lemma 9.7 *Let $\mu_n = \mathbf{E}[K_n]$ and $\sigma_n^2 = \mathbf{Var}[K_n]$, where K_n has the distribution ϖ_n defined above in (9.57). The following holds*

$$s_n \sim \exp\left(2\sqrt{cn} - \frac{7}{8} \log n + d + o(1)\right), \tag{9.59}$$

$$\mu_n = \frac{1}{4}\sqrt{\frac{n}{c}} \log \frac{n}{c} + o(\sqrt{n}), \tag{9.60}$$

$$\sigma_n^2 = O(n \log n) = o(\mu_n^2), \tag{9.61}$$

where $c = \pi^2/6 - 1$, $d = -\log 2 - \frac{3}{8} \log c - \frac{3}{4} \log \pi$.

Once the estimates of Lemma 9.7 are granted, the moments of order $\pm\frac{1}{2}$ of K_n follow by a standard argument based on concentration of the distribution ϖ_n as discussed in Chapter 5.

Lemma 9.8 *For large n*

$$\mathbf{E}[\sqrt{K_n}] = \mu_n^{1/2}(1 + o(1)) \tag{9.62}$$

$$\mathbf{E}[K_n^{-\frac{1}{2}}] = o(1), \tag{9.63}$$

where $\mu_n = \mathbf{E}[K_n]$.

Proof. We only prove (9.62) since (9.63) is obtained in a similar manner. The upper bound $\mathbf{E}[\sqrt{K_n}] \leq \sqrt{\mathbf{E}[K_n]}$ follows by Jensen's inequality (2.9) and the concavity of the function $\sqrt{x}$. The lower bound follows from concentration of the distribution. Chebyshev's inequality (2.7) and (9.61) of Lemma 9.7 entail, for any arbitrarily small $\varepsilon > 0$,

$$\Pr\{|K_n - \mu_n| > \varepsilon\mu_n\} \leq \frac{\mathbf{Var}[K_n]}{\varepsilon^2\mu_n^2} = \frac{\delta(n)}{\varepsilon^2},$$

where $\delta(n) \to 0$ as $n \to \infty$. Then

$$\mathbf{E}[\sqrt{K_n}] \geq \sum_{k \geq (1-\varepsilon)\mu_n} \sqrt{k}\Pr\{K_n \geq k\}$$

$$\geq (1-\varepsilon)^{\frac{1}{2}}\mu_n^{1/2}\Pr\{K_n \geq (1-\varepsilon)\mu_n\}$$

$$\geq (1-\varepsilon)^{\frac{1}{2}}\left(1 - \frac{\delta(n)}{\varepsilon^2}\right)\mu_n^{1/2}.$$

Hence for any $\eta > 0$ one has

$$\mathbf{E}[\sqrt{K_n}] > \mu_n^{1/2}(1 - \eta)$$

provided n is large enough. This completes the proof. ∎

In summary, r_n and s_n are related by

$$r_n = s_n\mathbf{E}[\sqrt{2\pi K_n}](1 + o(1))$$

$$= s_n\sqrt{2\pi\mu_n}(1 + o(1)),$$

by virtue of (9.58) and Lemma 9.8. This leads to

$$R_n = \log_2 r_n = \log_2 s_n + \frac{1}{2}\log_2\mu_n + \log_2\sqrt{2\pi} + o(1) \qquad (9.64)$$

$$= \frac{2}{\log 2}\sqrt{\left(\frac{\pi^2}{6} - 1\right)n} - \frac{5}{8}\log_2 n + \frac{1}{2}\log_2\log n + O(1). \qquad (9.65)$$

To complete the proof of our main result (9.64) we need to **prove Lemma 9.7**, which is discussed next. Let

$$\beta(z) = \sum_{k=0}^{\infty} \frac{k^k}{k!}e^{-k}z^k,$$

which by the Lagrange inversion formula is equal to

$$\beta(z) = \frac{1}{1 - T(ze^{-1})},$$

where $T(z)$ is the tree function discussed in Section 7.3.2 (e.g., see (7.31)). As a matter of fact, $\beta(z) = B(ze^{-1})$, where $B(z)$ is defined in (7.35).

The quantities s_n and $s_{n,k}$ of (9.56) have generating functions

$$S_n(u) = \sum_{k=0}^{\infty} s_{n,k} u^k, \qquad S(z, u) = \sum_{n=0}^{\infty} S_n(u) z^n.$$

Then since equation (9.66) involves convolutions of sequences of the form $k^k / k!$, we have

$$S(z, u) = \sum_{\mathcal{P}(n,k)} z^{1k_0 + 2k_1 + \cdots} \left(\frac{u}{e}\right)^{k_0 + \cdots + k_{n-1}} \frac{k_0^{k_0}}{k_0!} \cdots \frac{k_{n-1}^{k_{n-1}}}{k_{n-1}!}$$

$$= \prod_{i=1}^{\infty} \beta(z^i u). \tag{9.66}$$

The first moment $\mu_n = \mathbf{E}[K_n]$ and the second factorial moment $\mathbf{E}[K_n(K_n - 1)]$ are obtained as

$$s_n = [z^n] S(z, 1),$$

$$\mu_n = \frac{[z^n] S_u'(z, 1)}{[z^n] S(z, 1)},$$

$$\mathbf{E}[K_n(K_n - 1)] = \frac{[z^n] S_{uu}''(z, 1)}{[z^n] S(z, 1)},$$

where $S_u'(z, 1)$ and $S_{uu}''(z, 1)$ represent the first and the second derivative of $S(z, u)$ at $u = 1$.

We deal here with s_n that is accessible through its generating function

$$S(z, 1) = \prod_{i=1}^{\infty} \beta(z^i). \tag{9.67}$$

The behavior of the generating function $S(z, 1)$ as $z \to 1$ is an essential ingredient of the analysis. We know that the singularity of the tree function $T(z)$ at $z = e^{-1}$ is of the square-root type as discussed in Example 8.10 of Chapter 8 (cf. also [73]). Hence, near $z = 1$, $\beta(z)$ admits the singular expansion (cf. Example 8.10 of Chapter 8 or (8.140))

$$\beta(z) = \frac{1}{\sqrt{2(1-z)}} + \frac{1}{3} - \frac{\sqrt{2}}{24}\sqrt{1-z} + O(1-z).$$

We now turn to the infinite product asymptotics as $z \to 1^-$, with z real. Let $L(z) = \log S(z, 1)$ and $z = e^{-t}$, so that

$$L(e^{-t}) = \sum_{k=1}^{\infty} \log \beta(e^{-kt}). \qquad (9.68)$$

The Mellin transform technique discussed in this chapter provides an expansion for $L(e^{-t})$ around $t = 0$ (or equivalently $z = 1$) since the sum (9.68) is a *harmonic sum* paradigm discussed in this chapter. The Mellin transform $L^*(s) = \mathcal{M}[L(e^{-t}); s]$ of $L(e^{-t})$ computed by the harmonic sum formula for $\Re(s) \in (1, \infty)$ is

$$L^*(s) = \zeta(s)\Lambda(s),$$

where

$$\Lambda(s) = \int_0^{\infty} \log \beta(e^{-t}) t^{s-1} dt.$$

Now by the direct mapping Theorem 9.3 the expansion of $\beta(z)$ at $z = 1$ implies

$$\log \beta(e^{-t}) = -\frac{1}{2} \log t - \frac{1}{2} \log 2 + O(\sqrt{t}),$$

so that, collecting local expansions,

$$\Lambda(s) \asymp \left[\Lambda(1)\frac{1}{s-1} \right]_{s=1} + \left[\frac{1}{2}\frac{1}{s^2} - \frac{1}{2}\frac{\log 2}{s} \right]_{s=0}.$$

On the other hand, classical expansion of the zeta function gives (cf. [2, 424])

$$\zeta(s) \asymp \left[\frac{1}{s-1} + \gamma \right]_{s=1} + \left[-\frac{1}{2} - s \log \sqrt{2\pi} \right]_{s=0}.$$

Termwise multiplication then provides the singular expansion of $L^*(s)$:

$$L^*(s) \asymp \left[\frac{\Lambda(1)}{s-1} \right]_{s=1} + \left[-\frac{1}{4s^2} - \frac{\log \pi}{4s} \right]_{s=0}.$$

An application of the reverse mapping Theorem 9.2 allows us to come back to the original function

$$L(e^{-t}) = \frac{\Lambda(1)}{t} + \frac{1}{4}\log t - \frac{1}{4}\log \pi + O(\sqrt{t}), \qquad (9.69)$$

which (after using $z = 1 - t + O(t^2)$) translates into

$$L(z) = \frac{\Lambda(1)}{1-z} + \frac{1}{4}\log(1-z) - \frac{1}{4}\log \pi - \frac{1}{2}\Lambda(1) + O(\sqrt{1-z}). \qquad (9.70)$$

This computation is finally completed by the evaluation of $c := \Lambda(1)$:

$$
\begin{aligned}
c = \Lambda(1) &= -\int_0^1 \log(1 - T(x/e))\frac{dx}{x} \\
&= -\int_0^1 \log(1 - t)\frac{(1-t)}{t}dt \qquad (x = te^{1-t}) \\
&= \frac{\pi^2}{6} - 1.
\end{aligned}
$$

In summary, we just proved that as $z \to 1^-$

$$S(z, 1) = e^{L(z)} = a(1 - z)^{\frac{1}{4}} \exp\left(\frac{c}{1-z}\right)(1 + o(1)), \qquad (9.71)$$

where $a = \exp(-\frac{1}{4}\log \pi - \frac{1}{2}c)$. So far, the main estimate (9.71) has been established as z tends to 1 from the left, by real values, but by Theorem 9.4 we know that it is true for *complex t* only constrained in such a way that $-\frac{\pi}{2}+\epsilon \le \arg(t) \le \frac{\pi}{2} - \epsilon$, for any $\epsilon > 0$. Thus, the expansion (9.71) actually holds true as $z \to 1$ in a sector, say, $|\arg(1 - z)| < \frac{\pi}{4}$.

It remains to collect the information gathered on $S(z, 1)$ and recover $s_n = [z^n]S(z, 1)$ asymptotically. The inversion is provided by the Cauchy coefficient formula.

We shall use the following lemma.

Lemma 9.9 *For positive $A > 0$, and reals B and C, define $f(z) = f_{A,B,C}(z)$ as*

$$f(z) = \exp\left(\frac{A}{1-z} + B\log\frac{1}{1-z} + C\log\left(\frac{1}{z}\log\frac{1}{1-z}\right)\right). \qquad (9.72)$$

Then the nth Taylor coefficient of $f_{A,B,C}(z)$ satisfies asymptotically, for large n,

$$[z^n] f_{A,B,C}(z) = \exp\left(2\sqrt{An} + \frac{1}{2}\left(B - \frac{3}{2}\right)\log n + C \log\log\sqrt{\frac{n}{A}}\right.$$

$$\left. - \frac{1}{2}\log\left(4\pi e^{-A}/\sqrt{A}\right)\right)(1 + o(1)). \tag{9.73}$$

Proof. Problems of this kind have been considered by Wright [451] and others who, in particular, justify in detail that the saddle point method works in similar contexts. Therefore we merely outline the proof here (see Table 8.4 in Section 8.4). The starting point is Cauchy's formula

$$[z^n] f(z) = \frac{1}{2\pi i} \oint e^{h(z)} dz,$$

where

$$h(z) = \log f_{A,B,C}(z) - (n + 1)\log z.$$

In accordance with (SP1) of Table 8.4 one chooses a saddle point contour that is a circle of radius r defined by $h'(r) = 0$. Asymptotically, one finds

$$r = 1 - \sqrt{\frac{A}{n}} + \frac{B - A}{2n} + o(n^{-1}),$$

and

$$h(r) = 2A\sqrt{\frac{n}{A}} + B\log\left(\sqrt{\frac{n}{A}}\right) + C\log\log\left(\sqrt{\frac{n}{A}}\right) + \frac{1}{2}A + o(1).$$

The "range" $\delta = \delta(n)$ of the saddle point, where most of the contribution of the contour integral is concentrated asymptotically, is dictated by the order of growth of derivatives (cf. (SP2) of Table 8.4). Here, $h''(r) \approx n^{3/2}$, while $h'''(r) \approx n^2$, so that

$$\delta(n) = n^{-3/4}.$$

In accordance with requirement (SP3) of Table 8.4, tails are negligible since the function $\exp((1 - z)^{-1})$ decays very fast when going away from the real axis. In the central region, the local approximation (SP4) applies, as seen by expansions near $z = 1$. Thus requirements (SP1), (SP2), (SP3), and (SP4) of Table 8.4 are satisfied, implying, by (SP5) of Table 8.4

$$[z^{n-1}] f(z) = \frac{1}{\sqrt{2\pi |h''(r)|}} e^{h(r)}(1 + o(1)).$$

Some simple algebra, using

$$h''(r) = 2n\sqrt{n/A}\,(1 + o(1)),$$

yields the stated estimate (9.73). ∎

Now, the function $S(z, 1)$ is only known to behave like $f(z)$ of Lemma 9.9 in the vicinity of 1. In order to adapt the proof of Lemma 9.9 and legitimate the use of the resulting formula, we need to prove that $S(z, 1)$ decays fast away from the real axis. This somewhat technical Lemma 9.10 below is proved in [152] and we omit here its derivation.

Lemma 9.10 **(Concentration property)** *Consider the ratio*

$$q(z) = \prod_{j=1}^{\infty}\left|\frac{\beta(z^j)}{\beta(|z|^j)}\right|.$$

Then there exists a constant $c_0 > 0$ such that

$$q(re^{i\theta}) = O\left(e^{-c_0(1-r)^{-1}}\right),$$

uniformly, for $\frac{1}{2} \le r < 1$ and $|\arg(re^{i\theta} - 1)| > \frac{\pi}{4}$.

We are now finally ready to return to the estimate of s_n in Lemma 9.7. In the region $|\arg(z - 1)| < \frac{\pi}{4}$, the Mellin asymptotic estimates (9.69) and (9.71) apply. This shows that in this region,

$$S(z, 1) = e^{o(1)}f_{A,B,C}(z) \qquad (z \to 1),$$

where the function f is that of Lemma 9.9 and the constants A, B, C have the values assigned by (9.71):

$$A = c = \frac{\pi^2}{6} - 1, \quad B = -\frac{1}{4}, \quad C = 0.$$

In the complementary region, $|\arg(z - 1)| > \frac{\pi}{4}$, the function $S(z, 1)$ is exponentially smaller than $S(|z|, 1)$ by Lemma 9.10. From these two facts, the saddle point estimates of Lemma 9.9 are seen to apply, by a trivial modification of the proof of that lemma. This concludes the proof of Equation (9.59) in Lemma 9.7.

It remains to complete the evaluation of μ_n and σ_n^2, following the same principles as before. Start with $\mu_n = \mathbf{E}[K_n]$, with the goal of establishing the evaluation

(9.60) of Lemma 9.7. It is necessary to estimate $[z^n]S'_u(z, 1)$, with

$$S'_u(z, 1) = S(z, 1) \sum_{k=0}^{\infty} z^k \frac{\beta'(z^k)}{\beta(z^k)}.$$

Let

$$D_1(z) = \sum_{k=0}^{\infty} \alpha(z^k), \qquad \text{where} \qquad \alpha(z) = z \frac{\beta'(z)}{\beta(z)}.$$

Via the substitution $z = e^{-t}$, the function $D_1(e^{-t})$ falls under the harmonic sum so that its Mellin transform is

$$\mathcal{M}[D_1(e^{-t}); s] = \zeta(s)\mathcal{M}[\alpha(e^{-t}); s].$$

The asymptotic expansion

$$\alpha(e^{-t}) = \frac{1}{2t} - \frac{\sqrt{2}}{6}\frac{1}{\sqrt{t}} - \frac{1}{18} + O(\sqrt{t})$$

gives the singular expansion of the corresponding Mellin transform. This in turn yields the singular expansion of $\mathcal{M}[D_1(e^{-t}); s]$. Then the reverse mapping Theorem 9.2 gives back $D(e^{-t})$ at $t \sim 0$; hence,

$$D_1(z) = \frac{1}{2}\frac{1}{1-z}\log\frac{1}{1-z} + \frac{1}{2}\frac{\gamma}{1-z} - \frac{1}{6}\frac{\sqrt{2}\zeta(\frac{1}{2})}{\sqrt{1-z}} - \frac{1}{4}\log\frac{1}{1-z} + O(1),$$

where $\gamma = 0.577\ldots$ is the Euler constant. The combination of this last estimate and the main asymptotic form of $S(z, 1)$ in (9.71) yields

$$S'_u(z, 1) \sim \frac{1}{2}a\exp\left(\frac{c}{1-z} + \frac{3}{4}\log\frac{1}{1-z} + \log\log\frac{1}{1-z}\right),$$

where a is the same constant as in (9.71). As for $S(z, 1)$, the derivative $S'_u(z, 1)$ is amenable to Lemma 9.9, and this proves the asymptotic form of μ_n, as stated in (9.60) of Lemma 9.7.

Finally, we need to justify (9.61), which represents a bound on the variance of K_n. The computations follow the same steps as above, so we only sketch them briefly. One needs to estimate a second derivative,

$$\frac{S''_{uu}(z, 1)}{S(z, 1)} = D_2(z) + D_1^2(z),$$

where

$$D_2(z) = \sum_{k=0}^{\infty} z^{2k} \frac{\beta''(z^k)}{\beta(z^k)} - \left(\frac{z^k \beta'(z^k)}{\beta(z^k)} \right)^2.$$

The sum above is again a harmonic sum that is amenable to Mellin analysis, with the result that

$$D_2(z) = \frac{\zeta(2)}{2} \frac{1}{(1-z)^2} + O((1-z)^{-3/2}).$$

Then we appeal again to Lemma 9.9 to achieve the transfer to coefficients. Somewhat tedious calculations (that were assisted by the computer algebra system MAPLE) show that the leading term in $n \log^2 n$ of the second moment cancels with the square of the mean μ_n. Hence, the variance cannot be larger than $O(n \log n)$. This establishes the second moment estimate (9.61) of Lemma 9.7.

In summary, we proved that R_n (hence also the minimax redundancy R_n^*, which is within $O(\log n)$ from R_n) attains the following asymptotics

$$R_n = \frac{2}{\log 2} \sqrt{\left(\frac{\pi^2}{6} - 1 \right) n} - \frac{5}{8} \log_2 n + \frac{1}{2} \log_2 \log n + O(1)$$

as $n \to \infty$.

9.5   EXTENSIONS AND EXERCISES

9.1   Prove Entry 13 of Table 9.1.

9.2   Prove that

$$\mathcal{M}\left[e^{-x} - \sum_{j=0}^{k} (-1)^k \frac{x^j}{j!}; s \right] = \Gamma(s)$$

for $-k - 1 < \Re(s) < -k$. What is the Mellin transform of $e^{-x} - 2$, if it exists.

9.3   ⚠ (Louchard, Szpankowski, and Tang, 1999) Prove the following result.

Lemma 9.11   *Let $\{f_n\}_{n=0}^{\infty}$ be a sequence of real numbers. Suppose that its Poisson generating function $\tilde{F}(z) = \sum_{n=0}^{\infty} f_n \frac{z^n}{n!} e^{-z}$ is an entire function. Furthermore, let its Mellin transform $F(s)$ have the following factorization: $F(s) = \mathcal{M}[\tilde{F}(z); s] = \Gamma(s)\gamma(s)$, and assume $F(s)$ exists for*

$\Re(s) \in (-2, -1)$, while $\gamma(s)$ is analytic for $\Re(s) \in (-\infty, -1)$. Then

$$\gamma(-n) = \sum_{k=0}^{n} \binom{n}{k}(-1)^k f_k, \quad \text{for } n \geq 2.$$

9.4 ⚠ Define

$$f(x) = \sum_{k \geq 1} \frac{e^{kx}}{\left(e^{kx} - 1\right)^2}$$

$$g(x) = \sum_{k \geq 1} \frac{(-1)^{k-1}}{k\left(e^{kx} - 1\right)},$$

$$h(x) = \sum_{k \geq 1} \frac{1}{k\left(e^{2k\pi x} - 1\right)} .$$

Prove the following identities:

$$1 + 2\sum_{n \geq 1} e^{-\pi n^2 x} = \frac{1}{\sqrt{x}} + \frac{2}{\sqrt{x}} \sum_{n \geq 1} e^{-\pi n^2/x} \qquad \text{Jacobi's } \vartheta\text{-function,}$$

$$f(x) = \frac{1}{x^2}\frac{\pi^2}{6} - \frac{1}{2x} + \frac{1}{24} - \frac{4\pi^2}{x^2} f\left(\frac{4\pi^2}{x}\right) \qquad \text{Ramanujan [45]}$$

$$g(x) = \frac{\pi^2}{12x} - \frac{\log 2}{2} + \frac{x}{24} - g\left(\frac{2\pi^2}{x}\right) \qquad \text{Ramanujan [45]}$$

$$h(x) = \frac{\pi}{12}\left(\frac{1}{x} - x\right) + \frac{\log x}{2} + h\left(\frac{1}{x}\right) \qquad \text{Dedekind and Ramanujan [45].}$$

9.5 Consider the following functional equation

$$f(x) = 2f(x/2) + 1 - (1+x)e^{-x}, \quad x > 0.$$

Prove that for arbitrary positive M and $x \to \infty$

$$f(x) = x\left(\frac{1}{\log 2} + G(\log_2 x)\right) - 1 + O(x^{-M}),$$

where

$$G(x) = \frac{1}{\log^2 2} \sum_{k \neq 0} k \Gamma(-1 + 2\pi i k / \log 2) \exp(-2\pi i k x)$$

is a periodic function with period 1 and amplitude smaller than $1.6 \cdot 10^{-6}$.

9.6 Derive the following asymptotic expansion

$$\sum_{k=1}^{\log_2 n} 2^k \left(1 - (1 - 2^{-k})^n\right) = n(2 - C_1) - 2 - \frac{1}{2}C_2 + O(1/\log n)$$

where

$$C_1 = \sum_{l=0}^{\infty} 2^{-l} e^{-2^l},$$

$$C_2 = \sum_{l=0}^{\infty} 2^l e^{-2^l}$$

are constants.

Hint. Do **not** use the Mellin transform.

9.7 Using the generalized Euler-Maclaurin formula discussed in Example 9.4, analyze asymptotically

$$F_1(x) = \sum_{n \geq 1} e^{-\sqrt{nx}},$$

$$F_2(x) = \sum_{n \geq 1} \frac{\sqrt{nx}}{1 + n^2 x^2}.$$

9.8 Prove that under suitable conditions (state them!)

$$\sum_{n=1}^{\infty} (-1)^{n-1} f(nx) \sim \sum_{k=0}^{\infty} c_k (1 - 2^{1+\beta_k}) \zeta(-\beta_k) x^{\beta_k},$$

where as in Example 9.4 we assume that

$$f(x) = \sum_{k=0}^{\infty} c_k x^{\beta_k}, \qquad x \to 0,$$

where $-1 < \beta_0 < \beta_1 < \cdots$.

9.9 Generalize formula (9.37), that is, find an asymptotic expansion for

$$F_m(x) = \sum_{n\geq 1} \log^m n f(nx)$$

for $m \geq 1$.

9.10 Develop an asymptotic formula for

$$\sum_{n=1}^{\infty} (-1)^n f(nx) \log n.$$

Using this formula establish an asymptotic expansion of the sum from the introduction, namely,

$$H(x) = \sum_{k=1}^{\infty} (-1)^k e^{-k^2 x} \log k$$

for $x \to 0$.

9.11 Consider the function $\varphi(z)$ defined in Example 9.2, that is,

$$\varphi(z) = \prod_{j=0}^{\infty} \left(1 - e_d(z2^j)e^{-z2^j}\right),$$

where $e_d(z) = 1 + \frac{z^1}{1!} + \cdots + \frac{z^d}{d!}$. Prove that for all $z \in \mathcal{S}_\theta$ with $z \neq 0$ and $|\theta| < \pi/2$ we have uniformly $|\varphi(z)| < A$ for some constant $A > 0$.

9.12 Prove that

$$F(x) = \sum_{k=1}^{\infty} \left(1 - (1 - 2^{-k})^x\right)$$

attains the following asymptotics for $x \to \infty$:

$$F(x) \sim \log_2 x + \frac{\gamma}{\log 2} - \frac{1}{2} + \sum_{j=1}^{\infty} P_j(\log_2 x)x^{-j},$$

where P_j is a periodic function of period 1 (cf. Exercise 9.6).

9.13 Analyze asymptotically

$$\sum_{k\geq 0} (-1)^k \sqrt{k} \binom{2n}{n-k}.$$

9.14 ⚠ Let P_n be the number of integer partitions. In Example 7.5 we proved that its ordinary generating function is

$$P(z) = \prod_{k=1}^{\infty} (1 - z^k)^{-1}.$$

Show that

$$P_n \sim \frac{\exp(\pi \sqrt{2n/3})}{4\sqrt{3}n}.$$

Hint. The following steps are advised:

(a) Define

$$F(t) = \log P(e^{-t}) = \sum_{k=1}^{\infty} \log(1 - e^{kt})^{-1},$$

and then prove that the Mellin transform $F^*(s)$ of $F(t)$ is

$$F^*(s) = \zeta(s)\zeta(s+1)\Gamma(s).$$

(b) Using the reverse mapping Theorem 9.2 prove that

$$F(t) \sim \frac{\pi^2}{6t} + \log\sqrt{\frac{t}{2\pi}} - \frac{t}{24}, \qquad t \to 0,$$

which yields

$$P(z) = \frac{e^{-\pi^2/12}}{\sqrt{2\pi(1-z)}} \exp\left(\frac{\pi^2}{6(1-z)}\right)(1 + O(1-z)).$$

(c) Use the saddle point method (as described in Table 8.4 or in Lemma 9.9) to establish the result.

9.15 ⚠ Let Q_n be the number of integer partitions in distinct parts. Prove that its generating function is

$$Q(z) = \prod_{k=1}^{\infty} (1 + z^k),$$

and then establish

$$Q_n \sim \frac{\exp(\pi \sqrt{n/3})}{4 3^{1/4} n^{3/4}}.$$

9.16 ⚠ (Andrews, 1984) Let

$$G(z) = \prod_{n=1}^{\infty} (1 - z^n)^{-a_n} = 1 + \sum_{n=1}^{\infty} r_n z^n,$$

where $a_n \geq 0$ such that its Dirichlet series $D(s) = \sum_{n \geq 1} \frac{a_n}{n^s}$ is convergent for $\Re(s) > \alpha$ and

$$D(s) \asymp \left[\frac{A}{s - \alpha} \right]_{s=\alpha}.$$

Furthermore, assume that $D(s)$ is continuable to the left and of moderate growth; that is, $D(s) = O(|\Im(s)|^c)$ for some constant c. Let $z = e^{-t}$ and we assume that for $\arg(t) > \pi/4$.

$$\Re(G(e^{-t})) - G(e^{-\Re(t)}) \leq -C_2 \Re(t)^{-\varepsilon}$$

for sufficiently small $\Re(t)$ and arbitrarily small $\varepsilon > 0$. Generalize the approach presented in Exercise 14 to prove the following lemma due to G. Meinardus (cf. [13]).

Lemma 9.12 (Meinardus) *With the notations as above, the following holds*

$$r_n \sim C n^\kappa \exp\left(n^{\alpha/(1+\alpha)} (1 + 1/\alpha) \left(A\Gamma(\alpha + 1)\zeta(\alpha + 1) \right)^{1/(1+\alpha)} \right),$$

where

$$C = \frac{e^{D'(0)} \left(A\Gamma(\alpha + 1)\zeta(\alpha + 1) \right)^{(1-2D(0))/(2+2\alpha)}}{\sqrt{2\pi(1 + \alpha)}},$$

$$\kappa = \frac{D(0) - 1 - \alpha/2}{1 + \alpha}$$

as $n \to \infty$.

9.17 ⚠ Consider a trie built over n binary strings generated independently by a memoryless source. Let p and $q = 1 - p$ denote the probability of generating "1" and "0", respectively. In this exercise we are interested in finding the limiting distribution of the fill-up level F_n (see Section 1.1), that is, the maximal level, where the tree is still full (in other words, the tree contains all nodes up to level F_n). Let $f_n^k = \Pr\{F_n \geq k\}$. Verify that it satisfies the following recurrence for all $n \geq 0$ and $k \geq 0$

$$f_n^{k+1} = \sum_{i=0}^{n} \binom{n}{i} p^i q^{n-i} f_i^k f_{n-i}^k$$

with $f_0^0 = 0$ and $f_n^0 = 1$ for $n \geq 1$. Find asymptotics of f_n^k for all possible ranges of n and k. In particular, prove the result of Pittel [338] asserting that

$$\Pr\{F_n = k_n \text{ or } F_n = k_n + 1\} = 1 - o(1), \qquad n \to \infty,$$

where $k_n = \lfloor \log_Q n - \log_Q \log_Q \log n + o(1) \rfloor$ and $Q^{-1} = \min\{p, 1 - p\}$.

9.18 ▽ Consider the same trie as in Exercise 17 above. Consider now the shortest path s_n (see Section 1.1). Let $r_n^k = \Pr\{s_n \geq k\}$. Verify that it satisfies the following recurrence for $n \geq 0$ and $k \geq 0$

$$r_n^{k+1} = \sum_{i=0}^{n} \binom{n}{i} p^i q^{n-i} r_i^k r_{n-i}^k - \delta_{n,1} \delta_{k,0}$$

with $r_n^0 = 1$ for $n \geq 0$. In the above $\delta_{i,j}$ is the Kronecker delta (i.e., $\delta_{ij} = 1$ for $i = j$ and zero otherwise). Find asymptotics of r_n^k for all possible ranges of n and k.

9.19 ⚠ Show that the Mellin transform $f^*(s)$ of $f(x)$ in the strip of convergence can be represented as the following Hankel integral

$$f^*(s) = \frac{1}{2i \sin \pi s} \int_{\infty}^{0+} f(z)(-z)^{s-1} \, dz,$$

where the contour starts at infinity on the upper half-plane, encircles the origin, and proceeds back to infinity in the lower half-plane.

10

Analytic Poissonization and Depoissonization

In combinatorics and the analysis of algorithms a Poisson version of a problem (henceforth called *Poisson model* or *poissonization*) is often easier to solve than the original one, which is usually known under the name *Bernoulli model*. Poissonization is a technique that replaces the original input by a Poisson process. Poisson transform maps a sequence characterizing the Bernoulli model into a generating function of a complex variable characterizing the Poisson model. Once the problem is solved in the Poisson domain, one must *depoissonize* it in order to translate the results back to the original (i.e., Bernoulli) model. A large part of this chapter is devoted to various depoissonization results. As a matter of fact, analytic depoissonization can be viewed as an asymptotic technique applied to generating functions that are entire functions. We illustrate our analysis with numerous examples from combinatorics and the analysis of algorithms and data structures. These applications are among the most sophisticated that we discuss in this book.

Some algorithms (e.g., sorting and hashing) can be modeled by the "balls-and-urns" paradigm in which n elements are placed randomly into m bins. Questions arise such as how many urns are empty, how many balls are required to fill up all urns, and so forth. It is easy to see that the occupancies of urns are not independent (e.g., if all balls fall into one urn, then all the remaining urns are empty). To overcome this difficulty an interesting probabilistic technique called **poissonization** was suggested (see also Section 5.2). In this new Poisson model it is assumed that balls are "generated" according to a Poisson process N with mean n. Due to some unique properties of the Poisson process, discussed below, the streams of balls are now placed independently in every urn, thus overcoming the above-mentioned difficulty. Observe, however, that poissonization has its own problem since one must "extract" the original results from the Poisson model, that is, **depoissonize** the Poisson model. Throughout this chapter we shall refer to the original model as the *Bernoulli model*.

More generally, let g_n be a characteristic of the Bernoulli model of size n (e.g., g_n is the number of empty urns in the balls-and-urns model with n urns). The **Poisson transform** is defined as

$$\widetilde{G}(z) = \mathbf{E}[g_N] = \sum_{n \geq 0} g_n \frac{z^n}{n!} e^{-z}.$$

Observe that the input n becomes a Poisson variable N with mean z when $z \geq 0$. We later extend z to the whole complex plane. The Poisson transform was introduced by Gonnet and Munro [176] and extended in [343]. When z is complex, we also refer to it as **analytic poissonization**.

To further motivate our discussion, let us consider the following scenario. We have already seen in this book that many problems arising in the analysis of algorithms are mapped into certain recurrence equations or functional/differential equations (e.g., Sections 7.6.1, 8.7.3, and 9.4.1). Algorithms involved in splitting processes and digital trees are representative examples of this situation. Embedding this splitting process into a Poisson process often leads to more tractable functional/differential equations. Take the following general recurrence describing a splitting process, already discussed in Section 7.6.1,

$$x_n = a_n + \beta \sum_{k=0}^{n} \binom{n}{k} p^k q^{n-k} (x_k + x_{n-k}), \quad n \geq 2$$

with some initial values for x_0 and x_1, where a_n is a given sequence, and β is a constant. Let $X(z)$ and $A(z)$ be the exponential generating functions of x_n and a_n, respectively. By (7.17) of Table 7.3 we find

$$X(z) = \beta X(zp) e^{zq} + \beta X(zq) e^{zp} + A(z) - x_0 - x_1 z.$$

To further simplify it, we consider the Poisson transform $\widetilde{X}(z) = X(z)e^{-z}$, and then

$$\widetilde{X}(z) = \beta \widetilde{X}(zp) + \beta \widetilde{X}(zq) + \widetilde{A}(z) - x_0 e^{-z} - x_1 z e^{-z}. \qquad (10.1)$$

This is a linear additive functional equation that can be solved by consecutive iterations as shown in (7.67) of Section 7.6.1. But after solving it and eventually finding an asymptotic expansion of $\widetilde{X}(z)$, we still need a method to extract the coefficients x_n.

In general, if the Poisson transform $\widetilde{G}(z)$ can be computed exactly, then one can extract the coefficient $g_n = n! [z^n](\widetilde{G}(z)e^z)$. This is called *algebraic* or *exact depoissonization*. However, in most interesting situations (cf. Example 9.3) the Poisson transform $\widetilde{G}(z)$ satisfies a functional/differential equation like (10.1) that usually cannot be solved exactly. Nevertheless, one can find an asymptotic expansion of $\widetilde{G}(z)$ for $z \to \infty$ (e.g., using the Mellin transform discussed in the last chapter) on the real axis and can bound $\widetilde{G}(z)$ in the complex plane. Then one aims at finding an asymptotic expansion of g_n from the asymptotics of $\widetilde{G}(z)$. This we shall call **analytic depoissonization**. In Chapter 8 we discussed several methods to extract the coefficients g_n asymptotically as $n \to \infty$, from an asymptotic expansion of its generating function. In particular, the singularity analysis is designed to handle this situation when the analytic function is of moderate growth. Analytic depoissonization can be viewed as another asymptotic method, usually applied to functions of rapid growth.

Analytic depoissonization is well understood when the only singularities of the Poisson transform $\widetilde{G}(z)$ are poles or algebraic singularities. In this case, one can apply either Cauchy's residue theorem or the singularities analysis (see Section 8.3.2). The reader is referred to Gonnet and Munro [176], Poblete [341], and Poblete, Viola, and Munro [343] for such analyses. A new situation arises when the Poisson transform $\widetilde{G}(z)$ is an entire function and consequently does not have any singularity in the complex plane. In this case, one can rely only on partial information regarding $\widetilde{G}(z)$ and apply the saddle point method (see Section 8.4) to extract the asymptotics. This is exactly what we plan to do in this chapter.

Finally, one may ask *why poissonization*? Why not embed the Bernoulli model into another process? This seems to be a consequence of certain unique properties of the Poisson process that we briefly discuss below. In short, it is the only process that has stationary and independent increments, and which has no group arrivals. To be a little more precise, let us first define *superposition* and *thinning* or *splitting* processes of a renewal (point) process (details on renewal processes can be found in Durrett [117] or Ross [369]). Consider two (stationary) renewal processes, say N_1 and N_2. Then the superposition process $N = N_1 + N_2$ consists of all renewal points of both processes. To define the splitting or thinning process, take a single renewal process and for each point decide independently whether to omit it (thinning) or to direct it to one of two (or more) outputs (splitting). The following three properties are well known (cf. [369]), but the reader is asked to prove them in Exercise 1:

(P1) A superposition of renewal process is renewed if the underlying processes are Poisson.

(P2) A thinning or splitting process is Poisson with parameter zp, where p is the probability of thinning if the original process is Poisson with parameter z.

(P3) Let $N(t)$ denote Poisson arriving points in the interval $(0, t)$. Then

$$\Pr\{N(x) = k \mid N(t) = n\} = \binom{n}{k}(x/t)^k (1 - x/t)^{n-k},$$

where $x \leq t$. In other words, the points of Poisson process are uniformly and independently distributed in $(0, t)$ *conditioned* on n arrivals in this interval.

The last property is sometimes called "random occurrence of the conditional Poisson process."

This chapter discusses results obtained only recently that are still under vigorous investigations. Our goal is to present a readable account of poissonization and depoissonization. However, we also aspire to present rigorous derivations of depoissonization results so that this chapter can be used as a reliable reference on depoissonization. All depoissonization results presented in this chapter fall into the following general scheme: *If the Poisson transform has an appropriate growth in the complex plane, then an asymptotic expansion of the sequence can be expressed in terms of the Poisson transform and its derivatives evaluated on the real line.* Not surprisingly, actual formulations of depoissonization results depend on the nature of growth of the Poisson transform, and thus we have *polynomial* and *exponential* depoissonization theorems. Normalization (e.g., as in the central limit theorem) introduces another twist that led us to formulate the *diagonal* depoissonization theorems. We also have a short account on the *Dirichlet depoissonization* recently introduced by Clement, Flajolet, and Vallée (cf. [68]). In the application section we discuss the limiting distribution of the time to find a leader using the leader election algorithm, and then we analyze the depth of b-digital search trees for memoryless and Markovian sources. The latter application is one of the most complex in this book.

To the best of our knowledge, poissonization was introduced by Marek Kac [226], who half a century ago investigated the deviations between theoretical and empirical distributions. Aldous [4] gave a heuristic probabilistic principle for depoissonizing small probabilities rather than large expected values that is main focus of this chapter. Exact analytic depoissonization results for meromorphic functions were discussed in Gonnet and Munro [175, 176], Poblete [341], and Poblete, Viola and Munro [343]. Asymptotic analytic depoissonization for mero-morphic function can be found in [176, 343] while depoissonization for entire functions was initiated by Jacquet and Régnier, who analyzed the limiting dis-tributions of the depth and the size in tries [212, 213, 356]. Jacquet and Régnier introduced the basic idea of the analytical depoissonization for an implicit so-lution of a nonlinear functional equation. Jacquet and Szpankowski [217], and Rais et al. [350] extended these results and obtained the first general and simple version of the depoissonization result. The presentation in this chapter is based on two recent papers written together with my colleague P. Jacquet [218, 219]. We report here only basic depoissonization results and some of its generaliza-tions. We refer the reader to Jacquet and Szpankowski [218] for more advanced depoissonization results. Recently, poissonization was further popularized in the context of analysis of algorithms and combinatorial structures by Aldous [4], Ar-ratia and Tavaré [20], Clement, Flajolet, and Vallée [68], Gonnet [175], Gonnet and Munro [176], Holst [199], Jacquet and Régnier [212, 213], Jacquet and Sz-pankowski [214, 217, 218, 219], Janson and Szpankowski [223], Rais et al. [350], Fill et al. [124], Kirschenhofer et al. [254], Poblete [341], and Poblete et al. [343].

Finally, we must observe that any depoissonization result is in fact a Tauberian theorem for the Borel mean (cf. [188]), which is nothing else but the Poisson transform defined above. For an accessible account on modern development of Tauberian theorems the reader is referred to Bingham [50] and Postnikov [346].

10.1   ANALYTIC POISSONIZATION AND
THE POISSON TRANSFORM

We briefly discuss the poissonization technique. First, we review probabilistic poissonization, then discuss exact analytic poissonization, and finish this section with an overview of asymptotic poissonization.

10.1.1   Poisson Transform

Consider a combinatorial structure in which n objects are randomly distributed into some locations (e.g., one can think of n balls thrown into urns). The objects are not necessarily uniformly distributed among the locations (cf. digital trees). We call such a setting the Bernoulli model. Let X_n be a characteristic of the model (e.g., the number of throws needed to fill up all urns). Next, we define the Poisson model. Let N be a random variable distributed according to the Poisson distribu-tion with parameter $z \geq 0$, that is, $\Pr\{N = k\} = e^{-z}z^k/k!$, and let X_N be the above characteristic defined in the Poisson model in which the *deterministic* input (i.e., n) is replaced by the Poisson variable N. Then by definition

$$\widetilde{X}(z) := \mathbf{E}[X_N] = \sum_{n \geq 0} \mathbf{E}\left[X_N \mid N = n\right] e^{-z} \frac{z^n}{n!}$$

$$= \sum_{n \geq 0} \mathbf{E}[X_n] e^{-z} \frac{z^n}{n!}. \tag{10.2}$$

We now analytically continue $\widetilde{X}(z)$ to the whole complex plane, and throughout we make the following assumption:

(A) The sum in (10.2) converges absolutely for every z, that is, $\widetilde{X}(z)$ is an entire function of the complex variable z.

This defines the Poisson transform $\widetilde{X}(z)$ for all complex z.

In general, we use g_n as a generic notation for a sequence characterizing the Bernoulli model (e.g., $g_n = \mathbf{E}[X_n]$ or, in general, $g_n = \mathbf{E}[f(X_n)]$ for some function f). By $\widetilde{G}(z)$ or $\mathcal{P}(g_n; z)$ we denote the **Poisson transform** defined formally as

$$\widetilde{G}(z) = \mathcal{P}(g_n; z) := \sum_{n \geq 0} g_n \frac{z^n}{n!} e^{-z}.$$

Table 10.1 presents some common Poisson transforms. Items 1-4 are obvious. Item 5 (the additive (p, q)-splitting, where $q = 1 - p$) and Item 6 (i.e., multiplicative (p, q)-splitting) follow directly from (7.17) of Table 7.3, while Item 7 requires only trivial algebra (we return to it in Section 10.3.1). A more complete list of Poisson transforms can be found in [343].

TABLE 10.1. Some Poisson Transforms

Item	g_n	$\widetilde{G}(z)$
1	C	C
2	$n(n-1)\cdots(n-k+1)$	z^k
3	α^n	$e^{(\alpha-1)z}$
4	$n!$	$\frac{e^{-z}}{1-z}$
5	$\sum_{k=0}^{n} \binom{n}{k} p^k (1-p)^{n-k} (f_k + g_{n-k})$	$F(pz) + G((1-p)z)$
6	$\sum_{k=0}^{n} \binom{n}{k} p^k (1-p)^{n-k} f_k g_{n-k}$	$F(zp)G((1-p)z)$
7	$\sum_{k \geq 0} a_k (1 - (1-a_k)^n)$	$\sum_{k \geq 0} a_k (1 - e^{-za_k})$

An astute reader should notice that for Items 1 and 2 of Table 10.1 $g_n \sim \widetilde{G}(n)$ as $n \to \infty$ while this is not true for Items 3 and 4. In the latter case, the Poisson transform is either of rapid growth (Item 3) or not an entire function (so assumption (A) is not satisfied). For Items 1 and 2 the transforms are of a polynomial growth. This is not a coincidence, and in this chapter we systematically explore the relationship between g_n as $n \to \infty$ and $\widetilde{G}(z)$ as $z \to \infty$.

Example 10.1: *Depth and Size of a Trie*   Let us find the Poisson transform of the depth D_n for a trie built over n strings generated by biased memoryless source. We define

$$D_n(u) = \mathbf{E}[u^{D_n}] = \sum_{k=0}^{\infty} \Pr\{D_n = k\}u^k,$$

$$D(z, u) = \sum_{n=0}^{\infty} \mathbf{E}[u^{D_n}]\frac{z^n}{n!},$$

$$\widetilde{D}(z, u) = e^{-z}D(z, u).$$

We now derive a functional equation for $\widetilde{D}(z, u)$ directly from the properties of tries and the Poisson process (without going through a recurrence equation on D_n). We claim that

$$\widetilde{D}(z, u) = u(p\widetilde{D}(zp, u) + q\widetilde{D}(zq, u)) + (1 - u)(1 + z)e^{-z}.$$

To see this, we first observe that digital trees are recursive structures, so the Poisson transform of the depth of the left subtree is $D(zp, u)$ while for the right subtree it is $D(zq, u)$ (by property (P2) of the Poisson process). Since the subtrees are one level lower than the root, we have the factor u. But when computing the depth we either go left (with probability p) or right (with probability q), but never both ways. Finally, we know that $n = 0$ and $n = 1$ are special cases since for such n there is no left/right subtree recurrence. The probability of having $n = 0$ or $n = 1$ in the Poisson model is equal to $(1 + z)e^{-z}$. We must subtract from it $u(1 + z)e^{-z}$, which represents the subtree structures not accounted for in the recurrence. This establishes the above functional equation.

In Exercise 2 the reader is asked to prove that the Poisson transform $\widetilde{S}(z, u)$ of the trie size satisfies

$$\widetilde{S}(z, u) = u\widetilde{S}(zp, u)\widetilde{S}(zq, u) + (1 - u)(1 + z)e^{-z}$$

for $|u| \leq 1$.   ∎

10.1.2 Asymptotic Poissonization

In the previous section we computed the Poisson transform exactly. But there are sequences (e.g., $g_n = \log n$ and $g_n = n^\alpha$ for noninteger α) for which $\widetilde{G}(z)$ cannot be found in a closed form. The natural approach then is to compute $\widetilde{G}(z)$ asymptotically. We search for the asymptotics of $\widetilde{G}(z)$ as $z \to \infty$ in a linear cone

$$S_\theta = \{z : |\arg(z)| \le \theta, \ 0 < \theta < \pi/2\},$$

around the positive real axis. One expects to find asymptotic expansion $\widetilde{G}(z)$ in the following form

$$\widetilde{G}(z) = \sum_{i,j \ge 0} a_{ij} z^i g^{(j)}(z), \quad z \to \infty,$$

where $g^{(j)}(z)$ is the jth derivative of $g(z)$ which is the analytic continuation of g_n, if it exists; that is, $g_n = g(z)|_{z=n}$.

We first identify the coefficients a_{ij} formally, and later derive an error bound. We proceed as follows:

$$\widetilde{G}(z) = \sum_{n=0}^{\infty} g(n) \frac{z^n}{n!} e^{-z} = \sum_{n=0}^{\infty} \frac{z^n}{n!} e^{-z} \sum_{j=0}^{\infty} g^{(j)}(z) \frac{(n-z)^j}{j!}$$

$$= \sum_{n=0}^{\infty} \sum_{j=0}^{\infty} a_{nj} z^n g^{(j)}(z). \qquad (10.3)$$

We would like to formally set

$$a_{nj} = [z^n][y^j] \frac{1}{n!} \frac{(n-z)^j}{j!} e^{-z},$$

where y^j represents formally $g^{(j)}(z)$. This seemingly strange substitution becomes more familiar if we consider the following series

$$\sum_{n,j \ge 0} a_{n,j} z^n y^j = \sum_{n=0}^{\infty} e^{-z} \frac{z^n}{n!} \sum_{j=0}^{\infty} \frac{(n-z)^j}{j!} y^j = \sum_{n=0}^{\infty} e^{-z} \frac{z^n}{n!} e^{y(n-z)}$$

$$= \exp\left(z(e^y - 1) - zy\right).$$

Thus formally we define the coefficients a_{ij} as

$$\sum_{i=0}^{\infty} \sum_{j=0}^{\infty} a_{ij} x^i y^j = \exp(x(e^y - 1) - xy), \qquad (10.4)$$

that is, $a_{ij} = [x^i][y^j]\exp(x(e^y - 1) - xy)$. Actually, $a_{ij} = 0$ for $j < 2i$. To see this, let $f(x, y) = \exp(x(e^y - 1) - xy)$. Observe that $f(xy^{-2}, y)$ is analytic at $x = y = 0$, hence the Laurent series of $f(x, y)$ possesses only terms like $x^i y^{j-2i}$ with *nonnegative* powers. These nonzero coefficients a_{ij} have $j \geq 2i$.

In view of this, we can write formally $\widetilde{G}(z)$ in short as

$$\widetilde{G}(z) = \exp\left(z(e^y - 1) - zy\right), \quad \text{where} \quad y^j = g^{(j)}(z).$$

This is a convenient way of expressing analytic poissonization.

In some applications we really want to have an expansion around $w = \lambda z$, where λ is constant, so we extend a_{ij} to $a_{ij}(\lambda)$ defined as

$$\sum_{i=0}^{\infty} \sum_{j=0}^{\infty} a_{ij}(\lambda) x^i y^j = \exp(\lambda x(e^y - 1) - \lambda xy).$$

Observe that $a_{ij}(\lambda) = \lambda^i a_{ij}$. The same derivation as above leads to the following symbolic representation for $\widetilde{G}(z)$

$$\widetilde{G}(w) = \exp\left(w(e^y - 1) - wy\right) \quad \text{where} \quad y^j = g^{(j)}(w) \tag{10.5}$$

and $w = \lambda z$ with λ being a constant. In addition, we directly obtain derivatives of $\widetilde{G}(w)$ as

$$\widetilde{G}^{(k)}(w) = (e^y - 1)^k \exp\left(w(e^y - 1) - wy\right) \quad \text{where} \quad y^j = g^{(j)}(w). \tag{10.6}$$

So far we have used a formal series approach to find $\widetilde{G}(z)$. But we are really interested in an asymptotic expansion of $\widetilde{G}(z)$ as $z \to \infty$, thus we must derive an error bound. This is not difficult. However, one must first bound derivatives of an analytic function, and we discuss it next. This is of prime importance to the depoissonization results studied in the next section.

Lemma 10.1   *Let $\theta_0 < \pi/2$ and $D > 0$ be such that for all $z \in S_{\theta_0}$ the function $G(z)$ is analytic in this cone and*

$$|z| > D \quad \Longrightarrow \quad |G(z)| \leq B|z|^{\beta} \tag{10.7}$$

for some reals β and $B > 0$. Then for all $\theta < \theta_0$ there exist $B' > 0$ and $D' > D$ such that for all positive integers k and $z \in S_\theta$ the following holds:

$$|z| > D' \quad \Longrightarrow \quad |G^{(k)}(z)| \leq k!(B')^k |z|^{\beta-k}. \tag{10.8}$$

Proof. The proof follows directly from the Cauchy integral formula applied to derivatives. Indeed, from Cauchy's formula we know that

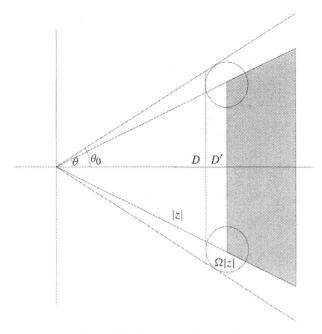

Figure 10.1. Illustration to Lemma 10.1.

$$G^{(k)}(z_0) = \frac{k!}{2\pi i} \oint \frac{G(z)dz}{(z - z_0)^{k+1}} \qquad (10.9)$$

for $z \in B(z_0, r) = \{z : |z - z_0| \leq r\} \subset S_\theta$, assuming $G(z)$ is analytic in $B(z_0, r)$ for some $r > 0$. If $|G(z)| \leq M$ for some $M > 0$, then clearly the above implies

$$|G^{(k)}(z)| \leq M k! r^{-k} \qquad (10.10)$$

for $z \in B(z_0, r)$, which is a generalization of the Cauchy bound. To prove our lemma, we place a circle of radius $r = \Omega|z|$ for some $\Omega > 0$ at the closest corner of the cone S_{θ_0} to the right of $\Re(z) = D$ as shown in Figure 10.1. Setting $M = B|z|^\beta$ in (10.10) we prove Lemma 10.1 with $B' = B\Omega^{-k}$. ∎

Now we can formulate our main asymptotic poissonization result.

Theorem 10.2 (Jacquet and Szpankowski, 1999) *Let $g(z)$ be an analytic continuation of $g(n) = g_n$ such that $g(z) = O(z^\beta)$ in a linear cone for some constant β. Then for every nonnegative integer m and complex $w = \lambda z$ for some* *constant λ the Poisson transform $\tilde{G}(z)$ as $z \to \infty$ has the following representation*

$$\tilde{G}(z) = \sum_{i=0}^{m} \sum_{j=0}^{i+m} a_{ij}(\lambda) z^i g^{(j)}(w) + O(z^{\beta-m-1}) \tag{10.11}$$

$$= g(w) + \sum_{k=1}^{m} \sum_{i=1}^{k} a_{i,k+i}(\lambda) z^i g^{(k+i)}(w) + O(z^{\beta-m-1}),$$

where

$$a_{ij}(\lambda) = [x^i][y^j] \exp(\lambda x(e^y - 1) - \lambda xy).$$

In particular, for $\lambda = 1$

$$\sum_{i=0}^{\infty} \sum_{j=0}^{\infty} a_{ij} x^i y^j = \exp(x(e^y - 1) - xy) \tag{10.12}$$

and $a_{ij} = 0$ for $j < 2i$.

Proof. In view of our previous discussion, it suffices to derive the error term. Using Lemma 10.1 we observe that $g^{(j)}(z) = O(z^{\beta-j})$ and using $a_{ij} = 0$ for $j < 2i$ we arrive at the following error term

$$\sum_{j=0}^{2m+1} a_{m+1,j}(\lambda) z^m g^{(j)}(w) = O(z^{\beta-m-1})$$

as $z \to \infty$, since $w = \lambda z$. This completes the proof. ∎

Here are the first few terms of the expansion of $\tilde{G}(z)$:

$$\tilde{G}(z) = g(z) + \frac{1}{2} z \, g^{(2)}(z) + \frac{1}{6} z \, g^{(3)}(z) + \left(\frac{1}{24} z + \frac{1}{8} z^2 \right) g^{(4)}(z)$$

$$+ \left(\frac{1}{120} z + \frac{1}{12} z^2 \right) g^{(5)}(z) + \left(\frac{1}{720} z + \frac{5}{144} z^2 + \frac{1}{48} z^3 \right) g^{(6)}(z)$$

$$+ \left(\frac{1}{5040} z + \frac{1}{90} z^2 + \frac{1}{48} z^3 \right) g^{(7)}(z) + \cdots .$$

If one wants the error term of order, say $O(z^{\beta-4})$ (we assume $g(z) = O(z^\beta)$), then we need only one line of the above expansion plus selected terms from the second line, that is,

$$\widetilde{G}(z) = g(z) + \frac{1}{2} z \, g^{(2)}(z) + \frac{1}{6} z \, g^{(3)}(z) + \left(\frac{1}{24} z + \frac{1}{8} z^2 \right) g^{(4)}(z)$$

$$+ \frac{1}{12} z^2 g^{(5)}(z) + \frac{1}{48} z^3 g^{(6)}(z) + O(z^{\beta-4}).$$

Example 10.2: *Poisson Transform of* $f_n = \log(n!)$ Let us consider $f_n = \log(n!)$. This sequence arises in the analysis of the entropy of the binomial distribution (cf. Example 10.7 below). Noting that $f(z) = \log \Gamma(z+1) = O(z \log z) = O(z^{1+\varepsilon})$, from Theorem 10.2 we obtain for any $\varepsilon > 0$

$$\widetilde{F}(z) = \sum_{i=0}^{m} \sum_{j=0}^{i+m} a_{ij}(\lambda) z^i f^{(j)}(w) + O(z^{\beta-m-1})$$

$$= f(z) + \frac{1}{2} z f^{(2)}(z) + \frac{1}{6} z f^{(3)}(z) + \frac{1}{8} f^{(4)}(z) + O(z^{\varepsilon-2}), \qquad m = 2.$$

To find the derivatives of $f(z) = \log \Gamma(z) = \log z + \log \Gamma(z)$ we use the following expansions as $z \to \infty$:

$$f(z) \sim \left(z + \frac{1}{2} \right) \ln z - z \frac{1}{2} + \sum_{m=1}^{\infty} \frac{B_{2m}}{2m(2m-1)z^{2m-1}} \qquad f^{(k)}(z) \sim \Psi^{(k-1)}(z+1)$$

$$= \frac{(-1)^k (k-2)!}{z^{k-1}} + \frac{(-1)^{k-1}(k-1)!(2k-1)}{2z^k} \qquad \text{for } k \geq 2$$

$$+ (-1)^k \sum_{l=1}^{\infty} \frac{B_{2l}(2l+k-2)!}{(2l)! \, z^{2l+k-1}},$$

where $\Psi(z) = \frac{d}{dz} \log \Gamma(z)$ is the psi function and B_n are Bernoulli numbers (see Table 8.1). ■

10.2 BASIC DEPOISSONIZATION

We now consider a sequence g_n and its Poisson transform $\widetilde{G}(z) = \mathcal{P}(g_n, z)$. Throughout, we assume that $\widetilde{G}(z)$ is an entire function. We first discuss exact or algebraic depoissonization in which the coefficients g_n are extracted exactly from its Poisson transform. Next, we focus on asymptotic depoissonization, which is the main goal of this chapter. We again use Cauchy's formula to find

$$g_n = \frac{n!}{2\pi i} \oint \frac{\tilde{G}(z)e^z}{z^{n+1}} dz = \frac{n!}{n^n 2\pi} \int_{-\pi}^{\pi} \tilde{G}(ne^{it}) \exp\left(ne^{it}\right) e^{-nit} dt. \quad (10.13)$$

All the asymptotic depoissonization results will follow from the above by a careful estimation of the integral using the saddle point method (see Section 8.4). Here we present, and give a simple proof of a basic asymptotic depoissonization result. Later in Section 10.3 we generalize it.

10.2.1   Algebraic Depoissonization

The exact or algebraic depoissonization is based on the exact extraction of g_n from its Poisson transform, that is,

$$g_n = n![z^n]\left(e^z \tilde{G}(z)\right).$$

In some cases this is quite simple. For example, applying directly Cauchy's formula to $\tilde{G}(z) = e^{-z}/(1-z)$ (see Item 4 in Table 10.1) we find $n![z^n]\left(e^z \tilde{G}(z)\right) = n!$, as required. Example 10.3 below illustrates a more complicated case. The reader is referred to Gonnet and Munro [176], Poblete, Viola, and Munro [343], and Hofri [197] for more interesting examples.

Example 10.3: *Ball-and-Urn Model*   Consider n balls (items) thrown randomly and uniformly into m urns (table of size m). What is the probability that precisely k specified urns are empty? Under the Poisson model the answer is immediate. Indeed, consider the Poisson process of balls with mean z. It is split into *m independent* Poisson processes of mean z/m entering each of m urns. All these processes are independent so that the probability $P_k(z)$ that exactly k specified urns are empty in the Poisson model is obviously equal to

$$P_k(z) = e^{-kz/m}(1 - e^{-z/m})^{m-k} = e^{-z}(e^{z/m} - 1)^{m-k}.$$

To find the answer to the original problem, we depoissonize $P_k(z)$ exactly, yielding

$$\Pr\{\text{number of empty urns} = k\} = n![z^n]\left(e^z P_k(z)\right) = [z^n]\left(n!(e^{z/m} - 1)^{m-k}\right)$$

$$= \frac{(m-k)!}{m^n}\left\{\begin{matrix} n \\ m-k \end{matrix}\right\},$$

where $\left\{\begin{smallmatrix} n \\ k \end{smallmatrix}\right\}$ denote the Stirling numbers of the second kind. The last line follows from Entry 9 of Table 7.2. The above formula can be directly obtained from the Bernoulli model. However, the derivation is much more troublesome.   ∎

The situation encountered in Example 10.3 above is rare. We usually cannot invert exactly the Poisson transform and we must resort to asymptotics. The rest of this chapter is devoted to asymptotic depoissonization. We start with a basic depoissonization (Section 10.2.2), then discuss in Section 10.3 some generalizations, and in Section 10.4 we deal with limiting distributions through depoissonization.

10.2.2   Asymptotic Depoissonization

Before we enter the realm of rigorous analysis, the reader should develop an intuition of how depoissonization works. We first propose a heuristic derivation. We refer to the definition of the Poisson transform, which we write as $\widetilde{G}(z) = \mathbf{E}[g_N]$, where N is a Poisson distributed random variable with mean $\mathbf{E}[N] = n$. Treating g_N as a function of N (and denoting it as $g(N)$), by Taylor's expansion we have

$$g(N) = g(n) + (N - n)g'(n) + \frac{1}{2}g''(n)(N - n)^2 + \cdots.$$

Taking the expectation we obtain

$$\widetilde{G}(n) = \widetilde{G}(z)|_{z=n} = \mathbf{E}[g(N)] = g(n) + \frac{1}{2}g''(n)n + \cdots$$

since $\mathbf{E}[N - n] = 0$ and $\mathbf{E}[N - n]^2 = n$. Solving the above for $g(n) = g_n$ we find that

$$g_n \approx \widetilde{G}(n) - \frac{1}{2}ng''(n) + \cdots = \widetilde{G}(n) + O(ng''(n)). \tag{10.14}$$

Provided that

$$ng''(n) = o(g(n)), \tag{10.15}$$

we expect to have $g_n \sim \widetilde{G}(n)$. To emphasize this, consider the following examples:

- Let $g(n) = n^\beta$. Then $\widetilde{G}(n) = n^\beta + O(n^{\beta-1})$, $g''(n) = O(n^{\beta-2})$, thus

$$g_n = \widetilde{G}(n) + O(ng''(n)) = \widetilde{G}(n) + O(n^{\beta-1}),$$

  which is true, as already seen in Table 10.1. This is proved in Theorem 10.3.
- Consider now $g(n) = \alpha^n$. This time $\widetilde{G}(z) = e^{z(\alpha-1)}$, $g''(n) = \alpha^n \log^2 \alpha$, and it is *not* true that $g_n \sim \widetilde{G}(n)$, as observed in Section 10.1.

- Now we assume $g_n = e^{n^\beta}$. In this case, it is harder to find the Poisson transform, but an extension of Theorem 10.2 suggests that $\widetilde{G}(z) \sim e^{z^\beta}$. We also have $g''(n) = O(n^{2\beta-2}e^{n^\beta})$. Observe that

$$g_n = \widetilde{G}(n) + O(ng''(n)) = \widetilde{G}(n) + O(n^{2\beta-1}e^{n^\beta})$$

and the error is small as long as $0 < \beta < \frac{1}{2}$. This is proved rigorously in Theorem 10.8.

The above derivation is hard to make rigorous since it is difficult to control the growth of the derivatives of a function while knowing only its behavior on the real axis. To circumvent this problem, we extend $\widetilde{G}(z)$ to the complex plane, where the growth of $g(z)$ and $\widetilde{G}(z)$ are easier to bound (Lemma 10.1).

Enriched by the above intuition, we now rigorously prove some depoissonization results. We start with a basic depoissonization that holds for sequences having a Poisson transform of a polynomial growth. As a matter of fact, we have to put certain restrictions on the growth of $\widetilde{G}(z)$ inside a cone S_θ (called conditions (I)) and another bound on the growth of $\widetilde{G}(z)$ outside the cone (called conditions (O)). The proof presented below will be modified throughout this chapter to obtain further generalizations but its main ingredients remain the same. The reader is advised to invest some time in studying the proof of Theorem 10.3 below.

Theorem 10.3   **(Jacquet and Régnier; Jacquet and Szpankowski)** *Let $\widetilde{G}(z)$ be the Poisson transform of a sequence g_n that is assumed to be an entire function of z. We postulate that in a linear cone S_θ ($\theta < \pi/2$) the following two conditions simultaneously hold:*

(I) *For $z \in S_\theta$ and some reals $B, R > 0, \beta$*

$$|z| > R \quad \Rightarrow \quad |\widetilde{G}(z)| \le B|z|^\beta, \tag{10.16}$$

(O) *For $z \notin S_\theta$ and $A, \alpha < 1$*

$$|z| > R \quad \Rightarrow \quad |\widetilde{G}(z)e^z| \le A \exp(\alpha|z|). \tag{10.17}$$

Then

$$g_n = \widetilde{G}(n) + O(n^{\beta-1}) \tag{10.18}$$

for large n.

Proof. The proof relies on the evaluation of Cauchy's formula (10.13) by the saddle point method. By Stirling's approximation $n! = n^n e^{-n} \sqrt{2\pi n}(1 + O(1/n))$; thus (10.13) becomes

$$g_n = \left(1 + O(n^{-1})\right) \sqrt{\frac{n}{2\pi}} \int_{-\pi}^{\pi} \tilde{G}(ne^{it}) \exp\left(n\left(e^{it} - 1 - it\right)\right) dt$$

$$= \left(1 + O(n^{-1})\right) (I_n + E_n),$$

where

$$I_n = \sqrt{\frac{n}{2\pi}} \int_{-\theta}^{\theta} \tilde{G}(ne^{it}) \exp\left(n\left(e^{it} - 1 - it\right)\right) dt,$$

$$E_n = \sqrt{\frac{n}{2\pi}} \int_{|t| \in [\theta, \pi]} \tilde{G}(ne^{it}) \exp\left(n\left(e^{it} - 1 - it\right)\right) dt$$

$$= \frac{n^n e^{-n} \sqrt{2\pi n}}{2\pi i} \int_{|t| \in [\theta, \pi]} \frac{\tilde{G}(z)e^z}{z^{n+1}} dz.$$

We estimate the above two integrals. We begin with the latter. Observe that by condition (O) (cf. (10.17)), we obtain

$$|E_n| \le A' \sqrt{2\pi n} e^{-(1-\alpha)n},$$

where A' depends only on A and R. Thus E_n decays exponentially to zero as $n \to \infty$, for $\alpha < 1$.

Now we turn our attention to the integral I_n, which is more intricate to handle. First, we replace t by $t/\sqrt{n}$ to find that

$$I_n = \frac{1}{\sqrt{2\pi}} \int_{-\theta\sqrt{n}}^{\theta\sqrt{n}} \tilde{G}(ne^{it/\sqrt{n}}) \exp\left(n\left(e^{it/\sqrt{n}} - 1 - it/\sqrt{n}\right)\right) dt. \qquad (10.19)$$

Let

$$h_n(t) = \exp\left(n\left(e^{it/\sqrt{n}} - 1 - it/\sqrt{n}\right)\right).$$

We need to estimate $h_n(t)$ in the interval $t \in [-\theta\sqrt{n}, \theta\sqrt{n}]$, and find the Taylor expansion of it in a smaller interval, say for $t \in [-\log n, \log n]$ (in order to apply Taylor's expansion of $h_n(t)$ around $t = 0$). The latter restriction is necessary since $\frac{t}{\sqrt{n}} = O(1)$ for $t \in [-\theta\sqrt{n}, \theta\sqrt{n}]$. Thus we split the integral I_n into two parts, I_n' and I_n'', such that

$$I'_n = \frac{1}{\sqrt{2\pi}} \int_{-\log n}^{\log n} \widetilde{G}(ne^{it/\sqrt{n}}) \exp\left(n\left(e^{it/\sqrt{n}} - 1 - it/\sqrt{n}\right)\right) dt,$$

$$I''_n = \frac{1}{\sqrt{2\pi}} \int_{t \in [-\theta\sqrt{n}, -\log n]} \widetilde{G}(ne^{it/\sqrt{n}}) \exp\left(n\left(e^{it/\sqrt{n}} - 1 - it/\sqrt{n}\right)\right) dt +$$

$$+ \frac{1}{\sqrt{2\pi}} \int_{t \in [\log n, \theta\sqrt{n}]} \widetilde{G}(ne^{it/\sqrt{n}}) \exp\left(n\left(e^{it/\sqrt{n}} - 1 - it/\sqrt{n}\right)\right) dt.$$

To estimate the second integral I''_n we observe that $|h_n(t)| \le e^{-\mu t^2}$ for $t \in [-\theta\sqrt{n}, \theta\sqrt{n}]$, where μ is a constant. By Lemma 8.15 (cf. (8.68)) and condition (I) we immediately see that $I''_n = O(n^\beta e^{-\mu \log^2 n})$, which decays faster than any polynomial.

Now we estimate I'_n. Observe first that for $t \in [-\log n, \log n]$ we can apply the Taylor expansion to obtain

$$h_n(t) = \exp\left(n\left(e^{it/\sqrt{n}} - 1 - it/\sqrt{n}\right)\right)$$

$$= e^{-t^2/2}\left(1 - \frac{it^3}{6\sqrt{n}} + \frac{t^4}{24n} - \frac{t^6}{72n} + O\left(\frac{\log^9 n}{n\sqrt{n}}\right)\right)$$

Furthermore, using (I) and Lemma 10.1 (with $D' = (1 + \Omega)R$ and $z \in \mathcal{S}_{\theta'}$ for $\theta' < \theta$) we have $|\widetilde{G}'(z)| \le B_1|z|^{\beta-1}$ and $|\widetilde{G}''(z)| \le B_2|z|^{\beta-2}$, for some constants B_1 and B_2. Thus we can expand $\widetilde{G}(ne^{it/\sqrt{n}})$ around $t = 0$ as follows

$$\widetilde{G}(ne^{it/\sqrt{n}}) = \widetilde{G}(n) + it\sqrt{n}\widetilde{G}'(n) + t^2\Delta_n(t),$$

where

$$|\Delta_n(t)| \le (B_1 + B_2)n^{\beta-1}$$

since for t in the vicinity of zero

$$\Delta_n(t) = -\widetilde{G}'(ne^{it/\sqrt{n}})e^{it/\sqrt{n}} - \widetilde{G}''(e^{it/\sqrt{n}})ne^{2it/\sqrt{n}}.$$

In summary, the integral I'_n can be written as

$$I'_n = \frac{1}{\sqrt{2\pi}} \int_{-\log n}^{\log n} e^{-t^2/2}\left(\widetilde{G}(n) + \widetilde{G}'(n)it\sqrt{n}\right)\left(1 - \frac{it^3}{6\sqrt{n}} + \frac{t^4}{24n} - \frac{t^6}{72n}\right) dt$$

$$(10.20)$$

$$+ \frac{1}{\sqrt{2\pi}} \int_{-\log n}^{\log n} e^{-t^2/2} \Delta_n(t) t^2 h_n(t) dt \tag{10.21}$$

$$+ \frac{1}{\sqrt{2\pi}} \int_{-\log n}^{\log n} e^{-t^2/2} \left(\widetilde{G}(n) + \widetilde{G}'(n) it \sqrt{n} \right) O\left(\frac{\log^9 n}{n\sqrt{n}} \right) dt. \tag{10.22}$$

To complete the proof we must estimate the above three integrals. From Lemma 8.15 and 10.1 we see that the first integral is equal to $\widetilde{G}(n) + O(n^{\beta-1})$. Using our estimate on $\Delta_n(t)$ we observe that the absolute value of the second integral is smaller than $(B_1 + B_2)n^{\beta-1}$. Finally, the last integral is $O(n^{\beta-\frac{3}{2}} \log^9 n)$. All together the error is $O(n^{\beta-1})$ and Theorem 10.3 is proved. ∎

Example 10.4: *Conditions (O) and (I) violated* Let $g_n = (-1)^n$; thus $\widetilde{G}(z) = e^{-2z}$ and condition (I) is true for any β and for any $\theta < \pi/2$. But in this case the condition (O) outside the cone S_θ does not hold because $|\widetilde{G}(z)e^z| = e^{|z|}$ for $\arg(z) = \pi$. Clearly, g_n is not asymptotically equivalent to $\widetilde{G}(n)$.

If $g_n = (1+t)^n$ for $t > 0$, then $\widetilde{G}(z) = e^{tz}$. Condition (O) holds for some θ such that $(1+t)\cos\theta < 1$. But condition (I) inside the cone S_θ does not hold because $\widetilde{G}(z)$ does not have a polynomial growth. As a matter of fact, g_n is not asymptotically equivalent to $\widetilde{G}(n)$. ∎

In some situations verifying condition (O) is quite troublesome. Fortunately, for some sequences having an analytic continuation of a polynomial growth, condition (O) is automatically satisfied, as shown below.

Theorem 10.4 (Jacquet and Szpankowski, 1999) *Let $g(z)$ be an analytic continuation of $g(n) = g_n$ such that $g(z) = O(z^\beta)$ in a linear cone S_{θ_0}. Then for some θ_0 and for all linear cones S_θ ($\theta < \theta_0$), there exist $\alpha < 1$ and $A > 0$ such that*

$$z \notin S_\theta \quad \Rightarrow \quad |\widetilde{G}(z)e^z| \le A e^{\alpha|z|},$$

where $\widetilde{G}(z)$ is the Poisson transform of g_n.

Proof. Let S_{θ_0} be the linear cone for which the polynomial bound over $g(z)$ holds. Let also $g^\star(s)$ be the Laplace transform of the function $g(x)$ of a real variable x, that is,

$$g^\star(s) = \int_0^\infty g(x) e^{-sx} dx$$

defined for $\Re(s) > 0$. It is well known (cf. [103]) that the inverse Laplace transform of $g^*(s)$ exists in $\Re(s) > 0$ and one can write

$$g(x) = \frac{1}{2\pi i} \int_{\delta-i\infty}^{\delta+i\infty} g^*(s)e^{sx}ds$$

with $\delta > 0$. In addition, $g^*(s)$ is absolutely integrable on the line of integration. Observe now that the exponential generating function $G(z) = \widetilde{G}(z)e^z$ of $g(n)$ can be represented in terms of $g^*(s)$ as follows

$$G(z) = \sum_{n=0}^{\infty} g(n)\frac{z^n}{n!} = \sum_{n=0}^{\infty} \frac{1}{2\pi i} \int_{\delta-i\infty}^{\delta+i\infty} g^*(s) \exp(ns)\frac{z^n}{n!}ds$$

$$= \frac{1}{2\pi i} \int_{\delta-i\infty}^{\delta+i\infty} g^*(s) \sum_{n=0}^{\infty} \exp(ns)\frac{z^n}{n!}ds = \frac{1}{2\pi i} \int_{\delta-i\infty}^{\delta+i\infty} g^*(s) \exp\left(ze^s\right)ds,$$

where the interchange of the integral and the summation is justified since both converge absolutely in their domains of definition. Thus

$$\widetilde{G}(z)e^z = \frac{1}{2\pi i} \int_{\delta-i\infty}^{\delta+i\infty} g^*(s)e^{e^s z}ds \tag{10.23}$$

for $\delta > 0$.

To take advantage of the above formula, we need an extension of the Laplace transform of a real variable to a complex variable, as we did in Section 9.3 for the Mellin transform. This fact is rather well known (e.g., [103, 219]), however, the reader is asked in Exercise 6 to derive such an extension. In particular, this exercise asks to show that if $g(z)$ is an analytic continuation of $g(x)$ in a cone S_θ, where $\theta < \theta_0$, then the inverse Laplace $g^*(s)$ of the function $g(z)$ (of complex z) exists in a *bigger* cone $S_{\theta+\pi/2}$ for all $\theta < \theta_0$ provided $g(z) = O(z^\beta)$ in the cone S_{θ_0} (i.e., $g(z)$ is of a polynomial growth).

In view of this we can write (10.23) as

$$\widetilde{G}(z)e^z = \frac{1}{2\pi i} \int_{\mathcal{L}_\varepsilon} g^*(s)e^{e^s z}ds, \tag{10.24}$$

where $\mathcal{L}_\varepsilon$ is a piece-linear curve that parallels the boundary of the cone $S_{\theta+\pi/2}$ at distance ε. That is, in $s = x + iy$ coordinates, the curve $\mathcal{L}_\varepsilon$ can be described as

$$y = \mathrm{sgn}(y)\tan(\pi/2 + \theta)(x - \varepsilon),$$

where $\mathrm{sgn}(y)$ is the sgn function (i.e., equal to 1 when $y \geq 0$ and -1 otherwise).

We also define $\mathcal{L}_0$ as a curve obtained from $\mathcal{L}_\varepsilon$ as $\varepsilon \to 0$, that is, having the description $y = \text{sgn}(y)x \tan(\pi/2 + \theta)$.

Using (10.24), we can now upper bound $\tilde{G}(z)e^z$ as follows

$$|\tilde{G}(z)e^z| \le \frac{1}{2\pi} \int_{\mathcal{L}_\varepsilon} |g^\star(s)| \exp(\Re(e^s z))ds$$

$$\le \frac{1}{2\pi} \int_{\mathcal{L}_\varepsilon} |g^\star(s)| \exp\left(\Re(e^s e^{i\theta})|z| \right) ds, \qquad (10.25)$$

since for $z \notin \mathcal{S}_\theta$ we have $\cos(\arg z) \le \cos\theta$ for $|\theta| \le \pi/2$. To complete the proof, we must show that $\Re(e^s e^{i\theta}) < \alpha$ for some $\alpha < 1$ and all $s \in \mathcal{L}_\varepsilon$. If this is true, we immediately obtain $|\tilde{G}(z)e^z| \le A e^{\alpha|z|}$ for $\alpha < 1$, where $A = \frac{1}{2\pi} \int_{\mathcal{L}_\varepsilon} |g^\star(s)|ds < \infty$.

We concentrate now on showing that $\Re(e^s e^{i\theta}) < \alpha < 1$. We study the image $\mathcal{I}_\varepsilon$ of $\mathcal{L}_\varepsilon$ under the function e^s, which is plotted in Figure 10.2. (In fact, in Figure 10.2 we assume that for $\varepsilon = 0.1$ the image $\mathcal{I}_\varepsilon$ has the following parametric description: $(\exp(-t/2+0.1)\cos(\pm t), \exp(-t/2+0.1)\sin(\pm t))$.) When $\varepsilon \to 0$, this image tends to the image $\mathcal{I}_0$ of $\mathcal{L}_0$. Observe that $\mathcal{I}_0$ is contained in the unit disk, and $\mathcal{I}_0$ touches the unit circle only at $s = 0$. We rewrite (10.24) as

$$\tilde{G}(z)e^z = \frac{1}{2\pi i} \int_{\mathcal{L}_\varepsilon} g^\star(s) \exp[e^s e^{i \arg(z)}|z|]ds$$

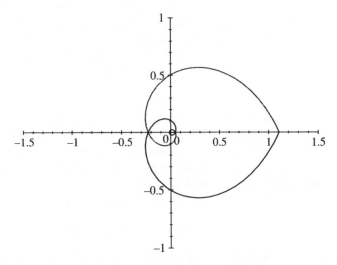

Figure 10.2. Image of $\mathcal{L}_\varepsilon$ with $\varepsilon = 0.1$ by e^s.

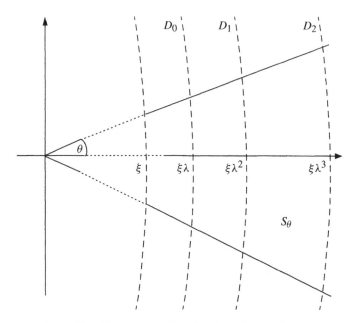

Figure 10.3. The linear cone S_θ and the increasing domains $\mathcal{D}_m$.

to see that the image $\mathcal{I}_\varepsilon$ of $\mathcal{L}_\varepsilon$ is rotated by the angle $\arg(z)$. Let us now fix an arbitrary $0 < \theta < \theta_0$, and consider for a moment only the image $\mathcal{I}_0$ of $\mathcal{L}_0$. Observe that if one rotates the image $\mathcal{I}_0$ by a non zero argument, then the new image has the *real part* strictly less than 1. In fact, the real part will be smaller than $1 - O(\theta^2)$. Finally, considering the image $\mathcal{I}_\varepsilon$ of $\mathcal{L}_\varepsilon$, we can easily choose ε so that the real part of the rotated images of $\mathcal{L}_\varepsilon$ remains smaller than some α such that $1 - O(\theta^2) < \alpha < 1$ for all arguments greater than θ. Thus, $\Re(e^s e^{i\theta}) < \alpha < 1$, which completes the proof. ∎

Before we illustrate the above theorem (see Example 10.7 below), we provide another approach to establish condition (O) that is useful when dealing with Poisson transforms expressed as linear functional equations. In this case conditions (I) and (O) can be verified relatively easily by applying mathematical induction over the **increasing domains**, which we define next.

Lee us consider the following general functional equation

$$\widetilde{G}(z) = \gamma_1(z)\widetilde{G}(zp) + \gamma_2(z)\widetilde{G}(zq) + t(z), \tag{10.26}$$

where $\gamma_1(z)$, $\gamma_2(z)$, and $t(z)$ are functions of z such that the above equation has a solution. The reader is referred to Fayolle et al. [119] (cf. also [379]) for conditions on these functions under which a solution of (10.26) exists. We further assume that $p + q = 1$, but in fact zp and zq could be replaced by more general functions that form a semigroup of substitutions under the operation of composition of functions. We recall that we analyzed a simple version of (10.26) in Section 7.6.1.

Let us define, for integers $m = 0, 1, \ldots,$ and a constant λ such that $0 < \max\{p, q\} \le \lambda^{-1} < 1$, a sequence of increasing domains (cf. Figure 10.3) $\mathcal{D}_m$ as

$$\mathcal{D}_m = \{z : \xi\delta \le |z| \le \xi\lambda^{m+1}\}$$

for some constant $\xi > 0$ and $\delta \le \min\{p, q\}$. Observe that

$$z \in \mathcal{D}_{m+1} - \mathcal{D}_m \quad \Rightarrow \quad pz, qz \in \mathcal{D}_m. \tag{10.27}$$

The last property is crucial for applying mathematical induction over m in order to establish appropriate bounds on $\tilde{G}(z)$ over the whole complex plane.

Theorem 10.5 *Consider the following functional equation (with $p + q = 1$)*

$$\tilde{G}(z) = \gamma_1(z)\tilde{G}(zp) + \gamma_2(z)\tilde{G}(zq) + t(z)$$

that is postulated to have an entire solution. Let for some positive $\beta, 0 < \theta < \pi/2$ and $0 < \eta < 1$ the following conditions hold for $|z| > \xi$:

(I) *For $z \in \mathcal{S}_\theta$*

$$|\gamma_1(z)|p^\beta + |\gamma_2(z)|q^\beta \le 1 - \eta, \tag{10.28}$$

$$|t(z)| \le B\eta|z|^\beta; \tag{10.29}$$

(O) *For $z \notin \mathcal{S}_\theta$ and for some $\alpha < 1$ the following three inequalities are true*

$$|\gamma_1(z)|e^{q\Re(z)} \le \frac{1}{3}e^{\alpha|z|q}, \tag{10.30}$$

$$|\gamma_2(z)|e^{p\Re(z)} \le \frac{1}{3}e^{\alpha|z|p}, \tag{10.31}$$

$$|t(z)|e^{\Re(z)} \le \frac{1}{3}e^{\alpha|z|}. \tag{10.32}$$

Then

$$g_n = \tilde{G}(n) + O(n^{\beta-1}), \tag{10.33}$$

where $g_n = [z^n]\left(n!e^z\tilde{G}(z)\right)$.

Proof. We apply Theorem 10.3. For this, we need to establish the following bounds

$$|\widetilde{G}(z)| \leq B|z|^{\beta} \tag{10.34}$$

for $z \in S_{\theta}$, and

$$|\widetilde{G}(z)e^z| \leq B'e^{\alpha|z|} \tag{10.35}$$

for $z \notin S_{\theta}$ with $\alpha < 1$, where $B, B' > 0$ are constants. It suffices to prove that under (10.28), (10.29), and (10.30)–(10.32) the above two conditions hold. We start with the proof of (10.34). We apply induction over the increasing domains $\mathcal{D}_m$ (Figure 10.3). Let us first consider $z \in S_{\theta}$. Define $\hat{\mathcal{D}}_m = \mathcal{D}_m \cap S_{\theta}$. We take B large enough such that for $m = 0$ the above inequalities hold for $|z| \leq \xi$. Let us now assume that inequality (10.34) is satisfied in $\hat{\mathcal{D}}_m$, and we prove that it also holds in a larger region, namely $\hat{\mathcal{D}}_{m+1}$, thus proving (10.34) in S_{θ}. For this it suffices to consider $z \in \hat{\mathcal{D}}_{m+1} - \hat{\mathcal{D}}_m$. By (10.27) we have $zp, zq \in \hat{\mathcal{D}}_m$, and we can invoke our induction hypothesis. Hence, taking into account (10.28) and (10.29) we conclude from equation (10.26) that

$$
\begin{aligned}
|\widetilde{G}(z)| &\leq |\gamma_1(z)||\widetilde{G}(zp)| + |\gamma_2(z)||\widetilde{G}(zq)| + |t(z)|, \\
&\leq B|\gamma_1(z)||z|^{\beta} p^{\beta} + B|\gamma_2(z)||z|^{\beta} q^{\beta} + |t(z)|, \\
&\leq B(1-\eta)|z|^{\beta} + B\eta|z|^{\beta} = B|z|^{\beta},
\end{aligned}
$$

which is the desired bound.

Now assume that $z \notin S_{\theta}$ and we aim at proving (10.35). We first observe that $|e^z| = e^{\Re(z)} \leq e^{\alpha|z|}$, where $\alpha \geq \cos\theta \geq \cos(\arg(z))$ The induction over the increasing domains can be applied as before, however, this time we consider $\overline{\mathcal{D}}_m = \mathcal{D}_m \cap \overline{S}_{\theta}$, where $\overline{S}_{\theta}$ is the complementary set to S_{θ}. Observe that

$$
\begin{aligned}
|\widetilde{G}(z)e^z| &\leq |\gamma_1(z)||\widetilde{G}(zp)e^{zp}||e^{zq}| + |\gamma_2(z)||\widetilde{G}(zq)e^{zq}||e^{zp}| + |t(z)e^z|, \\
&\leq B'|\gamma_1(z)|e^{\alpha|z|p}e^{q\Re(z)} + B'|\gamma_2(z)|e^{\alpha|z|q}e^{p\Re(z)} + |t(z)|e^{\Re(z)}, \\
&\leq B'e^{\alpha|z|},
\end{aligned}
$$

where the last inequality follows from (10.30)–(10.32) and

$$\Re(z) = |z|\cos(\arg(z)) \leq |z|\cos\theta \leq \alpha|z|.$$

This completes the proof.   ∎

Example 10.5: *Conflict Resolution Algorithm* We consider here a conflict resolution algorithm. Its description can be found in Capetenakis [65] and Tsybakov and Mikhailiov [426], but we provide here a brief review of the *interval searching algorithm*. The access to a slotted broadcast channel is controlled by a window based mechanism that works as follows: This window will be referred to as the *enabled interval* (EI). Let S_i denote the starting point for the ith EI, and t_i is the corresponding starting point for the *conflict resolution interval* (CRI), where CRI represents the number of slots needed to resolve a collision. Roughly speaking, at each step of the algorithm, we compute the endpoints of the EI based on the outcome of the channel. The parameters of interest are: the length of the conflict resolution interval T_n, the fraction of resolved interval W_n, and the number of resolved packets C_n. For example, the Poisson transform $\widetilde{C}(z)$ of C_n satisfies

$$\widetilde{C}(z) = \left(1 + (1 + z/2)e^{-z/2}\right)\widetilde{C}(z/2).$$

Indeed, if there is at most one packet in the first half of EI—which happens with probability $(1 + z/2)e^{-z/2}$ in the Poisson model—then we explore the second half of the interval represented by $\widetilde{C}(z/2)$. Otherwise, we explore the first half of the interval, and thus the first term in the above equation. This equation is of the form of (10.26), and thus falls under Theorem 10.5. Jacquet and Szpankowski [214] proved that for large $z \in \mathcal{S}_\theta$ the above equation admits asymptotically the following solution (cf. [214])

$$\widetilde{C}(z) = D + P(\log z) + O(1/z), \tag{10.36}$$

where

$$D = \exp\left(\frac{1}{\log 2}\int_0^\infty \frac{xe^{-x}\log x}{1 + (1 + x)e^{-x}}dx + \frac{1}{2}\log 2\right) \approx 2.505,$$

and $P(\log z)$ is our familiar fluctuating function with a small amplitude. This can be established by the method discussed in Chapter 9 and the reader is asked to prove it in Exercise 5. To depoissonize the solution, one must check conditions (I) and (O) of Theorem 10.3. But $\widetilde{C}(z) = O(1)$ in a cone $\mathcal{S}_\theta$, thus we need only verify (O) outside the cone $\mathcal{S}_\theta$. By Theorem 10.5 (e.g., see (10.30)) we need to show the existence of ξ such that for $|z| > \xi$ the following holds

$$(1 + (1 + |z|/2))e^{\Re(z)/2} \le e^{\alpha|z|/2}$$

which is clearly true for large enough ξ since $\alpha \ge \cos\theta$. By Theorem 10.5 we conclude that $C_n = D + P(\log n) + O(n^{-1})$. ∎

10.3 GENERALIZATIONS OF DEPOISSONIZATION

In this section we generalize our basic depoissonization Theorem 10.3 twofold:
(i) we derive a full asymptotic expansion for g_n, and (ii) we enlarge the class
of Poisson transforms $\tilde{G}(z)$ for which the depoissonization holds (i.e., we allow
for exponential rather than algebraic bounds on $\tilde{G}(z)$). We prove all of these results in a uniform manner in Section 10.3.3 using a general depoissonization tool
presented in Theorem 10.13.

10.3.1   A Full Asymptotic Depoissonization

Before we present a precise statement, let us formally derive a full asymptotic
expansion of g_n. We shall seek such an expansion in the following form

$$g_n = \sum_{i,j \geq 0} b_{ij} n^i \tilde{G}^{(j)}(n).$$

To identify formally the coefficients b_{ij} we proceed as follows. By Taylor's expansion

$$\tilde{G}(z) = \sum_{j \geq 0} \frac{(z-n)^j}{j!} \tilde{G}^{(j)}(n),$$

which yields

$$g_n = n![z^n](e^z \tilde{G}(z)) = \sum_j n![z^n] \frac{(z-n)^j e^z}{j!} \tilde{G}^{(j)}(n).$$

Hence, we formally set

$$b_{ij} = [n^i][z^n] \frac{n!(z-n)^j}{j!} e^z.$$

After interchanging indexes i and j, factorizing the operator $[z^n]$, noticing that
$\sum_i n^i[n^i]f(n,z) = f(n,z)$ for any analytic function $f(n,z)$, and formally setting $y^j = \tilde{G}^{(j)}(z)$, we obtain

$$\sum_{ij} b_{ij} n^i y^j = [z^n] \sum_j y^j \sum_i n^i[n^i] \frac{n!(z-n)^j e^z}{j!}$$

$$= [z^n] \sum_j \frac{n!(z-n)^j y^j e^z}{j!} = [z^n] n! \exp((z-n)y + z)$$

$$= (1+y)^n e^{-ny} = \exp(n \ln(1+y) - ny).$$

Thus

$$b_{ij} = [x^i][y^j] \exp(n \ln(1 + y) - ny).$$

We observe that $b_{ij} = 0$ for $j < 2i$. Indeed, as in the case of the coefficients a_{ij} discussed in Section 10.1, we let $f(x, y) = \exp(x \log(1 + y) - xy)$ and observe that $f(xy^{-2}, y)$ is analytic at $x = y = 0$. Hence, its Laurent expansion possesses only terms like $x^i y^{j-2i}$ with nonnegative exponents. This means that for $j < 2i$ the coefficients $b_{ij} = 0$.

We are now ready to formulate a general depoissonization result with polynomial bounds. We shall defer the proof until Section 10.3.3, where we present a general depoissonization tool that allows us to establish all results of this section in a uniform manner.

Theorem 10.6 **(Jacquet and Szpankowski, 1998)** *Let the following two conditions hold for some numbers A, B, R > 0 and α < 0, β, and γ:*

(I) *For $z \in S_\theta$, $0 < |\theta| < \pi/2$,*

$$|z| > R \quad \Rightarrow \quad |\widetilde{G}(z)| \leq B|z|^\beta \Psi(|z|), \tag{10.37}$$

where $\Psi(x)$ is a slowly varying function, that is, such that for fixed t $\lim_{x \to \infty} \frac{\Psi(tx)}{\Psi(x)} = 1$ (e.g., $\Psi(x) = \log^d x$ for some $d > 0$);

(O) *For all $z = \rho e^{i\vartheta}$ such that $z \notin S_\theta$ we have*

$$\rho = |z| > R \quad \Rightarrow \quad |\widetilde{G}(z)e^z| \leq A\rho^\gamma \exp[(1 - \alpha\vartheta^2)\rho]. \tag{10.38}$$

Then for every nonnegative integer m

$$g_n = \sum_{i=0}^{m} \sum_{j=0}^{i+m} b_{ij} n^i \widetilde{G}^{(j)}(n) + O(n^{\beta - m - 1} \Psi(n)) \tag{10.39}$$

$$= \widetilde{G}(n) + \sum_{k=1}^{m} \sum_{i=1}^{k} b_{i,k+i} n^i \widetilde{G}^{(k+i)}(n) + O(n^{\beta - m - 1} \Psi(n)),$$

where b_{ij} are as the coefficients of $\exp(x \log(1 + y) - xy)$ at $x^i y^j$, that is,

$$\sum_{i=0}^{\infty} \sum_{j=0}^{\infty} b_{ij} x^i y^j = \exp(x \log(1 + y) - xy), \tag{10.40}$$

with $b_{ij} = 0$ for $j < 2i$.

The first few terms of the above expansion are

$$g_n = \widetilde{G}(n) - \frac{1}{2}n\widetilde{G}^{(2)}(n) + \frac{1}{3}n\widetilde{G}^{(3)}(n) + \frac{1}{8}n^2\widetilde{G}^{(4)}(n) - \frac{1}{4}n\widetilde{G}^{(4)}(n)$$

$$- \frac{1}{6}n^2\widetilde{G}^{(5)}(n) - \frac{1}{48}n^3\widetilde{G}^{(6)}(n) + \frac{1}{5}n\widetilde{G}^{(5)}(n) + \frac{13}{72}n^2\widetilde{G}^{(6)}(n)$$

$$+ \frac{1}{24}n^3\widetilde{G}^{(7)}(n) + \frac{1}{384}n^4\widetilde{G}^{(8)}(n) - \frac{1}{6}n\widetilde{G}^{(6)}(n) - \frac{11}{60}n^2\widetilde{G}^{(7)}(n)$$

$$- \frac{17}{288}n^3\widetilde{G}^{(8)}(n) - \frac{1}{144}n^4\widetilde{G}^{(9)}(n) - \frac{1}{3840}n^5\widetilde{G}^{(10)}(n) + O(n^{\beta-6}).$$

We illustrate analytic depoissonization in two examples. The first one will lead us to the so-called Dirichlet depoissonization introduced recently by Flajolet.

Example 10.6:   *Dirichlet Depoissonization*   Let

$$g_n = \sum_{k=0}^{\infty}\left(1 - (1 - 2^{-k})^n\right) \tag{10.41}$$

whose Poisson transform is (Item 7 of Table 10.1)

$$\widetilde{G}(z) = \sum_{k=0}^{\infty}\left(1 - e^{-z2^{-k}}\right),$$

and the jth derivative is

$$\widetilde{G}^{(j)}(z) = (-1)^{j+1}\sum_{k=0}^{\infty} 2^{-jk}e^{-z2^{-k}}.$$

We are interested in a full asymptotic expansion of g_n using the depoissonization Theorem 10.6. For this we need to find the asymptotics of $\widetilde{G}^{(j)}(z)$ for z in a cone S_θ. But $\widetilde{G}^{(j)}(z)$ is a harmonic sum; hence by Theorem 9.1 the Mellin transforms are

$$\mathcal{M}[\widetilde{G}(z); s] = -\frac{\Gamma(s)}{1 - 2^s}, \qquad -1 < \Re(s) < 0,$$

$$\mathcal{M}[\widetilde{G}^{(j)}(z); s] = \frac{(-1)^{j+1}\Gamma(s)}{1 - 2^{s-j}}, \qquad 0 < \Re(s) < j, \quad j \geq 1.$$

Using the reverse mapping Theorem 9.2 we find for $x \to \infty$ (which by Theorem 9.5 can be extended to complex $z \to \infty$ in a cone S_θ with $|\theta| < \pi/2$)

$$\widetilde{G}(x) = \log_2 x + \frac{\gamma}{\log 2} + \frac{1}{2} + P_0(\log_2 x) + O(x^{-M}),$$

$$\tilde{G}^{(j)}(x) = \frac{(-1)^{j+1}}{x^j} P_j(\log_2 x) + O(x^{-M}), \quad j \geq 1$$

for any $M > 0$, where

$$P_0(\log_2 x) = \frac{1}{\log 2} \sum_{\ell \neq 0} \Gamma(2\pi i\ell/\log 2) e^{-2\pi i\ell \log_2 x}$$

$$P_j(\log_2 x) = \frac{1}{\log 2} \sum_{\ell=-\infty}^{\infty} \Gamma(j + 2\pi i\ell/\log 2) e^{-2\pi i\ell \log_2 x}, \quad j \geq 1.$$

Since $g(z) = O(\log z)$ inside the cone $\mathcal{S}_\theta$, by Theorem 10.4 and Theorem 10.6 we finally obtain

$$g_n = \log_2 n + \frac{\gamma}{\log 2} + \frac{1}{2} + P_0(\log_2 n)$$

$$+ \sum_{k=1}^{m} \sum_{i=1}^{k} (-1)^{k+i+1} b_{i,k+i} n^{-k} P_{k+i}(\log_2 n) + O(n^{-m-1} \log n) \quad (10.42)$$

for any $m \geq 1$, where b_{ij} are given by (10.40).

Clement, Fljaolet and Vellée [68] suggested to consider a more general sum, namely

$$g_n = \sum_{k \geq 0} a_k (1 - (1 - a_k)^n)$$

for a sequence $a_k \to 0$. From Item 7 of Table 10.1 we know that its Poisson transform is

$$\tilde{G}(z) = \sum_{k \geq 0} a_k (1 - e^{-za_k});$$

thus one can use depoissonization, as we did above, to find the full asymptotic expansion of g_n. Another approach, called the **Dirichlet depoissonization**, was proposed in [68]. It is based on the following observation. Define the following two Dirichlet series

$$\Lambda(s) = \sum_{k \geq 0} a_k^s, \qquad \Omega(s) = \sum_{k \geq 0} \left(\log \frac{1}{1 - a_k} \right)^s.$$

Formally, we have

$$\Omega(s) = \sum_{k \geq 0} \exp\left(s \log \log(1 - a_k)^{-1}\right)$$

$$= \sum_{k \geq 0} \exp\left(s \log\left(a_k + \frac{1}{2}a_k^2 + \frac{1}{3}a_k^3 + \cdots\right)\right)$$

$$= \sum_{k \geq 0} a_k^s \exp\left(s \log\left(1 + \frac{1}{2}a_k + \frac{1}{3}a_k^2 + \cdots\right)\right)$$

$$= \sum_{k \geq 0} a_k^s \exp\left(s \left(\frac{1}{2}a_k + \frac{5}{24}a_k^2 + \cdots\right)\right)$$

$$= \sum_{k \geq 0} a_k^s \left(1 + \frac{1}{2}sa_k + (\frac{5}{24}s + \frac{1}{8}s^2)a_k^2 + (\frac{1}{8}s + \frac{5}{48}s^2 + \frac{1}{48}s^3)a_k^3 + \cdots\right)$$

$$= \Lambda(s) + c_1(s)\Lambda(s + 1) + c_2(s)\Lambda(s + 2) + \cdots .$$

Thus the singularities of $\Omega(s)$ are those of $\Lambda(s)$ plus $\{0, -1, \ldots\}$. This was rigorously proved in [68] from where we quote the following result.

Lemma 10.7   (Dirichlet Depoissonization)   *Let $a_k \to 0$ with a_k positive. Define the above two Dirichlet series $\Lambda(s)$ and $\Omega(s)$. Then, if the Dirichlet series $\Lambda(s)$ of a_k is meromorphic in the whole complex plane and it is of "weak polynomial growth" (i.e., the growth is only controlled on certain lines parallel to the real axis tending to $\pm i\infty$), then the same properties hold for $\Omega(s)$. In addition, the singularities of Λ, Ω are related by*

$$\text{Sing}(\Omega) = \text{Sing}(\Lambda) + \{0, -1, -2, \cdots\},$$

and the singular expansions of $\Omega(s)$ and $\Lambda(s)$ are related by

$$\Omega(s) \asymp \Lambda(s) + c_1(s)\Lambda(s + 1) + c_2(s)\Lambda(s + 2) + \cdots, \tag{10.43}$$

where

$$c_j(s) = [x^j] \exp\left(\frac{s}{x} \log \frac{1}{1 - x}\right). \tag{10.44}$$

For g_n defined in (10.41) we have

$$\Omega(-s) \asymp \Lambda(-s) + \frac{s}{2}\Lambda(-s + 1) + \left(\frac{5s}{24} + \frac{s^2}{8}\right)\Lambda(-s + 2) + \cdots,$$

and (10.42) can be recovered.

Example 10.7: *Entropy of the Binomial Distribution*   We compute here a full asymptotic expansion of Shannon entropy h_n of the binomial distribution

$$\binom{n}{k} p^k q^{n-k} (q = 1 - p),$$

that is,

$$h_n = -\sum_{k=0}^{n} \binom{n}{k} p^k q^{n-k} \log\left(\binom{n}{k} p^k q^{n-k}\right).$$

We shall follow the approach suggested in Jacquet and Szpankowski [219], while the reader is asked in Exercise 10 to use the singularity analysis to obtain the same result (cf. Flajolet [129]). Observe that

$$h_n = -\ln(n!) - n(p \ln p + q \ln q)$$
$$+ \sum_{k=0}^{n} \ln(k!) \binom{n}{k} p^k q^{n-k} + \sum_{k=0}^{n} \ln((n-k)!) \binom{n}{k} p^k q^{n-k}.$$

We use analytic poissonization and depoissonization to estimate

$$g_n := \sum_{k=0}^{n} \ln(k!) \binom{n}{k} p^k q^{n-k} + \sum_{k=0}^{n} \ln((n-k)!) \binom{n}{k} p^k q^{n-k}.$$

Observe that the Poisson transform $\widetilde{G}(z)$ of g_n is

$$\widetilde{G}(z) = \widetilde{F}(zp) + \widetilde{F}(zq),$$

where

$$\widetilde{F}(z) = \sum_{n=0}^{\infty} \ln(n!) \frac{z^n}{n!} e^{-z}.$$

Thus

$$h_n = -\ln(n!) - n(p \ln p + q \ln q) + n![z^n]\left(e^z \widetilde{F}(zp) + e^z \widetilde{F}(zq)\right).$$

We cannot apply directly the depoissonization Theorem 10.6 since the Poisson transform $\widetilde{F}(z)$ is not given in a closed form. But, we can apply our previous analytic poissonization Theorem 10.2 to find $\widetilde{F}(z)$ for $z \to \infty$ and then use Theorem 10.6. In fact, in Example 10.2 we already computed $\widetilde{F}(z)$ asymptotically. Setting $\log(n!) = f_n = f(n) = \log \Gamma(n + 1)$ and using the symbolic notation proposed in Section 10.2 we can formally write for the kth derivative of $\widetilde{F}(z)$ as

$$\widetilde{F}^{(k)}(\lambda n) = (e^y - 1)^k \exp\left(n\lambda(e^y - 1) - n\lambda y\right), \qquad y^j = f^{(j)}(\lambda n),$$

where λ is equal to either p or q. Since f_n has an analytic continuation of a polynomial growth, we can apply Theorem 10.6 to obtain

$$n![z^n]\left(e^z \widetilde{F}(\lambda z)\right) = \sum_{i,j\geq 0} b_{ij} n^i \lambda^j (e^y - 1)^j \exp\left(n\lambda(e^y - 1) - n\lambda y\right),$$

$$= \exp\left(n \log\left(1 + \lambda(e^y - 1)\right) - \lambda n y\right), \qquad y^j = f^{(j)}(\lambda n),$$

where we used (10.40) in the last line. Introducing $c_{i,k}(\lambda)$ as follows

$$\exp\left(x \log\left(1 + \lambda(e^y - 1)\right) - \lambda x y\right) = \sum_{i=0}^{\infty} \sum_{k=2i}^{\infty} c_{i,k}(\lambda) x^i y^k,$$

we finally arrive at

$$h_n = -f(n) + f(np) + f(nq) - n(p \ln p + q \ln q)$$
$$+ \sum_{i=1}^{\infty} \sum_{k=2i}^{\infty} n^i \left(c_{i,k}(p) p^i f^{(k)}(np) + c_{i,k}(q) q^i f^{(k)}(nq)\right),$$

where $f^{(k)}(x)$ ($k \geq 1$) are computed in Example 10.2. This leads to

$$h_n \sim \frac{1}{2} \ln n + \frac{1}{2} + \frac{1}{2} \ln(2\pi pq) + \sum_{m=1}^{\infty} \frac{e_m}{n^m}$$

and e_m can be explicitly computed, as shown in [219].   ∎

10.3.2 Exponential Depoissonization

In some applications the assumption about the polynomial growth of $\widetilde{G}(z)$ is too restrictive (see Example 9.9.3 of Chapter 9). Here we extend Theorem 10.6 to Poisson transforms with exponential growth $\exp(z^\beta)$ for $0 < \beta < \frac{1}{2}$. The proof is again delayed until the last subsection. However, below we provide a sketch of the proof for the basic exponential depoissonization.

Theorem 10.8 *Let the conditions of Theorem 10.6 be satisfied with condition (I) replaced by*

$$|\widetilde{G}(z)| \leq A \exp(B|z|^\beta) \tag{10.45}$$

for some $0 < \beta < \frac{1}{2}$ and constants $A > 0$ and B. Then for every integer $m \geq 0$

$$g_n = \sum_{i=0}^{m} \sum_{j=0}^{i+m} b_{ij} n^i \widetilde{G}^{(j)}(n) + O(n^{-(m+1)(1-2\beta)} \exp(Bn^{\beta})) \qquad (10.46)$$

for large n.

Sketch of Proof. It seems appropriate to give here a sketch of the proof for the leading term (Section 10.3.3), namely,

$$g_n = \widetilde{G}(n) + O\left(n^{2\beta-1} \exp(Bn^{\beta})\right). \qquad (10.47)$$

This can be derived along the same lines as the proof of the basic depoissonization Theorem 10.3. As in the proof of Theorem 10.3, we can easily estimate the integrals E_n and I_n'', and therefore we are left with the integral I_n' (cf. (10.20)–(10.22)). To estimate I_n' we need an extension of Lemma 10.1 for functions $\widetilde{G}(z)$ with an exponential bound. We shall show that if (10.45) holds, then the kth derivative $\widetilde{G}^{(k)}(z)$ can be bounded as below

$$|\widetilde{G}^{(k)}(z)| \leq B_k |z|^{k(\beta-1)} \exp(B|z|^{\beta}). \qquad (10.48)$$

Indeed, using Cauchy's bound (10.10) with $r = \Omega|z|^{1-\beta}$ we obtain

$$|\widetilde{G}^{(k)}(z)| \leq \frac{\max_{|\omega|=\Omega|z|^{1-\beta}} |\widetilde{G}(z+\omega)|}{\Omega|z|^{k(1-\beta)}},$$

where $z + \omega \in S_\theta$, so that the exponential bound $|\widetilde{G}(z+\omega)| < A \exp(B|z+\omega|^{\beta})$ is still valid. Since the derivative of z^{β} is $O(z^{\beta-1})$ we have

$$\max_{|\omega|=\Omega|z|^{1-\beta}} |\widetilde{G}(z+\omega)| \leq A \exp(|B||z|^{\beta} + O(z^{\beta-1})|z|^{1-\beta}\Omega)$$

$$\leq A' \exp(|B||z|^{\beta})$$

for some constant A'. This proves (10.48).

Now we can consider the integral I_n', which we repeat here

$$I_n' = \frac{1}{\sqrt{2\pi}} \int_{-\log n}^{\log n} e^{-t^2/2} \left(\widetilde{G}(n) + \widetilde{G}'(n)it\sqrt{n}\right) \left(1 - \frac{it^3}{6\sqrt{n}} + \frac{t^4}{24n} - \frac{t^6}{72n}\right) dt$$

$$+ \frac{1}{\sqrt{2\pi}} \int_{-\log n}^{\log n} e^{-t^2/2} \Delta_n(t) t^2 h_n(t) dt$$

$$+ \frac{1}{\sqrt{2\pi}} \int_{-\log n}^{\log n} e^{-t^2/2} \left(\widetilde{G}(n) + \widetilde{G}'(n) it\sqrt{n} \right) O\left(\frac{\log^9 n}{n\sqrt{n}} \right) dt,$$

where we recall that

$$h_n(t) = \exp(n(e^{it/\sqrt{n}} - 1 - it/\sqrt{n})),$$
$$\Delta_n(t) = -\widetilde{G}'(ne^{it/\sqrt{n}})e^{it/\sqrt{n}} - \widetilde{G}''(e^{it/\sqrt{n}})ne^{2it/\sqrt{n}}.$$

Then (10.48) and the above yield $|\Delta_n(t)| = O(n^{2\beta-1} \exp(Bn^\beta))$. But the first part of the above integral is $\widetilde{G}(n) + O(n^{\beta-1} \exp(Bn^\beta))$ for $\beta > 0$, the second part is $O(n^{2\beta-1} \exp(Bn^\beta))$, and the last part is $O(n^{-3/2} \exp(Bn^\beta))$. This proves (10.47).   ■

Example 10.8: *Example 9.9.3 of Chapter 9 Revisited* In Example 9.9.3 of Chapter 9 we studied the sequence C_n satisfying the recurrence (9.31), that is,

$$C_n = 4 \sum_{k=2}^{n} \binom{n}{k} \frac{2^{-n}}{n-k+1} C_k, \quad n \geq 2$$

with $C_2 = 1$. We proved that its Poisson transform $\widetilde{C}(z)$ satisfies the following functional equation

$$\widetilde{C}(z) = \frac{8}{z}(1 - e^{-z/2})\widetilde{C}\left(\frac{z}{2}\right),$$

while $F(z) = \log z^2 \widetilde{C}(z)/2$ fulfills

$$F(2z) - F(z) = \log\left(\frac{1 - e^{-z}}{z}\right). \tag{10.49}$$

In (9.32) of Example 9.9.3 we proved that

$$\exp[F(z)] \sim \sqrt{z} 2^{-1/12} \exp\left(\frac{\gamma(1) + \gamma^2/2 - \pi^2/12}{\log 2}\right) \exp\left[-\frac{1}{2}\frac{\log^2(z)}{\log 2} + \Psi(\log_2 z)\right],$$

where

$$\Psi(\log_2 z) = \sum_{\substack{\ell=-\infty \\ \ell \neq 0}}^{\infty} \frac{1}{2\pi i \ell} \Gamma\left(1 - \frac{2\pi i \ell}{\log 2}\right) \varsigma\left(1 - \frac{2\pi i \ell}{\log 2}\right) e^{2\pi i \ell \log_2 z}.$$

Thus $\widetilde{C}(z) = O(z^{5/2} \exp(\log^2 z))$ in the cone $\mathcal{S}_\theta$ (by analytic continuation). By Theorem 10.8 we find

$$C_n \sim \frac{1}{2} n^{5/2} 2^{-1/12} \exp\left(\frac{\gamma(1) + \gamma^2/2 - \pi^2/12}{\log 2}\right) \exp\left[-\frac{1}{2} \frac{\log^2(j)}{\log 2} + \Psi(\log_2 n)\right]$$

as $n \to \infty$.   ∎

10.3.3   General Depoissonization Tool

Here we prove all the depoissonization results discussed so far. We start with a few lemmas followed by our main depoissonization tool Theorem 10.13.

It turns out that in the course of the proof we often have to deal with functions satisfying a certain property. We formalize it in the following definition.

Definition 10.9   *We say that a sequence of functions $f_n : I_n \to \mathbb{C}$ defined on subintervals I_n belongs to the class $\mathbb{D}_k(\omega)$ for ω real if there exist $D > 0$ such that for all integers $j \leq k$ and $x \in I_n$ we have $|f_n^{(j)}(x)| \leq Dn^{-j\omega}$.*

Example 10.9:   *Functions belonging to $\mathbb{D}_k(\omega)$*   Consider the following two functions:

1. Let f be infinitely differentiable on $[-1, 1]$, and define $f_n(x) = f(xn^{-\omega})$ on $I_n = [-n^\omega, n^\omega]$. Clearly, $f_n \in \mathbb{D}_k(\omega)$ for every $k \geq 0$.

2. Let F_n be an analytic function defined in $\{z : |z| \leq n^\omega\}$, $n \geq 1$, such that the sequence F_n is uniformly bounded. Then the restriction f_n of F_n to $I_n = [-\frac{1}{2}n^\omega, \frac{1}{2}n^\omega]$ belongs to $\mathbb{D}_k(\omega)$ for every $k \geq 0$.   ∎

The next lemma presents some simple properties of the class $\mathbb{D}_k(\omega)$. Its proof is quite simple and is left for the reader in Exercise 7.

Lemma 10.10

(i) *If f_n belongs to $\mathbb{D}_k(\omega)$, then f_n belongs to $\mathbb{D}_k(\omega')$ for all $\omega' \leq \omega$.*

(ii) *If f_n and g_n belong to class $\mathbb{D}_k(\omega)$, and if $H(x, y)$ is a function which is k times continuously differentiable, then $H(f_n, g_n)$ belongs to $\mathbb{D}_k(\omega)$. Consequently if f_n and $g_n \in \mathbb{D}_k(\omega)$, then $f_n + g_n$ and $f_n \times g_n \in \mathbb{D}_k(\omega)$.*

Observe that to prove our general asymptotic expansion, like the one in Theorem 10.6, we must study bounds and Taylor's expansions of the following function

$$h_n(t) = \exp(n(e^{it/\sqrt{n}} - 1 - it/\sqrt{n})). \tag{10.50}$$

One can interpret $h_n(t)$ as the kernel of the Cauchy integral (10.13) (cf. also (10.19)), thus it is not surprising that this function often appears in our depoissonization theorems. Its properties are discussed in the next lemma.

Lemma 10.11 *The following statements hold:*

(i) *For* $t \in [-\pi\sqrt{n}, \pi\sqrt{n}]$ *there exists* $\mu > 0$ *such that* $|h_n(t)| \le e^{-\mu t^2}$, *where* μ *is a constant.*

(ii) *For complex* t *such that* $|t| \le Bn^{\frac{1}{6}}$ *for some* $B > 0$, *the sequence of functions* $F_n(t) = h_n(t)e^{-t^2/2}$ *is bounded and belongs to* $\mathbb{D}_k(\frac{1}{6})$ *for any integer* $k \ge 0$.

Proof. Part (i) was already proved in Section 10.2.2. Part (ii) is a little more intricate. According to Lemma 10.10, it suffices to prove that the sequence of functions $\log F_n(t) \in \mathbb{D}_k(\frac{1}{6})$, and then refer to the fact that the sequence of exponentials of $\log F_n(t)$ still belongs to $\mathbb{D}_k(\frac{1}{6})$. Setting $r(x) = e^{ix} - 1 - ix - x^2/2$, we observe that the sequence of functions $r(t/\sqrt{n})$ belongs to $\mathbb{D}_k(\frac{1}{2})$ for $t = O(\sqrt{n})$ and any integer $k \ge 1$. Therefore, the ith derivative for $3 \le i \le k$ of $nr(t/\sqrt{n})$ is $O(n^{1-i/2})$, which is $O(n^{-i/6})$ for all $i \ge 3$. In particular, the third derivative is $O(n^{-1/2})$. But we also observe that by successive integrations the first derivative of $nr(t/\sqrt{n})$ is $O(n^{-1/6})$, the second derivative is $O(n^{-1/3})$, and $nr(t/\sqrt{n})$ is $O(1)$, because the first two derivatives of $r(t)$ are zero at $t = 0$ by the construction. Hence, part (ii) follows. ∎

Furthermore, when proving Theorems 10.6, we need an extension of Lemma 8.15, which is presented next.

Lemma 10.12 *For nonnegative* H *and* β, *let the sequence of complex functions* $F_n(x)$, *defined on* $x \in [-H \log n, H \log n]$, *belong to the class* $\mathbb{D}_k(\beta)$ *for any* $k \ge 1$. *Then*

$$\frac{1}{\sqrt{2\pi}} \int_{-H \log n}^{H \log n} F_n(x)e^{-x^2/2}dx = \sum_{i=0}^{k-1} \frac{1}{2^i i!} F_n^{(2i)}(0) + O(n^{-2k\beta}).$$

Proof. By Taylor's expansion $F_n(x) = \sum_{i=0}^{l-1} \frac{x^i}{i!} F^{(i)}(0) + \Delta_n(x) x^l$, where $\Delta_n(x) = O(n^{-l\beta})$ due to $F_n \in \mathbb{D}_l(\beta)$ for any l. Thus

$$\frac{1}{\sqrt{2\pi}} \int_{-H\log n}^{H\log n} F_n(x) e^{-x^2/2} dx = \sum_{i=0}^{l-1} \frac{F_n^{(i)}(0)}{i!} \frac{1}{\sqrt{2\pi}} \int_{-H\log n}^{H\log n} x^i e^{-x^2/2} dx$$

$$+ O(n^{-l\beta}).$$

Observe that $F_n^{(i)}(0) = O(n^{-i\beta})$ since $F_n \in \mathbb{D}_l(\beta)$ and $i \le l$. Furthermore, by Lemma 8.15 changing the limits of integration in the above to $\pm\infty$ introduces an error of order $O(e^{-(H\log n)^2/2})$ that decreases faster than any polynomial. Also, by (8.67) of Lemma 8.15 we know that for i even

$$\frac{1}{\sqrt{2\pi}} \int_{-\infty}^{\infty} x^i e^{-x^2/2} dx = \frac{(2m)!}{m! 2^m}, \qquad i = 2m,$$

and the integral is zero for i odd. After setting $k = 2l$ and rearranging the above sum, we prove the lemma. ∎

Finally, we are in a position to formulate the **depoissonization tool** theorem. To have our results apply also to the distributional problems discussed in the next section, we consider here a double-index sequence $g_{n,k}$ and a sequence of Poisson transforms

$$\widetilde{G}_k(z) = e^{-z} \sum_{n=0}^{\infty} g_{n,k} \frac{z^n}{n!}.$$

Theorem 10.13 **(Jacquet and Szpankowski, 1998)** *Let $\widetilde{G}_k(z)$ be the Poisson transform of a sequence $g_{n,k}$ that is assumed to be a sequence of analytical functions for $|z| \le n$. We postulate that the following three conditions simultaneously hold for some $H > 0$, γ, and integer $m \ge 0$:*

(I) *For $z \in S_\theta$, such that $|z| = n$: $|\widetilde{G}_n(z)| \le F(n)$ for a sequence $F(n)$,*

(L) *The sequence of function $f_n(t) = \widetilde{G}_n(ne^{it/\sqrt{n}})/F(n)$ defined for $t \in [-H\log n, H\log n]$ belongs to the class $\mathbb{D}_m(\gamma)$,*

(O) *For $z \notin S_\theta$ and $|z| = n$: $|\widetilde{G}_n(z)e^z| \le p(n)e^n F(n)$, where $p(n)$ decays faster than $n^{-\frac{1}{2}-(m+1)\min\{\gamma, \frac{1}{6}\}}$, that is,*

$$p(n) = o(n^{-\frac{1}{2}-(m+1)\min\{\gamma, \frac{1}{6}\}}).$$

Then

$$g_{n,n} = \widetilde{G}_n(n) + O(n^{-\min\{\gamma, \frac{1}{6}\}}) F(n), \tag{10.51}$$

and more generally for any integer $m \geq 0$

$$g_{n,n} = \sum_{i=0}^{m} \sum_{j=0}^{i+m} b_{ij} n^i \widetilde{G}_n^{(j)}(n) + O(n^{-(m+1)\min\{\gamma, 1/6\}}) F(n), \quad (10.52)$$

where b_{ij} are defined in (10.40).

Proof. The proof again relies on the Cauchy integral that we split into I_n and E_n as before, where I_n is the integral over t such that ne^{it} is inside the cone S_θ, and E_n is the integral that has ne^{it} outside the cone. That is:

$$I_n = \omega_n \sqrt{\frac{n}{2\pi}} \int_{-\theta}^{\theta} \widetilde{G}_n(ne^{it}) \exp\left(n\left(e^{it} - 1 - it\right)\right) dt,$$

$$E_n = \frac{n!}{2\pi i} \int_{|t| \in [\theta, \pi]} \frac{\widetilde{G}_n(z) e^z}{z^{n+1}} dz$$

$$= \omega_n \sqrt{\frac{n}{2\pi}} \int_{|t| \in [\theta, \pi]} \widetilde{G}_n(ne^{it}) \exp\left(n\left(e^{it} - 1 - it\right)\right) dt,$$

where $\omega_n = n! n^{-n} e^n (2\pi n)^{-1} = 1 + O(n^{-1})$ by the Stirling formula. We estimate the above two integrals. We begin with the latter. Observe that by condition (O) we have $|E_n|/F(n) \leq \sqrt{2\pi n} p(n) = o(n^{-k \min\{\gamma, \frac{1}{6}\}})$ for any $k \leq m + 1$, which is negligible when compared to the error term in (10.51).

The evaluation of I_n is more intricate. First, we replace t by $t/\sqrt{n}$ to get

$$I_n = \frac{\omega_n}{\sqrt{2\pi}} \int_{-\theta\sqrt{n}}^{\theta\sqrt{n}} \widetilde{G}_n(ne^{it/\sqrt{n}}) \exp\left(n\left(e^{it/\sqrt{n}} - 1 - it/\sqrt{n}\right)\right) dt$$

$$= \frac{\omega_n}{\sqrt{2\pi}} \int_{-\theta\sqrt{n}}^{\theta\sqrt{n}} \widetilde{G}_n(ne^{it/\sqrt{n}}) h_n(t) dt,$$

where $h_n(t)$ is defined in (10.50) and analyzed in Lemma 10.11. We further split the integral I_n into I'_n and I''_n as follows

$$I'_n = \frac{\omega_n}{\sqrt{2\pi}} \int_{-H\log n}^{H\log n} \widetilde{G}_n(ne^{it/\sqrt{n}}) \exp\left(n\left(e^{it/\sqrt{n}} - 1 - it/\sqrt{n}\right)\right) dt,$$

$$I''_n = \frac{\omega_n}{\sqrt{2\pi}} \int_{t \in [-\theta\sqrt{n}, -H\log n]} \widetilde{G}_n(ne^{it/\sqrt{n}}) \exp\left(n\left(e^{it/\sqrt{n}} - 1 - it/\sqrt{n}\right)\right) dt$$

$$+ \frac{\omega_n}{\sqrt{2\pi}} \int_{t \in [H\log n, \theta\sqrt{n}]} \widetilde{G}_n(ne^{it/\sqrt{n}}) \exp\left(n\left(e^{it/\sqrt{n}} - 1 - it/\sqrt{n}\right)\right) dt.$$

To estimate the second integral I_n'' we use the fact that $|h_n(t)| \le e^{-\mu t^2}$ with $\mu > 0$ for $t \in [-\theta\sqrt{n}, \theta\sqrt{n}]$, as discussed in Lemma 10.11. Thus by Lemma 8.15 and condition (I) we immediately obtain that $I_n''/F(n) = O(e^{-\mu H^2 \log^2 n})$, which decays faster than any polynomial.

Now we estimate I_n'. Observe first that since $h_n(t)e^{t^2/2} \in \mathbb{D}_m(\frac{1}{6})$ for any m, and we have $\widetilde{G}_n(ne^{it/\sqrt{n}})/F(n) \in \mathbb{D}_m(\gamma)$, therefore the product

$$R_n(t) = h_n(t)e^{t^2/2}\widetilde{G}_n(ne^{it/\sqrt{n}})/F(n) \in \mathbb{D}_m(\gamma_2),$$

where $\gamma_2 = \min\{\gamma, \frac{1}{6}\}$. In view of this and by Lemma 10.12 we obtain

$$I_n'/F(n) = \frac{\omega_n}{\sqrt{2\pi}} \int_{H\log n}^{H\log n} R_n(t)e^{-t^2/2}dt = \omega_n(F_n(0) + O(n^{-\gamma_2})),$$

and consequently $g_n = \widetilde{G}_n(n) + O(n^{-\gamma_2})F(n)$. This proves (10.51).

To prove our general result (10.52), we apply Lemma 10.12 for any k to obtain

$$g_{n,n}/F(n) = \omega_n \sum_{i=0}^{k-1} \frac{1}{2^i i!} R_n^{(2i)}(0) + O(n^{-2k\gamma_2}).$$

Computing explicitly the derivatives of $R_n(t)$ by their actual values (that involve the derivatives of $\widetilde{G}(z)$ at $z = n$), and noting that all odd powers of $n^{-\frac{1}{2}}$ disappear (since we are considering only derivatives of $R_n(t)$ of even order), multiplying by $\omega_n F(n)$, replacing ω_n by the Stirling expansion, we finally obtain

$$g_{n,n} = \sum_{i=0}^{m} \sum_{j=0}^{i+m} b_{ij} n^i \widetilde{G}_n^{(j)}(n) + O(n^{-(m+1)\min\{\gamma, \frac{1}{6}\}})F(n),$$

where we set $m + 1 = 2k$. Notice that the terms in the expansion do not involve m since the Stirling expansion and the expansion of Lemma 10.12 do not contain m.

∎

Using the depoissonization tool Theorem 10.13, we can now finally prove our remaining theorems by identifying the function $F(n)$ (see condition (I)) and finding the correct value for γ in condition (L) of Theorem 10.13.

We start with the proof of the generalized depoissonization **Theorem 10.6**. First of all, observe that conditions (I) and (O) of Theorem 10.6 imply conditions (I) and (O) of Theorem 10.13 when we set $F(n) = Bn^\beta \Psi(n)$. Thus to complete the proof we must verify condition (L) of Theorem 10.13 and find γ such that

$$f_n(t) = \widetilde{G}_n(ne^{it/\sqrt{n}})n^{-\beta}/\Psi(n) \in \mathbb{D}_m(\gamma)$$

for any integer $m \geq 1$. But by Lemma 10.1(ii), for all integer m and for all z belonging to a smaller cone $\mathcal{S}_{\theta'}$ with $\theta' < \theta$ there exists B_m such that $|\widetilde{G}_n^{(m)}(z)| \leq B_m |z|^{\beta-m} \Psi(|z|)$. After setting $z = ne^{it}/\sqrt{n}$ we see that the sequence of functions $f_n(t)$ belongs to $\mathbb{D}_m(\frac{1}{2})$ for all integer m. Now set $\gamma = \frac{1}{6}$. Then by Theorem 10.13 there is some $m' > m$ such that

$$g_{n,n} = \sum_{i=0}^{m'} \sum_{j=0}^{i+m'} b_{ij} n^i \widetilde{G}_n^{(j)}(n) + O(n^{\beta-m'\gamma}) \Psi(n).$$

The last delicate point is to obtain the correct error term in Theorem 10.6. But this can be achieved by setting $m' = \lfloor (m+1)/\gamma \rfloor$ for a given value m. This will lead to an error term equal to $O(n^{\beta-(m+1)})$. Indeed, the additional terms obtained are those with $i > m$ and $j > i + m$. The corresponding coefficients are $b_{ij} n^i \widetilde{G}_n^{(j)}(n)$ with $b_{ij} \neq 0$ for $j \geq 2i$ (we remind the reader that $b_{ij} = 0$ for $j < 2i$, as shown in Section 10.3.1). Then

$$n^{\beta+i-j} \leq n^{\beta+i-2i} = n^{\beta-i} \leq n^{\beta-(m+1)},$$

and this proves Theorem 10.6.

Now, we prove the exponential depoissonization **Theorem 10.8**. As before, the proof relies on Theorem 10.13 with $F(n) = \exp(Bn^\beta)$. Conditions (I) and (O) are again easy to verify, so we only need to check condition (L) of Theorem 10.13. That is, we must estimate the growth of $f_n(t) = \widetilde{G}(ne^{it}/\sqrt{n}) \exp(-Bn^\beta)$. But $f_n(t) \in \mathbb{D}(\frac{1}{2} - \beta)$ due to (10.48). To complete the proof, we must establish the error term. Let $\gamma = \min\{\frac{1}{2} - \beta, \frac{1}{6}\}$. An application of Theorem 10.13 leads to the error term $O\left(n^{-m'\gamma} \exp(Bn^\beta)\right)$ for some integer $m' \geq 0$. To establish the right error term $O(n^{-(m+1)(1-2\beta)} \exp(Bn^\beta))$ we follow the same approach as before. We set $m' = \lfloor (m+1)(1-2\beta)/\gamma \rfloor$, which introduces the additional terms $b_{ij} n^i \widetilde{G}^{(j)}(n)$ that contribute $O(n^{i-j(1-\beta)} \exp(Bn^\beta))$. But, since $b_{ij} \neq 0$ for $j \geq 2i$,

$$n^{i-j(1-\beta)} \exp(Bn^\beta) \leq n^{i-2i(1-\beta)} \exp(Bn^\beta) \leq n^{-i(1-2\beta)} \exp(Bn^\beta)$$
$$\leq n^{-m(1-2\beta)} \exp(Bn^\beta),$$

and this completes the proof of Theorem 10.8.

10.4 MOMENTS AND LIMITING DISTRIBUTIONS

Depoissonization techniques can also be used to derive limiting distributions. Extension to distributions requires us to investigate the double-index sequence $g_{n,k}$

(e.g, $g_{n,k} = \Pr\{X_n = k\}$ or $g_{n,k} = \mathbf{E}[e^{t X_n/\sqrt{V_k}}]$ for a sequence V_k). Its Poisson transform is denoted as

$$\widetilde{G}(z, u) = \mathcal{P}(g_{n,k}; z, u) = \sum_{n=0}^{\infty} \frac{z^n}{n!} e^{-z} \sum_{k=0}^{\infty} g_{n,k} u^k.$$

It is often more convenient to investigate the Poisson transform $\widetilde{G}_k(z) = \mathcal{P}(g_{n,k}; z)$ defined as

$$\widetilde{G}_k(z) = \sum_{n=1}^{\infty} g_{n,k} \frac{z^n}{n!} e^{-z},$$

and already discussed in Theorem 10.13. In this section we present the *diagonal depoissonization* result that allows us to extract asymptotically $g_{n,n}$ (e.g., $g_{n,n} = \mathbf{E}[e^{t X_n/\mathbf{Var}[X_n]}]$). We start, however, with the relationship between the Poisson and the Bernoulli moments of a random variable.

10.4.1   Moments

Let X_n be a sequence of discrete random variables and X_N its corresponding Poisson driven sequence, where N is a Poisson random variable with mean z. Let $\widetilde{G}(z, u) = \mathbf{E}u^{X_N} = \sum_{n=0}^{\infty} \mathbf{E}u^{X_n} \frac{z^n}{n!} e^{-z}$ be its Poisson transform. We introduce also the Poisson mean $\widetilde{X}(z)$ and the Poisson variance $\widetilde{V}(z)$ as

$$\widetilde{X}(z) = \widetilde{G}'_u(z, 1), \qquad \widetilde{V}(z) = \widetilde{G}''_u(z, 1) + \widetilde{X}(z) - \left(\widetilde{X}(z)\right)^2,$$

where $\widetilde{G}'_u(z, 1)$ and $\widetilde{G}''_u(z, 1)$ respectively denote the first and the second derivative of $\widetilde{G}(z, u)$ with respect to u at $u = 1$.

The next result presents a relationship between the Poisson mean $\widetilde{X}(z)$ and variance $\widetilde{V}(z)$, and the Bernoulli mean $\mathbf{E}[X_n]$ and variance $\mathbf{Var}[X_n]$.

Theorem 10.14   *Let $\widetilde{X}(z)$ and $\widetilde{V}(z) + \widetilde{X}^2(z)$ satisfy condition (O), and $\widetilde{X}(z)$ and $\widetilde{V}(z)$ satisfy condition (I) of Theorem 10.6 with $\beta \leq 1$ (e.g., $\widetilde{X}(z) = O(z^\beta \Psi(z))$, and $\widetilde{V}(z) = O(z^\beta \Psi(z))$ in a linear cone $\mathcal{S}_\theta$ and appropriate conditions (O) outside the cone, where $\Psi(z)$ is a slowly varying function). Then the following holds:*

$$\mathbf{E}[X_n] = \widetilde{X}(n) - \frac{1}{2} n \widetilde{X}''(n) + O(n^{\beta-2} \Psi(n)), \qquad (10.53)$$

$$\mathbf{Var}[X_n] = \widetilde{V}(n) - n[\widetilde{X}'(n)]^2 + O(n^{\beta-1} \Psi(n)) \qquad (10.54)$$

for large n.

Proof. The asymptotic expansion (10.53) directly follows from Theorem 10.6 for $m = 1$. To derive (10.54) observe that the Poisson transform of $\mathbf{E}[X_n^2]$ is $\widetilde{V}(z) + (\widetilde{X}(z))^2$, thus again by Theorem 10.6

$$
\begin{aligned}
\mathbf{E}[X_n^2] &= \widetilde{V}(n) + (\widetilde{X}(n))^2 - \frac{1}{2}n\widetilde{V}''(n) - n[\widetilde{X}'(n)]^2 \\
&\quad - n\widetilde{X}''(n)\widetilde{X}(n) + O(n^{\beta-2}\Psi(n)) \\
&= \widetilde{V}(n) + \left((\widetilde{X}(n))^2 - n\widetilde{X}''(n)\widetilde{X}(n) + O(n^{2\beta-2}\Psi(n))\right) \\
&\quad - n[\widetilde{X}'(n)]^2 + O(n^{\beta-1}\Psi(n)),
\end{aligned}
$$

where the last error term is a consequence of $n\widetilde{V}''(n) = O(n^{\beta-1}\Psi(n))$. Since $\mathbf{Var}[X_n] = \mathbf{E}[X_n^2] - \mathbf{E}[X_n]^2$, the result follows. ■

Example 10.10: *I.I.D. Random Variables* Let $Z_1, \ldots, Z_n$ be a sequence of independently and identically distributed random variables with generating function $P(u) = \mathbf{E}u^{Z_1}$ and mean μ and variance v. The generating function of $X_n = Z_1 + \ldots + Z_n$ is $G_n(u) = (P(u))^n$. Observe that $\mathbf{E}[X_n] = n\mu$ and $\mathbf{Var}[X_n] = nv$. If we consider the Poisson transform of X_n we obtain $\widetilde{G}(z, u) = \exp(z(P(u) - 1))$, and $\widetilde{X}(z) = z\mu$, $\widetilde{V}(z) = (\mu^2 + v)z$. Thus $\mathbf{Var}[X_n] = \widetilde{V}(n) - n[\widetilde{X}'(n)]^2$, as predicted by Theorem 10.14. ■

10.4.2 Limiting Distributions

When dealing with distributions, one must estimate $G_n(u)$ from the Poisson transform $\widetilde{G}(z, u)$. If u belongs to a compact set, then our previous depoissonization results can be directly applied. However, to unify our analysis we present below a general result called the *diagonal depoissonization*.

Theorem 10.15 **(Jacquet and Szpankowski, 1998)** *Let $\widetilde{G}_k(z)$ be a sequence of Poisson transforms of $g_{n,k}$, which are assumed to be a sequence of entire functions of z. Let the following two conditions hold for some $A > 0$, B, $R > 0$ and $\alpha < 1, 0, \beta,$ and γ:*

(I) *For $z \in S_\theta$ and*

$$|z| > R \quad \Rightarrow \quad |\widetilde{G}_n(z)| \leq Bn^\beta |\Psi(n)|, \tag{10.55}$$

  where $\Psi(x)$ is a slowly varying function.

(O) *For z outside the linear cone S_θ*

$$|z| = n \quad \Rightarrow \quad |\widetilde{G}_n(z)e^z| \leq An^\gamma \exp(\alpha n), \tag{10.56}$$

Then for large n

$$g_{n,n} = \widetilde{G}_n(n) + O(n^{\beta-1}\Psi(n)). \tag{10.57}$$

More generally, for every nonnegative integer m

$$g_{n,n} = \sum_{i=0}^{m}\sum_{j=0}^{i+m} b_{ij} n^i \widetilde{G}_n^{(j)}(n) + O(n^{\beta-m-1}\Psi(n)), \tag{10.58}$$

where $\widetilde{G}_n^{(j)}(n)$ denotes the jth derivative of $\widetilde{G}_n(z)$ at $z = n$.

Proof. The proof follows directly from the depoissonization tool Theorem 10.13 and is left for the reader (cf. Exercise 8). ∎

Using this we can prove the following two very useful corollaries.

Corollary 10.16 *Suppose $\widetilde{G}_k(z) = \sum_{n=0}^{\infty} g_{n,k}\frac{z^n}{n!}e^{-z}$, for k belonging to some set $\mathcal{K}$, are entire functions of z. For some constants A, B, R > 0, β and α < 1, let the following two conditions hold uniformly in $k \in \mathcal{K}$:*

(I) *For $z \in S_\theta$*

$$|z| > R \quad \Rightarrow \quad |\widetilde{G}_k(z)| \le B|z|^\beta, \tag{10.59}$$

(O) *For $z \notin S_\theta$*

$$|z| > R \quad \Rightarrow \quad |\widetilde{G}_k(z)e^z| \le A \exp(\alpha|z|). \tag{10.60}$$

Then uniformly in $k \in \mathcal{K}$

$$g_{n,k} = \widetilde{G}_k(n) + O(n^{\beta-1}) \tag{10.61}$$

and the error estimate does not depend on $\mathcal{K}$.

Proof. The above corollary is a direct consequence of our previous proofs. Nevertheless, we show below how it can be derived from Theorem 10.13. Indeed, let us assume the contrary (i.e., that the thesis of the corollary (10.61) does not hold). In other words, there is a subsequence (n_i, k_i) such that

$$\lim_{i\to\infty} \left| \frac{g_{n_i,k_i} - \widetilde{G}_{k_i}(n_i)}{n_i^{\beta-1}} \right| = \infty. \tag{10.62}$$

Observe that n_i cannot be bounded. Indeed, if the subsequence n_i would be bounded, then the uniform boundedness of $G_k(z)$ (by our assumption (I)) on any compact set implies that $g_{n,k}$ are uniformly bounded if n_i are uniformly bounded (it suffices to bound the integrand in $g_{n,k} = \frac{n!}{2i\pi} \oint G_k(z)e^z z^{-n-1} dz$), which contradicts (10.62). So assume now that n_i is unbounded and strictly increasing, and define for a nonnegative integer m

$$\tilde{H}_m(z) = \begin{cases} 0 & m \neq n_i \\ \tilde{G}_{k_i}(z) & m = n_i. \end{cases}$$

Then $h_{n_i,n_i} = g_{n_i,k_i}$ for all i. Clearly, $\tilde{H}_m(z)$ satisfies the assumptions of Theorem 10.13 (since it satisfies conditions (I) and (O) of the corollary, which imply conditions (I), (O), and (L) of Theorem 10.13). Thus,

$$h_{n_i,n_i} = g_{n_i,k_i} = \tilde{G}_{k_i}(n_i) + O(n_i^{\beta-1})$$

which is the desired contradiction.   ∎

The next corollary is a direct consequence of Corollary 10.16 but since it finds many useful applications, we formulate it separately.

Corollary 10.17 *Let $\tilde{G}(z, u)$ satisfy the hypothesis of Theorem 10.3, i.e., for some numbers $\theta < \pi/2$, A, B, $\xi > 0$, β, and $\alpha < 1$ (I) and (O) hold for all u in a set $\mathcal{U}$. Then*

$$G_n(u) = \tilde{G}(n, u) + O(n^{\beta-1}) \tag{10.63}$$

uniformly for $u \in \mathcal{U}$.

Proof. This directly follows from Corollary 10.16. Since the set $\mathcal{K}$ in Corollary 10.16 is arbitrary we can set $\mathcal{K} = \mathcal{U}$, and then $g_{n,u} = \sum_{l \geq 0} g_{n,l} u^l = G_n(u)$ and $\tilde{G}_u(z) = \tilde{G}(z, u)$. Thus (10.61) implies (10.63).   ∎

Example 10.11: *Depth in PATRICIA*   Let us consider the depth in a PATRICIA trie (Section 1.1). The Poisson transform $\tilde{D}(z, u)$ of the depth satisfies the following functional equation

$$\tilde{D}(z, u) = u(p\tilde{D}(zp, u) + q\tilde{D}(zq, u))$$
$$+ (1 - u)(p\tilde{D}(zp, u)e^{-qz} + q\tilde{D}(zq, u)e^{-pz}).$$

Due to the factors e^{-qz} and e^{-pz} in front of $\tilde{D}(z, u)$, the equation is more complicated than the ones we have seen so far. In fact, it does not have a simple explicit solution. Nevertheless, it falls under Theorem 10.5 with $\gamma_1(z) = up + (1 - u)pe^{-qz}$, $\gamma_2(z) = uq + (1 - u)qe^{-pz}$ and $t(z) = 0$. Furthermore, for $p \neq q$

the asymptotic behavior of $\widetilde{D}(z, u)$ inside a cone $\mathcal{S}_\theta$ ($z \to \infty$) is determined by the first two terms in the right side of the above equation. Using Mellin transforms Rais et al. [350] proved that for large z the Poisson depth is normal (the reader is asked to provide details in Exercise 9); that is, for z large in a cone $\mathcal{S}_\theta$, and $u = e^t$, where t is complex we have:

$$e^{-t\widetilde{X}(z)/\sigma(z)}\, \widetilde{D}(z, e^{t/\sigma(z)}) = e^{t^2/2}(1 + O(1/\sigma(z))), \qquad (10.64)$$

where $\widetilde{X}(z) = O(\log z)$ is the Poisson mean, and $\sigma^2(z) = \widetilde{V}(z) = O(\log z)$ is the Poisson variance.

One must now depoissonize (10.64) in order to establish the central limit law for the Bernoulli model. First, an easy application of Theorem 10.14 leads us to $\mathbf{E}[D_n] \sim \widetilde{X}(n)$ and $\mathbf{Var}[D_n] \sim \widetilde{V}(n)$. In fact, from Szpankowski [408] we know that $\mathbf{E}[D_n] = \frac{1}{h}\log n + O(1)$ and $\mathbf{Var}[D_n] = \frac{h_2 - h^2}{h^3}\log n + O(1)$, where $h = -p \log p - q \log q$ is the entropy, and $h_2 = p \log^2 p + q \log^2 q$. Thus we need only depoissonize (10.64) in order to obtain the limiting distribution. We first observe, as in [350], that $\widetilde{D}(z, u) = O(z^\varepsilon)$ for any $\varepsilon > 0$ inside a cone $\mathcal{S}_\theta$ ($\theta < \pi/2$). This can be proved by simple induction. With the help of Theorem 10.5 we verify conditions (I) and (O) of Corollary 10.17, which finally yields the following.

Theorem 10.18   (Rais, Jacquet and Szpankowski, 1993)   *For complex t*

$$e^{-t\mathbf{E}[D_n]/\sqrt{\mathbf{Var}[D_n]}}\mathbf{E}\left[e^{tD_n/\sqrt{\mathbf{Var}[D_n]}}\right] = e^{t^2/2}(1 + O(1/\sqrt{\log n})),$$

that is, $(D_n - \mathbf{E}[D_n])/\sqrt{\mathbf{Var}[D_n]}$ converges in distribution and in moments to the standard normal distribution.

Other examples will be discussed in the applications section below.   ∎

10.5   APPLICATIONS

We discuss three applications. First, we deal with the asymptotic distribution of the height in an incomplete trie associated with the leader election algorithm. Then we tackle the average and the variance of the depth in generalized b-digital search trees for two probabilistic models, namely, for biased memoryless and Markovian sources. The depth of b-DST was already discussed in Section 9.4.1 for the *unbiased* memoryless model. Extension to the biased memoryless model will require a new methodology, that of poissonization and depoissonization. Finally, further extension to the Markovian model will force us to consider systems of functional equations. This last application turns out to be the most sophisticated research problem of this book.

10.5.1   Leader Election Algorithm

This analysis is adopted from Fill *at al.* [124]. The following elimination process
has several applications, such as the election of a leader in a computer network.
A group of n people play a game to identify a winner by tossing fair coins. All
players who throw heads are losers; those who throw tails remain candidates and
flip their coins again. The process is repeated until a single winner is identified. If
at any stage all remaining candidates throw heads, the tosses are deemed incon-
clusive and all remaining players participate in the next round of coin tossing.

We investigate the distribution of the height of a random incomplete trie, the
discrete structure that underlies the elimination process described above. This binary *incomplete* tree can be described as follows. At the root of the tree we have
one node labeled with all participants. After all the participants flip their coins for
the first time, losers (if any) are placed in a *leaf* node that is attached to the root
as a right child, and all candidate winners are placed in a node that is attached as
a left child. Leaf nodes are terminal nodes that are not developed any further. The
process repeats recursively on every left child until a single winner is identified.
Figure 10.4 illustrates the discrete structure underlying the elimination process.

Our goal is to find the distribution of the height H_n of the associated incomplete
trie. We first use poissonization and then depoissonize the result to obtain the exact
and the limiting distribution of H_n. For each $n \geq 0$, let

$$G_n(u) := \sum_{k=0}^{\infty} \Pr\{H_n = k\} u^k$$

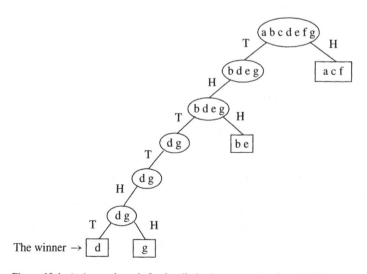

Figure 10.4. An incomplete trie for the elimination process starting with 7 players.

denote the probability generating function for the fixed-population height H_n (i.e., in the Bernoulli model). The starting point of our analysis is the following recurrence for $G_n(u)$

$$G_n(u) = u2^{-n} \sum_{k=1}^{n} \binom{n}{k} G_k(u) + uG_n(u)2^{-n}, \qquad (10.65)$$

which is valid for $n \geq 2$, with the boundary values $G_0(u) = G_1(u) = 1$. This is a standard equation for the depth in digital trees that we have seen before, except that the associated trie is incomplete so the term $G_{n-k}(u)$ does not appear. The additive term $G_n(u)2^{-n}$ accounts for the fact that when all players throw heads, then all of them still participate in the game (we must find a leader!).

We now consider the Poisson transform $\widetilde{G}(z, u) = \sum_{n=0}^{\infty} e^{-z} G_n(u) \frac{z^n}{n!}$ of $G_n(u)$, which satisfies the following functional equation

$$\widetilde{G}(z, u) = u(1 + e^{-z/2})\widetilde{G}\left(\frac{z}{2}, u\right) + e^{-z}\left[(1 + z)(1 - u) - ue^{z/2}\right]. \quad (10.66)$$

To handle this recurrence, we introduce

$$\widetilde{H}(z, u) = \frac{\widetilde{G}(z, u)}{1 - e^{-z}},$$

and obtain

$$\widetilde{H}(z, u) = u\widetilde{H}\left(\frac{z}{2}, u\right) + R(z, u), \qquad (10.67)$$

where

$$R(z, u) := \frac{(1 + z)(1 - u) - ue^{z/2}}{e^z - 1}.$$

The recurrence (10.67) can be solved by direct iteration, as we saw in Section 7.6.1, to obtain

$$\widetilde{H}(z, u) = \sum_{k=0}^{\infty} R\left(\frac{z}{2^k}, u\right) u^k. \qquad (10.68)$$

We now start the process of depoissonizing the last equation through Corollary 10.16. For this we need information about $\widetilde{G}(z, u)$ as $z \to \infty$ in a linear cone and u in a compact set. Certainly, the Mellin transform is the right tool since the sum in (10.68) is a harmonic sum and Theorem 9.1 applies. Simple algebra (with the help of Table 10.1) shows that the Mellin transform $\widetilde{H}^*(s, u)$ of $\widetilde{H}(z, u)$ is

$$\widetilde{H}^*(s, u) = \Gamma(s)\,\zeta(s) + \frac{(1 - u)\,\Gamma(s + 1)\,\zeta(s + 1)}{1 - 2^s u}$$

for $1 < \Re(s) < -\log|u|$. We further assume that $|u| < \frac{1}{2}$. Since we are concerned here with the distribution function of H_N, where N is a Poisson random variable, rather than its probability mass function, our interest centers on the transform

$$\frac{\widetilde{H}^*(s, u)}{1 - u} = \frac{\Gamma(s)\,\zeta(s)}{1 - u} + \frac{\Gamma(s + 1)\,\zeta(s + 1)}{1 - 2^s u}.$$

This expression can be immediately expanded in a power series in u yielding

$$[u^k]\left(\frac{\widetilde{H}^*(s, u)}{1 - u}\right) = \Gamma(s)\,\zeta(s) + 2^{ks}\Gamma(s + 1)\,\zeta(s + 1).$$

Inverting the transform gives

$$\frac{\widetilde{H}(z, u)}{1 - u} = \sum_{k=0}^{\infty} u^k \left[\frac{1}{e^z - 1} + \frac{z/2^k}{\exp(z/2^k) - 1}\right]$$

for $z > 0$. But

$$\sum_{k=0}^{\infty} \Pr\{H_{N(z)} \le k\} u^k = \frac{\widetilde{G}(z, u)}{1 - u} = \frac{(1 - e^{-z})\,\widetilde{H}(z, u)}{1 - u},$$

where $N(z)$ is a Poisson distributed random variable with mean z. For each $k = 0, 1, 2, \ldots$ and $z > 0$ we have

$$\Pr\{H_{N(z)} \le k\} = e^{-z} + (1 - e^{-z})\frac{z/2^k}{\exp(z/2^k) - 1}. \tag{10.69}$$

The relation (10.69) can be manipulated to get exact distribution. For any $z > 0$,

$$\Pr\{H_{N(z)} \le k\} = e^{-z}\left[1 + \frac{z}{2^k}\frac{e^z - 1}{\exp(z/2^k) - 1}\right]$$

$$= e^{-z}\left[1 + \frac{z}{2^k}\sum_{j=0}^{2^k - 1} \exp(jz/2^k)\right]$$

$$= e^{-z} \left[1 + \frac{1}{2^k} \sum_{j=0}^{2^k-1} \sum_{n=1}^{\infty} \frac{(j/2^k)^{n-1} z^n}{(n-1)!} \right]$$

$$= e^{-z} \left[1 + \sum_{n=1}^{\infty} \frac{z^n}{n!} \left(\frac{n}{2^{kn}} \sum_{j=0}^{2^k-1} j^{n-1} \right) \right].$$

We give the exact distribution in Theorem 10.19 below.

But we are really after the limiting distribution, if it exists, of the Bernoulli version of the height, that is, H_n. We use the depoissonization Corollary 10.16. Let

$$p_{k,n} := \Pr\{H_n \le k\} \tag{10.70}$$

and then by (10.69) the Poisson transform $\widetilde{P}_k(z)$ of $p_{k,n}$ is

$$\widetilde{P}_k(z) := e^{-z} + \left(1 - e^{-z} \right) \frac{z/2^k}{\exp(z/2^k) - 1}. \tag{10.71}$$

Although up until now the right side of (10.71) has arisen only as $\Pr\{H_{N(z)} \le k\}$ for real $z \ge 0$, note that it defines an entire function of the complex variable z (by analytic continuation or the extension of the Mellin transform to the complex plane as discussed in Section 9.3).

We must verify conditions (I) and (O) of Corollary 10.16. We may appeal to Theorem 10.5; however, for a change we check the conditions directly. To verify condition (I), we first observe that if $0 \ne z \in S_\theta$, then $|z| \le \Re z / \cos\theta$ and

$$|e^z - 1| \ge |e^z| - 1 = \exp(\Re z) - 1,$$

so that

$$\left| \frac{z}{e^z - 1} \right| \le (\sec\theta) \frac{\Re z}{\exp(\Re z) - 1} \le \sec\theta.$$

Thus for $0 \ne z \in S_\theta$ we have

$$|\widetilde{P}_k(z)| \le \left| e^{-z} \right| + \left| 1 - e^{-z} \right| \left| \frac{z/2^k}{\exp(z/2^k) - 1} \right|$$

$$\le \left| e^{-z} \right| + (\sec\theta) \left| 1 - e^{-z} \right|$$

$$\le \sec\theta + (1 + \sec\theta) \left| e^{-z} \right|$$

$$\le \sec\theta + (1 + \sec\theta) \exp(-\Re z) \le 1 + 2\sec\theta.$$

To verify condition (O), we first observe that if $z \notin S_\theta$, then $\Re z \leq |z| \cos\theta$. Thus for $0 \neq z \notin S_\theta$, we have from (10.71)

$$|\tilde{P}_k(z)e^z| \leq \left| 1 + \frac{z}{2^k} \times \frac{e^z - 1}{\exp(z/2^k) - 1} \right|$$

$$\leq 1 + |z| \times 2^{-k} \sum_{r=0}^{2^k - 1} \left| \exp(rz/2^k) \right|$$

$$\leq 1 + |z| \times 2^{-k} \sum_{r=0}^{2^k - 1} \exp\left(r2^{-k}|z| \cos\theta \right)$$

$$\leq 1 + |z| \int_0^1 \exp(x|z| \cos\theta) \, dx$$

$$= 1 + (\sec\theta) \left[\exp(|z| \cos\theta) - 1 \right]$$

$$\leq (\sec\theta) \exp(|z| \cos\theta).$$

This verifies both of the conditions in Corollary 10.16 with $\beta = 0$; hence

$$\Pr\{H_n \leq j\} = p_{j,n} = \tilde{P}_j(n) + O\left(n^{-1} \right) \tag{10.72}$$

holds uniformly for integers $j \geq 0$. But from (10.71)

$$\tilde{P}_j(n) = e^{-n} + \left(1 - e^{-n} \right) \frac{n/2^j}{\exp(n/2^j) - 1}.$$

Setting $j = \lfloor \log_2 n \rfloor + k$, we find that

$$\tilde{P}_j(n) = e^{-n} + \left(1 - e^{-n} \right) \frac{2^{\langle \log_2 n \rangle - k}}{\exp\left(2^{\langle \log_2 n \rangle - k} \right) - 1}$$

$$= \frac{2^{\langle \log_2 n \rangle - k}}{\exp\left(2^{\langle \log_2 n \rangle - k} \right) - 1} + O\left(e^{-n} \right) \tag{10.73}$$

holds, uniformly in all integers k ($\geq -\lfloor \log_2 n \rfloor$). We recall that $\langle \log_2 n \rangle = \log_2 n - \lfloor \log_2 n \rfloor$.

We have thus proved the following result.

Theorem 10.19 **(Fill, Mahmoud and Szpankowski, 1996)** *Consider an incomplete binary trie with $n \geq 1$, and let B_j denote the jth Bernoulli number.*

(i) *For any integer $k \geq 0$,*

$$\Pr\{H_n \leq k\} = \frac{n}{2^{kn}} \sum_{j=0}^{2^k-1} j^{n-1} = \sum_{j=0}^{n-1} \binom{n}{j} \frac{B_j}{2^{kj}}. \tag{10.74}$$

(ii) *Uniformly over all integers k,*

$$\Pr\{H_n \leq \lfloor \log_2 n \rfloor + k\} = \frac{2^{\langle \log_2 n \rangle - k}}{\exp(2^{\langle \log_2 n \rangle - k}) - 1} + O\left(\frac{1}{n}\right)$$

as $n \to \infty$.

From part (ii) of the above theorem we conclude that the limiting distribution of $H_n - \lfloor \log_2 n \rfloor$ does not exist (due to the term $\langle \log_2 n \rangle$). This is illustrated in Figure 10.5.

10.5.2 Depth in a b-Digital Search Tree for the Biased Memoryless Model

We discuss here the same data structures as in Section 9.4.1, that is, b-digital search trees. However, this time we analyze them under the *biased* memoryless model, that is, for $p \neq q = 1 - p$. We concentrate on the depth D_m in a b-DST built over a *fixed* number, say m, of independent binary strings. Let $\bar{B}_m^k$ be the

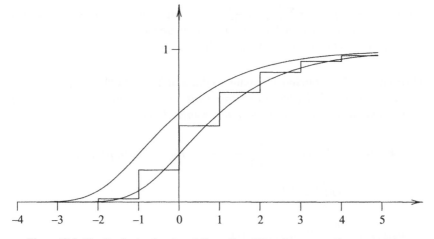

Figure 10.5. The distribution function of $H_{20} - \lfloor \log_2 20 \rfloor$ and the two continuous extremes.

expected number of strings at level k of a randomly built b-digital search tree. Since $\Pr\{D_m = k\} = \bar{B}_m^k / m$ (cf. Section 8.7.3), one can alternatively study the average $\bar{B}_m^k$ which is also called the *average profile*. Let $B_m(u) = \sum_{k \geq 0} \bar{B}_m^k u^k$ be the generating function of the average profile. As in Section 9.4.1, we argue that the average profile satisfies the following recurrence for all $m \geq 0$

$$B_{m+b}(u) = b + u \sum_{i=0}^{m} \binom{m}{i} p^i q^{m-i} \left(B_i(u) + B_{m-i}(u)\right), \tag{10.75}$$

with the initial conditions

$$B_i(u) = i \quad \text{for} \quad i = 0, 1, \ldots, b - 1.$$

For $b > 1$ this recurrence does not have a closed form solution. Therefore, we must apply an approach other than the one used in Section 9.4.1 for the unbiased model. We poissonize the model and study the Poisson transform

$$\tilde{B}(u, z) = \sum_{i=0}^{\infty} B_i(u) \frac{z^i}{i!} e^{-z} \tag{10.76}$$

of $B_m(u)$. We do so since the above recurrence transforms to a slightly more manageable differential-functional equation, namely

$$\left(1 + \frac{\partial}{\partial z}\right)^b \tilde{B}(u, z) = b + u \left(\tilde{B}(u, pz) + \tilde{B}(u, qz)\right),$$

where $(1 + \frac{\partial}{\partial z})^b f(z) := \sum_{i=0}^{b} \binom{b}{i} \frac{\partial^i f(z)}{\partial z^i}$. We shall study $\tilde{B}(u, z)$ for $z \to \infty$ in a cone around the real axis and u in a compact set around $u = 1$.

Our goal is to prove the following result.

Theorem 10.20 (Louchard, Szpankowski, and Tang, 1999) *Let D_m be the typical depth in a b-digital search tree built over m statistically independent strings generated by a biased memoryless source. Then*

$$\mathbf{E}[D_m] = \frac{1}{h} \log m + \frac{1}{h} \left(\frac{h_2}{2h} + \gamma - 1 - H_{b-1} - \Delta(b, p) + \delta_1(m, b)\right)$$

$$+ O\left(\frac{\log m}{m}\right)$$

$$\mathbf{Var}[D_m] = \frac{h_2 - h^2}{h^3} \log m + O(1),$$

where $h = -p \log p - q \log q$ is the entropy of the Bernoulli(p) distribution,
$h_2 = p \log^2 p + q \log^2 q$, $\gamma = 0.577\ldots$ is the Euler constant, while $H_{b-1} = \sum_{i=1}^{b-1} \frac{1}{i}$, $H_0 = 0$ are harmonic numbers. The constant $\Delta(b, p)$ can be computed as follows

$$\Delta(b, p) = \sum_{n=2b+1}^{\infty} \bar{f}_n \sum_{i=1}^{b} \frac{(i+1)b!}{(b-i)!n(n-1)\cdots(n-i-1)} < \infty,$$

where $\bar{f}_n$ is given recursively by

$$\begin{cases} f_{m+b} = m + \sum_{i=0}^{m} \binom{m}{i} p^i q^{m-i} (f_i + f_{m-i}), & m > 0, \\ f_0 = f_1 = \ldots = f_b = 0, \\ \bar{f}_{m+b} = f_{m+b} - m > 0, & m \geq 1. \end{cases} \tag{10.77}$$

Finally, $\delta_1(x, b)$ is a fluctuating function with a small amplitude (cf. (10.89)) when $(\log p)/(\log q)$ is rational, and $\lim_{x \to \infty} \delta_1(x, b) = 0$ otherwise.

The rest of this subsection is devoted to proving Theorem 10.20. We apply the Mellin transform and depoissonization methods. Observe that $B_m(1) = m$, $E[D_m] = B'_m(1)/m$, and $B''_m(1)/m = E[D_m(D_m - 1)] = \text{Var}[D_m] - E[D_m] + E[D_m]^2$. Thus

$$\text{Var}[D_m] = \frac{B''_m(1)}{m} + \frac{B'_m(1)}{m} - \left(\frac{B'_m(1)}{m}\right)^2.$$

Let $\tilde{X}(z) := \tilde{B}_u(1, z)$ and $\tilde{W}(z) := \tilde{B}_{uu}(1, z)$ be, respectively, the mean and the second factorial moment of the depth in the Poisson model. Using (10.76), we obtain

$$\left(1 + \frac{\partial}{\partial z}\right)^b \tilde{X}(z) = z + \tilde{X}(pz) + \tilde{X}(qz), \tag{10.78}$$

$$\left(1 + \frac{\partial}{\partial z}\right)^b \tilde{W}(z) = 2\left(\tilde{X}(pz) + \tilde{X}(qz)\right) + \left(\tilde{W}(pz) + \tilde{W}(qz)\right). \tag{10.79}$$

We plan to apply the Mellin transform to find the asymptotics of $\tilde{X}(z)$ and $\tilde{W}(z)$ as $z \to \infty$ (first as $x = \Re(z) \to \infty$ and then we analytically continue to the complex plane). Due to the appearance of the derivative in (10.78) and (10.79), we rather do not directly find the Mellin transforms $X(s)$ and $Y(s)$ of $\tilde{X}(z)$ and $\tilde{W}(z)$, but shall seek $X(s)$ and $Y(s)$ in the following forms

$$X(s) = \mathcal{M}[\tilde{X}(z); s] = \Gamma(s)\gamma(s), \tag{10.80}$$

$$Y(s) = \mathcal{M}[\tilde{W}(z); s] = \Gamma(s)\beta(s), \tag{10.81}$$

where $\gamma(s)$ and $\beta(s)$ are to be computed. They exist in the proper strips, as proved below.

Lemma 10.21 *The following is true:*

(i) $X(s)$ *exists for* $\Re(s) \in (-b-1, -1)$, *and* $Y(s)$ *is defined for* $\Re(s) \in (-2b-1, -1)$.

(ii) *Furthermore,* $\gamma(-1-i) = 0$ *for* $i = 1, \ldots, b-1$, $\gamma(-1-b) = (-1)^{b+1}$, *and* $\beta(-1-i) = 0$ *for* $i = 1, \ldots, b$, *and* $\gamma(s)$ *has simple poles at* $s = -1, 0, 1, \ldots$.

Proof. By recurrence (10.75), we have $B_i(u) = i$ for $i = 0, 1, \ldots, b$, and thus $B_i(u) = b + (i-b)u$ for $i = 1+b, \ldots, 2b$. Taking derivatives, we obtain $\frac{\partial B_i(u)}{\partial u} = 0$ for $i = 0, 1, \ldots, b$ and $\frac{\partial B_i(u)}{\partial u} = i - b$ for $i = b, 1+b, \ldots, 2b$. Furthermore, the second derivative becomes $\frac{\partial^2 B_i(u)}{\partial u^2} = 0$ for $i = 0, 1, \ldots, 2b$. Hence, for $z \to 0$

$$\widetilde{X}(z) = \left(z^{b+1}/(b+1)! + 2z^{b+2}/(b+2)! + 3z^{b+3}/(b+3)! + O(z^{b+4}) \right) e^{-z},$$

$$= z^{b+1}/(b+1)! + O(z^{b+2}) \ as \ z \to 0$$

$$\widetilde{W}(z) = O(z^{2b+1}) \ as \ z \to 0.$$

On the other hand, for $z \to \infty$ we conclude from (10.78) and (10.79) that $\widetilde{X}(z) = O(z \log z)$ and $\widetilde{W}(z) = O(z \log^2 z)$. Thus the first part of the lemma is proven. Part (ii) directly follows from Lemma 10.22 below and (10.86). ∎

Lemma 10.22 *Let* $\{f_n\}_{n=0}^{\infty}$ *be a sequence of real numbers having the Poisson transform* $\widetilde{F}(z) = \sum_{n=0}^{\infty} f_n \frac{z^n}{n!} e^{-z}$, *which is an entire function. Furthermore, let its Mellin transform* $F(s)$ *have the following factorization*

$$F(s) = \mathcal{M}[\widetilde{F}(z); s] = \Gamma(s)\gamma(s).$$

Assume that $F(s)$ *exists for* $\Re(s) \in (-2, -1)$, *and that* $\gamma(s)$ *is analytic for* $\Re(s) \in (-\infty, -1)$. *Then*

$$\gamma(-n) = \sum_{k=0}^{n} \binom{n}{k}(-1)^k f_k, \quad for \ n \geq 2. \tag{10.82}$$

Proof. Let sequence $\{g_n\}_{n=0}^{\infty}$ be such that $\widetilde{F}(z) = \sum_{n=0}^{\infty} g_n \frac{z^n}{n!}$, that is,

$$g_n = \sum_{k=0}^{n} \binom{n}{k} (-1)^{n-k} f_k, \quad n \geq 0.$$

Define for some fixed $M \geq 2$, the function $\widetilde{F}_M(z) = \sum_{n=0}^{M-1} g_n \frac{z^n}{n!}$. Due to our assumptions, we can continue $F(s)$ analytically to the whole complex plane except $s = -2, -3, \ldots$. In particular, for $\Re(s) \in (-M, -M+1)$ we have

$$F(s) = \mathcal{M}[\widetilde{F}(z) - \widetilde{F}_M(z); s].$$

(The above is true since a polynomial in z, such as $\widetilde{F}_M(z)$, can only shift the fundamental strip of the Mellin transform but cannot change its value, as we saw in Section 9.2.) As $s \to -M$, due to the factorization $F(s) = \Gamma(s)\gamma(s)$, we have

$$F(s) = \frac{1}{s+M} \frac{(-1)^M}{M!} \gamma(-M) + O(1);$$

thus by the inverse Mellin transform, we obtain

$$\widetilde{F}(z) - \widetilde{F}_M(z) = \frac{(-1)^M}{M!} \gamma(-M) z^M + O(z^{M+1}) \quad \text{as} \quad z \to 0. \qquad (10.83)$$

But

$$\widetilde{F}(z) - \widetilde{F}_M(z) = \sum_{i=M}^{\infty} g_n \frac{z^n}{n!} = g_M \frac{z^M}{M!} + O(z^{M+1}). \qquad (10.84)$$

Comparing (10.83) and (10.84) shows that

$$\gamma(-M) = (-1)^M g_M = \sum_{k=0}^{M} \binom{M}{k} (-1)^k f_k \quad \text{for } M \geq 2$$

which completes the proof ∎

Now we are in a position to compute the Mellin transforms of $\widetilde{X}(z)$ and $\widetilde{W}(z)$. From (10.78)–(10.81) we obtain

$$\sum_{i=0}^{b} \binom{b}{i} (-1)^i \gamma(s-i) = (p^{-s} + q^{-s})\gamma(s),$$

$$\sum_{i=0}^{b} \binom{b}{i} (-1)^i \beta(s-i) = 2(p^{-s} + q^{-s})\gamma(s) + (p^{-s} + q^{-s})\beta(s),$$

and, by Lemma 10.21, $\gamma(s)$ exists at least for $\Re(s) \in (-b-1, -1)$, while $\beta(s)$ is well defined in the strip $\Re(s) \in (-2b-1, -1)$. To simplify the above, we define, for any function $g(s)$,

$$\widehat{g}(s) = \sum_{i=1}^{b} \binom{b}{i} (-1)^{i+1} g(s-i), \qquad (10.85)$$

provided $g(s-1), \ldots, g(s-b)$ exist. Then

$$\gamma(s) = \frac{1}{1-p^{-s}-q^{-s}} \sum_{i=1}^{b} \binom{b}{i} (-1)^{i+1} \gamma(s-i)$$

$$= \frac{1}{1-p^{-s}-q^{-s}} \widehat{\gamma}(s) \qquad (10.86)$$

$$\beta(s) = \frac{1}{1-p^{-s}-q^{-s}} \widehat{\beta}(s) + \frac{2(p^{-s}+q^{-s})}{(1-p^{-s}-q^{-s})^2} \widehat{\gamma}(s). \qquad (10.87)$$

Let s_k, $k = 0, \pm 1, \pm 2, \ldots$, be roots of $1 - p^{-s} - q^{-s} = 0$. Observe that $s_0 = -1$. We need to know precisely the locations of the roots s_k. Fortunately, we already studied this problem in Lemma 8.21. At $s = s_k$ we have

$$\frac{1}{1-p^{-s}-q^{-s}} = -\frac{1}{h(s_k)} \frac{1}{s-s_k} + \frac{h_2(s_k)}{2h^2(s_k)} + O(s-s_k), \qquad (10.88)$$

where

$$h(t) = -p^{-t} \log p - q^{-t} \log q,$$

$$h_2(t) = p^{-t} \log^2 p + q^{-t} \log^2 q.$$

Expanding $\Gamma(s)\widehat{\gamma}(s)$ around $s = s_k$, we find that

$$\Gamma(s)\widehat{\gamma}(s) = \Gamma(s_k)\widehat{\gamma}(s_k) + \left(\Gamma(s_k)\widehat{\gamma}'(s_k) + \Gamma'(s_k)\widehat{\gamma}(s_k)\right)(s-s_k) + O((s-s_k)^2).$$

Therefore, since $X(s) = \Gamma(s)\gamma(s) = \frac{1}{1-p^{-s}-q^{-s}}\Gamma(s)\widehat{\gamma}(s)$, we obtain

$$X(s) = -\frac{1}{s-s_k}\frac{\Gamma(s_k)}{h(s_k)}\widehat{\gamma}(s_k) + \frac{h_2(s_k)}{2h^2(s_k)}\Gamma(s_k)\widehat{\gamma}(s_k)$$

$$- \frac{1}{h(s_k)}\left(\Gamma(s_k)\widehat{\gamma}'(s_k) + \Gamma'(s_k)\widehat{\gamma}(s_k)\right) + O(s-s_k).$$

In a similar manner, from (10.87) we have

$$Y(s) = \frac{2\Gamma(s_k)\widehat{\gamma}(s_k)}{h^2(s_k)}\frac{1}{(s-s_k)^2} + \left(\frac{2\Gamma'(s_k)\widehat{\gamma}(s_k)}{h^2(s_k)} - \frac{\Gamma(s_k)}{h(s_k)}\widehat{\beta}(s_k)\right)$$

$$- 2\Gamma(s_k)\frac{h_2(s_k) - h^2(s_k)}{h^3(s_k)}\widehat{\gamma}(s_k) - \frac{2\Gamma(s_k)\widehat{\gamma}'(s_k)}{h^2(s_k)}\right)\frac{1}{s-s_k} + O(1).$$

On the other hand, from (10.88) and (10.86) at $s = s_0 = -1$, we find that

$$\gamma(s) = -\frac{1}{h}\frac{1}{s+1} + \frac{h_2}{2h^2} - \frac{\widehat{\gamma}'(-1)}{h} + O(s+1),$$

$$\beta(s) = \frac{2}{h^2}\frac{1}{(s+1)^2} + \left(-2\frac{h_2 - h^2}{h^3} + 2\widehat{\gamma}'(-1)\frac{1}{h^2}\right)\frac{1}{s+1} + O(1).$$

In the above, we used the fact that $\widehat{\gamma}(-1) = 1$ and $\widehat{\beta}(-1) = 0$, which directly follows from Lemma 10.21. But $\Gamma(s) = -\frac{1}{s+1} + (\gamma - 1) + O(s+1)$; hence the Laurent expansion of $X(s)$ at $s = -1$ is

$$X(s) = \Gamma(s)\gamma(s) = \frac{1}{h}\frac{1}{(s+1)^2} - \left(\frac{h_2}{2h^2} - \frac{1}{h}\widehat{\gamma}'(-1) + \frac{\gamma - 1}{h}\right)\frac{1}{s+1} + O(1).$$

We now use the reverse mapping Theorem 9.2 to find asymptotically $\widetilde{X}(z)$ and $\widetilde{W}(z)$ for $z \to \infty$ in a cone $\mathcal{S}_\theta$ (we again use either analytic continuation or the extension to the complex plane of the Mellin transform as discussed in Section 9.3). However, to estimate the error term we note that $\gamma(s)$ has additional simple poles at $s = 0, 1, \ldots$. The pole at $s = 0$ is a double pole of $X(s) = \Gamma(s)\gamma(s)$, and thus its contribution to $\widetilde{X}(z)$ is $O(\log z)$. Putting everything together, we finally arrive at

$$\widetilde{X}(z) = \frac{1}{h}z\log z + \left(\frac{h_2}{2h^2} - \frac{1}{h}\widehat{\gamma}'(-1) + \frac{\gamma - 1}{h}\right)z + \sum_{k \neq 0}\frac{\Gamma(s_k)\widehat{\gamma}(s_k)}{h(s_k)}z^{-s_k}$$

$$+ O(\log z). \tag{10.89}$$

Similarly, at $s = -1$,

$$Y(s) = -\frac{2}{h^2}\frac{1}{(s+1)^3} + \frac{2}{h}\left(\frac{h_2 - h^2}{h^2} - \frac{1}{h}\widehat{\gamma}'(-1) + \frac{\gamma - 1}{h}\right)\frac{1}{(s+1)^2}$$

$$+ O\left(\frac{1}{s+1}\right).$$

In addition, there is a double pole at $s = 0$, and hence by the reverse mapping Theorem 9.2 and Lemma 8.21 we obtain

$$
\widetilde{W}(z) = \frac{1}{h^2} z \log^2 z + \frac{2}{h} \left(\frac{h_2 - h^2}{h^2} - \frac{1}{h} \widehat{\gamma}'(-1) + \frac{\gamma - 1}{h} \right) z \log z
$$

$$
+ 2 \sum_{k \neq 0} \frac{\Gamma(s_k) \widehat{\gamma}(s_k)}{h^2(s_k)} z^{-s_k} \log z + O(z)
$$

for $z \to \infty$, where $O(z)$ error term comes from the term $O((s+1)^{-1})$. This formula will allow us to infer the asymptotics of the variance of D_m.

The final step is depoissonization. We shall use Theorem 10.3. To verify conditions (I) and (O) we need a slight extension of Theorem 10.5. Condition (I) follows from (10.89), thus we need only verify (O). We consider only $\widetilde{X}(z)$. Let $X(z) = \widetilde{X}(z)e^z$. Then the functional equation (10.78) transforms into

$$
X^{(b)}(z) = X(zp)e^{zq} + X(zq)e^{zp} + ze^z,
$$

where $X^{(b)}(z)$ denotes the bth derivative of $X(z)$. Observe that the above equation can be integrated to yield

$$
X(z) = \underbrace{\int_0^z \int_0^{w_1} \cdots \int_0^{w_{b-1}}}_{b \text{ times}} \left(X(w_1 p)e^{w_1 q} \right.
$$

$$
\left. + X(w_1 q)e^{w_1 p} + w_1 e^{w_1} \right) dw_b \cdots dw_2 dw_1, \qquad (10.90)
$$

where the integration is along lines in the complex plane.

We now prove that $|X(z)| \leq e^{\alpha |z|}$ for $z \notin S_\theta$ for $\alpha < 1$. We use induction over *increasing domains* as discussed in Theorem 10.5. That is, for all positive integers $m \geq 1$ and constants $\xi, \delta > 0$, let

$$
\mathcal{D}_m = \{ z = \rho e^{i\vartheta} : \rho \in [\xi \delta, \xi v^{-m}], \; 0 \leq \vartheta < 2\pi \},
$$

where $\max\{p, q\} \leq v < 1$ and $\delta \leq \min\{p, q\}$. As we have already seen, such a representation enjoys the property that if $z \in \mathcal{D}_{m+1} - \mathcal{D}_m$ then $zp, zq \in \mathcal{D}_m$ provided that $|z| \geq \xi$, which is assumed to hold. To carry out the induction, we first define $\bar{\mathcal{D}}_m = \mathcal{D}_m \cap \bar{S}_\theta$, where $\bar{S}_\theta$ denotes points in the complex plane outside S_θ. Since $X(z)$ is bounded for $z \in \bar{\mathcal{D}}_1$, the initial step of induction holds. Let us now assume that for some $m > 1$ and for $z \in \bar{\mathcal{D}}_m$ we have $|X(z)| \leq e^{\alpha |z|}$ with $\alpha < 1$. We intend to prove that $|X(z)| \leq e^{\alpha |z|}$ for $z \in \bar{\mathcal{D}}_{m+1}$. Indeed, let $z \in \bar{\mathcal{D}}_{m+1}$. If also $z \in \bar{\mathcal{D}}_m$, then the proof is complete. Let us assume that $z \in \bar{\mathcal{D}}_{m+1} - \bar{\mathcal{D}}_m$. Since then $zp, zq \in \bar{\mathcal{D}}_m$, we can use our induction hypothesis together with the integral equation (10.90) to obtain the following estimate for $|z| > \xi$, where ξ is sufficiently large:

$$|X(z)| \leq |z|^{b+1} \left(e^{|z|(pa+q\cos\theta)} + e^{|z|(qa+p\cos\theta)} + e^{|z|\cos\theta} \right).$$

Let us now define $1 > \alpha > \cos\theta$ such that the following three inequalities simultaneously hold:

$$|z|^b e^{|z|(pa+q\cos\theta)} \leq \frac{1}{3} e^{\alpha|z|},$$

$$|z|^b e^{|z|(qa+p\cos\theta)} \leq \frac{1}{3} e^{\alpha|z|},$$

$$|z|^{b+1} e^{|z|\cos\theta} \leq \frac{1}{3} e^{\alpha|z|}.$$

Then for $z \in \bar{D}_{m+1}$ we have $|X(z)| \leq e^{\alpha|z|}$, as needed to verify condition (O) of the depoissonization lemma.

In view of the above, we can apply the depoissonization Theorem 10.3 to $\tilde{X}(z)$ and $\tilde{W}(z)$ and obtain

$$\mathbf{E}[D_m] = \frac{\tilde{X}(m)}{m} + O\left(\frac{\log m}{m}\right) = \frac{1}{h}\log m + \frac{h_2}{2h^2} - \frac{1}{h}\hat{\gamma}'(-1) + \frac{\gamma-1}{h}$$

$$+ \sum_{k\neq 0} \frac{\Gamma(s_k)\hat{\gamma}(s_k)}{h(s_k)} m^{-1-s_k} + O\left(\frac{\log m}{m}\right),$$

and

$$\mathbf{E}[D_m(D_m - 1)] = \frac{\tilde{W}(m)}{m} + O(1) = \frac{1}{h^2}\log^2 m$$

$$+ 2\frac{1}{h}\left(\frac{h_2 - h^2}{h^2} - \frac{1}{h}\hat{\gamma}'(-1) + \frac{\gamma-1}{h}\right)\log m$$

$$+ 2\sum_{k\neq 0} \frac{\Gamma(s_k)\hat{\gamma}(s_k)}{h^2(s_k)} m^{-1-s_k}\log m + O(1).$$

After computing $\mathbf{E}[D_m]^2$ we arrive at

$$\mathbf{Var}[D_m] = E D_m(D_m - 1) + E D_m - (E D_m)^2 = \frac{h_2 - h^2}{h^3}\log m$$

$$+ 2\sum_{k\neq 0} \frac{\Gamma(s_k)\hat{\gamma}(s_k)}{h(s_k)}\left(\frac{1}{h(s_k)} - \frac{1}{h}\right) m^{-1-s_k}\log m + O(1).$$

If $\Re(s_k) = -1$ for all k, then by Lemma 8.21 one can prove that $h(s_k) = h$ (see Exercise 19). If $\Re(s_k) > -1$, then $m^{-1-s_k} \log m = o(1)$, and $\delta_1(x, b) = o(1)$ as $x \to \infty$. From Lemma 8.21 we know that $\Re(s_k) = -1$ whenever $(\log p)/(\log q)$ is rational, and otherwise $\Re(s_k) > -1$.

To complete the proof of Theorem 10.20 we evaluate the constant $\widehat{\gamma}'(-1)$. In passing we point out that the second-order term of $\mathbf{E}[D_m]$ plays an important role in some applications (e.g., the computation of the average code redundancy for the standard Lempel-Ziv scheme [299], and for k-resilient Lempel-Ziv codes, introduced recently by Storer and Reif [399] and analyzed in [361]). Therefore, knowing its value, or providing a numerical algorithm to compute it, is of prime interest. But the second-order term depends on $\widehat{\gamma}'(-1)$, which also can be expressed as

$$\widehat{\gamma}'(-1) = \sum_{i=1}^{b} \binom{b}{i} (-1)^{i+1} \gamma'(-1-i),$$

where $\gamma(s)\Gamma(s) = \mathcal{M}[\tilde{X}(z); s]$. Define the sequence $\{f_m\}_{m=0}^{\infty}$ from $\tilde{X}(z) = \sum_{m=0}^{\infty} f_m \frac{z^m}{m!} e^{-z}$. Since clearly $f_m = \mathbf{E}[D_m]$, it satisfies recurrence (10.77) of Theorem 10.20.

Assume first that $b > 1$. To compute $\widehat{\gamma}'(-1)$ we must find $\gamma(s)$ in terms of computable quantities, such as f_n. We proceed as follows:

$$\gamma(s) = \frac{1}{\Gamma(s)} \mathcal{M}\left[\sum_{n=b+1}^{\infty} f_n \frac{z^n}{n!} e^{-z}; s \right] = \sum_{n=b+1}^{\infty} \frac{f_n}{n!} \frac{\Gamma(s+n)}{\Gamma(s)}$$

$$= \sum_{n=b+1}^{\infty} \frac{f_n}{n!} s(s+1) \cdots (s+n-1).$$

We assume that $\Re(s) \in (-b-1, -1)$ to ensure the existence of the Mellin transform and the convergence of the above series. Then one easily derives

$$\gamma'(s) = \sum_{n=b+1}^{\infty} \frac{f_n}{n!} s(s+1) \cdots (s+n-1) \sum_{i=0}^{n-1} \frac{1}{s+i}, \qquad s \notin \{-2, -3, \ldots, -b\}.$$

After some algebra, we arrive at the following:

$$\gamma'(-k) = (-1)^k \sum_{n=b+1}^{\infty} \frac{f_n}{n!} k! (n-k-1)!, \quad \text{for} \quad k = 2, \ldots, b,$$

$$\gamma'(-b-1) = (-1)^b H_{b+1} + (-1)^{b+1} \sum_{n=b+2}^{\infty} \frac{f_n}{n!} (b+1)! (n-b-2)! .$$

To estimate $\widehat{\gamma}'(-1)$ we find, after some tedious algebra, that

$$\widehat{\gamma}'(-1) = \sum_{i=1}^{b} \binom{b}{i}(-1)^{i+1}\gamma'(-i-1) - \frac{1}{b} - \frac{b}{b+1} + A + \Delta(b, p),$$

where

$$\Delta(b, p) = \sum_{n=b+2}^{\infty} \bar{f}_n \sum_{i=1}^{b} \frac{(i+1)b!}{(b-i)!n(n-1)\cdots(n-i-1)},$$

$$A = \sum_{n=b+2}^{\infty} (n-b) \sum_{i=1}^{b} \frac{(i+1)b!}{(b-i)!n(n-1)\cdots(n-i-1)}. \quad (10.91)$$

We recall that $\bar{f}_{m+b} = f_{m+b} - m$ and f_m satisfies (10.77). The above series converge since the summands are $O(\log n/n^2)$. Finally, observe that $\bar{f}_{m+b} = 0$ for $m = 1, 2, \ldots, b$ and $\bar{f}_i > 0$ for $i > 2b$; hence

$$\Delta(b, p) = \sum_{n=b+2}^{\infty} \bar{f}_n \sum_{i=1}^{b} \frac{(i+1)b!}{(b-i)!n(n-1)\ldots(n-i-1)}$$

$$= \sum_{n=2b+1}^{\infty} \bar{f}_n \sum_{i=1}^{b} \frac{(i+1)b!}{(b-i)!n(n-1)\cdots(n-i-1)}.$$

It can be proved that $A = H_b + b(1+b)^{-1}$ (the reader is asked to verify this in Exercise 20); hence $\widehat{\gamma}'(-1) = H_{b-1} + \Delta(b, p)$ as presented in Theorem 10.20.

The case $b = 1$ was already discussed in Section 8.7.3, where we found

$$\Delta(1, p) = \theta = -\sum_{k=1}^{\infty} \frac{p^{k+1}\log p + q^{k+1}\log q}{1 - p^{k+1} - q^{k+1}}.$$

This finally completes the proof of Theorem 10.20 for all b.

10.5.3 Depth in a Digital Search Tree with a Markovian Source

We again analyze the depth in a digital search tree with $b = 1$. However, this time the strings are emitted by a Markovian source. More precisely, we assume that m sequences are independently generated by irreducible and aperiodic Markov sources over a finite alphabet $\mathcal{A} = \{1, 2, \ldots, V\}$ with a positive transition matrix $\mathbf{P} = \{p_{ii}\}_{i,j=1}^{V}$ (i.e., $p_{ii} > 0$ for $i, \in \mathcal{A}$) and a stationary vector $\boldsymbol{\pi} = (\pi_1, \ldots, \pi_V)$.

We build a digital search tree from m strings. Let I_i be the depth of insertion for the ith string, that is, the path length from the root to the node containing the

ith string. As before, the typical depth D_m is defined as

$$\Pr\{D_m = k\} = \frac{1}{m} \sum_{i=1}^{m} \Pr\{I_i = k\},$$

and the *average* profile $\bar{B}_m^k$ represents the *average* number of nodes at level k of the tree. Observe that $\bar{B}_0^k = 0$ for all $k \geq 0$, and for $m \geq 1$

$$\Pr\{D_m = k\} = \frac{\bar{B}_m^k}{m}.$$

This and the definition of the typical depth immediately imply that

$$\Pr\{I_{m+1} = k\} = \bar{B}_{m+1}^k - \bar{B}_m^k, \tag{10.92}$$

with $\Pr\{I_0 = 0\} = 1$ and $\Pr\{I_0 = k\} = 0$ for all $k \geq 1$. Throughout, we shall work with generating functions of the above quantities, that is,

$$D_m(u) = E[u^{D_m}] = \sum_{k \geq 0} \Pr\{D_m = k\}u^k, \qquad D_0(u) = 1,$$

$$I_m(i) = E[u^{I_m}] = \sum_{k \geq 0} \Pr\{I_m = k\}u^k, \qquad I_0(u) = 1,$$

$$B_m(u) = \sum_{k \geq 0} \bar{B}_m^k u^k \qquad\qquad\qquad B_0(u) = 0$$

for a complex u such that $|u| < 1$, and the Poisson transforms

$$\tilde{D}(z, u) = \sum_{m \geq 0} D_m(u) \frac{z^m}{m!} e^{-z},$$

$$\tilde{B}(z, u) = \sum_{m \geq 0} B_m(u) \frac{z^m}{m!} e^{-z},$$

$$\tilde{I}(z, u) = \sum_{m \geq 0} I_m(u) \frac{z^m}{m!} e^{-z}.$$

Observe that from (10.92), we have

$$\frac{\partial \tilde{I}(z, u)}{\partial z} + \tilde{I}(z, u) = \frac{\partial \tilde{B}(z, u)}{\partial z}.$$

Since $D_m(u) = B_m(u)/m$, we can recover all results on the depth of insertion I_m and on the typical depth, from the average profile $\bar{B}_m^k$. Therefore, hereafter we concentrate on the analysis of the average profile.

To start the analysis, we derive a system of recurrence equations for the generating function of the average profile. Let $B_m^i(u)$ for $i \in \mathcal{A}$ be the ordinary generating function of the average profile when all sequences start with symbol i. Let also $\mathbf{p} = (p_1, \ldots, p_V)$ be the initial probability vector of the underlying Markov chain, that is, $\Pr\{X_0 = i\} = p_i$. (For the stationary Markov chain we have $\mathbf{p} = \pi$.) The probability that the first subtree of the root contains j_1 strings, the second subtree has j_2 strings, and so on until the Vth subtree stores j_V strings is equal to

$$\binom{m}{j_1, \ldots, j_V} p_1^{j_1} \cdots p_V^{j_V}.$$

But the ith subtree is a digital search tree itself of size j_i containing only those strings that start with symbol i. Hence, its average profile generating function must be $B_{j_i}^i(u)$. This leads to the following recurrence equation, assuming $B_0(u) = 0$

$$B_{m+1}(u) = 1 + u \sum_{|\mathbf{j}|=m} \binom{m}{\mathbf{j}} p_1^{j_1} \cdots p_V^{j_V} \left(B_{j_1}^1(u) + \cdots + B_{j_V}^V(u) \right), \quad (10.93)$$

where $\mathbf{j} = (j_1, \ldots, j_V)$, $|\mathbf{j}| = j_1 + \cdots + j_V$ and for simplicity we write $\binom{m}{\mathbf{j}} = \binom{m}{j_1, \ldots, j_V}$. Clearly, we can set up similar recurrences for the subtrees. That is,

$$B_{m+1}^i(u) = 1 + u \sum_{|\mathbf{j}|=m} \binom{m}{\mathbf{j}} p_{i1}^{j_1} \cdots p_{iV}^{j_V} \left(B_{j_1}^1(u) + \cdots + B_{j_V}^V(u) \right), \quad \text{for all } i \in \mathcal{A},$$

$$(10.94)$$

where $B_0^i(u) = 0$ for $i \in \mathcal{A}$.

Before we present the main result, let us establish some notation. For a complex s we define

$$Q(s) = I - P(s), \quad \text{where} \quad P(s) = \{p_{ij}^{-s}\}_{i,j=1}^V.$$

where I is the identity matrix. In addition, we use the standard notation for the entropy of a Markov source, (cf. Example 6.2):

$$h = -\sum_{i=1}^V \pi_i \sum_{j=1}^V p_{ij} \ln p_{ij},$$

and for a probability vector $\mathbf{p} = (p_1, \ldots, p_V)$ we define

$$h_{\mathbf{p}} = -\sum_{i=1}^{V} p_i \ln p_i.$$

We also write $\mathbf{p}(s) = [\pi_1^{-s}, \pi_2^{-s}, \cdots, \pi_V^{-s}]$, which becomes π when $s = -1$.

Finally, for the matrix $\mathsf{P}(s)$ we define the principal left eigenvector $\pi(s)$, and the principal right eigenvector $\psi(s)$ associated with the largest eigenvalue $\lambda(s)$ as

$$\pi(s)\mathsf{P}(s) = \lambda(s)\pi(s),$$

$$\mathsf{P}(s)\psi(s) = \lambda(s)\psi(s),$$

where $\pi(s)\psi(s) = 1$. Observe that $\pi(-1) = \pi = (\pi_1, \ldots, \pi_V)$, $\psi(-1) = \psi = (1, \ldots, 1)$, and $\lambda(-1) = 1$. For a function, vector, or matrix $A(s)$ we write $\dot{A}(s) = \frac{d}{ds}A(s)$ and $\ddot{A}(s) = \frac{d^2}{ds^2}A(s)$. In Exercise 22 the reader is asked to check that

$$\dot{\lambda}(-1) = \pi\dot{\mathsf{P}}(-1)\psi = h, \tag{10.95}$$

$$\ddot{\lambda}(-1) = \pi\ddot{\mathsf{P}}(-1)\psi + 2\dot{\pi}(-1)\dot{\mathsf{P}}(-1)\psi - 2\dot{\lambda}(-1)\dot{\pi}(-1)\psi. \tag{10.96}$$

We now formulate the main result, which we derive throughout the rest of this section.

Theorem 10.23 (Jacquet, Szpankowski, and Tang, 2000) *Let a digital search tree be built from m strings generated independently by Markov stationary sources with the transition matrix* $\mathsf{P} = \{p_{ij}\}_{i,j=1}^{V}$.

(i) [TYPICAL DEPTH] *For large m the following holds*

$$\mathbf{E}[D_m] = \frac{1}{\dot{\lambda}(-1)}\left(\ln m + \gamma - 1 + \dot{\lambda}(-1) + \frac{\ddot{\lambda}(-1)}{2\dot{\lambda}^2(-1)} - \vartheta - \pi\dot{\psi}(-1) + \delta_1(\ln m)\right)$$

$$+ O\left(\frac{\ln m}{m}\right), \tag{10.97}$$

$$\mathbf{Var}[D_m] = \frac{\ddot{\lambda}(-1) - \dot{\lambda}^2(-1)}{\dot{\lambda}^3(-1)}\ln m + O(1). \tag{10.98}$$

Here $\vartheta = \pi\dot{\mathbf{x}}(-2)$ *and*

$$\dot{\mathbf{x}}(-2) := \sum_{i=1}^{\infty}\left(\mathsf{Q}^{-1}(-2)\cdots\mathsf{Q}^{-1}(-i)(\mathsf{Q}^{-1}(s))'|_{s=-i-1}\mathsf{Q}^{-1}(-i-2)\cdots\right)\mathsf{K}, \tag{10.99}$$

$$K = \left(\prod_{i=0}^{\infty} Q^{-1}(-2-i)\right)^{-1} \psi, \tag{10.100}$$

where $\psi = (1, \ldots, 1)$. The function $\delta_1(x)$ is a fluctuating function with a small amplitude when

$$\frac{\ln p_{ij} + \ln p_{1i} - \ln p_{1j}}{\ln p_{11}} \in \mathbb{Q}, \qquad i, j = 1, 2, \ldots, V, \tag{10.101}$$

where $\mathbb{Q}$ is the set of rational numbers. If (10.101) does not hold, then $\lim_{x \to \infty} \delta_1(x) = 0$.

(ii) [DEPTH OF INSERTION] *The depth of insertion I_m behaves asymptotically as D_m, that is,*

$$\mathbf{E}[I_m] = \frac{1}{\dot{\lambda}(-1)} \left(\ln m + \gamma + \dot{\lambda}(-1) + \frac{\ddot{\lambda}(-1)}{2\dot{\lambda}^2(-1)} - \vartheta - \pi \dot{\psi}(-1) + \delta_2(\ln m) \right)$$
$$+ O\left(\frac{\ln m}{m}\right)$$

$$\mathbf{Var}[I_m] = \mathbf{Var}[D_m] + O(1),$$

where $\delta_2(x)$ is a fluctuating function with the same property as $\delta_1(x)$.

The rest of this section is devoted to proving Theorem 10.23. The generating function $B_m(u)$ of the average profile satisfies (10.93) with the initial vector $\mathbf{p} = \pi$. Observe that the conditional generating functions $B_m^i(u)$ fulfill the system of recurrence equations (10.94). We shall first deal with (10.94). There is no easy way to solve these recurrences. Therefore, we transform them to the Poisson model. Let

$$\widetilde{B}^i(z, u) = \sum_{n=1}^{\infty} B_n^i(u) \frac{z^n}{n!} e^{-z}, \quad i \in \mathcal{A}$$

be the Poisson transform of $B_m^i(u)$. In addition, we shall write $\widetilde{B}_z^i(z, u) := \frac{\partial}{\partial z} \widetilde{B}^i(z, u)$ for the derivative of $\widetilde{B}^i(z, u)$ with respect to z. After some simple algebra, we have the following poissonized differential-functional equations of recurrences (10.93) and (10.94)

$$\widetilde{B}_z(z, u) + \widetilde{B}(z, u) = u[\widetilde{B}^1(u, \pi_1 z) + \cdots + \widetilde{B}^V(u, \pi_V z)] + 1,$$

and

$$\widetilde{B}_z^i(z, u) + \widetilde{B}^i(z, u) = u[\widetilde{B}^1(u, p_{i1} z) + \cdots + \widetilde{B}^V(u, p_{iV} z)] + 1 \quad \text{for all } i \in \mathcal{A}.$$

We now concentrate on the evaluation of the first two moments of the depth. Thus we need the first two derivatives of $\widetilde{B}(z, u)$ with respect to u at $u = 1$. Noting that $\widetilde{B}^i(z, 1) = z$, we have

$$\widetilde{B}_{zu}(z, 1) + \widetilde{B}_u(z, 1) = z + [\widetilde{B}_u^1(1, \pi_1 z) + \cdots + \widetilde{B}_u^V(1, \pi_V z)],$$

$$\widetilde{B}_{zu}^i(z, 1) + \widetilde{B}_u^i(z, 1) = z + [\widetilde{B}_u^1(1, p_{11}z) + \cdots + \widetilde{B}_u^V(1, p_{1V}z)], \quad i \in \mathcal{A},$$

and

$$\widetilde{B}_{zuu}(z, 1) + \widetilde{B}_{uu}(z, 1) = 2[\widetilde{B}_u^1(1, \pi_1 z) + \cdots + \widetilde{B}_u^V(1, \pi_V z)]$$
$$+ [\widetilde{B}_{uu}^1(1, \pi_1 z) + \cdots + \widetilde{B}_{uu}^V(1, \pi_V z)],$$

$$\widetilde{B}_{zuu}^i(z, 1) + \widetilde{B}_{uu}^i(z, 1) = 2[\widetilde{B}_u^1(1, p_{11}z) + \cdots + \widetilde{B}_u^V(1, p_{1V}z)] \qquad i \in \mathcal{A}$$
$$+ [\widetilde{B}_{uu}^1(1, p_{11}z) + \cdots + \widetilde{B}_{uu}^V(1, p_{1V}z)]$$

Our goal is to solve asymptotically (as $z \to \infty$ in a cone around $\Re(z) > 0$) the above two sets of functional equations using Mellin transforms. A direct solution through the Mellin transform does not work well, as we saw in the previous section. Therefore, we factorize the Mellin transforms as follows

$$B_i^*(s) := \mathcal{M}[B_u^i(z, 1); s] = \Gamma(s)x_i(s), \quad i \in \mathcal{A}$$
$$B^*(s) := \mathcal{M}[B_u(z, 1); s] = \Gamma(s)x(s),$$
$$C_i^*(s) := \mathcal{M}[B_{uu}^i(z, 1); s] = \Gamma(s)v_i(s), \quad i \in \mathcal{A}$$
$$C^*(s) := \mathcal{M}[B_{uu}(z, 1); s] = \Gamma(s)v(s),$$

where $x_i(s)$, $x(s)$, $v_i(s)$ and $v(s)$ are unknown. The lemma below establishes the existence of the above Mellin transforms. The reader is asked to prove it in Exercise 23.

Lemma 10.24 *The Mellin transforms $B_i^*(s)$, $B^*(s)$ and $C_i^*(s)$, $C^*(s)$ exist for $\Re(s) \in (-2, -1)$. In addition,*

$$x_i(-2) = 1, \quad v_i(-2) = 0, \qquad i \in \mathcal{A},$$

and $x(-2) = 1$, $v(-2) = 0$.

Using properties of Mellin transforms (cf. (M3) in Chapter 9) we obtain

$$-(s - 1)B^*(s - 1) + B^*(s) = B_1^*(s)\pi_1^{-s} + \cdots + B_V^*(s)\pi_V^{-s}, \qquad (10.102)$$

$$-(s-1)B_i^*(s-1) + B_i^*(s) = B_1^*(s)p_{11}^{-s} + \cdots + B_V^*(s)p_{1V}^{-s}, \quad i \in \mathcal{A},$$

$$(10.103)$$

and

$$-(s-1)C^*(s-1) + C^*(s) = 2[B_1^*(s)\pi_1^{-s} + \cdots + B_V^*(s)\pi_V^{-s}] \quad (10.104)$$

$$+ [C_1^*(s)\pi_1^{-s} + \cdots + C_V^*(s)\pi_V^{-s}], \quad (10.105)$$

$$-(s-1)C_i^*(s-1) + C_i^*(s) = 2[B_1^*(s)p_{11}^{-s} + \cdots + B_V^*(s)p_{1V}^{-s}] \quad (10.106)$$

$$+ [C_1^*(s)p_{11}^{-s} + \cdots + C_V^*(s)p_{1V}^{-s}], \quad i \in \mathcal{A}. \quad (10.107)$$

Now set

$$\mathbf{x}(s) = (x_1(s), \ldots, x_V(s))^T,$$
$$\mathbf{v}(s) = (v_1(s), \ldots, v_V(s))^T,$$
$$\mathbf{b}(s) = (B_1^*(s), \ldots, B_V^*(s))^T,$$
$$\mathbf{c}(s) = (C_1^*(s), \ldots, C_V^*(s))^T,$$

where T denotes transpose. Using $\Gamma(s) = (s-1)\Gamma(s-1)$, we can rewrite the above equations as

$$\mathbf{x}(s) - \mathbf{x}(s-1) = \mathsf{P}(s)\mathbf{x}(s),$$
$$\mathbf{v}(s) - \mathbf{v}(s-1) = 2\mathsf{P}(s)\mathbf{x}(s) + \mathsf{P}(s)\mathbf{v}(s).$$

Thus

$$\mathbf{x}(s) = \mathsf{Q}^{-1}(s)\mathbf{x}(s-1) = \left(\prod_{i=0}^{\infty} \mathsf{Q}^{-1}(s-i)\right)\mathsf{K}, \quad (10.108)$$

$$\mathbf{v}(s) = 2\mathsf{Q}^{-1}(s)\mathsf{P}(s)\mathbf{x}(s) + \mathsf{Q}^{-1}(s)\mathbf{v}(s-1). \quad (10.109)$$

We recall that $\mathsf{Q} = \mathsf{I} - \mathsf{P}$, I is the identity matrix, and K is defined in (10.100). The formula for K follows from Lemma 10.24 (i.e., $\mathbf{x}(-2) = (1, \ldots, 1)^T$) and (10.108).

Thus far we have obtained the Mellin transforms of the conditional generating functions $\tilde{B}_i(z, 1)$. In order to obtain the composite Mellin transform $B^*(s)$ and $C^*(s)$ of $\tilde{B}_u(z, 1)$ and $\tilde{B}_{uu}(z, 1)$, we use (10.102) and (10.104) and, after some algebra, we finally obtain

$$B^*(s) = \mathbf{p}(s)\mathbf{b}(s) + \Gamma(s)x(s-1), \tag{10.110}$$

$$C^*(s) = 2\mathbf{p}(s)\mathbf{b}(s) + \mathbf{p}(s)\mathbf{c}(s) + \Gamma(s)v(s-1), \tag{10.111}$$

where $\mathbf{p}(s) = (\pi_1^{-s}, \ldots, \pi_V^{-s})$. We shall see that the dominant asymptotics of $B^*(s)$ and $C^*(s)$ follow from the asymptotics of $\mathbf{b}(s)$ and $\mathbf{c}(s)$, and these depend on the singularities of $Q(s)$, which we discuss next.

We now prove the following lemma that characterizes the location of singularities of $Q(s)$. It extends the Jacquet-Schachinger Lemma 8.21 to Markovian sources.

Lemma 10.25 (Tang 1996) *Let* $Q(s) = I - P(s)$ *and* $P(s) = \{p_{ij}^{-s}\}_{i,j \in A}$. *Let* s_l *denote the singularities of* $Q(s)$, *where* $l \in \mathbb{Z}$ *is an integer. Then:*

(i) *Matrix* $Q(s)$ *is nonsingular for* $\Re(s) < -1$, *and* $s_0 = -1$ *is a simple pole.*

(ii) *Matrix* $Q(s)$ *has simple poles on the line* $\Re(s) = -1$ *if and only if*

$$\frac{\ln p_{ij} + \ln p_{1i} - \ln p_{1j}}{\ln p_{11}} \in \mathbb{Q}, \quad i, j \in A, \tag{10.112}$$

where $\mathbb{Q}$ *is the set of rational numbers. These poles have the following representation*

$$s_l = -1 + l\theta i,$$

where

$$\theta = \frac{n_1}{n_2} \left| \frac{2\pi}{\ln p_{11}} \right|,$$

for integers n_1, n_2 *such that* $\left\{ \left| \frac{n_1}{n_2 \ln p_{11}} (\ln p_{ij} + \ln p_{1i} - \ln p_{1j}) \right| \right\}_{ij=1}^{V}$ *is a set of relative primes.*

(iii) *Finally,*

$$Q(-1 + l\theta i) = \mathsf{E}^{-l} Q(-1) \mathsf{E}^{l},$$

where $\mathsf{E} = \mathrm{diag}(1, e^{\theta_{12}i}, \ldots, e^{\theta_{1V}i})$ *is the diagonal matrix with* $\theta_{ik} = -\theta \ln p_{ik}$.

Proof. Observe that for $\Re(s) < -1$,

$$|1 - p_{ii}^{-s}| \geq 1 - |p_{ii}^{-s}| > 1 - p_{ii} = \sum_{j \neq i} p_{ij} \geq \sum_{j \neq i} |p_{ij}^{-s}|; \tag{10.113}$$

hence $Q(s)$ is a strictly diagonally dominant matrix, and therefore nonsingular (cf. [200, 327]). We proceed with the proof of part (ii) of the lemma. For $b \neq 0$ such that $Q(-1 + bi)$ is singular, let $\mathbf{x} = [x_1, x_2, \ldots, x_V]^T \neq 0$ be a solution of $Q(-1 + bi)\mathbf{x} = 0$, where

$$
Q(-1+bi) =
\begin{bmatrix}
1 - p_{11}e^{\xi_{11}i} & -p_{12}e^{\xi_{12}i} & \cdots & -p_{1V}e^{\xi_{1V}i} \\
-p_{21}e^{\xi_{21}i} & 1 - p_{22}e^{\xi_{22}i} & \cdots & -p_{2V}e^{\xi_{2V}i} \\
\vdots & \vdots & \ddots & \vdots \\
-p_{j1}e^{\xi_{j1}i} & -p_{j2}e^{\xi_{j2}i} & \cdots & -p_{jV}e^{\xi_{jV}i} \\
\vdots & \vdots & \ddots & \vdots \\
-p_{V1}e^{\xi_{V1}i} & -p_{V2}e^{\xi_{V2}i} & \cdots & 1 - p_{VV}e^{\xi_{VV}i}
\end{bmatrix}
$$

with $\xi_{ik} = -b \ln p_{ik}$. Without loss of generality, suppose

$$
|x_1| = \max\{|x_1|, |x_2|, \ldots, |x_V|\} \neq 0
$$

(since $Q(-1 + bi)$ is singular). Then

$$
(1 - p_{11}e^{\xi_{11}i})x_1 - p_{12}e^{\xi_{12}i}x_2 - \cdots - p_{1V}e^{\xi_{1V}i}x_V = 0,
$$

implies that

$$
1 - p_{11}e^{\xi_{11}i} = p_{12}e^{\xi_{12}i}x_2/x_1 + \cdots + p_{1V}e^{\xi_{1V}i}x_V/x_1.
$$

But as in (10.113)

$$
|1 - p_{11}e^{\xi_{11}i}| \geq 1 - p_{11},
$$

and

$$
|p_{12}e^{\xi_{12}i}x_2/x_1 + \ldots + p_{1V}e^{\xi_{1V}i}x_V/x_1| \leq p_{12} + \cdots + p_{1V} = 1 - p_{11}.
$$

Thus

$$
1 - p_{11}e^{\xi_{11}i} = 1 - p_{11},
$$

$$
p_{12}e^{\xi_{12}i}x_2/x_1 + \cdots + p_{1V}e^{\xi_{1V}i}x_V/x_1 = p_{12} + \cdots + p_{1V}.
$$

This implies that

$$
e^{\xi_{11}i} = e^{\xi_{1i}i}x_i/x_1 = 1,
$$

and $|x_i| = |x_j|$ for any $i, j = 1, 2, \ldots, V$, so that $e^{\xi_{jj}i} = 1$ for all j. Define now ξ_i such that $x_i/x_1 = e^{-\xi_{1i}i} = e^{\xi_i i}$. Then

$$- p_{j1}e^{\xi_{j1}i} - p_{j2}e^{\xi_{j2}i}e^{\xi_2 i} - \cdots - p_{j(j-1)}e^{\xi_{j(j-1)}i}e^{\xi_{j-1}i}$$
$$+ (1 - p_{jj})e^{\xi_j i} - \cdots - p_{jv}e^{\xi_{jv}i}e^{\xi_v i} = 0$$

for any $1 \le j \le V$. Note that since

$$-p_{j1} - p_{j2} - \cdots - p_{j(j-1)} + 1 - p_{jj} - \cdots - p_{jv} = 0$$

we must have $e^{\xi_{ji}i}e^{\xi_i i}e^{-\xi_j i} = 1$, and thus

$$e^{\xi_{ji}i} = e^{(\xi_j - \xi_i)i}.$$

Hence, $-b(\ln p_{ji} + \ln p_{1j} - \ln p_{1i}) = 2\pi n_{ji}$ for some integer n_{ji}, and as a consequence $(\ln p_{ij} + \ln p_{1i} - \ln p_{1j})/\ln p_{11}$ is rational for any $i, j = 1, 2, \ldots, V$.

To prove the second part of (ii), suppose b is such that $|\frac{b}{2\pi}(\ln p_{ji} + \ln p_{1j} - \ln p_{1i})|$ are integers for any $i, j = 1, 2, \ldots, V$. Then

$$Q(-1 + bi) =$$

$$\begin{bmatrix} 1 - p_{11} & -p_{12}e^{(\xi_1 - \xi_2)i} & \cdots & -p_{1v}e^{(\xi_1 - \xi_v)i} \\ -p_{21}e^{(\xi_2 - \xi_1)i} & 1 - p_{22} & \cdots & -p_{2v}e^{(\xi_2 - \xi_v)i} \\ \vdots & \vdots & \ddots & \vdots \\ -p_{i1}e^{(\xi_i - \xi_1)i} & -p_{i2}e^{(\xi_i - \xi_2)i} \cdots & -p_{iv}e^{(\xi_i - \xi_v)i} & \vdots \\ \vdots & \vdots & \ddots & \vdots \\ -p_{v1}e^{(\xi_v - \xi_1)i} & -p_{v2}e^{(\xi_v - \xi_2)i} & \cdots & 1 - p_{vv} \end{bmatrix}$$

$$= [\text{diag}(1, e^{-\xi_2}, e^{-\xi_3}, \ldots, e^{-\xi_v})]^{-1}Q(-1)\text{diag}(1, e^{-\xi_2}, e^{-\xi_3}, \ldots, e^{-\xi_v}).$$

Since $Q(-1)$ is singular, so is $Q(-1 + bi)$. Hence $s = -1 + bi$ is a pole of $Q(s)$ if and only if $|\frac{b}{2\pi}(\ln p_{ji} + \ln p_{1j} - \ln p_{1i})|$ are integers for any $i, j = 1, 2, \ldots, V$. Since $\{|\frac{\theta}{2\pi}(\ln p_{ij} + \ln p_{1i} - \ln p_{1j})|\}_{ij=1}^{V}$ is a set of relative primes, we have $b = l\theta$ for some integer l. Thus part (ii) is proved. Part (iii) follows from our previous analysis. ∎

Observe that for the memoryless case we have $p_{ji} = \pi_i$, and condition (10.112) becomes $\frac{\ln \pi_i}{\ln \pi_j} \in \mathbb{Q}$ for all i, j. This agrees with Lemma 8.21.

Now, we are prepared to prove (10.97) and (10.98) of Theorem 10.23. We first consider the mean. The starting point is (10.108), which we rewrite as

$$x(s) = Q^{-1}x(s-1) = \sum_{k=0}^{\infty} P^k(s)x(s-1).$$

We denote by $\lambda(s), \mu_2(s), \ldots, \mu_V(s)$ the eigenvalues of $P(s)$. By the Perron-Frobenius Theorem 4.14 we know that $|\lambda(s)| > |\mu_1(s)| \geq \cdots \geq |\mu_V(s)|$ (recall that we assumed that $p_{ii} > 0$ for $i \in \mathcal{A}$). The corresponding left eigenvectors are $\pi(s), \pi_2(s), \ldots, \pi_V(s)$ while the right eigenvectors are $\psi(s), \psi_2(s), \ldots, \psi_V(s)$. We write here $\langle \mathbf{x}, \mathbf{y} \rangle$ for the scalar product of the vectors $\mathbf{x}$ and $\mathbf{y}$.

Using the spectral representation of $P(s)$ (see Table 4.1) we obtain

$$P^k(s)x(s-1) = \lambda^k(s)\langle \pi(s), x(s-1) \rangle \psi(s) + \sum_{i=2}^{V} \mu_i^k(s)\langle \pi_i(s), x(s-1) \rangle \psi_i(s).$$

Thus $\mathbf{b}(s) = \Gamma(s)x(s)$ becomes

$$\mathbf{b}(s) = \frac{\Gamma(s)\langle \pi(s), x(s-1) \rangle \psi(s)}{1 - \lambda(s)} + \sum_{i=2}^{V} \frac{\Gamma(s)\langle \pi_i(s), x(s-1) \rangle \psi_i(s)}{1 - \mu_i(s)}. \quad (10.114)$$

To obtain the leading order asymptotics of $B^*(s) = p(s)\mathbf{b}(s) + \Gamma(s)x(s-1)$ (cf. (10.110)), we need Laurent's expansion of the above around the roots of $\lambda(s) = -1$. Observe that the second term of (10.114) contributes $o(m)$ since $\lambda(s)$ is the largest eigenvalue; hence we ignore this term. To simplify the presentation, we deal here only with the root $s_0 = -1$. The other singularities are handled in the same way as in the previous section. We use our previous expansions for $x(s-1)$ and $\Gamma(s)$ together with

$$\frac{1}{1 - \lambda(s)} = \frac{-1}{\dot{\lambda}(-1)} \frac{1}{s+1} + \frac{\ddot{\lambda}(-1)}{2\dot{\lambda}^2(-1)} + O(s+1),$$

$$\psi(s) = \psi + \dot{\psi}(-1)(s+1) + O((s+1)^2).$$

This finally leads to

$$B^*(s) = \frac{-1}{\dot{\lambda}(-1)} \frac{1}{(s+1)^2}$$
$$+ \frac{1}{s+1} \left(\frac{\langle \pi, \dot{x}(-2) \rangle}{\dot{\lambda}(-1)} - \frac{\gamma - 1}{\dot{\lambda}(-1)} + \frac{\langle p(-1), \dot{\psi}(-1) \rangle}{\dot{\lambda}(-1)} + \frac{\ddot{\lambda}(-1)}{2\dot{\lambda}^2(-1)} - 1 \right)$$
$$+ O(1).$$

After finding the inverse Mellin transform and depoissonizing, we obtain (10.97) for the mean $E[D_m]$.

Finally, we turn our attention to the second factorial moment and the variance. We need to study $c(s) = \Gamma(s)v(s)$, where $v(s) = 2Q^{-1}(s)P(s)x(s) + Q^{-1}(s)v(s-1)$. As before, we obtain

$$c(s) = \frac{2\Gamma(s)\langle\pi(s), x(s-1)\rangle\langle\pi(s), P(s)\psi(s)\rangle\psi(s)}{(1-\lambda(s))^2} + O\left((1-\lambda(s))^{-1}\right).$$

A calculation similar to that for $b(s)$ leads to

$$c(s) = \frac{-2}{\dot\lambda^2(-1)}\frac{1}{(s+1)^3} + \frac{1}{(s+1)^2}\left(\frac{\ddot\lambda(-1)}{2\dot\lambda^3(-1)}\right.$$
$$\left. + 2\frac{\gamma - 1 - \langle\pi, \dot x(-2)\rangle - \langle p(-1), \dot\psi(-1)\rangle - \dot\lambda(-1)}{\dot\lambda^2(-1)}\right) + O\left(\frac{1}{s+1}\right).$$

This is sufficient to prove (10.98) after some tedious algebra that was done using MAPLE. Details of the proof can be found in Jacquet, Szpankowski, and Tang [220].

10.6 EXTENSIONS AND EXERCISES

10.1 Prove properties (P1)–(P3) of the Poisson process.

10.2 Let S_n be the size of a trie built over n independent strings generated by a memoryless source. Define $S_n(u) = E[u^{S_n}]$ and let $\widetilde S(z, u)$ be the Poisson transform of $S_n(u)$. Prove that

$$\widetilde S(z, u) = u\widetilde S(zp, u)\widetilde S(zq, u) + (1-u)(1+z)e^{-z}.$$

10.3 Let $g_n = n(n-1)\cdots(n-k+1)$. First show that $\widetilde G(z) = z^k$. Next directly prove the general depoissonization result (as in Theorem 10.6) in this case, and show that the coefficients b_{ij} defined in (10.40) are equal to

$$b_{ij} = \frac{1}{j!}s_2(j, i),$$

where $s_2(n, k)$ are related to Stirling's number of first kind, and defined via

$$\sum_{n,k} s_2(n, k)t^n u^k / n! = e^{-tu}(1 + t)^u$$

(cf. Comtet [71]).

10.4   Consider

$$g_{n,k} = 1 + \frac{n!}{(k + \log k)^n}.$$

Find its Poisson transform and decide whether the analytic depoissoniza-tion will work in this case. Explain your answer. If you cannot justify it, carefully read Theorem 10.13 and see if you can apply it to this case.

10.5   ⚠ (Jacquet and Szpankowski, 1989) Consider the functional equation discussed in Example 10.5, that is,

$$\tilde{C}(z) = \left(1 + (1 + z/2)e^{-z/2}\right)\tilde{C}(z/2).$$

Establish asymptotics (10.36) of $\tilde{C}(z)$ as $z \to \infty$ in a cone $\mathcal{S}_\theta$ around the real axis. In particular, compute the constant D that appears in the asymptotic expansion of $\tilde{C}(z)$.

10.6   Consider the Laplace transform $g^\star(s)$ of a function $g(x)$. Extend the defi-nition of the Laplace transform to the complex plane (assuming $g(x)$ can be analytically continued to $g(z)$). (The reader may use arguments similar to those we adopted in Section 9.3, where we extended Mellin transforms to the complex plane.) In particular, prove that if $g(z)$ is an analytic con-tinuation of $g(x)$ in a cone $\mathcal{S}_\theta$, then the inverse Laplace transform $g^\star(s)$ of the function $g(z)$ exists in a *bigger* cone $\mathcal{S}_{\theta + \pi/2}$ for all $\theta < \theta_0$, *provided* that $g(z) = O(z^\beta)$ in the cone $\mathcal{S}_{\theta_0}$.

10.7   Prove Lemma 10.10.

10.8   Provide details of the proof of the diagonal depoissonization Theo-rem 10.15.

10.9   ⚠ (Rais, Jacquet, and Szpankowski, 1993) Prove that the depth in a PA-TRICIA trie in the Poisson model is normally distributed, that is, establish (10.64).

10.10   ⚠ (Flajolet 1999) Use singularity analysis to derive the asymptotic ex-pansion for Shannon entropy h_n of the binomial distribution, that is,

$$h_n = -\sum_{k=0}^{n} \binom{n}{k} p^k q^{n-k} \log\left(\binom{n}{k} p^k q^{n-k}\right).$$

10.11   Let

$$h_n(t) = \exp(n(e^{it/\sqrt{n}} - 1 - it/\sqrt{n}))$$

(cf. Lemma 10.11). Prove that for all nonnegative integers k there exist ν_k such that for all $t \in [-\log n, \log n]$,

$$h_n(t) = e^{-t^2/2}\left(1 + \sum_{i=3}^{2k}\sum_{j=1}^{k} \xi_{ij} t^i n^{j-i/2} + O(\nu_k \log^{3(k+1)} n^{-(k+1)/2})\right),$$

where the coefficients ξ_{ij} are defined as

$$\sum_{ij\geq 0} \xi_{ij} x^i y^j = \exp(y(e^{ix} - 1 - ix + \frac{1}{2}x^2)).$$

10.12   ⚠   (Jacquet and Szpankowski, 1998) The purpose of this exercise is to extend the general depoissonization Theorem 10.6 to the polynomial cones $\mathcal{C}(D, \delta)$ defined as follows:

$$\mathcal{C}(D, \delta) = \{z = x + iy : |y| \leq Dx^\delta, \quad 0 < \delta \leq 1, \quad D > 0\}.$$

Observe that when $\delta = 1$ the polynomial cone becomes a linear cone. Prove the following extension of Theorem 10.6.

Theorem 10.26   *Consider a polynomial cone $\mathcal{C}(D, \delta)$ with $1/2 < \delta \leq 1$. Let the following two conditions hold for some numbers $A, B, R > 0$ and $\alpha > 0$, β, and γ:*
(I) *For $z \in \mathcal{C}(D, \delta)$*

$$|z| > R \quad \Rightarrow \quad |\widetilde{G}(z)| \leq B|z|^\beta \Psi(|z|),$$

where $\Psi(x)$ is a slowly varying function, that is, for each fixed t $\lim_{x\to\infty} \frac{\Psi(tx)}{\Psi(x)} = 1$ (e.g., $\Psi(x) = \log^d x$ for some $d > 0$).
(O) *For all $z = \rho e^{i\theta}$ with $\theta \leq \pi$ such that $z \notin \mathcal{C}(D, \delta)$*

$$\rho = |z| > R \quad \Rightarrow \quad |\widetilde{G}(z)e^z| \leq A\rho^\gamma \exp[(1 - \alpha\theta^2)\rho].$$

Then for every nonnegative integer m

$$g_n = \sum_{i=0}^{m} \sum_{j=0}^{i+m} b_{ij} n^i \widetilde{G}^{(j)}(n) + O(n^{\beta-(m+1)(2\delta-1)}\Psi(n))$$

$$= \widetilde{G}(n) + \sum_{k=1}^{m} \sum_{i=1}^{k} b_{i,k+i} n^i \widetilde{G}^{(k+i)}(n) + O(n^{\beta-(m+1)(2\delta-1)}\Psi(n)),$$

where b_{ij} are the coefficients defined in (10.40).

10.13 ⚠ Prove the central limit theorem for the Poisson model. More specifically, establish the following result.

Theorem 10.27 *Let X_N be a characteristic of the Poisson model with $\widetilde{G}(z, u) = \mathbf{E} u^{X_N}$, mean $\widetilde{X}(z)$, and variance $\widetilde{V}(z)$. Let the following hold for $z \to \infty$ in a cone S_θ and for u belonging to a neighborhood $\mathcal{U}$ of $u = 1$ in the complex plane:*

$$\log \widetilde{G}(z, u) = O(A(z)),$$

where $A(z)$ is such that

$$\lim_{z \to \infty} \frac{A(z)}{\widetilde{V}^{3/2}(z)} = 0$$

and furthermore $\lim_{z \to \infty} \widetilde{V}(z) = \infty$. Then for complex τ

$$\widetilde{G}\left(z, e^{\tau/\sqrt{\widetilde{V}(z)}}\right) e^{-\tau \widetilde{X}(z)/\sqrt{\widetilde{V}(z)}} = e^{\frac{\tau^2}{2}} \left(1 + O\left(\frac{A(z)}{\widetilde{V}^{3/2}(z)}\right)\right).$$

Thus $\bar{X}_N = (X_N - \widetilde{X}(z))/\sqrt{\widetilde{V}(z)}$ converges in distribution and in moments to the standard normal distribution.

10.14 ⚠ (Jacquet and Szpankowski, 1998) Prove the following extension of the depoissonization tool Theorem 10.13.

Theorem 10.28 *Let $\widetilde{G}_k(z)$ be a sequence of Poisson transforms of $g_{n,k}$, and each $\widetilde{G}_k(z)$ is assumed to be an entire function of z. Let $\log \widetilde{G}_k(z)$ exist in a cone S_θ. We suppose that there exists $\frac{1}{2} \le \beta < \frac{2}{3}$ such that the following two conditions hold:*

(I) *For all $z \in S_\theta$ such that for $|z| = n$*

$$|\log \widetilde{G}_n(z)| \leq Bn^\beta$$

for some constant $B > 0$.

(O) *For all $z \notin S_\theta$ such that $|z| = n$:*

$$|\widetilde{G}_n(z)e^z| \leq \exp(n - An^\alpha)$$

for some $\alpha > \beta$. Then for all $\varepsilon > 0$:

$$g_{n,n} = \widetilde{G}_n(n) \exp[-\frac{n}{2}(L'_n(n))^2] \left(1 + O(n^{3\beta - 2 + \varepsilon})\right),$$

where $L_n(z) = \log \widetilde{G}_n(z)$ and $L'_n(z) = \widetilde{G}'_n(z)/\widetilde{G}_n(z)$ is the first derivative of $L_n(z)$.

10.15 ⚠ Consider again the leader election algorithm discussed in Section 10.5.1. In this exercise the reader is asked to consider the mean $\mathbf{E}[H_n]$ and the variance $\mathbf{Var}[H_n]$ of the height H_n. Prove the following result.

Theorem 10.29 *Define $L := \ln 2$ and $\chi_k := 2\pi ik/L$.*

(i) *(Prodinger, 1993) The average height $\mathbf{E}[H_n]$ satisfies*

$$\mathbf{E}[H_n] = \log_2 n + \frac{1}{2} - \delta_1(\log_2 n) + O\left(\frac{1}{n}\right),$$

where $\delta_1(\cdot)$ is a periodic function of magnitude $\leq 2 \times 10^{-5}$, given by

$$\delta_1(x) := \frac{1}{L} \sum_{\mathbb{Z} \setminus \{0\}} \zeta(1 - \chi_k) \Gamma(1 - \chi_k) e^{2\pi ikx}.$$

(ii) *(Fill, Mahmoud and Szpankowski, 1996) The variance $\mathbf{Var}[H_n]$ of the height satisfies*

$$\mathbf{Var}[H_n] = \frac{\pi^2}{6\log^2 2} + \frac{1}{12} - \frac{2\gamma_1}{\log^2 2} - \frac{\gamma^2}{\log^2 2} + \delta_2(\log n)$$

$$+ O\left(\frac{\log n}{n}\right) = 3.116695\ldots + \delta_2(\log n) + O\left(\frac{\log n}{n}\right).$$

Here the constants $(-1)^k \gamma_k / k!$, $k \geq 0$, are the Stieltjes constants defined as

$$\gamma_k := \lim_{m \to \infty} \left(\sum_{i=1}^{m} \frac{\log^k i}{i} - \frac{\log^{k+1} m}{k+1} \right).$$

In particular, $\gamma_0 = \gamma = 0.577215\ldots$ is Euler's constant and $\gamma_1 = -0.072815\ldots$. The periodic function $\delta_2(\cdot)$ has magnitude $\leq 2 \times 10^{-4}$.

10.16 ⚠ (Janson and Szpankowski, 1997) Extend the analysis of the leader election algorithm to biased coins, that is, assume that the probability of throwing a head is $p \neq \frac{1}{2}$.

10.17 ▽ Consider again the leader election algorithm discussed in Section 10.5.1 but this time the process stops when at most b leaders are selected. In other words, consider a b-incomplete trie version of the regular incomplete trie. Find the mean of the height (cf. Grabner [167]), the variance, and the limiting distribution, if it exists.

10.18 ⚠ (Kirschenhofer, Prodinger, and Szpankowski, 1996) Consider the generalized probabilistic counting, as discussed in Example 9.2 of Chapter 9. The quantity $R_{n,d}$ determines the number of rounds before the algorithm terminates. Let $G_n(u) = \mathbf{E}u^{R_{n,d}}$, and $\widetilde{G}(z, u)$ be the probability generating function and its Poisson transform, respectively. Observe (cf. [254]) that

$$\widetilde{G}(z, u) = u f_d(z/2) \widetilde{G}(z/2, u) + (u - 1)(f_d(z/2) - 1),$$

where $f_d(z) = 1 - e_d(z)e^{-z}$ and $e_d(z) = 1 + \frac{z^1}{1!} + \cdots + \frac{z^d}{d!}$ is the truncated exponential function. Then prove the following result.

Theorem 10.30 (Kirschenhofer, Prodinger and Szpankowski, 1996)
For any integer m

$$\Pr\{R_{n,d} \leq \log_2 n + m - 1\} = 1 - \varphi \left(2^{-m - \langle \log_2 n \rangle} \right) + O(n^{-1}),$$

where

$$\varphi(z) = \prod_{j=0}^{\infty} f_d(z2^j) = \prod_{j=0}^{\infty} \left(1 - e_d(z2^j)e^{-z2^j}\right)$$

and $\langle \log n \rangle = \log n - \lfloor \log n \rfloor$.

10.19 Let

$$h(t) = -p^{-t} \log p - q^{-t} \log q$$

and s_k be a root of $1 - p^s - q^{-s} = 0$ for $k \in \mathbb{Z}$. Prove that if $\Re(s_k) = -1$, then $h(s_k) = h$, where h is the entropy of *Bernoulli*(p).

10.20 Prove that the constant A defined in (10.91) simplifies to $A = H_b + b(1+b)^{-1}$.

10.21 ⚠ (Louchard, Szpankowski, and Tang, 1999) Consider again the b-digital search tree model in Section 10.5.2 and prove the following limiting distribution for the depth D_m.

Theorem 10.31 *Let* $G_m(u)$ *be the probability generating function of* D_m *(i.e.,* $G_m(u) = \mathbf{E}[u^{D_m}]$*),* $\mu_m = E D_m$*, and* $\sigma_m = \sqrt{\mathrm{Var}\, D_m}$*. Then for complex* τ

$$e^{-\tau \mu_m / \sigma_m} G_m(e^{\tau/\sigma_m}) = e^{\frac{\tau^2}{2}} \left(1 + O\left(\frac{1}{\sqrt{\log m}}\right)\right).$$

Thus the limiting distribution of $\frac{D_m - \mu_m}{\sigma_m}$ *is normal, and it converges in moments to the appropriate moments of the standard normal distribution. Also, there exist positive constants* A *and* $\alpha < 1$ *(that may depend on* p *and* b*) such that uniformly in* k*, and for large* m

$$\mathbf{Pr}\left\{\left|\frac{D_m - c_1 \log m}{\sqrt{c_2 \log m}}\right| > k\right\} \le A\alpha^k,$$

where $c_1 = 1/h$ *and* $c_2 = (h_2 - h^2)/h^3$.

10.22 Prove (10.95) and (10.96).

10.23 Prove Lemma 10.24.

10.24 ⚠ (Jacquet, Szpankowski, and Tang, 2000) Consider a digital search tree under the Markovian model as in Section 10.5.3 and prove the following limiting distribution for the depth D_m

$$\frac{D_m - \mathbf{E}[D_m]}{\sqrt{\mathbf{Var}\,D_m}} \to N(0, 1),$$

where $N(0, 1)$ represents the standard normal distribution and $\mathbf{E}[D_m]$ and $\mathbf{Var}[D_m]$ are computed in Theorem 10.23.

10.25 ⚠ (Jacquet, Szpankowski, and Tang, 2000) Extend Theorem 10.23 to a *nonstationary* Markov model.

10.26 Extend Theorem 10.23 to b-DST under the Markovian model.

10.27 ⚠ (Jacquet, Szpankowski, and Tang, 2000) Using the analysis from Section 10.5.3, establish the mean, the variance, and the limiting distribution for the phrase length in the Lempel-Ziv'78 scheme, when a string of length n is generated by a Markovian source.

10.28 ▽ Consider the digital search tree model in Section 10.5.3 with $b = 1$ and where strings are generated by m independent Markov sources. Let L_m be the total path length, that is, the sum of all depths. Prove that $E[L_m] = m\mathbf{E}[D_m]$ and $\mathbf{Var}[L_m] \sim m\mathbf{Var}[D_m]$, where $\mathbf{E}[D_m]$ and $\mathbf{Var}[D_m]$ were computed in Theorem 10.23. Then show that $(L_m - \mathbf{E}[D_m])/\mathbf{Var}[D_m]$ tends to the standard normal distribution. Finally, prove similar results for the path lengths for regular tries and PATRICIA tries.

10.29 ▽ Consider the Lempel-Ziv'78 scheme, where a string of length n is generated by a Markovian source. Use the same notation as in Section 10.5.3. In particular, let

$$c_2 = \frac{\ddot{\lambda}(-1) - \dot{\lambda}^2(-1)}{\dot{\lambda}^3(-1)},$$

where $\lambda(s)$ is the principal eigenvalue of $\mathbf{I} - \mathbf{P}(s)$. Prove that, under certain conditions, the number of phrases $(M_n - \mathbf{E}[M_n])/\mathbf{Var}[M_n]$ tends to the standard normal distribution with

$$\mathbf{E}[M_n] \sim \frac{nh}{\log n},$$

$$\mathbf{Var}[M_n] \sim \frac{c_2 h^3 n}{\log^2 n},$$

where h is the entropy rate of the Markov source. This is an open and a *very* difficult problem.

Bibliography

[1] J. Abrahams, "Code and Parse Trees for Lossless Source Encoding, *Proc. of Compression and Complexity of SEQUENCE'97*, Positano, IEEE Press, 145–171, 1998.

[2] M. Abramowitz, and I. Stegun, *Handbook of Mathematical Functions*, Dover, New York, 1964.

[3] A. Aho, J. Hopcroft, and J. Ullman, *The Design and Analysis of Computer Algorithms*, Addison-Wesley, Reading, MA, 1974.

[4] D. Aldous, *Probability Approximations via the Poisson Clumping Heuristic*, Springer Verlag, New York, 1989.

[5] D. Aldous, and P. Shields, A Diffusion Limit for a Class of Random-Growing Binary Trees, *Probab. Th. Rel. Fields*, 79, 509–542, 1988.

[6] D. Aldous, M. Hofri, and W. Szpankowski, Maximum Size of a Dynamic Data Structure: Hashing with Lazy Deletion Revisited, *SIAM J. Computing*, 21, 713–732, 1992.

[7] K. Alexander, Shortest Common Superstring of Random Strings, *J. Applied Probability*, 33, 1112–1126, 1996.

[8] P. Algoet, and T. Cover, A Sandwich Proof of the Shannon-McMillan-Beriman Theorem, *Annals of Probability*, 16, 899–909, 1988.

[9] N. Alon, and J. Spencer, *The Probabilistic Method*, John Wiley & Sons, New York, 1992.

[10] N. Alon, and A. Orlitsky, A Lower Bound on the Expected Length of One-to-One Codes, *IEEE Trans. Information Theory*, 40, 1670–1672, 1994.

[11] M. Alzina, W. Szpankowski, and A. Grama, 2D-Pattern Matching Image and Video Compression *IEEE Trans. Image Processing*, 2000.

[12] C. Anderson, Extreme Value Theory for a Class of Discrete Distributions With Applications to Some Stochastic Processes, *J. Applied Probability*, 7, 99–113, 1970.

[13] G. Andrews, *The Theory of Partitions*, Cambridge University Press, Cambridge, 1984.

[14] G. Andrews, R. Askey, and R. Roy, *Special Functions*, Cambridge University Press, Cambridge, 1999.

[15] T. Apostol, *Introduction to Analytic Number Theory*, Springer Verlag, New York, 1976.

[16] T. Apostol, *Modular Functions and Dirichlet Series in Number Theory*, Springer-Verlag, New York, 1978.

[17] A. Apostolico, The Myriad Virtues of Suffix Trees, in *Combinatorial Algorithms on Words*, Springer-Verlag, New York, 1985.

[18] A. Apostolico, and W. Szpankowski, Self-Alignments in Words and Their Applications, *J. Algorithms*, 13, 446–467, 1992.

[19] A. Apostolico, M. Atallah, L. Larmore, and S. McFaddin, Efficient Parallel Algorithms for String Editing and Related Problems, *SIAM J. Computing*, 19, 968–988, 1990.

[20] R. Arratia, and S. Tavaré, Independent Processes Approximations for Random Combinatorial Structures. *Advances in Mathematics*, 104, 90–154, 1994.

[21] R., Arratia, and M. Waterman, Critical Phenomena in Sequence Matching, *Annals of Probability*, 13, 1236–1249, 1985.

[22] R. Arratia, and M. Waterman, The Erdős-Rényi Strong Law for Pattern Matching with Given Proportion of Mismatches, *Annals of Probability*, 17, 1152–1169, 1989.

[23] R. Arratia, and M. Waterman, A Phase Transition for the Score in Matching Random Sequences Allowing Deletions, *Annals of Applied Probability*, 4, 200–225, 1994.

[24] R., Arratia, L. Gordon, and M. Waterman, An Extreme Value Theory for Sequence Matching, *Annals of of Statistics*, 14, 971–993, 1986.

[25] R. Arratia, L. Gordon, and M. Waterman, The Erdős-Rényi Law in Distribution for Coin Tossing and Sequence Matching, *Annals of Statistics*, 18, 539–570, 1990.

[26] J. Ashley, and P. Siegel, A Note on the Shannon Capacity of Run-Length-Limited Codes, *IEEE Trans. Information Theory*, 33, 601–605, 1987.

[27] M. Atallah (Editor), *Algorithms and Theory of Computation Handbook*, CRC Press, Boca Raton, FL, 1998.

[28] M. Atallah, P. Jacquet, and W. Szpankowski, Pattern Matching with Mismatches: A Randomized Algorithm and Its Analysis, *Random Structures & Algorithms*, 4, 191–213, 1993.

[29] M. Atallah, Y. Génin, and W. Szpankowski, Pattern Matching Image Compression: Algorithmic and Empirical Results, *IEEE Trans. Pattern Analysis and Machine Intelligence*, 21, 618-627, 1999.

[30] K. Azuma, Weighted Sums of Certain Dependent Random Variables, *Tôhoku Math. J.*, 19, 357–367, 1967.

[31] R. Baeza-Yates, A Trivial Algorithm Whose Analysis is Not: A Continuation, *BIT*, 29, 378–394, 1989.

[32] C. Banderier, P. Flajolet, G. Schaeffer, and M. Soria, Planar Maps and Airy Phenomena, *Proc. ICALP'2000*, Geneva, Lecture Notes in Computer Science, No. 1853, 388–402, 2000.

[33] A. Barbour, L. Holst, and S. Janson, *Poisson Approximation*, Clarendon Press, Oxford, 1992.

[34] A. Barron, *Logically Smooth Density Estimation*, Ph.D. Thesis, Stanford University, Stanford, CA, 1985.

[35] H. Bateman, *Higher Transcendental Functions*, Vols. I–III (Ed. A. Erdélyi), Robert. E. Krieger Publishing, Malabar, FL, 1985.

[36] E. Bender, Central and Local Limit Theorems Applied to Asymptotic Enumeration, *J. Combinatorial Theory, Ser. A*, 15, 91–111, 1973.

[37] E. Bender, Asymptotic Methods in Enumeration, *SIAM Review*, 16, 485–515, 1974.

[38] E. Bender, and J. Goldman, Enumerative Uses of Generating Functions, *Indiana University Mathematical Journal*, 753–765, 1971.

[39] C. Bender and S. Orszag, *Advanced Mathematical Methods for Scientists and Engineers*, McGrew-Hill, New York, 1978.

[40] E. Bender and B. Richmond, Central and Local Limit Theorems Applied to Asymptotic Enumeration II: Multivariate Generating Functions, *J. Combinatorial Theory, Ser. A*, 34, 255–265, 1983.

[41] J. Bentley, Multidimensional Binary Trees Used for Associated Searching, *Commun. ACM*, 18, 509–517, 1975.

[42] T. Berger, *Rate Distortion Theory: A Mathematical Basis for Data Compression*, Englewood Cliffs, NJ, Prentice-Hall, 1971.

[43] B. Berger, The Fourth Moment Methods, *SIAM J. Computing*, 26, 1188–1207, 1997.

[44] B.C. Berndt, *Ramanujan's Notebooks. Part I*, Springer-Verlag, New York, 1985.

[45] B.C. Berndt, *Ramanujan's Notebooks. Part II*, Springer-Verlag, New York, 1989.

[46] J. D. Biggins, The First and Last Birth Problems for a Multitype Age-Dependent Branching Process, *Adv. in Applied Probability*, 8, 446–459, 1976.

[47] J. D. Biggins, Chernoff's Theorem in the Branching Random Walk, *J. Applied Probability*, 14, 630–636, 1977.

[48] P. Billingsley, *Convergence of Probability Measures*, John Wiley & Sons, New York, 1968.

[49] P. Billingsley, *Probability and Measures*, Second Edition, John Wiley & Sons, New York, 1986.

[50] N. H. Bingham, Tauberian Theorems and the central Limit Theorem, *Annals of Probability*, 9, 221-231, 1981.

[51] N. Bleistein, and R. Handelsman, *Asymptotic Expansions of Integrals*, Dover Publications, New York, 1986.

[52] A. Blum, T. Jiang, M. Li, J. Tromp, and M. Yannakakis, Linear Approximation of Shortest Superstring, *J. ACM*, 41, 630–647, 1994

[53] A. Blum, T. Jiang, M. Li, J. Tromp, and M. Yannakakis, Linear Approximation of Shortest Superstring, *J. ACM*, 41, 630–647, 1994

[54] A. Blumer, A. Ehrenfeucht, and D. Haussler, Average Size of Suffix Trees and DAWGS, *Discrete Applied Mathematics*, 24, 37–45, 1989.

[55] B. Bollobás, *Random Graphs*, Academic Press, London 1985.

[56] S. Boucheron, G. Lugosi, and P. Massart, A Sharp Concentration Inequality with Applications, *Random Structures & Algorithms*, 16, 277–292, 2000.

[57] R. Boyer, and J. Moore, A Fast String Searching Algorithm, *Comm. ACM*, 20, 762–772, 1977.

[58] R. Bradley, Basic Properties of Strong Mixing Conditions, in *Dependence in Probability and Statistics* (Eds. E. Eberlein and M. Taqqu), 165–192, 1986.

[59] G. Brassard, and P. Bratley, *Algorithmics. Theory and Practice*, Prentice Hall, Englewood Cliffs, NJ, 1988.

[60] F. Bruss, M. Drmota and G. Louchard, The Complete Solution of the Competitive Rank Selection Problem, *Algorithmica*, 22, 413–447, 1998.

[61] R. Burkard and U. Fincke, Probabilistic Asymptotic properties of some Combinatorial Optimization Problems, *Discrete Applied Mathematics*, 12, 21–29, 1985.

[62] E. R. Canfield, Central and Local Limit Theorems for the Coefficients of Polynomials of Binomial Type, *J. Combinatorial Theory, Ser. A*, 23, 275–290, 1977.

[63] E. R. Canfield, From Recursions to Asymptotics: On Szekeres' Formula for the Number of Partitions, *Electronic J. of Combinatorics*, 4, RR6, 1997.

[64] J. Capetanakis, Tree Algorithms for Packet Broadcast Channels. *IEEE Trans. Information Theory*, 25, 505–515, 1979.

[65] J. Capetanakis, Generalized TDMA: The Multiaccessing Tree Protocol, *IEEE Trans. Communications*, 27, 1476–1484, 1979.

[66] S-N. Choi, and M. Golin, Lopsided trees: Algorithms, Analyses and Applications, *Proc. the 23rd International Colloquium on Automata Languages and Programming* (ICALP '96), 538–549, July 1996.

[67] V. Chvatal, and D. Sankoff, Longest Common Subsequence of Two Random Sequences, *J. Applied Probability.*, 12, 306–315, 1975.

[68] J. Clement, P. Flajolet, and B. Vallée, Dynamic Sources in Information Theory: A General Analysis of Trie Structures, *Algorithmica*, 29, 307–369, 2001.

[69] E. Coffman, and G. Lueker, *Probabilistic Analysis of Packing and Partitioning Algorithms*, John Wiley & Sons, New York, 1991.

[70] E. Coffman, L. Flatto, P. Jelenković, and B. Poonen, Packing Random Intervals on-Line, *Algorithmica*, 22, 448–475, 1998.

[71] L. Comtet, *Advanced Combinatorics*, D. Reidel Publishing Company, Boston, 1974.

[72] L. Colussi, Z. Galil, and R. Giancarlo, On the Exact Complexity of String Matching, *Proc. 31st Annual IEEE Symposium on the Foundations of Computer Science*, 135–143. IEEE, 1990.

[73] R. Corless, G. Gonnet, D. Hare, D. Jeffrey, and D. Knuth, On the Lambert W Function, *Adv. Computational Mathematics*, 5, 329–359, 1996.

[74] T. Cormen, C. Leiserrson, and R. Rivest, *Introduction to Algorithms*, MIT Press, New York, 1990.

[75] T.M. Cover and J.A. Thomas, *Elements of Information Theory*, John Wiley & Sons, New York, 1991.

[76] M. Cramer, A Note Concerning the Limit Distribution of the Quicksort Algorithm, *Theoretical Informatics and Applications*, 30, 195–207, 1996.

[77] M. Crochemore, and W. Rytter, *Text Algorithms*, Oxford University Press, New York, 1995.

[78] I. Csiszár, and J. Körner, *Information Theory: Coding Theorems for Discrete Memoryless Systems*, Academic Press, New York, 1981.

[79] I. Csiszàr and P. Shields, Redundancy Rates for Renewal and Other Processes, *IEEE Trans. Information Theory*, 42, 2065–2072, 1996.

[80] V. Dančik, and M. Paterson, Upper Bounds for the Expected Length of a Longest Common Subsequence of Two Binary Sequences, *Random Structures & Algorithms*, 6, 449–458, 1995.

[81] H. Daniels, Saddlepoint Approximations in Statistics, *Annals of Math. Stat.*, 25, 631–650, 1954.

[82] B. Davies, *Integral Transforms and Their Applications*, Springer-Verlag, New York, 1978.

[83] H. David, *Order Statistics*, John Wiley & Sons, New York, 1980.

[84] N. G. De Bruijn, *Asymptotic Methods in Analysis*, Dover Publications, New York, 1958.

[85] N. G. De Bruijn, D. E. Knuth, and S. O. Rice, The Average Height of Planted Trees, in *Graph Theory and Computing*, (Ed. R. C. Read), 15–22, Academic Press, 1972.

[86] N. G. De Bruijn, and P. Erdős, Some Linear and some Quadratic Recursion Formulas I, *Indagationes Mathematicae*, 13, 374–382, 1952.

[87] N. G. De Bruijn, and P. Erdős, Some Linear and some Quadratic Recursion Formulas II, *Indagationes Mathematicae*, 14, 152–163, 1952.

[88] J. Deken, Some Limit results for the Longest Common Subsequences, *Discrete Mathematics*, 26, 17–31, 1979.

[89] H. Delange, Généralisation du Théoréme d'Ikehara, *Ann. Sc. ENS*, 213–242, 1954.

[90] A. Dembo, and O. Zeitouni, *Large Deviations Techniques*, Jones and Bartlett Publishers, Boston, 1993.

[91] A. Dembo, and I. Kontoyiannis, The Asymptotics of Waiting Times Between Stationary Processes, Allowing Distortion, *Annals of Applied Probability*, 9, 413–429, 1999.

[92] G. Derfel, and F. Vogl, Divide-and-Conquer Recurrences—Classification of Asymptotics, *Aquations Math.*, 1999.

[93] Y. Derriennic, Un Théorème Ergodique Presque Sous Additif, *Annals of Probability*, 11, 669–677, 1983.

[94] L. Devroye, A Probabilistic Analysis of the Height of Tries and the Complexity of Triesort, *Acta Informatica*, 21, 229–237, 1984.

[95] L. Devroye, The Expected Length of the Longest Probe Sequence When the Distribution Is Not Uniform, *J. Algorithms*, 6, 1–9, 1985.

[96] L. Devroye, A Note on the height of Binary Search Trees, *J. ACM.*, 33, 489–498, 1986.

[97] L. Devroye, Branching Processes in the Analysis of the Heights of Trees, *Acta Informatica*, 24, 277–298, 1987.

[98] L. Devroye, A Study of Trie-Like Structures Under the Density Model, *Annals of Applied Probability*, 2, 402–434, 1992.

[99] L. Devroye, A Note on the Probabilistic Analysis of Patricia Tries, *Random Structures & Algorithms*, 3, 203–214, 1992.

[100] L. Devroye, and L. Laforest, An Analysis of Random d-Dimensional Quadtrees, *SIAM J. Computing*, 19, 821–832, 1992.

[101] L. Devroye, W. Szpankowski, and B. Rais, A Note of the Height of Suffix Trees, *SIAM J. Computing*, 21, 48–53, 1992.

[102] L. Devroye, and B. Reed, On the Variance of the Height of Random Binary Search Trees, *SIAM. J. Computing*, 24, 1157–1162, 1995.

[103] G. Doetsch, *Handbuch der Laplace Transformation,* Vols. 1–3, Birkhäuser-Verlag, Basel, 1955.

[104] M. Drmota, Systems of Functional Equations, *Random Structures & Algorithms*, 10, 103–124, 1997.

[105] M. Drmota, An Analytic Approach to the Height of Binary Search Trees, *Algorithmica*, 29, 89–119, 2001.

[106] M. Drmota, An Analytic Approach to the Height of Binary Search Trees. II, preprint.

[107] M. Drmota, and U. Schmid, The Analysis of the expected Successful Operation Time of Slotted ALOHA, *IEEE Trans. Information Theory*, 39, 1567–1577, 1993.

[108] M. Drmota, and M. Soria, Marking in Combinatorial Constructions: Generating Functions and Limiting Distributions, *Theoretical Computer Science*, 144, 67–100, 1995.

[109] M. Drmota, and R. Tichy, *Sequences, Discrepancies, and Applications*, Springer Verlag, Berlin, Heidelberg, 1997.

[110] W. Eddy, and M. Schervish, How Many Comparisons Does Quicksort Use, *J. Algorithms*, 19, 402–431, 1995.

[111] R. Ellis, Large Deviations for a General Class of Random Vectors, *Annals of Probability*, 1–12, 1984.

[112] P. Erdős, and A. Rényi, On Random Graphs I, *Publicationes Mathematicæ*, 6, 385–287, 1959.

[113] P. Erdős, and A. Rényi, On the Evolution of Random Graphs, *Matematikai Kutató Intézetének Kőzleményei*, 5, 17–61, 1960.

[114] P. Erdős, and J. Spencer, *Probabilistic Methods in Combinatorics*, Academic Press, New York, 1974.

[115] A. Erdélyi, *Asymptotic Expansions*, Dover, New York, 1956.

[116] R. Estrada, and R. Kanwal, *Asymptotic Analysis: A Distributional Approach*, Birkhäuser, New York, 1994.

[117] R. Durrett, *Probability: Theory and Examples*, Wadsworth, Belmont, 1991.

[118] G. Fayolle, P. Flajolet, M. Hofri, and P. Jacquet, Analysis of Stack Algorithm for Random Multiple-access Communication, *IEEE Trans. Information Theory*, 31, 244–254, 1985.

[119] G. Fayolle, P. Flajolet, and M. Hofri, On a Functional Equation Arising in the Analysis of a Protocol for a Multi-Access Broadcast Channel, *Adv. Applied Probability*, 18, 441–472, 1986.

[120] M. Feder, N. Merhav, and M. Gutman, Universal Prediction of Individual Sequences, *IEEE Trans. Information Theory*, 38, 1258–1270, 1992.

[121] M. V. Fedoryuk, Asymptotic Methods in Analysis, in *Analysis I: Integral Representations and Asymptotic Methods* (Ed. R.V. Gamkrelidze), Springer-Verlag, Berlin, 1989.

[122] W. Feller, *An Introduction to Probability Theory and its Applications*, Vol.I, John Wiley & Sons, New York, 1970.

[123] W. Feller *An Introduction to Probability Theory and its Applications*, Vol.II, John Wiley & Sons, New York, 1971.

[124] J. Fill, H. Mahmoud, and W. Szpankowski, On the Distribution for the Duration of a Randomized Leader Election Algorithm. *Annals of Applied Probability*, 6, 1260–1283, 1996.

[125] P. Flajolet, On the Performance Evaluation of Extendible Hashing and Trie Search, *Acta Informatica*, 20, 345–369, 1983.

[126] P. Flajolet, Mathematical Methods in the Analysis of Algorithms and Data Structures. In *Trends in Theoretical Computer Science*, (Ed. E. Börger), Computer Science Press. Rockville, 225–304, 1988.

[127] P. Flajolet, Analytic Analysis of Algorithms, *Lectures Notes in Computer Science*, Vol. 623, (Ed. W. Kuich), 186–210, Springer-Verlag, 1992.

[128] P. Flajolet, and G. Louchard, Analytic Variations on the Airy Distributions, *Algorithmica*, 2002.

[129] P. Flajolet, Singularity Analysis and Asymptotics of Bernoulli Sums, *Theoretical Computer Science*, 215, 371–381, 1999.

[130] P. Flajolet, D. Gardy, and L. Thimonier, Birthday Paradox, Coupon Collectors, Caching Algorithms, and Self-Organizing Search, *Discrete Applied Mathematics*, 39, 207–229, 1992.

[131] P. Flajolet, and M. Golin, Mellin Transforms and Asymptotics: The Mergesort Recurrence, *Acta Informatica*, 31, 673–696, 1994.

[132] P. Flajolet, X. Gourdon, and P. Dumas, Mellin Transforms and Asymptotics: Harmonic sums, *Theoretical Computer Science*, 144, 3–58, 1995.

[133] P. Flajolet, X. Gourdon, and C. Martinez, Patterns in Random Binary Search Trees, *Random Structures and Algorithms*, 11, 223–244, 1997.

[134] P. Flajolet, P. Grabner, P. Kirschenhofer, H. Prodinger, and R. Tichy, Mellin Transforms and Asymptotics, Digital Sums, *Theoretical Computer Science*, 123, 291–314, 1994.

[135] P. Flajolet, P. Grabner, P. Kirschenhofer, and H. Prodinger, On Ramanujan's Q-function, *J. Comp. and Appl. Math.*, 58, 103–116, 1995.

[136] P. Flajolet, P. Kirschenhofer, and R. Tichy, Deviations from Uniformity in Random Strings, *Probab. Th. Rel. Fields*, 80, 139–150, 1988.

[137] P. Flajolet, D. E. Knuth, and B. Pittel, The First Cycles in an Evolving Graph, *Discrete Mathematics*, 75, 167–215, 1989.

[138] P. Flajolet, and G. Martin, Probabilistic counting algorithms for data base applications. *J. Computer and System Sciences*, 31, 182–209, 1985.

[139] P. Flajolet, and A. Odlyzko, The Average Height of Binary Trees and Other Simple Trees, *J. Computer and System Sciences*, 25, 171–213, 1982.

[140] P. Flajolet, and A. Odlyzko, Singularity Analysis of Generating Functions, *SIAM J. Discrete Mathematics*, 3, 216–240, 1990.

[141] P. Flajolet, P. Poblete, and A. Viola, On the Analysis of Linear Probing Hashing, *Algorithmica*, 22, 490–515, 1998.

[142] P. Flajolet, and H. Prodinger, Register Allocation for Unary-Binary Trees, *SIAM J. Computing*, 15, 629–640, 1986.

[143] P. Flajolet, and C. Puech, Partial Match Retrieval of Multidimensional Data, *J. ACM*, 371–407, 1986.

[144] P. Flajolet, M. Régnier, and D. Sotteau, Algebraic Methods for Trie Statistics, *Annals of Discrete Mathematics*, 25, 145–188, 1985.

[145] P. Flajolet, and B. Richmond, Generalized Digital Trees and Their Difference-Differential Equations, *Random Structures and Algorithms*, 3, 305–320, 1992.

[146] P. Flajolet, and N. Saheb, The Complexity of Generating an Exponentially Distributed Variate, *J. Algorithms*, 1986.

[147] P. Flajolet, and R. Sedgewick, Digital search trees revisited, *SIAM J. Computing*, 15, 748–767, 1986.

[148] P. Flajolet, and R. Sedgewick, Mellin Transforms and Asymptotics: Finite Differences and Rice's Integrals. *Theoretical Computer Science*, 144, 101–124, 1995.

[149] P. Flajolet, and R. Sedgewick, *Analytical Combinatorics*, in preparation; see also INRIA TR-1888 1993, TR-2026 1993 and TR-2376 1994.

[150] P. Flajolet, and M. Soria, Gaussian Limiting Distributions for the Number of Components in Combinatorial Structures, *J. Combinatorial Theory, Ser. A*, 53, 165–182, 1990.

[151] P. Flajolet, and M. Soria, General Combinatorial Schemas: Gaussian Limit Distributions and Exponential Tails, *Discrete Mathematics*, 114, 159–180, 1993.

[152] P. Flajolet, and W. Szpankowski, Analytic Variations on Redundancy Rates of Renewal Processes, INRIA TR-3553, November 1998.

[153] E. Fredkin, Trie Memory, *Commun. ACM*, 3, 490–499, 1960.

[154] M. Freeman, and D. E. Knuth, Recurrence Relations Based on Minimization, *J. Mathematical Analysis and Applications*, 48, 534–559, 1974.

[155] J. Frenk, M. van Houweninge, and A. Rinnooy Kan, Asymptotic Properties of the Quadratic Assignment Problem, *Mathematics of Operations Research*, 10, 100–116, 1985.

[156] A. Frieze, and C. McDiarmid, Algorithmic Theory of Random Graphs, *Random Structures & Algorithms*, 10, 1997.

[157] A. Frieze, and W. Szpankowski, Greedy Algorithms for the Shortest Common Superstring that Are Asymptotically Optimal, *Algorithmica*, 21, 21–36, 1998.

[158] D. Foata, La Série Génératice Exponentielle dans les Problémes d'énumeration, S.M.S. Montreal University Press, Montreal, 1974.

[159] W. Ford, *Studies on Divergent Series and the Asymptotic Developments of Functions Defined by Maclaurin Series*, Chelsea Publishing Company, New York, 1960.

[160] I. Fudos, E. Pitoura, and W. Szpankowski, On Pattern Occurrences in a Random Text, *Information Processing Letters*, 57, 307–312, 1996.

[161] J. Galambos, *The Asymptotic Theory of Extreme Order Statistics*, Robert E. Krieger Publishing Company, Malabar, FL, 1987.

[162] Z. Galil, and R. Giancarlo, Data Structures and Algorithms for Approximate String Matching, *J. Complexity*, 4, 33–72, 1988.

[163] R. Gallager, *Information Theory and Reliable Communication*, John Wiley & Sons, New York, 1968.

[164] R. Gallager, Variations on the Theme by Huffman, *IEEE Trans. Information Theory*, 24, 668–674, 1978.

[165] J. Gärtner, On Large Deviations from the Invariant Measure, *Theory of Probability and Applications*, 22, 24–39, 1977.

[166] G. Gasper, and M. Rahman, *Basic Hypergeometric Series*, Cambridge University Press, Cambridge, 1990.

[167] P. Grabner, Searching for Losers, *Random Structures and Algorithms*, 4, 99–110, 1993.

[168] I. S. Gradshteyn, and I. M. Ryzhik, *Table of Integrals, Series, and Products*, Fourth Edition, Academic Press, New York, 1980.

[169] R. L. Graham, D. E. Knuth, and O. Patashnik, *Concrete Mathematics*, Addison-Wesley, Reading, MA 1994.

[170] A. Greenberg, P. Flajolet, and R. Ladner, Estimating the Multiplicities of Conflicts to Speed their Resolution in Multiaccess Channels, *J. ACM*, 34, 289–325, 1987.

[171] D.H. Greene, and D.E. Knuth, *Mathematics for the Analysis of Algorithms*, Birkhauser, Boston, 1990.

[172] J. Griggs, P. Halton, and M. Waterman, Sequence Alignments with Matched Sections, *SIAM J. Algebraic and Discrete Methods*, 7, 604–608, 1986.

[173] J. Griggs, P. Halton, A. Odlyzko, and M. Waterman, On the Number of Alignments of *k* Sequences, *Graphs and Combinatorics*, 6, 133–146, 1990.

[174] M. Golin, Limit Theorems for Minimum-Weight Triangulation, Other Euclidean Functionals and Probabilistic Recurrence Relations, *Seventh Annual ACM-SIAM Symposium on Discrete Algorithms* (SODA96), 252–260, 1996.

[175] G. H. Gonnet, The Expected Length of the Longest Probe Sequence in Hash Code Searching *J. ACM*, 28, 289–304, 1981.

[176] G. H. Gonnet, and J. Munro, The Analysis of Linear Probing Sort by the Use of a New Mathematical Transform, *J. Algorithms*, 5, 451–470, 1984.

[177] G. H. Gonnet, and R. Baeza-Yates, *Handbook of Algorithms and Data Structures*, Addison-Wesley, Workingham, 1991.

[178] I. Goulden, and D. Jackson, *Combinatorial Enumerations*, John Wiley, New York, 1983.

[179] R. M. Gray, *Probability, Processes, and Ergodic Properties*, Springer-Verlag, New York, 1988.

[180] R. M. Gray, *Entropy and Information Theory*, Springer-Verlag, New York, 1988.

[181] R. M. Gray, *Source Coding Theory*, Kluwer Academic Press, Boston, 1990.

[182] L. Guibas, and A. Odlyzko, Maximal Prefix-Synchronized Codes, *SIAM J. Applied Mathematics*, 35, 401–418, 1978.

[183] L. Guibas, and A. Odlyzko, Periods in Strings, *J. Combinatorial Theory Ser. A*, 30, 19–43, 1981.

[184] L. Guibas and A. W. Odlyzko, String Overlaps, Pattern Matching, and Nontransitive Games, *J. Combinatorial Theory Ser. A*, 30, 183–208, 1981.

[185] M. Habib, C. McDiarmid, J. Ramiriez, and B. Reed (Eds.), *Probabilistic Methods for Algorithmic Discrete Mathematics*, Springer Verlag, New York, 1998.

[186] M. Hall, Jr., *Combinatorial Theory*, John Wiley & Sons, New York, 1986.

[187] G. H. Hardy, *A Course of Pure Mathematics*, Cambridge University Press, Cambridge, 1952.

[188] G. H. Hardy, *Divergent Series*, Chelsea Publishing Company, New York, 1991.

[189] G. H. Hardy, J. E. Littlewood, and G. Pólya, *Inequalities*, Second Edition, Cambridge University Press, Cambridge, 1952.

[190] G. Hardy, and E. Wright, *An Introduction to the Theory of Numbers*, Clarendon Press, Oxford, 1979.

[191] B. Harris, and L. Schoenfeld, Asymptotic Expression for the Coefficients of Analytic Functions, *Illinois J. Mathematics*, 264–277, 1968.

[192] W. K. Hayman, A Generalization of Stirling's Formula, *J. Reine Angew. Math*, 196, 67–95, 1956.

[193] P. Hennequin, Combinatorial Analysis of Quicksort Algorithm, *Theoretical Informatics and Applications*, 23, 317–333, 1989.

[194] P. Hennequin, *Analyse en Moyenne d'Algorithmes, Tri Rapide at Arbres de Recherche*, Ph.D. Thesis, Ecole Politechnique, Palaiseau 1991.

[195] P. Henrici, *Applied and Computational Complex Analysis*, Vols. 1–3, John Wiley & Sons, New York, 1977.

[196] E. Hille, *Analytic Function Theory*, Vols. I and II, Chelsea Publishing Company, New York, 1962.

[197] M. Hofri, *Analysis of Algorithms. Computational Methods and Mathematical Tools*, Oxford University Press, New York, 1995.

[198] M. Hofri, and P. Jacquet, Saddle Points in Random Matrices: Analysis of Knuth Search Algorithm, *Algorithmica*, 22, 516–528, 1998.

[199] L. Holst, On Birthday, Collector's, Occupancy and other Classical Urn Problems, *International Statistical Review*, 54, 15–27, 1986.

[200] R. Horn, and C. Johnson, *Matrix Analysis*, Cambridge University Press, Cambridge, 1990.

[201] H-K. Hwang, *Théorèmes Limites Pour les Structures Combinatoires et les Fonctions Arithmétiques*, Thèse de Doctorat de l'Ecole Polytechnique, Palaiseau, 1994.

[202] H-K. Hwang, Asymptotic Expansions for Stirling Numbers of the First Kind, *J. Combinatorial Theory, Ser. A*, 71, 343–351, 1995.

[203] H-K. Hwang, Large Deviations for Combinatorial Distributions I: Central Limit Theorems, *Annals of Applied Probability*, 6, 297–319, 1996.

[204] H-K. Hwang, Limit Theorems for Mergesort, *Random Structures & Algorithms*, 8, 319–336, 1996.

[205] H-K. Hwang, Asymptotic of Divide-and-Conquer Recurrences: Batcher's Sorting Algorithm and a Minimum Euclidean Matching Heuristic, *Algorithmica*, 22, 529–546, 1998.

[206] H-K. Hwang, On Convergence Rates in the Central Limit Theoremsfor Combinatorial Structures, *European J. Combinatorics*, 19, 329–343, 1998.

[207] H-K. Hwang, Large Deviations for Combinatorial Distributions II: Local Limit Theorems, *Annals of Applied Probability*, 8, 163–181, 1998.

[208] H-K. Hwang, Uniform Asymptotics of Some Abel Sums Arising in Coding Theory, *Theoretical Computer Science*, 2001.

[209] F. Hubalek, On the Variance of the Internal Path Length of Generalized Digital Trees - The Mellin Convolution Approach, *Theoretical Computer Science*, 242, 143–168, 2000.

[210] E. L. Ince, *Ordinary Differential Equations*, Dover, New York, 1956.

[211] P. Jacquet, *Contribution de l'Analyse d'Algorithmes a l'Evaluation de Protocoles de Communication*, Thèse Université de Paris Sud-Orsay, Paris, 1989.

[212] P. Jacquet, and M. Régnier, Limiting Distributions for Trie Parameters, *Lecture Notes in Computer Science*, 214, 196–210, 1986.

[213] P. Jacquet, and M. Régnier, Normal Limiting Distribution of the Size of Tries, *Proc. Performance'87*, 209–223, North Holland, Amsterdam, 1987.

[214] P. Jacquet, and W. Szpankowski, Ultimate Characterizations of the Burst Response of an Interval Searching Algorithm: A Study of a Functional Equation, *SIAM J. Computing*, 18, 777–791, 1989.

[215] P. Jacquet, and W. Szpankowski, Analysis of Digital Tries with Markovian Dependency, *IEEE Trans. Information Theory*, 37, 1470–1475, 1991.

[216] P. Jacquet, and W. Szpankowski, Autocorrelation on Words and Its Applications. Analysis of Suffix Trees by String-Ruler Approach, *J. Combinatorial Theory Ser. A*, 66, 237–269, 1994.

[217] P. Jacquet, and W. Szpankowski, Asymptotic Behavior of the Lempel-Ziv Parsing Scheme and Digital Search Trees, *Theoretical Computer Science*, 144, 161–197, 1995.

[218] P. Jacquet, and W. Szpankowski, Analytical Depoissonization and Its Applications, *Theoretical Computer Science*, 201, 1–62, 1998.

[219] P. Jacquet, and W. Szpankowski, Entropy Computations Via Analytic Depoissonization, *IEEE Trans. on Information Theory*, 45, 1072–1081, 1999.

[220] P. Jacquet, W. Szpankowski, and J. Tang, Average Profile of the Lempel-Ziv Parsing Scheme for a Markovian Source, *Algorithmica*, 2002.

[221] S. Janson, D. E. Knuth, T. Łuczak, and B. Pittel, The Birth of the Giant Component, *Random Structures & Algorithms*, 4, 231–358, 1993.

[222] S. Janson, and D. E. Knuth, Shellsort with Three Increments, *Random Structures & Algorithms*, 10, 125–142, 1997.

[223] S. Janson, and W. Szpankowski, Analysis of an Asymmetric Leader Election Algorithm, *Electronic J. Combinatorics*, 4, R17, 1997.

[224] A. Jonassen, and D. E. Knuth, A Trivial Algorithm Whose Analysis Isn't, *J. Computer and System Sciences*, 16, 301–322, 1978.

[225] C, Jordan, *Calculus of Finite Differences*, Chelsea Publishing Company, New York, 1960.

[226] M. Kac, On the Deviations between Theoretical and Empirical Distributions, *Proc. Natl. Acad. Sci. USA*, 35, 252–257, 1949.

[227] N. Kamarkar, R. Karp, G. Lueker, and A. Odlyzko, Probabilistic Analysis of Optimum Partitioning, *J. Applied Probability*, 23, 626–645, 1986.

[228] S. Karlin, and F. Ost, Some Monotonicity Properties of Schur Powers of Matrices and Related Inequalities, *Linear Algebra and Its Applications*, 68, 47–65, 1985.

[229] S. Karlin, and F. Ost, Counts of Long Aligned Word Matches Among Random Letter Sequences, *Adv. in Applied Probability*, 19, 293–351, 1987.

[230] S. Karlin, and F. Ost, Maximal Length of Common Words Among Random Letter Sequences, *Annals of Probability*, 16, 535–563, 1988.

[231] S. Karlin, A. Dembo, and T. Kawabata, Statistical Composition of High-Scoring Segments from Molecular Sequences, *Annals of Statistics*, 18, 571–581, 1990.

[232] S. Karlin, and A. Dembo, Limit Distributions of Maximal Segmental Score Among Markov-Dependent Partial Sums, *Adv. Applied Probability*, 24, 113–140, 1992.

[233] R. Karp, The Probabilistic Analysis of Some Combinatorial Search Algorithms. In *Algorithms and Complexity*, ed. J.F. Traub, Academic Press, New York, 1976.

[234] R. Karp, A Characterization of the Minimum Cycle Mean in a Digraph, *Discrete Mathematics*, 23, 309–311, 1978.

[235] R. Karp, An Introduction to Randomized Algorithms, *Discrete Applied Mathematics*, 34, 165–201, 1991.

[236] R. Kemp, *Fundamentals of the Average Case Analysis of Particular Algorithms*, Wiley-Teubner, Stuttgart, 1984.

[237] J.C. Kieffer, A Unified Approach to Weak Universal Source Coding, *IEEE Tans. Information Theory*, 24, 340–360, 1978.

[238] J. C. Kieffer, Sample Converses in Source Coding Theory Relative to a Fidelity Criterion, *IEEE Trans. Information Theory*, 37, 257–262, 1991.

[239] J. C. Kieffer, Sample Converses in Source Coding Theory, *IEEE Trans. Information Theory*, 37, 263–268, 1991.

[240] J.F.C. Kingman, *Subadditive Processes*, in Ecole d'Eté de Probabilités de Saint-Flour V-1975, Lecture Notes in Mathematics, 539, Springer-Verlag, Berlin, 1976.

[241] J.F.C. Kingman, The First Birth Problem for an Age-Dependent Branching Process, *Annals of Probability*, 3, 790–801, 1975.

[242] P. Kirschenhofer, and H. Prodinger, Further Results on Digital Search Trees, *Theoretical Computer Science*, 58, 143–154, 1988.

[243] P. Kirschenhofer, and H. Prodinger, On Some Applications of Formulæ of Ramanujan in the Analysis of Algorithms, *Mathematika*, 38, 14–33, 1991.

[244] P. Kirschenhofer, and H. Prodinger, Approximate Counting: An Alternative Approach, *Informatique Théorique et Applications/Theoretical Informatics and Applications*, 25, 43–48, 1991.

[245] P. Kirschenhofer, and H. Prodinger, Multidimensional Digital Searching— Alternative Data Structures, *Proc. Random Graphs'91*, John Wiley & Sons, New York, 1992.

[246] P. Kirschenhofer, and H. Prodinger, A Result in Order Statistics Related to Probabilistic Counting, *Computing*, 51, 15–27, 1993.

[247] P. Kirschenhofer, C. Martinez, and H. Prodinger, Analysis of an Optimized Search Algorithm for Skip Lists *Theoretical Computer Science*, 144, 199–220, 1995.

[248] P. Kirschenhofer, H. Prodinger, and C. Martinez, Analysis of Hoare's FIND Algorithms with Median-of-Three Partition, *Random Structures & Algorithms*, 10, 143–156, 1997.

[249] P. Kirschenhofer, H. Prodinger, and J. Schoissengeier, Zur Auswertung Gewisser Reihen mit Hilfe Modularer Funktionen, *Zahlentheoretische Analysis 2* (Ed. E. Hlawka), Springer Lecture Notes in Math. 1262, 108–110, 1987.

[250] P. Kirschenhofer, H. Prodinger, and W. Szpankowski, On the Variance of the External Path in a Symmetric Digital Trie *Discrete Applied Mathematics*, 25, 129–143, 1989.

[251] P. Kirschenhofer, H. Prodinger, and W. Szpankowski, On the Balance Property of PATRICIA Tries: External Path Length Viewpoint, *Theoretical Computer Science*, 68, 1–17, 1989.

[252] P. Kirschenhofer, H. Prodinger, and W. Szpankowski, Multidimensional Digital Searching and Some New Parameters in Tries, *International J. Foundation of of Computer Science*, 4, 69–84, 1993.

[253] P. Kirschenhofer, H. Prodinger, and W. Szpankowski, Digital Search Trees Again Revisited: The Internal Path Length Perspective, *SIAM J. Computing*, 23, 598–616, 1994.

[254] P. Kirschenhofer, H. Prodinger, and W. Szpankowski, Analysis of a Splitting Process Arising in Probabilistic Counting and Other Related Algorithms, *Random Structures & Algorithms*, 9, 379–401, 1996.

[255] T. Kløve, Bounds for the Worst Case Probability of Undetected Error, *IEEE Trans. Information Theory*, 41, 298–300, 1995.

[256] C. Knessl, A Note on the Asymptotic Behavior of the Depth of Tries, *Algorithmica*, 22, 547–560, 1998.

[257] C. Knessl, and J.B. Keller, Partition Asymptotics from Recursions, *SIAM J. Applied Mathematics*, 50, 323–338, 1990.

[258] C. Knessl, and W. Szpankowski, Quicksort algorithm again revisited, *Discrete Mathematics and Theoretical Computer Science*, 3, 43–64, 1999.

[259] C. Knessl, and W. Szpankowski, Limit Laws for Heights in Generalized Tries and PATRICIA Tries, *Proc. LATIN'2000*, Punta del Este, Lecture Notes in Computer Science, No. 1776, 298–307, 2000.

[260] C. Knessl, and W. Szpankowski, Asymptotic Behavior of the Height in a Digital Search Tree and the Longest Phrase of the Lempel-Ziv Scheme, *SIAM J. Computing*, 30, 923-964, 2000.

[261] C. Knessl, and W. Szpankowski, A Note on the Asymptotic Behavior of the Height in *b*-Tries for *b* Large, *Electronic J. Combinatorics*, 7, R39, 2000.

[262] D. E. Knuth, Mathematical Analysis of Algorithms, *Information Processing 71*, Proc. of the IFIP Congress, 19–27, Ljublijna, 1971.

[263] D. E. Knuth, Optimum Binary Search Trees, *Acta Informatica*, 1, 14–25, 1971.

[264] D. E. Knuth, Big Omicron and Big Omega and Big Theta, *SIGACT News*, 18–24, 1976.

[265] D. E. Knuth, The Average Time for Carry Propagation, *Indagationes Mathematicae*, 40, 238–242, 1978.

[266] D. E. Knuth, *Stable Marriage and Its Relation to Other Combinatorial Problems*, American Mathematical Society, Providence, 1997.

[267] D. E. Knuth, *The Art of Computer Programming. Fundamental Algorithms,* Vol. 1, Third Edition, Addison-Wesley, Reading, MA, 1997.

[268] D. E. Knuth, *The Art of Computer Programming. Seminumerical Algorithms*. Vol. 2, Third Edition, Addison Wesley, Reading, MA, 1998.

[269] D. E. Knuth, *The Art of Computer Programming. Sorting and Searching*, Vol. 3, Second Edition, Addison-Wesley, Reading, MA, 1998.

[270] D. E. Knuth, Linear Probing and Graphs, *Algorithmica*, 22, 561–568, 1998.

[271] D. E. Knuth *Selected Papers on the Analysis of Algorithms*, Cambridge University Press, Cambridge, 2000.

[272] D. E. Knuth, J. Morris, and V. Pratt, Fast Pattern Matching in Strings, *SIAM J. Computing*, 6, 189–195, 1977.

[273] D. Knuth, and A. Schönhage, The Expected Linearity of a Simple equivalence Algorithm, *Theoretical Computer Science*, 6, 281–315, 1978.

[274] D. Knuth, An Analysis of Optimal Caching, *J. Algorithms*, 6, 181–199, 1985.

[275] A. Konheim, and D.J. Newman, A Note on Growing Binary Trees, *Discrete Mathematics*, 4, 57–63, 1973.

[276] I. Kontoyiannis, and Y. Suhov, Prefix and the Entropy Rate for Long-Range Sources, in *Probability, Statistics, and Optimization* (Ed. F.P. Kelly), John Wiley, New York, 1993.

[277] I. Kontoyiannis, Second-Order Noiseless Source Coding Theorems, *IEEE Trans. Information Theory*, 43, 1339–1341, 1997.

[278] I. Kontoyiannis, An Implementable Lossy Version of the Lempel-Ziv Algorithm — Part I: Optimality for Memoryless Sources, *IEEE Trans. Information Theory*, 45, 2285–2292, 1999.

[279] I. Kontoyiannis, Pointwise Redundancy in Lossy Data Compression and Universal Lossy Data Compression, *IEEE Trans. Inform. Theory*, 46,136–152, 2000.

[280] T. W. Körner, *Fourier Analysis*, Cambridge University Press, Cambridge 1988.

[281] M. Kuczma, B. Choczewski, and R. Ger, *Iterative Functional Equations*, Cambridge University Press, New York, 1990.

[282] L. Kuipers, and H. Niederreiter, *Uniform Distribution of Sequences*. John Wiley, New York, 1974.

[283] H. Lauwerier, *Asymptotics Methods*, Mathematisch Centrum, Amsterdam, 1977.

[284] A. Lempel, and J. Ziv, On the Complexity of Finite Sequences, *IEEE Trans. Information Theory*, 22, 75–81, 1976.

[285] A. Lesek (Ed.), *Computational Molecular Biology, Sources and Methods for Sequence Analysis*, Oxford University Press, Oxford, 1988.

[286] L. Levin, Average Case Complete Problems, *SIAM J. Computing*, 15, 285–286, 1986.

[287] M. Li, and P. Vitànyi, *An Introduction to Kolmogorov Complexity and Its Applications*, Springer-Verlag, New York, 1993.

[288] T. Liggett, *Interacting Particle Systems*, Springer-Verlag, New York, 1985.

[289] M. Lothaire, *Combinatorics on Words*, Addison-Wesley, Reading, MA, 1982.

[290] G. Louchard, The Brownian Motion: A Neglected Tool for the Complexity Analysis of sorted Tables Manipulations, *RAIRO Theoretical Informatics*, 17, 365–385, 1983.

[291] G. Louchard, Brownian Motion and Algorithm Complexity, *BIT*, 26, 17–34, 1986.

[292] G. Louchard, Exact and Asymptotic Distributions in Digital and Binary Search Trees, *RAIRO Theoretical Inform. Applications*, 21, 479–495, 1987.

[293] G. Louchard, Random Walks, Gaussian Processes and List Structures, *Theoretical Computer Science*, 53, 99–124, 1987

[294] G. Louchard, Probabilistic Analysis of Adaptive Sampling, *Random Structures & Algorithms*, 10, 157–168, 1997.

[295] G. Louchard, and R. Schott, Probabilistic Analysis of Some Distributed Algorithms, *Random Structures & Algorithms*, 2, 151–186, 1991.

[296] G. Louchard, and W. Szpankowski, A Probabilistic Analysis of a String Editing Problem and its Variations, *Combinatorics, Probability and Computing*, 4, 143–166, 1994.

[297] G. Louchard, and W. Szpankowski, Average Profile and Limiting Distribution for a Phrase Size in the Lempel-Ziv Parsing Algorithm, *IEEE Trans. Information Theory*, 41, 478–488, 1995.

[298] G. Louchard, and W. Szpankowski, On the Average Redundancy Rate of the Lempel-Ziv Code, *IEEE Trans. Information Theory*, 43, 2–8, 1997.

[299] G. Louchard, W. Szpankowski, and J. Tang, Average Profile of Generalized Digital Search Trees and the Generalized Lempel-Ziv Algorithm, *SIAM J. Computing*, 935–954, 1999.

[300] L. Lovász, *Combinatorial Problems and Exercises*, North Holland, Amsterdam, 1993.

[301] G. Lueker, Some Techniques for Solving Recurrences, *Computing Surveys*, 12, 419–436, 1980.

[302] G. Lueker, Optimization Problems on Graphs with Independent Random Edge Weights, *SIAM J. Computing*, 10, 338–351, 1981.

[303] T. Łuczak, and W. Szpankowski, A Suboptimal Lossy Data Compression Based in Approximate Pattern Matching, *IEEE Trans. Information Theory*, 43, 1439–1451, 1997.

[304] F.J. MacWilliams, and N.J.A. Sloane, *The Theory of Error-Correcting Codes*, North-Holland, Amsterdam, 1977.

[305] H. Mahmoud, *Evolution of Random Search Trees*, John Wiley & Sons, New York, 1992.

[306] H. Mahmoud, and B. Pittel, On the Most Probable Shape of a Search Tree Grown from a Random Permutation, *SIAM J. Algebraic and Discrete Methods*, 5, 69–81, 1984.

[307] H. Mahmoud, and B. Pittel, On the Joint Distribution of the Insertion Path length and the Number of Comparisons in Search Trees, *Discrete Applied Mathematics*, 20, 243–251, 1988.

[308] H. Mahmoud, and B. Pittel, Analysis of the Space of Search Trees Under the Random Insertion Algorithm, *J. Algorithms*, 10, 52–75, 1989.

[309] U. Manber, *Introduction to Algorithms*, Addison-Wesley, Reading, MA, 1989.

[310] K. Marton, and P. Shields, The Positive-Divergence and Blowing-up Properties, *Israel J. Math*, 80, 331–348, 1994.

[311] P. Mathys, and P. Flajolet, Q-ary Collision Resolution Algorithms in Random Access Systems with Free and Blocked Channel Access, *IEEE Trans. Information Theory*, 31, 217–243, 1985.

[312] C. McDiarmid, On the Method of Bounded Differences, in *Surveys in Combinatorics* (Ed. J. Siemons), vol 141, 148–188, London Mathematical Society Lecture Notes Series, Cambridge University Press, Cambridge, 1989.

[313] C.J. McDiarmid, and R. Hayward, Large Deviations for Quicksort, *J. Algorithms*, 21, 476–507, 1996.

[314] K. Mehlhorn, and S. Näher, *A Platform for Combinatorial and Geometric Computing: LEDA*, Cambridge University Press, Cambridge, 1999.

[315] A. Meir, and J. W. Moon, On the Altitude of Nodes in Random Trees, *Canadian J. Mathematics*, 30, 997–1015, 1978.

[316] N. Merhav, Universal Coding with Minimum Probability of Codeword Length Overflow, *IEEE Trans. Information Theory*, 37, 556–563, 1991.

[317] N. Merhav, and J. Ziv, On the Amount of Statistical Side Information Required for Lossy Data Compression, *IEEE Trans. Information Theory*, 43, 1112–1121, 1997.

[318] L. Milne-Thomson, *The Calculus of Finite Differences*, Chelsea Publishing Company, New York, 1981.

[319] D. Mitrinović, J. Pečarić, and A. Fink, *Classical and New Inequalities in Analysis*, Kluwer Academic Publishers, Dordrecht, 1993.

[320] L. Moser, and M. Wyman, Asymptotic Expansions, *Canadian J. Mathematics*, 225–233, 1956.

[321] L. Moser, and M. Wyman, Asymptotic Expansions II, *Canadian J. Mathematics*, 194–209, 1957.

[322] R. Motwani, and P. Raghavan, *Randomized Algorithms*, Cambridge University Press, Cambridge, 1995.

[323] E. Myeres, An $O(ND)$ Difference Algorithm and Its Variations, *Algorithmica*, 1, 251–266, 1986.

[324] A. H. Nayfeh, *Introduction to Perturbation Techniques*, John Wiley & Sons, New York, 1981.

[325] C. Newman, Chain Lengths in Certain Random Directed Graphs, *Random Structures & Algorithms*, 3, 243–254, 1992.

[326] P. Nicodéme, B. Salvy, and P. Flajolet, Motif Statistics, *European Symposium on Algorithms*, Lecture Notes in Computer Science, No. 1643, 194–211, 1999.

[327] B. Noble, and J. Daniel, *Applied Linear Algebra*, Prentice Hall, Englewood Cliffs, NJ, 1988.

[328] N. E. Nörlund, *Vorlesungen über Differenzenrechnung*, Chelsea Publishing Company, New York, 1954.

[329] A. Odlyzko, Periodic Oscillation of Coefficients of Power Series that Satisfy Functional Equations, *Advances in Mathematics*, 44, 180–205, 1982.

[330] A. Odlyzko, Asymptotic Enumeration, (Eds. R. Graham, M. Götschel, and L. Lovász), 1063–1229, Elsevier Science, Amsterdam, 1995.

[331] F. Olver, *Asymptotics and Special Functions*, Academic Press, New York, 1974.

[332] D. Ornstein, and P. Shields, Universal Almost Sure Data Compression, *Annals of Probability*, 18, 441–452, 1990.

[333] D. Ornstein, and B. Weiss, Entropy and Data Compression Schemes, *IEEE Trans. Information Theory*, 39, 78–83, 1993.

[334] A. Panholzer, and H. Prodinger, Average-Case Analysis of Priority Trees: A Structure for Priority Queue Administration, *Algorithmica*, 22, 600–630, 1998.

[335] C. Papadimitriou, and K. Steiglitz, *Combinatorial Optimization: Algorithms and Complexity*, Prentice-Hall, Englewood Cliffs, NJ, 1982.

[336] M. Petkovšek, H. Wilf, and D. Zeilberger, $A = B$, A K Peters, Wellesley, 1996.

[337] B. Pittel, Asymptotic Growth of a Class of Random Trees, *Annals of Probability*, 18, 414–427, 1985.

[338] B. Pittel, Paths in a Random Digital Tree: Limiting Distributions, *Adv. in Applied Probability*, 18, 139–155, 1986.

[339] B. Pittel, and H. Rubin, How Many Random Questions Are Necessary to Identify n Distinct Objects?, *J. Combinatorial Theory, Ser. A*, 55, 292–312, 1990.

[340] V. Pless, W. Huffman (Eds.), *Handbook of Coding Theory*, Vols. I and II, Elsevier, Amsterdam, 1998.

[341] P. Poblete, Approximating Functions by Their Poisson Transform, *Information Processing Letters*, 23, 127–130, 1986.

[342] P. Poblete, and J. Munro, The Analysis of a Fringe Heuristic for Binary Search Trees, *J. Algorithms*, 6, 336–350, 1985.

[343] P. Poblete, A. Viola, and J. I. Munro, The Diagonal Poisson Transform and its Application to the Analysis of a Hashing Scheme, *Random Structures & Algorithms*, 10, 221–256, 1997.

[344] G. Pólya, and R. C. Read, *Combinatorial Enumeration of Groups, Graphs, and Chemical Compounds*, Springer Verlag, New York, 1987.

[345] A. D. Polyanin, and A. V. Manzhirov, *Handbook of Integral Equations*, CRC Press, Boca Raton, FL, 1998.

[346] A. G. Postnikov, *Tauberian Theory and Its Applications, Proc. Steklov Institute of Mathematics*, Vol. 144, American Mathematical Society, Providence, 1980.

[347] H. Prodinger, How to Select a Loser, *Discrete Mathematics*, 120, 149–159, 1993.

[348] A. P. Prudnikov, Y. A. Brychkov, and O. I. Marichev, *Integrals and Series*, Vols. I-III, Gordon and Breach Science Publishers, New York, 1986.

[349] B. Prum, F. Rodolphe, and E. Turckheim, Finding Words with Unexpected Frequencies in Deoxyribonucleic Acid Sequence, *J.R. Stat. Soc. B*, 57, 205–220, 1995.

[350] B. Rais, P. Jacquet, and W. Szpankowski, Limiting Distribution for the Depth in Patricia Tries, *SIAM J. Discrete Mathematics*, 6, 197–213, 1993.

[351] S. Rachev, and L. Rüschendorf, Probability Metrics and Recursive Algorithms, *Adv. Applied Probability*, 27, 770–799, 1995.

[352] B. Reed, How Tall Is a Tree, *Proc. STOC*, 479–483, Portland, 2000.

[353] M. Régnier, Analysis of Grid File Algorithms, *BIT*, 25, 335–357, 1985.

[354] M. Régnier, Trie Hashing Analysis, *Proc. 4th International Conference on Data Engineering*, 377–387, Los Angeles, 1988.

[355] M. Régnier, A Limiting Distribution of Quicksort, *RAIRO Theoretical Informatics and Applications*, 23, 335–343, 1989.

[356] M. Regnier, and P. Jacquet, New Results on the Size of Tries, *IEEE Trans. Information Theory*, 35, 203–205, 1989.

[357] M. Régnier, and W. Szpankowski, Complexity of Sequential Pattern Matching Algorithms, *Proc. Randomization and Approximate Techniques in Computer Science, RANDOM'98*, LCNS No. 1518, 187–199, Barcelona, 1998.

[358] M. Régnier, and W. Szpankowski, On the Approximate Pattern Occurrences in a Text, *Proc. Compression and Complexity of SEQUENCE'97*, IEEE Computer Society, 253–264, Positano, 1997.

[359] M. Régnier, and W. Szpankowski, On Pattern Frequency Occurrences in a Markovian Sequence, *Algorithmica*, 22, 631–649, 1998.

[360] A., Rényi, On Measures of Entropy and Information. *Proc. 4th Berkeley Symp. on Math. Statist. and Prob.*, 547–561, 1961.

[361] Y. Reznik, and W. Szpankowski, On the Average Redundancy Rate of the Lempel-Ziv Code with K-Error Protocol, *Proc. Data Compression Conference*, 373–382, Snowbird, 2000.

[362] W.T. Rhee, and M. Talagrand, Martingales Inequalities of NP-complete Problems, *Math. Oper. Res.*, 12, 177–181, 1987.

[363] R. Remmert, *Theory of Complex Functions*, Springer Verlag, New York, 1991.

[364]   J. Rissanen, A Universal Data Compression System, *IEEE Trans. Information Theory*, 29, 656–664, 1983.

[365]   J. Rissanen, Complexity of Strings in the Class of Markov Sources, *IEEE Trans. Information Theory*, 30, 526–532, 1984.

[366]   J. Rissanen, Universal Coding, Information, Prediction, and Estimation, *IEEE Trans. Information Theory*, 30, 629–636, 1984.

[367]   J. Rissanen, *Stochastic Complexity in Statistical Inquiry*, World Scientific, Singapore, 1998.

[368]   L.V. Romanovski, Optimization of Stationary Control of a Discrete Deterministic Process, *Cybertnetics*, 3, 52–62, 1967.

[369]   S. Ross, *Stochastic Processes*, John Wiley & Sons, New York, 1983.

[370]   U. Rösler, A Limit Theorem for Quicksort, *Theoretical Informatics and Applications*, 25, 85–100, 1991.

[371]   U. Rösler, A Fixed Point Theorem for Distributions, *Stochastic Processes and Their Applications*, 42, 195–214, 1992.

[372]   H. J. Ryser, *Combinatorial Mathematics*, Mathematical Association of America, Providence, 1963.

[373]   V. Sachov, *Combinatorial Methods in Discrete Mathematics*, Cambridge University Press, Cambridge, 1996.

[374]   V. Sachov, *Probabilistic Methods in Combinatorial Analysis*, Cambridge University Press, Cambridge, 1997.

[375]   D. Sankoff, and J. Kruskal (Eds.), *Time Warps, String Edits, and Macromolecules: The Theory and Practice of Sequence Comparison*, Addison-Wesley, Reading, MA, 1983.

[376]   S. Savari, Redundancy of the Lempel-Ziv Incremental Parsing Rule, *IEEE Trans. Information Theory*, 43, 9–21, 1997.

[377]   W. Schachinger, *Beiträge zur Analyse von Datenstrukturen zur Digitalen Suche*, Dissertation Technische Universität Wien, Vienna, 1993.

[378]   W. Schachinger, On the Variance of a Class of Inductive Valuations of Data Structures for Digital Search, *Theoretical Computer Science*, 144, 251–275, 1995.

[379]   U. Schmid, The Average CRI-length of a Tree Collision Resolution Algorithm in Presence of Multiplicity-Dependent Capture Effects, *Proc. ICALP 92*, Vienna, 223–234, 1992.

[380]   U. Schmid, Random Trees in Queueing Systems with Deadlines, *Theoretical Computer Science*, 144, 277–314, 1995.

[381]   R. Sedgewick, The Analysis of Quicksort Programs, *Acta Informatica*, 7, 327–355, 1977.

[382]   R. Sedgewick, Data Movement in Odd-Even Merging, *SIAM J. Computing*, 239–272, 1978.

[383]   R. Sedgewick, and P. Flajolet, *An Introduction to the Analysis of Algorithms*, Addison-Wesley, Reading, MA, 1995.

[384]   J. Seidler, *Information Systems and Data Compression*, Kluwer Academic Publishers, Boston, 1997.

[385] L. Shepp, Covering the Circle with Random Arcs, *Israel J. Math.*, 11, 328–345, 1972.

[386] P. Shields, Entropy and Prefixes, *Annals of Probability*, 20, 403–409, 1992.

[387] P. Shields, Waiting Times: Positive and Negative Results on the Wyner-Ziv Problem, *J. Theoretical Probability*, 6, 499–519, 1993.

[388] P. Shields, *The Ergodic Theory of Discrete Sample Paths*, American Mathematical Society, Providence, 1996.

[389] A. N. Shiryayev, *Probability*, Springer-Verlag, New York, 1984.

[390] Y. Shtarkov, Universal Sequential Coding of Single Messages, *Problems of Information Transmission*, 23, 175–186, 1987.

[391] Y. Shtarkov, T. Tjalkens, and F.M. Willems, Multi-alphabet Universal Coding of Memoryless Sources, *Problems of Information Transmission*, 31, 114–127, 1995.

[392] N. J. Sloane, *The Encyclopedia of Integer Sequences*, Academic Press, New York, 1995.

[393] Special Volume on Mathematical Analysis of Algorithms, Dedicated to D. E. Knuth (Eds. H. Prodinger and W. Szpankowski), *Theoretical Computer Science*, 144, 1995.

[394] Special Issue on Average-Case Analysis of Algorithms (Eds. P. Flajolet, and W. Szpankowski), *Random Structures & Algorithms*, 10, 1997.

[395] Special Issue on Average-Case Analysis of Algorithms, Dedicated to Philippe Flajolet on the Occasion of His 50th Birthday (Eds. H. Prodinger, and W. Szpankowski), *Algorithmica*, 22, 1998.

[396] R. Stanley, *Enumerative Combinatorics*, Vol. I, Wadsworth & Brooks/Cole Advanced Books & Software, Monterey, CA, 1986.

[397] R. Stanley, *Enumerative Combinatorics*, Vol. II, Cambridge University Press, Cambridge, 1999.

[398] J-M. Steyaert, and P. Flajolet, Patterns and Pattern-Matching in Trees: An Analysis, *Information and Control*, 58, 19–58, 1983.

[399] J. Storer, and J. Reif, Error Resilient Optimal Data Compression, *SIAM J. Computing*, 26, 934–939, 1997.

[400] V. Strassen, Asymptotische Abschätzungen in Shannon's Informationstheorie, *Trans. Third Prague Conference on Information Theory*, 689–723, 1962.

[401] J. M. Steele, An Efron-Stein Inequality for Nonsymmetric Statistics, *Annals of Statistics*, 14, 753–758, 1986.

[402] J. M. Steele, *Probability Theory and Combinatorial Optimization*, SIAM, Philadelphia, 1997.

[403] Y. Steinberg, and M. Gutman, An Algorithm for Source Coding Subject to a Fidelity Criterion, Based on String Matching, *IEEE Trans. Information Theory*, 39, 877–886, 1993.

[404] W. Szpankowski, Solution of a Linear Recurrence Equation Arising in the Analysis of Some Algorithms, *SIAM J. Algebraic and Discrete Methods*, 8, 233–250, 1987.

[405] W. Szpankowski, On a Recurrence Equation Arising in the Analysis of Conflict Resolution Algorithms, *Stochastic Models*, 3, 89–114, 1987.

[406] W. Szpankowski, Some Results on V-ary Asymmetric Tries, *J. Algorithms*, 9, 224–244, 1988.

[407] W. Szpankowski, The Evaluation of an Alternating Sum with Applications to the Analysis of Some Data Structures, *Information Processing Letters*, 28, 13–19, 1988.

[408] W. Szpankowski, Patricia Tries Again Revisited, *J. ACM*, 37, 691–711, 1990.

[409] W. Szpankowski, A Characterization of Digital Search Trees From the Successful Search Viewpoint, *Theoretical Computer Science*, 85, 117–134, 1991.

[410] W. Szpankowski, On the Height of Digital Trees and Related Problems, *Algorithmica*, 6, 256–277, 1991.

[411] W. Szpankowski, Asymptotic Properties of Data Compression and Suffix Trees, *IEEE Trans. Information Theory*, 39, 1647–1659, 1993.

[412] W. Szpankowski, A Generalized Suffix Tree and Its (Un)Expected Asymptotic Behaviors, *SIAM J. Computing*, 22, 1176–1198, 1993.

[413] W. Szpankowski, On Asymptotics of Certain Sums Arising in Coding Theory, *IEEE Trans. Information Theory*, 41, 2087–2090, 1995.

[414] W. Szpankowski, Combinatorial Optimization Problems for Which Almost Every Algorithm Is Asymptotically Optimal, *Optimization*, 33, 359–367, 1995.

[415] W. Szpankowski, On Asymptotics of Certain Recurrences Arising in Universal Coding, *Problems of Information Transmission*, 34, 142–146, 1998.

[416] W. Szpankowski, Asymptotic Average Redundancy of Huffman (and Other) Block Codes, *IEEE Trans. Information Theory*, 46, 2434–2443, 2000.

[417] L. Takács, The Asymptotic Distribution of the Total Heights of Random Rooted Trees, *Acta Scientifica Mathematica. (Szeged)*, 57, 613–625, 1993.

[418] M. Talagrand, Concentration of Measure and Isoperimetric Inequalities in Product Spaces, *Pub. Math. IHES*, 81, 73–205, 1995.

[419] M. Talagrand, A New Look at Independence, *Annals of Applied Probability*, 6, 1–34, 1996.

[420] K.H. Tan and P. Hadjicostas, Some Properties of a Limiting Distribution in Quicksort, *Statistics & Probability Letters*, 25, 87–94, 1995.

[421] J. Tang, *Probabilistic Analysis of Digital Search Trees*, Ph.D. Thesis, Purdue University, 1996.

[422] N. Temme, *Special Functions*, John Wiley & Sons, New York, 1996.

[423] G. Tenenbaum, *Introduction to Analytic and Probabilistic Number Theory*, Cambridge University Press, Cambridge, 1995.

[424] E. C. Titchmarsh, *The Theory of Functions*, Oxford University Press, Oxford, 1991.

[425] E. C. Titchmarsh, and D. Heath-Brown, *The Theory of the Riemann Zeta-Functions*, Oxford University Press, Oxford, 1988.

[426] B. Tsybakov, and V. Mikhailov, Random Multiple Packet Access: Part-and-Try Algorithm. *Prob. Information Transmission*, 16, 305–317, 1980.

[427] E. Ukkonen, A Linear-Time Algorithm for Finding Approximate Shortest Common Superstrings, *Algorithmica*, 5, 313–323, 1990.

[428] B. Vallée, Gauss' Algorithm Revisited, *J. Algorithms*, 12, 556–572, 1991.

[429] B. Vallée, Opérateurs de Ruelle-Mayer Généralisés et Analyse des Algorithmes d'Euclide et de Gauss, *Acta Arithmetica*, 8, 101–144, 1997.

[430] B. Vallée, Dynamics of the Binary Euclidean Algorithm: Functional Analysis and Operators, *Algorithmica*, 22, 660–685, 1998.

[431] B. Vallée, Dynamical Sources in Information Theory: Fundamental intervals and Word Prefixes, *Algorithmica*, 29, 262–306, 2001.

[432] V. N. Vapnik, *Statistical Learning Theory*, John Wiley & Sons, New York, 1998.

[433] J. Vitter, and W. Chen, *Design and Analysis of Coalesced Hashing*, Oxford University Press, New York, 1987.

[434] J. Vitter, and P. Flajolet, Average-Case Analysis of Algorithms and Data Structures, *Handbook of Theoretical Computer Science* (Ed. J. van Leewen), 433–524, Elsevier Science Publishers, Amsterdam, 1990.

[435] J. Vitter, and P. Krishnan, Optimal Prefetching via Data Compression, *J. ACM*, 43, 771–793, 1996.

[436] J. Vuillemin, A Unifying Look at Data Structures, *Commun. ACM*, 23, 229–239, 1980.

[437] E.H. Yang, and J. Kieffer, Simple Universal Lossy Data Compression Schemes Derived From Lempel-Ziv algorithm, *IEEE Trans. Information Theory*, 42, 239–245, 1996.

[438] E.H. Yang, and J. Kieffer, On the Redundancy of the Fixed–Database Lempel-Ziv Algorithm for Φ-Mixing Sources, *IEEE Trans. Information Theory*, 43, 1101–1111, 1997.

[439] E.H. Yang, and J. Kieffer, On the Performance of Data Compression Algorithms Based upon String Matching, *IEEE Trans. Information Theory*, 44, 47–65, 1998.

[440] E.H. Yang, and Z. Zhang, The Shortest Common Superstring Problem: Average Case Analysis for Both Exact Matching and Approximate Matching, *IEEE Trans. Information Theory*, 45, 1867–1886, 1999.

[441] A. Yao, On Random 2-3 Trees, *Acta Informatica*, 9, 159–170, 1978.

[442] A. Yao, An Analysis of $(h, k, 1)$ Shellsort, *J. Algorithms*, 14–50, 1980.

[443] B.L. van der Waerden, On the Method of Saddle Points, *Appl. Sci. Res.*, B., 2, 33–45, 1950.

[444] W. Wasow, *Asymptotic Expansions for Ordinary Differential Equations*, Dover, New York, 1987.

[445] M. Waterman, *Introduction to Computational Biology*, Chapman & Hall, London, 1995.

[446] B. Weide, Random Graphs and Graph Optimization Problems, *SIAM J. Computing*, 9, 552–557, 1980.

[447] H. Wilf, *Generatingfunctionology*, Academic Press, Boston, 1990.

[448] E. Whittaker, and G. Watson, *A Course of Modern Analysis*, Cambridge University Press, Cambridge, 1927.

[449] J. Wolfowitz, The Coding of Messages Subject to Chance Errors, *Illinois J. Mathematics*, 1, 591–606, 1961.

[450] R. Wong, *Asymptotic Approximations of Integrals*, Academic Press, Boston, 1989.

[451] E. M. Wright, The Coefficients of Certain Power Series, *J. London Mathematical Society*, 7, 256–262, 1932.

[452] A. Wyner, and J. Ziv, Some Asymptotic Properties of the Entropy of a Stationary Ergodic Data Source with Applications to Data Compression, *IEEE Trans. Information Theory*, 35, 1250–1258, 1989.

[453] A. Wyner, and J. Ziv, Fixed Data Base Version of the Lempel-Ziv Data Compression Algorithm, *IEEE Trans. Information Theory*, 37, 878–880, 1991.

[454] A. Wyner, and J. Ziv, The Sliding Window Lempel-Ziv Algorithm Is Asymptotically Optimal, *Proc. IEEE*, 82, 872–877, 1994.

[455] A. Wyner, and J. Ziv, Classification with Finite Memory, *IEEE Trans. Information Theory*, 42, 337–347, 1996.

[456] A.J. Wyner, The Redundancy and Distribution of the Phrase Lengths of the Fixed-Database Lempel-Ziv Algorithm, *IEEE Trans. Information Theory*, 43, 1439–1465, 1997.

[457] Z. Zhang, and E. Yang, An On-Line Universal Lossy Data Compression Algorithm via Continuous Codebook Refinement — Part II: Optimality for Phi-Mixing Source Models, *IEEE Trans. Information Theory*, 42, 822–836, 1996.

[458] J. Ziv, Compression, Test of Randomness, and Estimating the Statistical Model of Individual Sequences, *SEQUENCES*, 366–373, Springer-Verlag, New York, 1990.

[459] J. Ziv, Back from Infinity: A Constrained Resources Approach to Information Theory, *IEEE Information Theory Society Newsletter*, 48, 30–33, 1998.

[460] J. Ziv, and A. Lempel, A Universal Algorithm for Sequential Data Compression, *IEEE Trans. Information Theory*, 23, 3, 337–343, 1977.

[461] J. Ziv, and A. Lempel, Compression of Individual Sequences via Variable-rate Coding, *IEEE Trans. Information Theory*, 24, 530–536, 1978.

[462] J. Ziv, and N. Merhav, A Measure of Relative Entropy Between Individual Sequences with Application to Universal Classification, *IEEE Trans. Information Theory*, 39, 1270–1279, 1993.

[463] A. Zygmund, *Trigonometric Series*, Cambridge University Press, New York, 1959.

Index

NEMIROVSKY AND YUDIN • Problem Complexity and Method Efficiency in Optimization *(Translated by E. R. Dawson)*

PACH AND AGARWAL • Combinatorial Geometry

PLESS • Introduction to the Theory of Error-Correcting Codes, Third Edition

ROOS AND VIAL • Ph. Theory and Algorithms for Linear Optimization: An Interior Point Approach

SCHEINERMAN AND ULLMAN • Fractional Graph Theory: A Rational Approach to the Theory of Graphs

SCHRIJVER • Theory of Linear and Integer Programming

SZPANKOWSKI • Average Case Analysis of Algorithms on Sequences

TOMESCU • Problems in Combinatorics and Graph Theory *(Translated by R. A. Melter)*

TUCKER • Applied Combinatorics, Second Edition

WOLSEY • Integer Programming

YE • Interior Point Algorithms: Theory and Analysis

Printed and bound by CPI Group (UK) Ltd, Croydon, CR0 4YY

27/10/2024

14580331-0004